Springer Proceedings in Business and Economics

Springer Proceedings in Business and Economics brings the most current research presented at conferences and workshops to a global readership. The series features volumes (in electronic and print formats) of selected contributions from conferences in all areas of economics, business, management, and finance. In addition to an overall evaluation by the publisher of the topical interest, scientific quality, and timeliness of each volume, each contribution is refereed to standards comparable to those of leading journals, resulting in authoritative contributions to the respective fields. Springer's production and distribution infrastructure ensures rapid publication and wide circulation of the latest developments in the most compelling and promising areas of research today.

The editorial development of volumes may be managed using Springer Nature's innovative EquinOCS, a proven online conference proceedings submission, management and review system. This system is designed to ensure an efficient timeline for your publication, making Springer Proceedings in Business and Economics the premier series to publish your workshop or conference volume.

This book series is indexed in SCOPUS.

Anna Rumyantseva • Vladimir Plotnikov •
Alexey Minin • Hod Anyigba
Editors

Challenges and Solutions in the Digital Economy and Finance

Proceedings of the 5th International Scientific Conference on Digital Economy and Finances (DEFIN 2022), St. Petersburg 2022

Springer

Editors
Anna Rumyantseva
St. Petersburg University of Management Technologies and Economics
St. Petersburg, Russian Federation

Vladimir Plotnikov
St. Petersburg State University Of Economics
St. Petersburg, Russian Federation

Alexey Minin
MHP Management- und IT-Beratung GmbH
Ludwigsburg, Germany

Hod Anyigba
Nobel International Business School
Acca, Ghana

ISSN 2198-7246 ISSN 2198-7254 (electronic)
Springer Proceedings in Business and Economics
ISBN 978-3-031-14409-7 ISBN 978-3-031-14410-3 (eBook)
https://doi.org/10.1007/978-3-031-14410-3

This Springer imprint is published by the registered company Springer Nature Switzerland AG
The registered company address is: Gewerbestrasse 11, 6330 Cham, Switzerland

Contents

The original version of the book was revised: The author's names changes updated in several chapters. The correction to the book is available at https://doi.org/10.1007/978-3-031-14410-3_59

Contributors

T. Afanasyeva St. Petersburg University of Management Technologies and Economics, St. Petersburg, Russian Federation

Irina Gareevna Akhmetova Kazan State Power Engineering University, Kazan, Russian Federation

Anna-Maria Arias St. Petersburg University of Management Technologies and Economics, St. Petersburg, Russian Federation

L. N. Babkina North-Western Institute of Management, Branch of the Russian Presidential Academy of National Economy and Public Administration Attached to the President of the Russian Federation, St. Petersburg, Russian Federation

Dinar R. Baetova Federal State-Funded Educational Institution of Higher Education Omsk State Agrarian University Named After Pyotr A. Stolypin, Omsk, Russian Federation

M. M. Balog Pskov State University, Pskov, Russian Federation

Alexander S. Baranov Herzen State Pedagogical University, St. Petersburg, Russian Federation

Sofiia A. Baranova Herzen State Pedagogical University, St. Petersburg, Russian Federation

Marina Ivanovna Barabanova St. Petersburg State University of Economics, St. Petersburg, Russian Federation

Snezhana N. Bareyko Luga Les LLC Group of Companies, St. Petersburg, Russian Federation

Irina Sergeevna Bazen Volksuniversiteit Enschede, Enschede, The Netherlands

Jacob Cornelis Bazen School of Business, Building & Technology, Saxion University of Applied Sciences, Enschede, The Netherlands

Tatyana S. Bazhaeva Leningrad State University Named After A.S. Pushkin, St. Petersburg, Russian Federation

Anna M. Belyatskaya State Institute of Economics, Finance, Law and Technology, Gatchina, Russian Federation

Tatyana I. Bezdenezhnykh St. Petersburg State University of Economics, St. Petersburg, Russian Federation

Olga Bezgatcheva St. Petersburg University of Management Technologies and Economics, St. Petersburg, Russian Federation

Oleg A. Blinov Federal State-Funded Educational Institution of Higher Education Omsk State Agrarian University Named After Pyotr A. Stolypin, Omsk, Russian Federation

Valentina Vladimirovna Bologova National Research University "Moscow Power Engineering Institute", Moscow, Russian Federation

R. S. Bolshakov Irkutsk State Transport University, Irkutsk, Russian Federation

Larisa Bondar St. Petersburg University of Management Technologies and Economics, St. Petersburg, Russian Federation
St. Petersburg Branch of the Archive of RAS, SPbB ARAS, St. Petersburg, Russian Federation

Anastasiya A. Bordyugovskaya State Institute of Economics, Finance, Law and Technology, Gatchina, Russian Federation

Tatiana A. Borisova St. Petersburg University of Management Technologies and Economics, St. Petersburg, Russian Federation

Olga Boyarskaya Gzhel State University, Electroizolyator, Russian Federation

A. A. Burykin Admiral Makarov State University of Maritime and Inland Shipping, St. Petersburg, Russian Federation
St. Petersburg University of Management and Economics Technologies, St. Petersburg, Russian Federation
Russian Academy of Transport, Moscow, Russian Federation

A. B. Chernykh St. Petersburg University of Management Technologies and Economics, St. Petersburg, Russian Federation

S. E. Demidova Financial University Under the Government of the Russian Federation, Moscow, Russian Federation

Mikhail Georgievich Djanelidze Institute for Regional Economy Studies, Russian Academy of Science, St. Petersburg, Russian Federation

Timofey N. Dovbush St. Petersburg Institute of ARSUJ, RLA of Ministry of Justice of the Russian Federation, St. Petersburg, Russian Federation

Maria P. Efremova St. Petersburg State University of Economics (UNECON), St. Petersburg, Russian Federation

Svetlana Anatolyevna Fatova St. Petersburg State Institute of Film and Television, St. Petersburg, Russian Federation

Maria Fedyukovskaya St. Petersburg University of Management Technologies and Economics, St. Petersburg, Russian Federation

Alexander Fedyukovsky Peter the Great St. Petersburg Polytechnic University (SPbPU), St. Petersburg, Russian Federation

Inga G. Filippova St. Petersburg University of Management Technologies and Economics, St. Petersburg, Russian Federation

Artur A. Fomin National Research University "Higher School of Economics", Moscow, Russian Federation

Nailya Gareeva Kazan Innovative University Named After V.G. Timiryasov, Nizhnekamsk, Russian Federation

Anastasia S. Gerasimova Peter the Great St. Petersburg Polytechnic University (SPbPU), St. Petersburg, Russian Federation

Zhanna Viktorovna Gornostaeva Kadyrov Chechen State University, Grozny, Russian Federation

M. N. Grachev PPF Meridian JSC, St. Petersburg, Russian Federation

Maria Viktorovna Grudina Russian Customs Academy St. Petersburg Branch Named After Vladimir Bobkov, St. Petersburg, Russian Federation

Alexander Vyacheslavovich Gudkov Glinka Nizhny Novgorod State Conservatoire, Nizhny Novgorod, Russian Federation

Iurii Nikolaevich Guzov St. Petersburg State University, St. Petersburg, Russian Federation

Yaroslav Iurievich Guzov LLC "Geological Center of St. Petersburg State University", St. Petersburg, Russian Federation

Grigory Khoteevich Humenyuk St. Petersburg University of Management Technologies and Economics, St. Petersburg, Russian Federation

Nurtdinov Ilgiz Irekovich Kazan State Medical University, Kazan, Russian Federation

Dinara Iskhakova Kazan National Research Technological University, Kazan, Russian Federation

Sergey Sayarovich Ivanov National Research Lobachevsky State University of Nizhny Novgorod, Nizhny Novgorod, Russian Federation

Elena V. Ivanova St. Petersburg University of Management Technologies and Economics, St. Petersburg, Russian Federation

Elena Ivleva St. Petersburg University of Management Technologies and Economics, St. Petersburg, Russian Federation

Rinas V. Kashbrasiev Kazan (Volga Region) Federal University, Kazan, Russian Federation

Natalya Leonidovna Ketoeva National Research University "Moscow Power Engineering Institute", Moscow, Russian Federation

E. S. Khatsenko Chair of Youth Department Murmansk Region, Murmansk, Russian Federation

Maria Andreevna Kiseleva National Research University "Moscow Power Engineering Institute", Moscow, Russian Federation

Kirill V. Kliyev St. Petersburg University of Management Technologies and Economics, St. Petersburg, Russian Federation

S. V. Kolesnichenko Admiral Makarov State University of Maritime and Inland Shipping, St. Petersburg, Russian Federation
St. Petersburg University of Management and Economics Technologies, St. Petersburg, Russian Federation

Elena Kolesnik St. Petersburg State University of Economics, St. Petersburg, Russian Federation

Vladislava V. Kosareva St. Petersburg University of Management Technologies and Economics, St. Petersburg, Russian Federation

Kseniya A. Kozhukhina University at Interparliamentary Assembly of EurAsEC, St. Petersburg, Russian Federation

E. S. Kuznetsova Murmansk State Technical University, Murmansk, Russian Federation

Vladimir Aleksandrovich Kunin St. Petersburg University of Management Technologies and Economics, St. Petersburg, Russian Federation

Viktor Valerievich Lavrov St. Petersburg State Institute of Film and Television, St. Petersburg, Russian Federation

Marina V. Lazareva St. Petersburg University of Management Technologies and Economics, St. Petersburg, Russian Federation

Natalia A. Lazareva St. Petersburg University of Management Technologies and Economics, St. Petersburg, Russian Federation

Gregory V. Lepesh St. Petersburg State Economic University, St. Petersburg, Russian Federation

Veronika V. Lizovskaya St. Petersburg University of Management Technologies and Economics, St. Petersburg State University of Industrial Technologies and Design, St. Petersburg, Russian Federation

Mikhail Pavlovich Loginov Ural State University of Economics, Yekaterinburg, Russian Federation

Nelly Lugert St. Petersburg University of Management Technologies and Economics, St. Petersburg, Russian Federation

Irina V. Makarova Administration of the Governor of Perm Krai, Perm, Russian Federation

Elizaveta Igorevna Makovkina Ural State University of Economics, Yekaterinburg, Russian Federation

E. V. Malovetskaya Irkutsk State Transport University, Irkutsk, Russian Federation

Dmitry Andreevich Makarov St. Petersburg University of Management Technologies and Economics, St. Petersburg, Russian Federation

Mikhail Yurevich Makarov St. Petersburg University of Management Technologies and Economics, St. Petersburg, Russian Federation

Nikolay Petrovich Mayurov St. Petersburg University of Management Technologies and Economics, St. Petersburg, Russian Federation

Yulia V. Meleshko Belarusian National Technical University, Minsk, Republic of Belarus

Victor Aleksandrovich Menshov National Research University "Moscow Power Engineering Institute", Moscow, Russian Federation

Galina Stepanovna Merzlikina Volgograd State Technical University, Volgograd, Russian Federation

Natalia Meshkova Lugansk Vladimir Dahl State University, Lugansk, Ukraine

Yuriy V. Mishalchenko St. Petersburg Institute of ARSUJ, RLA of Ministry of Justice of the Russian Federation, St. Petersburg, Russian Federation

Maria Yu Mishalchenko St. Petersburg Institute of ARSUJ, RLA of Ministry of Justice of the Russian Federation, St. Petersburg, Russian Federation

Artem A. Moldovan St. Petersburg State University, St. Petersburg, Russian Federation

Vitaly A. Mordovets St. Petersburg University of Management Technologies and Economics, St. Petersburg, Russian Federation

G. Morunova St. Petersburg State University of Economics, St. Petersburg, Russian Federation

Vladimir V. Okrepilov Institute for Regional Economic Studies, Russian Academy of Sciences, St. Petersburg, Russian Federation

Aleksandr L. Pastukhov St. Petersburg University of Management Technologies and Economics, St. Petersburg, Russian Federation

Dariia Andreevna Pavlova St. Petersburg State University of Economics, St. Petersburg, Russian Federation

Vladimir Plotnikov St. Petersburg State University of Economics, St. Petersburg, Russian Federation

Nikolay Aleksandrovich Polyakov St. Petersburg State University, St. Petersburg, Russian Federation

Elena G. Pozdeeva Peter the Great St. Petersburg Polytechnical University, St. Petersburg, Russian Federation

Valeriy I. Prasolov Financial University, Moscow, Russian Federation

Svetlana Pshenichnikova St. Petersburg State University of Economics, St. Petersburg, Russian Federation

Yuliya Ivanovna Rastova St. Petersburg State University of Economics, St. Petersburg, Russian Federation

Natalia Nikolaevna Reshetnikova Rostov State University of Economics (RINH), Rostov-on-Don, Russian Federation
Don State Technical University, Rostov-on-Don, Russian Federation

Roman A. Romashov Murmansk Arctic State University, Murmansk, Russian Federation

Anna Rumyantseva St. Petersburg University of Management Technologies and Economics, St. Petersburg, Russian Federation

Marina Vladimirovna Rybkina St. Petersburg State University of Economics, St. Petersburg, Russian Federation

Vitaly D. Sattarov Perm Institute of the Federal Penal Service, Perm, Russian Federation

Nina Aleksandrovna Seliverstova St. Petersburg State Institute of Film and Television, St. Petersburg, Russian Federation

Elena A. Sereda St. Petersburg University of Management Technologies and Economics, St. Petersburg, Russian Federation

Tatsiana V. Serhiyevich Belarusian National Technical University, Minsk, Republic of Belarus

Elena E. Sharafanova St. Petersburg State University of Economics, St. Petersburg, Russian Federation

Nina Shashina St. Petersburg Institute of Economics and Management, St. Petersburg, Russian Federation

Natalia Nikolaevna Shestakova Institute for Regional Economy Studies, Russian Academy of Science, St. Petersburg, Russian Federation

Olga D. Shipunova Peter the Great St. Petersburg Polytechnical University, St. Petersburg, Russian Federation

Alexey D. Shmatko Institute for Regional Economic Studies Russian Academy of Sciences, St. Petersburg, Russian Federation

Daria Georgievna Shuvalova National Research University "Moscow Power Engineering Institute", Moscow, Russian Federation

Tatiana Borisovna Sidneva Glinka Nizhny Novgorod State Conservatoire, Nizhny Novgorod, Russian Federation

Elena Sintsova St. Petersburg University of Management Technologies and Economics, St. Petersburg, Russian Federation

O. V. Skotarenko Murmansk Arctic State University, Murmansk, Russian Federation
Military Academy of Logistical Support Named After General of the Army A. V. Khrulyov, St. Petersburg, Russian Federation

Margarita Borisovna Skvortsova Institute for Regional Economy Studies Russian Academy of Science, St. Petersburg, Russian Federation

Sergey Yu Solodovnikov Belarusian National Technical University, Minsk, Republic of Belarus

Anna Starodubova Kazan National Research Technological University, Kazan, Russian Federation

Mikhail Mikhailovich Strelnik St. Petersburg State University of Economics (UNECON), St. Petersburg, Russian Federation

Ekaterina A. Sysoeva National Research University "Moscow Power Engineering Institute", Moscow, Russian Federation

Olga Tarutko St. Petersburg University of Management Technologies and Economics, St. Petersburg, Russian Federation

Natalia Taskaeva National Research University Moscow State University of Civil Engineering, Moscow, Russian Federation

Irina A. Tkacheva St. Petersburg University of Management Technologies and Economics, St. Petersburg, Russian Federation

Elena Tsyplakova Pushkin Leningrad State University, St. Petersburg, Russian Federation

Olga Dmitrovna Ugolnikova St. Petersburg University of Management Technologies and Economics, St. Petersburg, Russian Federation

Elena V. Ushakova St. Petersburg University of Management Technologies and Economics, St. Petersburg, Russian Federation

Oleg Pavlovich Ushchekin Kazan State Power Engineering University, Kazan, Russian Federation

Natalya Vitalievna Usova Ural State University of Economics, Ekaterinburg, Russian Federation

Ruslan Yakovlevich Vakulenko Nizhny Novgorod State Linguistic University Named After N. A. Dobrolyubov, Nizhny Novgorod, Russian Federation

Yuliya S. Valeeva Kazan State Power Engineering University, Kazan, Russian Federation

Natalia V. Vasilenko St. Petersburg University of Management Technologies and Economics, St. Petersburg, Russian Federation

Olga O. Vasileva Herzen State Pedagogical University of Russia, St. Petersburg, Russian Federation

Iana Verkhovskaia St. Petersburg University of Management Technologies and Economics, St. Petersburg, Russian Federation
St. Petersburg State University of Industrial Technologies and Design, St. Petersburg, Russian Federation

Kunin Vladimir St. Petersburg University of Management Technologies and Economics, St. Petersburg, Russian Federation

Ksenia Sergeevna Volodina National Research University "Moscow Power Engineering Institute", Moscow, Russian Federation

Natalia Voronina Moscow Aviation Institute (National Research University), Moscow, Russian Federation

Ludmila V. Yakovleva St. Petersburg University of Management Technologies and Economics, St. Petersburg, Russian Federation

I. Yumasheva St. Petersburg University of Management Technologies and Economics, St. Petersburg, Russian Federation

Anastasia A. Zagorenko Federal State-Funded Educational Institution of Higher Education Omsk State Agrarian University Named After Pyotr A. Stolypin, Omsk, Russian Federation

Irina G. Zaiceva St. Petersburg State University of Economics, St. Petersburg, Russian Federation

Elena Sergeevna Zakharchenko Don State Technical University, Rostov-on-Don, Russian Federation
Rostov Branch, Russian Customs Academy, Rostov-on-Don, Russian Federation

Meri T. Zargaryan National Research University "Moscow Power Engineering Institute", Moscow, Russian Federation

Vladimir V. Zemskov Financial University, Moscow, Russian Federation

Marina Vianorovna Zolotova National Research Lobachevsky State University of Nizhny Novgorod, Nizhny Novgorod, Russian Federation

Elena M. Zvyagina St. Petersburg University of Management Technologies and Economics, St. Petersburg, Russian Federation

Teaching Students Professionally Based Foreign Language Communication in the Era of Digitalization

Irina A. Tkacheva, Marina V. Lazareva, and Ludmila V. Yakovleva

Introduction

In the modern world a foreign language is a great part of professional communication, that is why it is very necessary for the university foreign languages professors to reckon the timeliness of the student's professional competence with reference to a foreign language problem.

It is proved that foreign-language skills are acquired in the most effective way if professional knowledge is closely combined with communicative competence. Therefore, from a methodical point of view, it is important to consider communicative training of students in a foreign language in the context of a professionally oriented approach, which is related to the needs of students in learning the language for future professional activities. Such training means professional orientation of the content of educational materials and the training activities that form specific skills.

The solution to this problem leads to the goal of the study, which is to consider effective approaches to students' achievement of foreign-language communicative competence in the framework of professionally oriented training (Odinokaya et al., 2020). Research priorities also include a detailed study of the possibilities of creating methodical materials with a competently selected stagewise system of tasks and exercises that contribute to the quality training of students in foreign language communication, considering the specifics of the future profession.

I. A. Tkacheva (✉) · M. V. Lazareva · L. V. Yakovleva
St. Petersburg University of Management Technologies and Economics, St. Petersburg, Russian Federation
e-mail: i.tkacheva@spbacu.ru; l.jakovleva@spbacu.ru

A. Rumyantseva et al. (eds.), *Challenges and Solutions in the Digital Economy and Finance*, Springer Proceedings in Business and Economics,
https://doi.org/10.1007/978-3-031-14410-3_1

Materials and Methods

Nowadays, Russian education pays a great attention to the study of problems related to the development and implementation of new methods and techniques in foreign language teaching, as well as practical skills acquisition by students in the process of professional foreign language communication.

For teachers at the St. Petersburg University of Management Technologies and Economics such creative initiatives help to implement a practice-based approach in foreign language teaching, taking into account the future specialty. They are also mandatory in the conditions of modern requirements of the GEF and informatization of society.

The research methods are represented by a set of different approaches of theoretical and applied scientific knowledge that contribute to the implementation of analytical tools. Methods of comparative analysis and synthesis, grouping of actual data are used (Bylieva et al., 2020).

The interdisciplinary approach used in the study allows us to consider the interaction between teachers and students in the framework of the scientific study of the problem related to the qualitative development of students' professionally based speech when they achieve foreign language communicative competence.

The results of diagnostics of the level of formation of students' speech competencies, as well as the results of a graduates' survey about practical skills in the process of professional foreign language communication were the research material.

The results of the study can be further used in the professional training of students - future specialists of non-linguistic profiles in foreign language classes.

Results and Discussion

Professional communication in a foreign language is understood as communication carried out within the framework of professional activities.

In line with this consideration, the discipline "Foreign Language" requires an interdisciplinary connection with major disciplines to ensure the formation of the necessary professional competencies of a future specialist.

The detailed analysis of the works of scientists on pedagogy, methodology (for example, the use of interactive technologies in the educational process) (Young Oh et al., 2020), as well as their own experience of teaching professionally oriented communication of students in foreign language classes lead to the conclusion that the content of teaching a foreign language should include:

- areas of communicative activity of students, with the future specialty.
- foreign language material in the aspect of the main regulators of the student's work.
- practical knowledge of a foreign language based on the formed skills and abilities.

Table 1 The results of the students' testing

Type of speech activity	What was tested
Listening	Multiple choice, determination of correspondence, open-ended questions for the development of critical thinking on what was heard
Speaking	Interview, interactive conversation, monologue, business game, description of graphs and tables
Reading	Semantic, productive, functional
Writing	CV letter, essay, article, description graphic information

Table 2 The objectivity of the analysis

Level of foreign language proficiency	Number of students 170 (100%)	Objectivity of the correct answer (1)
A2	0	–
B1	25 (7.1%)	0.76
B2-	30 (14%)	0.77
B2	49 (27%)	0.76
C1	56 (51%)	0.80

– national and cultural features of the foreign language sphere in the realities of a foreign language.

University has not only the responsibility of making students know, but also must construct new knowledge that could be of utility to face the current challenges human beings must deal with (Zamora-Polo & Sánchez-Martín, 2019; Tverdokhlib, 2016).

The training of students of various directions (non-linguistic) by means of a foreign language at the university includes the formation of complex skills that contribute to the development of foreign language professional interaction in situational realities of a business nature. Consequently, a foreign language acts here as a means of personal development of students, as well as increasing professional competence, which is an important condition for successful professional activity in the future.

To prove that a diagnostic of the speech competences of graduates was carried out, in which 170 pcople took part.

Test tasks checked the level of competence formation in all types of speech activity. The digital methods available at the university made it possible to conduct analytics clearly and quickly on the following parameters (Table 1).

Table 2 shows the objectivity of the correct answer depending on the level of preparedness of students.

The results show that a more professionally prepared student has a greater probability and objectivity of the correct answer to all tasks, when compared with a less prepared student.

It is reasonable to note that it is the competent selection of a professional lexicon that is the most effective way for students to develop foreign language communication.

Learning foreign languages, a student creates not only a new language system, but he also programs the implementation of speech utterances (Rubtsova et al., 2020; Gorter et al., 2021). Learning foreign languages a student creates not only a new language system, but he also programs the implementation of speech utterances. Great opportunities for this are provided by the educational technologies of modern computer systems used in foreign language classes. They develop not only personal characteristics, but also allow students to accumulate knowledge independently. Training based on the performance of relevant tasks has a beneficial effect on the qualitative mastery of key foreign language competencies.

With the use of IT, teachers could spend less time on organizational matters related to documentation preparation and more time on working with students as to monitor their progress (Huk, 2021).

Computer technology also reduces the amount of time required to check and evaluate student work and does not even require the personal participation of the teacher. Correcting the course of work with the help of computer technology allows you to pay attention to the interests and capabilities of each student in the educational process.

Educational complexes with the use of modern electronic educational technologies in the context of the actualization of the problem disproportionately increase the motivation of students to learn a foreign language, activate speech-thinking activity, improve the skills and abilities of foreign language communication.

In this regard, it is advisable to consider the process of teaching professionally oriented communication of students in the format of two successive stages. The first one is aimed at the formation of lexical skills within the framework of communicative topics. The second stage is directly focused on the application of the studied lexical material in a foreign language, with the specific vector of the chosen profession.

For the first stage, a system of exercises aimed at teaching professional vocabulary is in demand. Since the process of forming lexical skills is communicative in nature, it is necessary to take the relationship of the vocabulary that students must learn with the topics of oral utterances. The topics of students' statements when teaching a professional foreign language can be divided into the following groups:

- economic structure in the countries of the language being studied (banking office work and the financial system).
- structural components of business (types of companies, management, marketing, etc.).
- management (in the field of higher education and the media, tourism, hotel services, etc.).
- daily business communication (contacts, receptions, presentations, telephone conversations, etc.).

- employment (job search, interview, professional duties, etc.).
- professional contacts and business correspondence.

One should note as well that in the methodology of teaching foreign languages, a step-by-step process of forming lexical skills is generally accepted, where there are the following levels of work on lexical material:

1. Familiarization and primary consolidation of lexical material.
2. Simulators for the use of vocabulary in the communicative process.
3. The use of lexical skills in speech activity.

In English-language pedagogical discourse, the pedagogical situation is also a rather vague concept and is referred to as educational, pedagogical, learning, etc. situation (Dobrova, 2021; Maldonado, 2022). For a strong formation of students' skills and abilities in the lexical formulation of a professionally oriented statement, it is necessary continuous work with lexical material in all classroom classes.

The analysis of the system of exercises aimed at the formation of lexical skills leads to the conclusion that there are many ways and techniques for organizing effective learning activities with students. You should focus on the exercises that are most useful for enhancing the speech-cogitative activity of students. They must also comply with the psychological, didactic, methodical, and other principles of teaching (Ylimaki & Brunderman, 2022).

At various levels of formation and development of lexical skills, appropriate types of exercises are used.

The second stage is the process of developing communication skills in a foreign language, with the specifics of the future profession. There are three levels of formation of foreign language communicative competence of students:

I-level—reproductive. It does not involve the speech creativity of students. It is characterized by a lack of independence in the choice of language material to formulate an utterance.

II-level—reproductive and productive. This level provides elements of creative and independent activity of students. It manifests itself in the expansion of vocabulary and its use in new situations.

III-level—the level of productive professionally oriented speech. It is characterized by the fact that the student has already have a certain language and speech experience, can express his personal attitude to facts and events and give his own assessment.

Based on the results of the survey on the sources of the formation of speech competence, graduate students named the following: the digital environment of the university (22%), reading professional literature (37%), fiction (10%), communication in a foreign language (15%), all indicators (16%). The results are presented in Fig. 1.

It is important for the teacher to pay attention the conditions and features of teaching a foreign language to students of various areas of training (non-linguistic). They are expressed in the following aspects: various levels of pre-university training

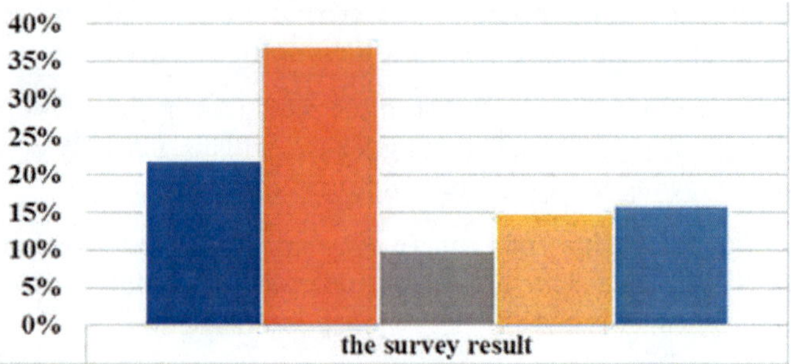

	the survey result
the digital environment of the university	22%
reading professional literature	37%
readung fiction	10%
communication in a foreign language	15%
all indicators	16%

Fig. 1 The results of the survey on the sources of the formation of speech competence

of students; a small number of hours in the discipline "Foreign language" according to the curriculum; often low level of motivation of students to learn a foreign language and others.

At the same time, it is important to recall that a prerequisite for the formation of professional competence of students is the development of communicative competence, which is reflected in the new Federal State Educational Standards.

The dimensions of continuous assessment and development of procedural knowledge skills have an extremely small presence in the definitions of pedagogical differentiation (Kanellopoulou & Darra, 2021; Klasnic et al., 2021).

In the line of the solution of the task set, the skillful application of interactive teaching methods acquires great significance. They provide a real opportunity for a qualitative increase in students' level of language proficiency and the achievement of the competence necessary for the use of a foreign language in professional communication and intercultural communication (Levandovski et al., 2021).

Foreign language teachers have in their arsenal a wide selection of such teaching methods and technologies to expand the range of communicative activities. It is logical to use them in the classroom as part of the second stage—the process of developing communicative skills in a foreign language, with the specifics of the future profession (Gerasimova et al., 2021).

At the reproductive level: brainstorming, role-playing games of a problem orientation, the method of situational analysis, training in small groups of cooperation and others.

At the level of reproductive and productive utterance: brainstorming, role-playing games of a problem orientation, collage method, learning in cooperation and others.

At a productive level, examples of the greatest efficiency and effectiveness can be: discussions, project method, information technology, collaborative learning, non-traditional activities (interviews, videos, excursions, musicals, performances, holidays) and others.

The approach described above to studying the problem of teaching students professionally oriented foreign language communication allows you to create a methodological system of tasks and exercises, which is educational complex. It is based on two levels of education using modern electronic educational technologies:

- Stage I of training—a tutorial book for updating basic foreign language knowledge with further improvement of practical skills and abilities at an advanced level.
- Stage II of education—a series of tutorial books in a foreign language for classroom and independent work of students of various areas and profiles of education ("Economics", "Applied Informatics", "Psychology", "Management", "Tourism and hotel service", "State and municipal administration", "Advertising and public relations" and others).

As a result, we state that the educational complex, consisting of methodological developments with a well-chosen step-by-step system of tasks and exercises, contributes to a more effective implementation of the process of teaching students' foreign language communication, with the specifics of the future profession. The use of modern electronic educational technologies, of course, serves as a powerful factor in strengthening the motivation to learn a language and qualitatively improving the level of students' communicative competence.

Conclusion

Summarizing all this, it can be argued that the prospects of this study have the further development of methodical materials aimed at improving the practical skills of foreign professional communication of future specialists. In our opinion, this should lead to a qualitative change in the process of teaching foreign language to students at non-linguistic universities (Odinokaya et al., 2020).

New training standards pose new challenges and tasks for university professors. Using a creative approach with methodically competent and reasonable use of modern e-learning resources and skillful use of interactive learning methods, we will be able to form a socially active initiative creative personality with a high level of proficiency in a foreign language, which is proved by the results of the diagnosis and survey of graduates.

References

Bylieva, D. S., Lobatyuk, V. V., & Fedyukovsky, A. A. (2020, September). Ways of sociotechnical integration of scientists and volunteers in citizen science. In *IOP conference series: Materials science and engineering* (Vol. 940, No. 1, p. 012150). IOP Publishing.

Dobrova, V. V. (2021). Pedagogical situations: Interpretation difference. *Vestnik of Samara State Technical University Psychological and Pedagogical Sciences, 18*(4), 73–82. https://doi.org/10.17673/vsgtu-pps.2021.4.6

Gerasimova, A. S., Sereda, E., & Rubtsova, S. (2021, October). Transcreation as a creative tool of translation. In *International conference on professional culture of the specialist of the future* (pp. 331–338). Cham: Springer. https://doi.org/10.1007/978-3-030-89708-6_28.

Gorter, D., Cenoz, J., & der Worp, K. V. (2021). The linguistic landscape as a resource for language learning and raising language awareness. *Journal of Spanish Language Teaching, 8*, 161–181. https://doi.org/10.1080/23247797.2021.2014029

Huk, T. (2021). *From education 1.0 to education 4.0–challenges for the contemporary school.* Mihaela Cvek, Mateja Pšunder, 36. https://doi.org/10.15804/tner.2021.66.4.03

Kanellopoulou, E.-M., & Darra, M. (2021). Defining pedagogical differentiation in higher education. *International Journal of Education, 13*(4), 139–154. https://doi.org/10.5296/ije.v13i4.19209

Klasnic, I., Duranovic, M., & Vidic, T. (2021). Flipped learning–Pedagogic dilemmas. *Contemporary Educational Researches Journal, 11*(4), 138–146. https://doi.org/10.18844/cerj.v11i4.5463

Levandovski, A. R., Passos, M. M., de Araújo, R. N., & de Mello Arruda, S. (2021). Pedagogical course projects: An interpretation through the knowledge matrix. *Revista Prática Docente, 6*(3), e099. https://doi.org/10.23926/RPD.2021.v6.n3.e099.id1322

Maldonado, K. (2022). Pedagogical strategies for the learning of medicine students. *Technium Social Sciences Journal, 27*, 65–69. https://doi.org/10.47577/tssj.v27i1.5339

Odinokaya, M., Andreeva, A., Mikhailova, O., Petrov, M., & Pyatnitsky, N. (2020). Modern aspects of the implementation of interactive technologies in a multidisciplinary university. In *E3S web of conferences* (Vol. 164, p. 12011). EDP Sciences. https://doi.org/10.1051/e3sconf/202016412011.

Odinokaya, M. A., Karpovich, I. A., Mikhailova, O. J., Piyatnitsky, A. N., & Klímová, B. (2020, September). Interactive technology of pedagogical assistance as a means of adaptation of foreign first-year students. In *IOP conference series: Materials science and engineering* (Vol. 940, No. 1, p. 012130). IOP Publishing. https://doi.org/10.1088/1757-899X/940/1/012130.

Rubtsova, A. V., Almazova, N. I., Bylieva, D. S., & Krylova, E. A. (2020, September). Constructive model of multilingual education management in higher school. In *IOP conference series: Materials science and engineering* (Vol. 940, No. 1, p. 012132). IOP Publishing. https://doi.org/10.1088/1757-899X/940/1/012132.

Tverdokhlib, T. (2016). Forms of teaching pedagogical disciplines in Orthodox Religious Educational Institutions of Ukraine (the end of 19th–the beginning of 20th century). *International Letters of Social and Humanistic Sciences, 71*, 40–46. https://doi.org/10.18052/www.scipress.com/ILSHS.71.40

Ylimaki, R. M., & Brunderman, L. A. (2022). Going deeper into curriculum and pedagogical activity. In *Evidence-based school development in changing demographic contexts* (pp. 73–79).-Cham: Springer. https://doi.org/10.1007/978-3-030-76837-9_6.

Young Oh, E., Song, D., & Hong, H. (2020). Interactive computing technology in anti-bullying education: The effects of conversation-bot's role on K-12 students' attitude change toward bullying problems. *Journal of Educational Computing Research, 58*(1), 200–219. https://doi.org/10.1177/0735633119839177

Zamora-Polo, F., & Sánchez-Martín, J. (2019). Teaching for a better world. Sustainability and sustainable development goals in the construction of a change-maker university. *Sustainability, 11*(15), 4224. https://doi.org/10.3390/su11154224

Possibility of Digital Technology Use to Improve the Efficiency of Economic Entities Interaction

Vladimir Aleksandrovich Kunin and Mikhail Mikhailovich Strelnik

Introduction

The modern stage of economic development is characterized by the presence of two opposite tendencies: the intention of economic entities to cooperate and the isolation of their economic activities. Latest trend is clearly manifested at the macro level in the policy of some countries to protect national markets, and it prevents development of interaction with foreign companies and destroys already established relations increasing business entities risks and affecting the key factors of competitiveness negatively: strategic positioning, operational positioning, efficiency, and innovation potential (Kunin & Tarutko, 2018).

The following forms of business entities interaction are distinguished in the theory and practice of economic activity: cooperation, coordination, and collaboration. The theory of cooperation appeared in the nineteenth century (founders are Ch. Fourier, R. Owen, W. King) and since then has repeatedly transformed in the process of the economic environment, society and the state management changes (Palladina & Voronina, 2014). The International Cooperative Alliance (ICA) defines a cooperative as an autonomous association of persons united voluntarily to meet their common economic, social, and cultural needs and aspirations through a jointly owned and democratically controlled enterprise (International Cooperative Alliance, 2022). The differences between the terms of cooperation, coordination and collaboration are as follows:

V. A. Kunin (✉)
St. Petersburg University of Management Technologies and Economics, St. Petersburg, Russian Federation

M. M. Strelnik
St. Petersburg State University of Economics (UNECON), St. Petersburg, Russian Federation

A. Rumyantseva et al. (eds.), *Challenges and Solutions in the Digital Economy and Finance*, Springer Proceedings in Business and Economics,
https://doi.org/10.1007/978-3-031-14410-3_2

- coordination is the harmonization of joint work, combining of various parts or groups into a single whole to achieve common goals of inter-organizational relationships together (inter-organizational relationships—IOR) (Castañer & Oliveira, 2020), while cooperation is a process of joint work of independent economic entities to achieve a common (agreed) goal in the long term.
- collaboration means voluntary help others to achieve the common goal of the IOR or a particular goal of the participant. Collaboration is usually strategic in nature (Wankmüller & Reiner, 2020; Weaver, 2012).

The development of inter-organizational relationships processes is happening amid the rapid spread of breakthrough digital technologies transforming the methods and technologies of business doing (Makarov et al., 2020) and can become an important tool for strengthening and expanding effective interaction of business entities. Digital Transformation can be understood as the changes in all aspects of human life due to digital technology (Stolterman & Fors, 2004). A study of scientific publications on digital transformation shows the lack of a clear definition of digital transformation areas (Van Veldhoven, 2021). It should be noted that the engine of changes is innovation causing industrial revolutions.

In total, four industrial revolutions are identified by different features: the first revolution is characterized by "mechanization", the second revolution by "mass production", the third revolution by "computerization", the fourth revolution is characterized by "cyber-physical systems" and "Internet of Things" (Sikandar et al., 2021). Industry 4.0 is a term to describe the changes due to the fourth industrial revolution. This concept combines the strengths of traditional industries with cutting edge internet technologies. It embraces a set of technologies that support effective and accurate decision-making in real time through the introduction of various Information and Communication Technologies and the convergence with the existing manufacturing technologies (Schmidt et al., 2015). Digitalization represents a paramount process started some decades ago, but which received a strong acceleration by Industry 4.0 and now directly impacts all the process and manufacturing sectors (Branca et al., 2020). It is important to underline that digitalization is not a simple transfer from "analogic" to digital data and documents. It is rather the networking between the business processes, the creation of efficient interfaces, and the integrated data exchange and management (Bogner et al., 2016).

This paper examines the peculiarities of modern digital technologies use for effective business entities interaction and formulates the principal directions of such high-tech interaction.

Digitalization transformed economic activity by dramatically reducing the cost of data collecting, storing, and processing, and due to a significant increase in computing power (Wysokińska, 2021). The types of digital technologies proposed by different authors is given as a comparison in Table 1.

The differences existing in the understanding of the areas of application and the practical significance of various digital technologies, along with the importance of developing the process of business entities interaction, determine the relevance of

Table 1 The types of digital technologies (DT)

The types of DT by C.Vite, R. Morbiducci:	The types of DT by L.V. Tashenova etc.:
Autonomous robot	Internet of things
Simulations	Big data
Integrated systems	Blockchain
Internet of things	Cyberphysical systems
Cybersecurity	
Cloud computing	
Additive manufacturing	
Augmented reality	

Source: made by the authors based on (Vite & Morbiducci, 2021) and (Tashenova et al., 2019)

the research carried out in this article aimed at identifying the prospects of digital technology use to improve the efficiency of economic entities interaction.

Materials and Methods

The methods of comparative analysis, synthesis, grouping, and fuzzy logic are used in the article. Comparative analysis provides an opportunity to identify and group the essential features of digital technologies in the process of interaction between economic entities and to determine possible restrictions of their use.

The fuzzy logic method allows to obtain the probability distribution of digital technology choice based on the distribution of fuzzy sets. The probability distributions of the choice of digital technology by risk accepting people and risk avoiding people were obtained. The decision-making on the choice of digital technology by decision-makers (subjects) is examined in the article.

Results

There is little systematization of digital technologies in scientific publications which makes it difficult to determine their significant properties and areas of practical application. The classification system of digital technologies proposed in this paper is given in Table 2. This classification does not include information and telecommunication digital technologies (chat bots, internet platforms, social networks etc.), since there is no doubt that they can be applied in the interaction of business entities.

This authors' classification is based on three criteria: the level of digital technology specialization, the types of economic activity, the level of interaction between business entities. The authors propose to distinguish general digital technologies that can be applied as a basis for other technologies and special ones, limited by a certain field of application. The application of the type of economic activity as the

Table 2 The classification of digital technologies

№	Digital technology name	Level of digital technology specialization	Types of economic activity	Level of interaction between economic entities
1.	Autonomous robots	Special technology	P; C; S; W	B2B; B2C; B2G
2.	Simulations	Special technology	S; W	
3.	Integrated systems	Special technology	P; C; S; W	B2B; B2C; B2G
4.	Internet of things	Special technology	P; C; S; W	B2B; B2C
5.	Cybersecurity	Special technology	P; C; S; W	B2B; B2C; B2G
6.	Cloud computing	General technology	P; C; S; W	B2B; B2C; B2G
7.	Additive manufacturing	Special technology	P	B2B
8.	Augmented reality	Special technology	C; S; W	B2B; B2C
9.	Big data	General technology	P; C; S; W	B2B; B2C
10.	Machine learning	General technology	P; C; S; W	
11.	Blockchain	General technology	P; C; S; W	B2B; B2C; B2G
12.	Cyberphysical systems	General technology	P; C; S; W	B2B; B2C; B2G

Source: made by the authors.* The types of economic activities: production (P), commerce (C), provision of services (S), performance of work (W). The level of interaction between business entities: Business to Business (B2B), Business to Consumer (B2C), Business to Government (B2G)

classification criterion involves the use of digital technology in the interaction of entities engaging the certain economic activity (production, commerce, provision of services, performance of work). The level of interaction of economic entities presupposes the allocation of the following levels: interaction between businesses; interaction between business and consumer; the relationship between business and government.

Cloud computing, big data, machine learning, blockchain, cyberphysical systems are classified as general technologies in the article. It is possible to get mobile access to big data stored on a digital platform or to machine learning tools with the help of cloud computing. Cloud computing along with blockchain can be used for mobile control of business processes implementation, the passage of transactions etc. Such complex application of digital technologies is reasonable to recommend in practices when implementing such form of interaction between business and the government as public-private partnership (Kunin & Semenov, 2021). Some digital technologies are special, for example, additive manufacturing, which is the process of joining materials to make parts from 3D model data, usually layer upon layer, as opposed to subtractive manufacturing and formative manufacturing methodologies (ISO/ASTM 52900-15, 2015).

The following digital technologies can be used in the interaction of economic entities engaged in production, commerce, provision of services (works): autonomous robots, integrated systems, internet of things, cybersecurity, cloud computing, big data, machine learning, blockchain, cyberphysical systems. The finished goods that can be commercialized are not the result of simulator use. Therefore, the possibility of using this technology should be considered in the interaction of entities providing services or performing works. The use of additive technologies is possible only in the interaction of industrial entrepreneurship entities. Entities operating in the field of commercial entrepreneurship, providing services and performing works can apply augmented reality. The authors of the article suppose that the application of big data and machine learning is expensive, but large enterprises can usually afford it. However, when large and small enterprises interact under the sub-contracting scheme, the large enterprise (subcontractor) working together with small enterprises can provide them with the necessary support in the use of digital technologies. At the same time, big data and machine learning is equally important to apply in the field of commerce, provision of services (performance of work) where business entities carry out active interactions with the external environment which leads to the generation of big data.

The use of additive manufacturing technology is possible in B2B interaction. Internet of things, simulators are possible in the event of B2B and B2C interactions. The rest of digital technologies can be applied at all levels of interaction between economic entities: B2B; B2C; B2G.

An important consequence of digital technologies use in business entities interaction is the potential of significant efficiency increasing of economic activity and the competitiveness of the participants of such interaction. The use of digital technologies allows interacting subjects to increase labor productivity and the quality of goods, services or works performed. This leads to revenue increase of the interaction participants. Revenue increase can be accompanied by fixed costs increase and will be effective if the growth rate of fixed costs is less than the critical value determined by the following formula:

$$\kappa_{\mathrm{f}}^{(crit)} = \frac{bc}{f} - (c - 1), \tag{1}$$

where

$\kappa_{\mathrm{f}}^{(crit)}$—critical value of fixed costs growth rate.

b—revenue growth rate.

c—basic position determined by the ratio of revenue to revenue at the break-even point.

f—a parameter determined by the change in the share of variable costs in revenue, and equal to the ratio of marginal profit before and after the revenue change (Kunin, 2015).

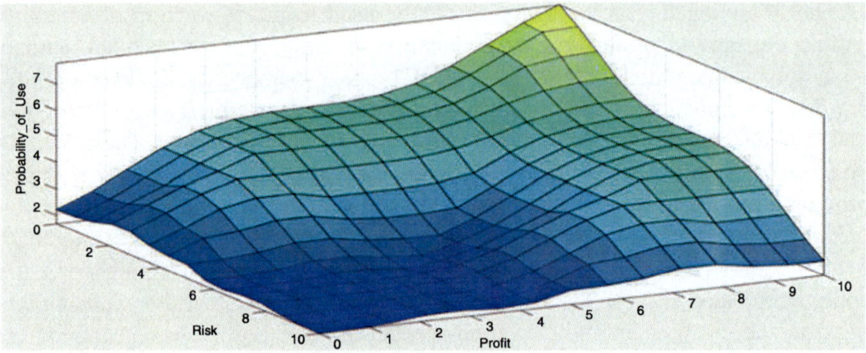

Fig. 1 The distributions of probability of digital technologies choice by a risk avoiding person. Source: authors' development

If condition one (1) is met, it means that revenue growth is accompanied by sales profit and economic profitability increase, therefore, the efficiency of economic activity is going up.

Based on Fuzzy Logic Toolbox (MATLAB extension package for designing fuzzy logic systems), an expert system can be developed that determines the probability of digital technology choice by economic entity deciding of particular digital technology to be used for interacting with other entities.

The following factor indicators are considered:

- the risk level of digital technology application.
- the profit from the digital technology use.

It should be noted that risk can appear because of management decision. (Strelnik, 2014). The resulting indicator is the probability of deciding on the digital technology choice. The following assumptions are made:

- the range of factor indicators values is from 0 to 10 (where 0 is the minimum value, 10 is the maximum value).
- the distribution of factor indicator function is normal.

The distributions of probability of digital technologies choice by a risk avoiding person is presented in Fig. 1.

It should be noted that if the person making a decision on digital technology use avoids taking risks, the probability of digital technology choice will be minimal with high risk level of digital technology application, as well as with minimal values of profit. As profit rises and risk level declines, the probability of choice of a given digital technology grows.

The distributions of probability of digital technologies choice by a risk accepting person is shown in Fig. 2.

If the decision maker accepts risk, the probability of digital technology choice will grows in proportion to rising profits from the use of digital technology in order to build relationships of interaction with other business entities, even if the risk level

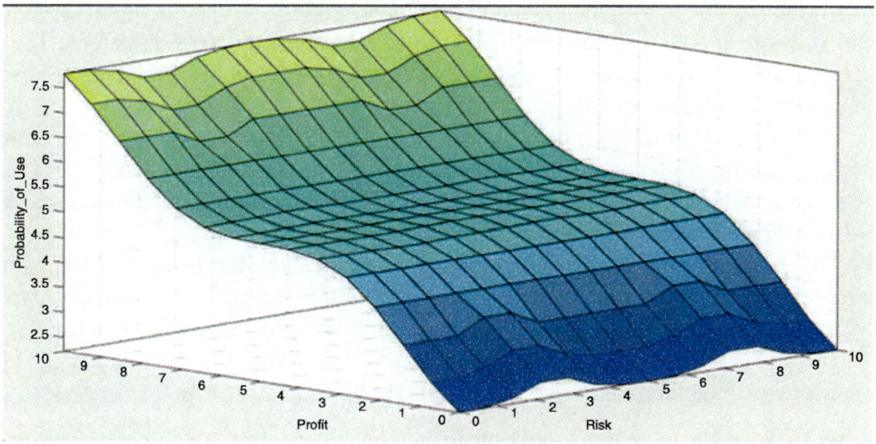

Fig. 2 The distributions of probability of digital technologies choice by a risk accepting person. Source: authors' development

of digital technology application rises. The probability of digital technology use declines if the profit from the digital technology use decreases. The advantage of fuzzy logic use in expert system consists in the universal validity of its application and individuality in setting a type of the distribution of factor indicator, choice of performance indicators and generation of selection rules used by the decision-maker.

Generation of databases on assessments of the probability of choice of a particular digital technology for different scenarios of attitudes towards risk of decision makers and subjects of different economic activities with different scales of economic activity allows:

- to reveal the prospects of the digital technologies, use in business interaction.
- to determine key areas of technological support for such kind of interaction.

Discussions

The economic entities are increasingly using modern digital technologies in internal business processes. At the same time, when several economic entities interact, the barriers impeding the effective use of digital technologies for the development of interaction may arise. The classification of digital technologies which makes it possible to determine their properties in the case of their application in economic entities interaction is presented in the article. Thus, it becomes possible to assess the restrictions of the use of digital technologies proactively.

In this regard, the importance of the development of interaction technologies based on the use of cloud technologies and low code platforms for wide access, providing the relative ease of use and mobility of access should be emphasized. Such platforms are able to provide the possibility of easy mobile access to digital services and software products for a wide range of users of interacting organizations. They can provide access to databases and software for their processing, including the use of machine learning methods.

The digital technology choice depends on the criteria applied by a decision maker. Apart from technology, digital transformation includes the activities of an actor (for example, a manager) involved in promoting transformation processes (Mikhaylova, 2019; Nadkarni & Prügl, 2021).

Today, the connection between cloud technologies and the internet of things is possible which opens the possibility of combining classical artificial intelligence with advanced approaches of machine learning (Alberternst et al., 2021).

There is the possibility of augmented reality technology use in commerce. The application of augmented reality as exemplified by food and beverages allows:

- to track products in the supply chain helping to identify goods counterfeiting, especially in international markets.
- to inform consumers about product characteristics (Penco et al., 2021).

Digital technologies have also become widespread in government management. For example, since 2014, the concept of electronic residency ("e-Residency") has been implemented in Estonia. This concept implies the possibility for other countries citizens to receive an e-resident ID card giving access to government services: online registration of an e-business, opening of a bank account etc. Thus, the government can expand its capabilities by attracting new residents - potential consumers of state services (individuals and legal entities) (Mikhaylova, 2019).

The probability of digital technology choice in business entities interaction, depending on two factors is considered in the article: the risk level of digital technology implementation and profit from the digital technology use. The authors admit the existence of other factors that can affect the digital technology choice in economic entities interaction. For example: the required level of user competence; the amount of investment required for digital technologies introduction; the possibility of digital technologies use by all interacting subjects etc. Nevertheless, it is possible to consider these parameters as additional variables in Fuzzy Logic Toolbox, setting the appropriate distribution laws for their values, and get the probability distribution for digital technology choice based on the criteria.

New technologies lead to changes in the factors of production which, in their turn, can contribute to the emergence of new innovations. For future studies, the most relevant option is to consider the evolution of digital technologies in the aspect of economic systems transformation.

Conclusion

The classification system presented in the article makes it possible to determine the essential features of digital technologies use in of economic entities interaction in accordance with the following criteria:

- the level of digital technology specialization.
- the types of economic activity.
- the level of interaction between economic entities.

There are restrictions of digital technology use in business entities interaction. The use of fuzzy logic for the expert system makes it possible to determine the probability of decision making on the choice of digital technology in interacting with other economic entities based on such factor indicators as the risk level of digital technology application and profit from the digital technology use.

The use of digital technology is reasonable not only at the level of internal business processes, but also in economic entities interaction. At the same time, the complex application of digital technologies and cloud computing and low code platforms aimed at providing easy mobile access for users of interacting entities, helps to overcome the barriers and increase the efficiency of interaction between economic entities with different scales of economic activity.

References

Alberternst, S., Anisimov, A., Antakli, A., Duppe, B., Hoffmann, H., Meiser, M., et al. (2021). Orchestrating heterogeneous devices and AI services as virtual sensors for secure cloud-based IoT applications. *Sensors, 21*(22), 7509. https://doi.org/10.3390/s21227509

ASTM I (2015). *ASTM52900-15 standard terminology for additive manufacturing—general principles—terminology*. ASTM International, West Conshohocken, PA, 3(4), 5.

Bogner, E., Voelklein, T., Schroedel, O., & Franke, J. (2016). Study based analysis on the current digitalization degree in the manufacturing industry in Germany. *Procedia Cirp, 57*, 14–19.

Branca, T. A., Fornai, B., Colla, V., Murri, M. M., Streppa, E., & Schröder, A. J. (2020). The challenge of digitalization in the steel sector. *Metals, 10*(2), 288. https://doi.org/10.3390/met10020288

Castañer, X., & Oliveira, N. (2020). Collaboration, coordination, and cooperation among organizations: Establishing the distinctive meanings of these terms through a systematic literature review. *Journal of Management, 46*(6), 965–1001. https://doi.org/10.1177/0149206320901565

International Cooperative Alliance. (2022). *What is a cooperative?* Retrieved from https://www.ica.coop/en/cooperatives/what-is-a-cooperative

Kunin, V. A. (2015). Business performance management in conditions of economic instability, UchenyezapiskiMezhdunarodnogobankovskogoinstituta. *In the International Banking Institute, 11*(2), 87–97.

Kunin, V. A., & Semenov, A. V. (2021). On the issue of ways to increase the social orientation of public-private entrepreneurship. *Problems of the Modern Economy*, 2(78), 93–98. (In Russ.)

Kunin, V. A., & Tarutko, O. A. (2018). The system of indicators of the competitiveness of entrepreneurial structures. *Problems of the Modern Economy, 1*(65), 65–68.

Makarov, M., Ivleva, E., Shashina, N., & Shashina, E. (2020, April). Transforming entrepreneurship factors and technologies in the digital economy. In *Proceedings of the III International Scientific and Practical Conference «Digital Economy and Finances»(ISPC–DEF 2020)*. https://doi.org/10.2991/aebmr.k.200423.005.

Mikhaylova, A. (2019). The role of innovations in providing economic security: The Estonian case. *Contemporary Europe, 7*, 136–147. https://doi.org/10.15211/soveurope72019136147

Nadkarni, S., & Prügl, R. (2021). Digital transformation: A review, synthesis and opportunities for future research. *Management Review Quarterly, 71*, 233–341. https://doi.org/10.1007/s11301-020-00185-7

Palladina, M. I., & Voronina, N. P. (2014). The origins of the cooperative theory, cooperatives and the development of the cooperative movement. *State and Law, 4*, 83–95.

Penco, L., Serravalle, F., Profumo, G., & Viassone, M. (2021). Mobile augmented reality as an internationalization tool in the "Made In Italy" food and beverage industry. *Journal of Management and Governance, 25*, 1179. https://doi.org/10.1007/s10997-020-09526-w

Schmidt, R., Möhring, M., Härting, R. C., Reichstein, C., Neumaier, P., & Jozinović, P. (2015, June). Industry 4.0-potentials for creating smart products: empirical research results. In *International conference on business information systems* (pp. 16–27). Cham: Springer.

Sikandar, H., Vaicondam, Y., Khan, N., Qureshi, M. I., & Ullah, A. (2021). Scientific Mapping of Industry 4.0 Research: A bibliometric analysis. *International Journal of Interactive Mobile Technologies, 15*(18). https://doi.org/10.3991/ijim.v15i18.25535

Stolterman, E., & Fors, A. C. (2004). Information technology and the good life. In *Information systems research* (pp. 687–692). Boston, MA: Springer.

Strelnik, M. (2014). Approving the ISDWIR Method of Risk Measurement in Making Risk Management Decision//Aprobación del método de medición del riesgo SIIPDR en el manejo de asunción de riesgos. *Revista de Métodos Cuantitativos para la Economía y la Empresa*, 17, áginas-42.

Tashenova, L. V., Babkin, A. V., & Mamrayeva, D. G. (2019). Digital transformation of industrial production in the context of Industry 4.0, Bulletin of Karaganda University. *Economy Series, 96, 4*, 154–162.

Van Veldhoven, Z. (2021). A scoping review of the digital transformation literature using scientometric analysis. In *2021: 24th International conference on business information systems*. https://doi.org/10.52825/bis.v1i.49.

Vite, C., & Morbiducci, R. (2021). Optimizing the sustainable aspects of the design process through building information modeling. *Sustainability, 13*(6), 3041. https://doi.org/10.3390/su13063041

Wankmüller, C., & Reiner, G. (2020). Coordination, cooperation and collaboration in relief supply chain management. *Journal of Business Economics, 90*(2), 239–276. https://doi.org/10.1007/s11573-019-00945-2

Weaver, B. (2012). *Coordination, cooperation, and collaboration: Defining the C3 framework.*

Wysokińska, Z. (2021). A review of the impact of the digital transformation on the global and European economy. *Comparative Economic Research. Central and Eastern Europe, 24*(3), 75–92.

Mutual Investment Funds Under Financial Instability

Olga Bezgatcheva, Anna Rumyantseva, and Elena Tsyplakova

Introduction

The dynamics of the social and economic development of the Russian Federation in recent years has been showing that investments play a crucial role in maintaining and increasing the country's economic potential.

The complexity of the investment concept ultimately comes down to activities that allow to overcome the technological gap and increase the future ability of an individual enterprise and the national economy to manufacture products.

The country lives in a new economic reality. These are not only the coronavirus pandemic, the unstable situation in the world energy markets, the decline in business activity, but also the difficult foreign policy situation.

All these factors make it necessary to look for new sources of investment, to attract extrabudgetary funds for the development of the national economy more actively.

Such a financial intermediary as investment funds is proposed to be considered as a driver for the recovery of the economic, investment activity of the country.

Today financial intermediaries can be called a driver for the recovery of the national economic and investment activity. Of course, the dominant place among them is occupied by banks. But not the last place belongs to investment funds (Bezgatcheva, 2020).

O. Bezgatcheva (✉) · A. Rumyantseva
St. Petersburg University of Management Technologies and Economics, St. Petersburg, Russian Federation
e-mail: o.bezgacheva@spbacu.ru

E. Tsyplakova
Pushkin Leningrad State University, St. Petersburg, Russian Federation

A. Rumyantseva et al. (eds.), *Challenges and Solutions in the Digital Economy and Finance*, Springer Proceedings in Business and Economics,
https://doi.org/10.1007/978-3-031-14410-3_3

It is analyzed how efficiently mutual investment funds operate in the financial market as one of the sources of investment in the real and financial sectors of the economy.

As a rule, investors in mutual funds are legal entities and individuals who do not have sufficient qualifications to select investment objects and further manage them.

To form one's own effective portfolio, it is necessary to understand how to reduce risks and not lose in profitability, which is what non-professional investors lack. MIFs undertake the solution of these tasks, successfully diversifying risks and demonstrating investment income to their investors.

Attracting funds from non-professional investors by mutual funds to the investment sphere by purchasing shares and bonds of various market entities can be considered not only as a source of investment for these entities, but also as one of the ways to protect the personal savings of people, private business from inflation and earning an income.

Methods and Materials

The article used research methods typical for this work, namely, an analysis of the forms and possibilities of investing mutual funds in the real and financial sectors of the Russian economy; generalization, based on information, statistical and reporting data of the experience of mutual investment funds activity in conditions of financial instability and self-isolation as a result of the COVID-19 pandemic. The research apparatus chosen by the authors made it possible to study and generalize the experience of the fund's activity, to draw interesting and well-founded conclusions. Namely, to find out that mutual investment funds have great opportunities to raise finances from unqualified investors to the economy; the potential of funds has not yet been exhausted; there is a tendency towards an increase in investor portfolios and the volume of funds invested in both the financial and real sectors of the economy.

Kharisma Novita Sari, Moch Sulchan, & Mimah Mutamimah (2021) analyzed the differences in the performance of Islamic mutual funds with mutual funds based on the Sharpe and Jensen method for the 2010–2020 period in the Indonesian Capital Market. Syed Feroz Aziz (2019) explained the performance of mutual fund and mutual fund manager in Pakistan. Javier Gil-Bazo, Peter Hoffmann, & Sergio Mayordomo (2020) using data on Spanish mutual funds showed that bank-affiliated funds provide funding support to their parent company via purchases of bonds in the primary market.

Barbara Abou Tanos (2021) investigated how culture affects US global mutual funds' financial performance. Nikolai Roussanov, Hongxun Ruan, & Yanhao Wei (2021) estimated a structural model with costly investor search and learning about fund skill; it is found that marketing is nearly as important as performance and fees in determining fund size. Kenneth Khang & Thomas W. Miller (Khang & Miller, 2021) examined the activity level of portfolio managers and how this activity affects portfolio performance. Ed deHaan, Yang Song, Chloe Xie, & Christina Zhu (2021) improved our understanding of why investors make poor mutual fund choices and how price dispersion persists among homogeneous index funds.

Results

The impetus for the flow of savings of unqualified investors to the capital market in recent years has been, first, higher interest yield on securities compared to interest rates on deposits. Due to the COVID-19 pandemic, the policy of the Central Bank was aimed at easing monetary conditions that caused a decrease in interest rates on deposits to 4.2% per annum in December 2020 (−1.4 percentage points to December 2019) (Bank of Russia, 2020). Recently rates have started to rise gradually and have reached the 2019 level before the COVID-19 pandemic. But inflation is also increasing, devaluing income and savings, so the trend towards an overflow of funds from the population to the stock market continues to this day.

Second, the financial and information literacy of users of financial services has increased.

Third, the mutual funds introduce new products based on the use of information technology. "Going online has played a critical role in developing the market and attracting people to invest," says Andrey Makarov, Head of Sales at Sberbank Asset Management. "In the spring, before the start of the COVID-19 pandemic, we brought our open-end mutual funds to Sberbank Online. The result exceeded our expectations: more than 60% of investments in our mutual funds people make online" (Maltseva & Dranishnikova, 2020).

Fourth, the risks of investing in mutual funds for large businesses are next lower order than investing in public-private partnership projects.

Based the Report of the Central Bank of the Russian Federation for 2020, the following was the result of this:

- a twofold decrease in the growth of household deposits compared to 2019.
- a significant flow of funds from retail investors to the stock market.

And as a confirmation of these trends:

- growth in the number of mutual funds (see Table 1).
- growth in the number of operating mutual funds in the financial market (see Table 1).
- growth in the value of investment portfolios of clients within the framework of trust management. In 2020 it increased by 33.4% and exceeded 1.5 trillion rubles.

Table 1 Number of mutual funds by type, units (InvestFunds, 2021)

Mutual investment funds	As of 8 December 2021	Change in the number of funds for the period from 1 Jan 2020 to 8 Dec 2021
Exchange-traded	255	+90
Open end	170	+8
Interval	109	−2
Closed end	9	−28
All funds	543	+68

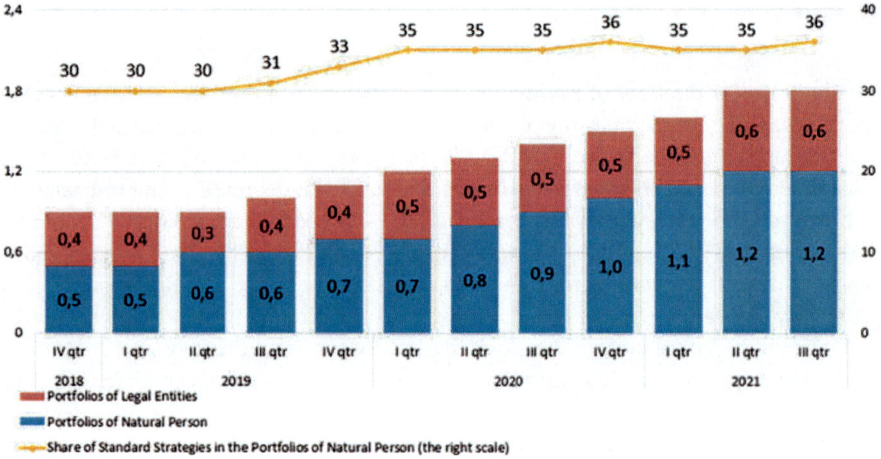

Fig. 1 Dynamics of the value of portfolios in trust and the share of standard strategies, trillion roubles

– stability and profitability of management companies (the value of net assets of mutual funds increased by more than 12% and exceeded 5 trillion rubles) (Bank of Russia (2020).

In addition, the financial stability of mutual funds has allowed them to participate in state programs and national projects since 2020 and made it possible to expand active operations throughout the country.

According to the reported data of the Central Bank as of 30 June 2021, the number of trust management clients with funded accounts in the Russian Federation was 4 people per 1000 population, the average size of a funded account is 1891 thousand rubles. At the same time, the following regions have the largest data indicators: in Moscow—11 people and 3960 thousand rubles; in St. Petersburg—6 people and 3333 thousand rubles; in the Moscow Region—5 people and 1467 thousand rubles; in the Novosibirsk Region—4 people and 1604 thousand rubles; in the Samara Region—5 people and 1519 thousand rubles; in the Murmansk Region—5 people and 1058 thousand rubles, respectively.

The value of portfolios in trust has doubled in recent years (Fig. 1).

In terms of categories of mutual funds, at the end of 2020 the largest were closed end combined and open-end funds of market financial instruments (68.3% and 13.2% of the total value of net assets, respectively).

The largest type of investment fund are closed-end funds, focused on working with legal entities and large private clients (83.7% of the total value of the net assets of mutual funds). It is on these funds that the Ministry of Economic Development and Trade of Russia places hopes on the diversification of the Russian economy with the active involvement of private Russian money.

In 2020 two closed-end mutual investment funds began to work to attract funds from large businesses and individuals to national projects that should act as co-investors on an automatic basis in all projects of the Russian Direct Investment Fund (Tkachev & Yushkov, 2019). This is another evidence of the importance attached to the development of this source of attracting extrabudgetary funds to the national economy and co-financing of national projects.

In 2020 the largest number of investors came to the segment of exchange-traded mutual funds. The number of owners of investment units of exchange-traded investment funds for the year increased to 1358.5 thousand units (+1287.9 thousand), most of which are retail investors.

As a result, the total number of owners of investment units of mutual funds increased by 92%, up to 3637.9 thousand owners.

The given statistical data testify to the potential possibilities of mutual funds to attract additional resources to the economy.

This trend indicates that potential investors have available funds, they are confident in funds, and they can invest their funds in more profitable assets (Table 2).

As a rule mutual funds choose conservative or balanced strategies that are based on investing in stable assets such as bonds and shares of large companies; government and municipal bonds. Both more and less risky assets can be added to the portfolio, depending on the investor's decisions and the horizon for achieving the financial goal, but the risks remain balanced. Funds actively monitor the market situation that allows them to quickly get rid of low-profit and risky instruments.

As of the end of the third quarter of 2021, the volume of investments of individuals in bonds reached 3.1 trillion roubles (+ 14% YTD). The growth rates of investments in bonds were significantly lower than the rates of investments in shares (+ 57% YTD), however, bonds are perceived by many investors as an alternative to bank deposits and are held until maturity. In addition, unlike stocks, a significant portion of bonds is purchased on the primary market, so individuals' funds act as a direct source of borrowed resources for the economy.

The structure of individuals' investments in bonds of various issuers is presented in Table 3.

A rather curious fact calls attention to itself. In the segment of debt instruments, the demand of mutual funds (retail investors) for classic bank bonds and subordinated bonds expanded (see Table 4).

Investments in subordinated and securitized bonds can be associated with additional risks for unprepared investors due to the long maturity of these securities that significantly exceeds the traditional horizon for bank deposits.

But this is a stable income for investors, and the placement of subordinated bonds on the market for banks has a double effect: it has a positive effect on their own capital and increases their "long" resources that they invest through their clients in the national economy, whose projects comply with the principles of sustainable development (for example, Sberbank, VTB, Gazprombank and others).

Table 2 Indicators of yield on portfolios of individuals in the Top-30 standard strategies as of 30 September 2021 (Bank of Russia, 2021)

Indicator	Number of strategies, units	Number of clients ind., thousands of units	Ind. portfolio size, bln rub.	Share of Top-30 portfolio size, %	Income (3 months), %	Income (12 months), %	Sharpe ratio
Whole market	544	626.8	428.2	–	–	–	
Top-30 standard strategies	30	414.2	303.6	100	0.8	2.8	1.0
Including by investment objects							
Bonds of residents	6	264.3	76.4	25	−0.2	0.5	0.2
Russian shares	3	34.6	32.5	11	10.1	37.7	2.7
Foreign bonds	7	19.9	58.7	19	−0.5	−5.7	−0.8
Foreign shares	4	12.9	20.1	7	−6.0	3.7	0.3
Structured products	10	82.4	115.9	38	1.0	3.1	0.8
Including by types of investors							
For qualified investors	10	10.5	99.6	33	0.1	3.9	0.7
For unqualified investors	20	403.7	204.0	67	1.1	2.7	0.9
Of which IIA strategies	8	322.7	87.7	29	2.0	5.0	1.3
Including by the method of investment							
Direct investments	15	27.4	110.2	36	−0.5	3.1	0.8
Investments in mutual funds, structured bonds or ETF	15	386.8	193.4	64	1.4	2.7	1.0
Benchmarks							

Table 3 Structure of individuals' investments in bonds, in %% (Bank of Russia, 2021)

	IV quarter of 2018	IV quarter of 2019	IV quarter of 2020	III quarter of 2021
Bonds of foreign issuers	24	21	24	23
Bonds of quasi-nonresidents	24	17	18	18
Bonds of other residents	6	10	12	13
Bonds of credit institutions	19	28	27	30
Government, sub-federal and municipal bonds	27	24	20	17

Table 4 Structure of individuals' investments in bonds of residents by type, in %% (Bank of Russia, 2021)

Types of bonds	IV quarter of 2018	IV quarter of 2019	IV quarter of 2020	III quarter of 2021
Unsecured	87	79	69	60
Subordinated	6	4	10	20
Structured and investment	7	16	21	18
Securitization	–	1	–	2

Discussion

The range of trust management strategies with concern for the environment and the social sphere is expanding more and more. These strategies give the investor the opportunity to build a long-term capital through responsible investing. The essence of this approach is that funds are invested not only in bonds, but also in shares of stable, responsible, and growing companies that follow the international ESG principles (Bezgatcheva & Rumyantseva, 2021).

Inputs are invested in stocks of companies that have shown the best dynamics in the field of sustainable development and corporate social responsibility through the purchase of units of mutual funds.

Using the example of individual mutual investment funds, let's see what products they provide to their investors.

Sberbank (2021)—Responsible Investment ETF. It is an exchange-traded fund investing in shares that are part of the Moscow Exchange Index—RUIE Vector of sustainable development of Russian issuers, full profitability. The fund is aimed at long-term capital growth by investing in shares of companies with sufficient liquidity, high growth potential and following the principles of ESG (Environmental, Social, Governance).

The structure of the fund's investments by sector:

- communications—8.2%
- Energy carriers—26.5%
- Electricity 18.0%

- Finance—14.1%
- natural resources 26.6%, etc.

Sberbank (2021)—Balanced Bond Fund OEMF (Sber Asset Management JSC managing company) is aimed at investing in shares and bonds of predominantly Russian issuers to ensure a balance between an acceptable level of risk and expected return.

There are three offered scenarios of the average annual return on investment for at least 3 years: Positive +20.8%; Neutral +12.7%; Negative +5.0%.

The structure of the fund's investments by sector:

- bonds—31.7%
- Energy carriers—27.4%
- Finance—13.6%
- natural resources 11.6%
- chemical industry—1.7%, etc.

Sberbank (2021)—Ilya Muromets Bond Fund OEMF (Sber Asset Management JSC managing company). The funds are invested in ruble-denominated government, sub-federal and corporate bonds of Russian issuers. There are three offered scenarios of profitability when investing for at least 3 years: Positive +11.1%; Neutral +9.1%; Negative +7.1%.

The structure of the fund's investments by sector:

- sovereign Eurobonds—6.2%
- Finance—21.4%
- Telecommunications—13.3%
- Mechanical engineering—7%.

Sberbank—Dobrynya Nikititch Equity Fund OEMF (Sber Asset Management JSC managing company) is aimed at generating income on an investment horizon of one year or more by investing in shares of predominantly Russian issuers with sufficient liquidity and high growth potential. There are three offered scenarios of profitability when investing for at least 3 years: Positive +29.8%; Neutral +14.6%; Negative +1.1%.

The structure of the fund's investments by sector:

- Energy carriers—36.0%
- Finance—22.4%
- natural resources—19.8%
- communications—12.0%
- cash—2.5%, etc.

Conclusion

In conditions of financial and political instability more and more attention is paid to the search for extrabudgetary sources of investment in the economy, and primarily in the domestic market. The investments of mutual funds are one of these sources.

Mutual funds are one of the largest intermediaries in the financial market. The potential of the funds has not been exhausted; it continues to grow. They use a variety of strategies. They introduce new products based on the use of information technologies and online services that allows them to significantly increase their presence in the stock market. They expand the range of environmental and social trust management strategies that follow the international ESG principles, thereby contributing to the country's progress towards the development of a sustainable economy.

By accumulating available funds of the general population and private business (from several thousand to millions of roubles), they not only increase the opportunities for investment in the national economy, but also contribute to the strengthening of the national currency, "binding" the available funds of the population.

While maintaining financial instability in the domestic and foreign markets in the coming years, it is possible to talk about further strengthening the positions of investment funds and expanding their participation in the investment activities of the economy.

References

Abou Tanos, B. (2021). Culture and mutual funds' performance. *Finance Research Letters*, pp. 102466. https://doi.org/10.1016/j.frl.2021.102466.

Aziz, S. F. (2019). Performance of mutual funds and mutual fund manager in Pakistan. *South Asian Journal of Social Studies and Economics*, 1–9. https://doi.org/10.9734/sajsse/2019/v4i230122.

Bank of Russia. (2020). *Annual report of the Bank of Russia for 2020*. Retrieved from https://www.cbr.ru/Collection/Collection/File/32268/ar_2020.pdf

Bank of Russia. (2021). *Review of key indicators of professional participants in the securities market no 3. III quarter of 2021*. Retrieved from https://cbr.ru/Collection/Collection/File/39609/review_secur_21Q3.pdf

Bezgatcheva, O. L. (2020). A new level of financial relations between banks and customers with the development of digital banking. *Proceedings on Digital economy and finance: materials of the IV International scientific and practical conference on 18–19 March, 2021*.

Bezgatcheva, O. L., & Rumyantseva, A. Y. (2021). Development of ESG banking in Russia. *Russian Science Magazine Economics and Management, 27*(10).

DeHaan, E., Song, Y., Xie, C., & Zhu, C. (2021). Obfuscation in mutual funds. *Journal of Accounting and Economics, 72*(2–3), 101429. https://doi.org/10.1016/j.jacceco.2021.101429

Gil-Bazo, J., Hoffmann, P., & Mayordomo, S. (2020). Mutual funding. *The Review of Financial Studies, 33*(10), 4883–4915. https://doi.org/10.1093/rfs/hhz111

InvestFunds (2021). *Mutual investment funds market statistics*. Retrieved from https://investfunds.ru/funds-statistics/

Khang, K., & Miller, T. W. (2021). Mutual fund performance components: An application to asset allocation mutual funds. *Applied Economics*, 1–16. https://doi.org/10.1080/00036846.2021.2000583.

Maltseva, A., & Dranishnikova, M. (2020). *Digital revolution: What can be done without get-ting up from the couch.* Newspaper Vedomosti of 25 December, 2020. Retrieved from https://www.vedomosti.ru/partner/articles/2020/12/25/852187-tsifrovaya-revolyutsiya.

Roussanov, N., Ruan, H., & Wei, Y. (2021). Marketing mutual funds. *The Review of Financial Studies, 34*(6), 3045–3094. https://doi.org/10.1093/rfs/hhaa095

Sari, K. N., Sulchan, M., & Mutamimah, M. (2021). Performance comparison of mutual funds and Sharia mutual funds. *Bukhori: Kajian Ekonomi dan Keuangan Islam, 1*(1), 65–77. https://doi.org/10.35912/bukhori.v1i1.600

Sberbank. (2021). *Responsible Investment Fund (SBRI ETF).* Retrieved from https://www.sber-am.ru/individuals/etf/etf-esg/

Tkachev, I., & Yushkov, M. (2019). *RBC: Business was offered to invest in national pro-jects through analogs of trusts*. Retrieved from https://www.rbc.ru/economics/06/12/2019/5dea85689a7947223633c7b0

International Legal and Economic Aspects of Regulation of Digital Technologies and Artificial Intelligence

Yuriy V. Mishalchenko, Timofey N. Dovbush, and Maria Yu Mishalchenko

Introduction

The rapid development of digital technologies requires the international community to take coordinated actions to legally regulate this area. Currently, artificial intelligence is changing many aspects in the daily life of society and individuals. National government bodies should respond to the digital technologies development and establish national rules for the artificial intelligence development.

In 2021 more than 40 economically developed countries had their own strategies for the artificial intelligence development, including the transformation of the public sector by increasing the automation of several administrative and management processes to increase the efficiency of the state machinery (Bremer, 2019).

Despite the generation of strategies for the artificial intelligence development by national states, there is actually no comprehensive legal regulation at the international level at the present time. At this stage a number of international organizations are trying to lay the foundations for the artificial intelligence regulation that can be included in the legislation of nation states.

Thus, as part of an international conference in Glasgow in November 2021, several national states and international automotive concerns signed an agreement to ban cars with an internal combustion engine in the main world markets by 2025, and in the rest of the world until 2040. Within the framework of this agreement, it is also planned to develop unmanned electric vehicles (Epoch Times, 2021).

The artificial intelligence development affects the labor market, since the automation of a number of production processes leads to the displacement of a person from a number of industrial sectors that leads to the need to change the approach to

Y. V. Mishalchenko (✉) · T. N. Dovbush · M. Y. Mishalchenko
St. Petersburg Institute of ARSUJ, RLA of Ministry of Justice of the Russian Federation, St. Petersburg, Russian Federation

A. Rumyantseva et al. (eds.), *Challenges and Solutions in the Digital Economy and Finance*, Springer Proceedings in Business and Economics,
https://doi.org/10.1007/978-3-031-14410-3_4

social relations regulation. A few years ago, Elon Musk said that artificial intelligence could be the main threat to humanity (Higgins, 2017).

In 2015 Stephen Hawking and several researchers in the field of artificial intelligence signed an open letter on the need to ban the use of artificial intelligence technologies in the field of weapons (Matney, 2015).

UN High Commissioner for Human Rights Michelle Bachelet voiced concerns about the impact of some artificial intelligence technologies on human rights and called for a ban on the use of artificial intelligence technologies that can prevent a person from exercising their rights and that are incompatible with generally recognized human rights. She also proposed to introduce a moratorium on the use of biometric recognition systems in public places without ensuring the protection of personal data by state authorities and the absence of discrimination in the operation of these systems (ENDS, 2021).

Her report states that, in accordance with Articles 2 and 17 of the International Covenant on Civil and Political Rights, states have not only the obligation not to violate the fundamental human rights to privacy, but also the obligation to protect a person from such invasions, as well as from discrimination within their jurisdiction (Human Rights Council, 2020).

The UN High Commissioner for Human Rights pointed out that artificial intelligence technologies are increasing intrusion into the privacy of people as many companies seek access to personal data to improve their business processes.

Modern technologies for accumulating and analyzing the data obtained create opportunities for influencing a person by both state bodies and private organizations that increases the risks of illegal interference in people's private lives. In addition, artificial intelligence technologies can significantly affect human rights even without the use of personal data, for example, to identify and predict his future actions, as well as in a situation of automated decision-making that may affect human rights.

Thus, the use of biometric recognition technologies by police authorities significantly affects the privacy of a person, since biometric identification of a person allows the state to track the presence of people in public places, nullifying the right to freedom of expression, freedom of movement and freedom of peaceful assembly and rallies.

Technologies that allow to determine the emotional and mental state of a person can negatively affect the realization of human rights; they undermine the human right to privacy and a fair trial. The collection and analysis of information about employees and candidates for filling vacancies and automatic decision-making can lead to a violation of the labor rights of an employee.

The artificial intelligence development significantly affects the legal sciences development right now. Thus, in the European Court of Human Rights a special algorithm is used to analyze judicial practice and prospects for litigation; it can predict the outcome of a case with a probability of up to 80% (Aletras et al., 2016).

Meanwhile, the lack of unified international regulation in this area may lead to the need to introduce many amendments to national legislation that will deprive the national legal system of the predictability of its development. At the same time, the stability of the legal system will not allow the legislation to keep up with the digital technologies development that will lead to the lack of actual legal regulation of social relations.

Materials and Methods

The above-mentioned shows that the world community needs to start preparing a comprehensive international act on the use and limits of the artificial intelligence systems development right now. Within the framework of this international act, it is necessary to legitimize the artificial intelligence concept and to develop international mechanisms for monitoring the artificial intelligence development.

As part of the development of this regulatory act, it is necessary to provide for the limits of regulation of various spheres of society, culture, and science advancement by artificial intelligence. It is also necessary to develop general principles on which artificial intelligence technologies should be based.

Using the comparative law method, we can see that in many countries of the world national legal acts that regulate the digital intelligent systems development are being enacted.

Thus, in the Russian Federation it is planned to create universal intelligent systems that can be integrated to regulate various spheres of public life, within the framework of the Digital Economy of the Russian Federation national program.

In order to develop artificial intelligence in Russia, the Decree of the President of the Russian Federation dated October 10, 2019, No 490 approved the National Strategy for the Artificial Intelligence Development until 2030 in the Russian Federation.

As part of this strategy implementation, it is planned to support scientific research in order to ensure the rapid artificial intelligence development and to design software that uses artificial intelligence technologies, as well as to increase the availability and quality of data necessary for the artificial intelligence technologies development and to increase the level of providing the Russian market of artificial intelligence technologies with qualified personnel and the level of public awareness of the possible areas of use of such technologies.

Within the framework of the departmental digital transformation program, the Ministry of Foreign Affairs of the Russian Federation plans to develop an intelligent system in the period from 2021–2023 that will analyze data arrays on foreign policy issues; in addition, this system is planned to be used for the analysis of personal data and interaction with applicants.

It should be noted that now the UK, the USA and a number of other countries are using an artificial intelligence system for screening visa applications, communicating with citizens, creating various models for analyzing the distribution of the population in developing countries.

In the European Union countries, the advisory resolution of the European Parliament dated February 16, 2017 "Civil Law Rules on Robotics" No. 2015/2103 (INL) (European Parliament, 2017) is of interest.

Within the framework of this resolution, an attempt is made to comprehensively approach the artificial intelligence regulation by legitimizing the basic principles of regulation, the legal responsibility of developers of artificial intelligence systems, the legal status of intelligent systems and ethical recommendations for the use of these systems.

In addition, in April 2018, the European Union Artificial Intelligence Strategy was enacted. It involves encouraging the intelligent systems development by increasing annual investments and creating research centers, supporting the development of these systems, providing access to all users to the necessary resources of the European Union, developing the artificial intelligence systems for key sectors of the European economy (European Commission, 2019a).

In June 2019, the European High-Level Expert Group on Artificial Intelligence drew up two advisory documents:

1. Ethics Guidelines for Trustworthy Artificial Intelligence. This document names seven fundamental requirements that artificial intelligence systems must comply with (the possibility of human supervision of artificial intelligence systems; technical reliability and security; confidentiality of the data received; transparency of the artificial intelligence system; accessibility for all users and all social groups; ensuring the well-being of society and the environment; controllability) (European Commission, 2019b).

2. Policy and Investment Recommendations for Trustworthy Artificial Intelligence that should ensure the growth of the European economy competitiveness (European Commission, 2021a).

In November 2019, the Report on Liability for Artificial Intelligence and Other Emerging Digital Technologies was released in the European Union by the Expert Group on Liability and New Technologies. Within the framework of this report, the principles of legal responsibility for the actions of artificial intelligence are set out and an explanation of the main legal concepts in this area is given (European Commission, 2021b).

In February 2020, the European Commission published the White Paper on Artificial Intelligence, where it indicated that in addition to the benefits that artificial intelligence has, making production processes safer, it can also cause damage (European Commission, 2020).

The damage caused can be both material (destruction of property, loss of equipment) and intangible (disclosure of personal data, restriction of the rights of employees, etc.).

Using the method of legal hermeneutics, we can argue that the legal base being developed in the above countries in the area under consideration seeks to reduce potential risks and is aimed at protecting the fundamental rights of the individual, as well as at security issues.

Results

Artificial intelligence will gradually perform those operations that were previously performed only by people that in the future will lead to the fact that individuals and legal entities will face decisions that are made by artificial intelligence and that will be quite difficult to dispute.

Moreover, artificial intelligence systems will help track the actions of people and be able to predict their behavior that may allow government bodies to massively monitor citizens, and employers to track the actions of their employees.

At the same time, it must be considered that the prejudgement of artificial intelligence systems can have much greater consequences than the mistakes made by ordinary people. The autonomy and non-transparency of artificial intelligence systems can make it difficult to verify the decision taken under the European Union legislation in the field of protection of fundamental rights. Regulatory authorities may not have the means to verify the validity of an artificial intelligence decision and affected individuals may find it difficult to exercise their right to a fair trial.

The lack of clear security criteria can lead to a situation of legal uncertainty for organizations that sell products with artificial intelligence systems, because government bodies may find themselves in a situation where they lack the technical capabilities to test systems and it can reduce the overall level of security and jeopardize the competitiveness of individual industries.

Discussion

Despite the problems discussed above, the countries of the European Union are trying to approve provisions on liability for harm caused by artificial intelligence at the level of national legislation.

For example, Article 7 of the German Road Traffic Act (StVG) provides for the liability of the owner of a motor vehicle even if the vehicle is equipped with artificial intelligence systems (Bundesamt für Justiz, 2021).

Thus, when a traffic accident occurs without the fault of the driver due to the shortcomings of the artificial vehicle control system, the owner of the vehicle remains directly responsible for the requirements of the victims.

There is a similar legal norm in France in Decree No. 2018–211 of March 28, 2018, on experiments with automated vehicles on public roads.

We believe that the artificial intelligence regulation at the level of regional state associations is unproductive since it is necessary to develop and implement global standards.

Realizing the importance of developing common approaches in this area, the Second Conference on Artificial Intelligence was held in September 2021 with the participation of representatives of the OECD, the World Bank, and a number of other international organizations. During this event the need for comprehensive regulation of artificial intelligence at the global level was stated (Globalpolicy, 2021).

As part of meeting the need for global regulation of artificial intelligence, under the UNESCO auspices recommendations on the problems of artificial intelligence that should become one of the first international legal documents on the ethical aspects of artificial intelligence are being developed.

UNESCO notes that in the field of artificial intelligence it is necessary to focus on building a "knowledge society" and achieving sustainable development goals (SDGs) (UNESCO, 2019).

Moreover, UNESCO recommends approaching artificial intelligence in terms of ensuring human rights, openness, accessibility, participation of all interested groups (UNESCO, 2019). In addition to these principles, UNESCO notes the need to develop ethical principles, as well as the need to respect gender and social minorities equality, to pay special attention to overcoming the "digital divide", especially in relation to African countries.

The great importance attached by UNESCO to the issues of artificial intelligence can be seen from the activities that have been carried out on this issue:

the UNESCO global conference "Principles of Artificial Intelligence: Towards a Humanistic Approach" in March 2019 that states that artificial intelligence systems should not replace human intelligence.

the UNESCO International Conference on Artificial Intelligence and Education in Beijing (UNESCO, 2019) that published the final Beijing Consensus (UNESCO, 2019).

XV World Conference on Intellectual Capital "Artificial intelligence and the next generation of competences: how digital and artificial intelligence will impact jobs and competences profiles?"

By the end of 2021, the UNESCO General Conference is expected to approve the developed and agreed ethical principles in the field of artificial intelligence.

Back in 2019 a recommendation on artificial intelligence was approved within the OECD framework. It indicates the main principles of responsible governance to ensure sustainable development and inclusive growth, respect for the rule of law, democratic values and diversity, and, if necessary, human intervention into the operation of the system to guarantee a fair and honest society, as well as transparency and responsible disclosure of information about artificial intelligence systems, the reliability and safety of the operation of artificial intelligence systems throughout their cycle, and the responsibility of organizations developing or operating artificial intelligence systems.

To implement the stated principles, the OECD recommends that national governments promote public and private investments in the field of artificial intelligence, develop modern technologies and digital infrastructure, promote the acquisition of knowledge and skills in working with artificial intelligence systems by people, and carry out cross-border cooperation to achieve progress in this area.

Within the framework of international cooperation, the OECD recommendation emphasises the need to develop international technical standards for compatibility and uniform regulation of artificial intelligence systems.

In June 2019 at G20 summit the Osaka Declaration stating that artificial intelligence should be focused on human interests was enacted.

The COVID-19 pandemic has played an important role in the rapid development of artificial intelligence technologies. Various investment projects in the field of artificial intelligence were able to raise more than $73 billion in the fourth quarter of 2020, which is $15 billion more than in 2019.

In 2021, the volume of the global software market in the field of artificial intelligence exceeded $51.5 billion, increasing at 21.3% compared to 2020.

Conclusion

Thus, the artificial intelligence development has a significant impact on the global economy and requires the adoption of comprehensive urgent measures for the legal and economic regulation of these technologies at the international level.

As a solution to these problems, it seems appropriate to create a special international body under the UN, the main task of which will be to analyze the consequences of the use of artificial intelligence and develop a comprehensive international treaty on the use of artificial intelligence.

The activities of this international body in the future may lead to the creation of an international union in the field of artificial intelligence that will contribute to the economic prosperity of the participants in this association and guarantee the safety of the parties conducting research in the field of artificial intelligence. Within the framework of this association, it is possible to develop uniform regulatory norms that will consider the interests of the parties as much as possible.

In the context of the complication of international economic conditions in the field of artificial intelligence, relevant mechanisms to prevent and promptly resolve conflict situations between states will be created.

The creation of supranational integration associations in the field of artificial intelligence will make it possible to create a legal international mechanism that ensures stable rules for the functioning of artificial intelligence systems and non-discriminatory conditions for all interested parties to work with these systems.

References

Bremer, C. (2019). *Forty-two countries adopt new OECD principles on artificial intelligence.* Retrieved from https://www.oecd.org/science/forty-two-countries-adopt-new-oecd-principles-on-artificial-intelligence.htm

Epoch Times. (2021). *Bis 2040: Ausstieg aus Verbrennungsmotoren – Deutschland hat nicht unterzeichnet.* Retrieved from https://www.epochtimes.de/wirtschaft/bis-2040-ausstieg-aus-verbrennungs motoren-deutschland-hat-nicht-unterzeichnet-a3641485.html

Higgins, T. (2017). Elon musk lays out worst-case scenario for AI threat. *The Wall Street Journal*. Retrieved from https://www.wsj.com/articles/elon-musk-warns-nations-governors-of-looming-ai-threat-calls-for-regulations-1500154345

Matney, L. (2015). *Hawking, musk warn of 'virtually inevitable' AI arms race*. Retrieved from https://techcrunch.com/2015/07/27/artificially-assured-destruction/

ENDS. (2021). *Artificial intelligence risks to privacy demand urgent action – Bachelet*. Retrieved from https://www.ohchr.org/EN/NewsEvents/Pages/DisplayNews.aspx?NewsID=27469&LangID=E

Human Rights Council. (2020). *Impact of new technologies on the promotion and protection of human rights in the context of assemblies, including peaceful protests*. Report of the United Nations High Commissioner for Human Rights. Retrieved from https://undocs.org/en/A/HRC/44/24

Aletras, N., Tsarapatsanis, D., Preoţiuc-Pietro, D., & Lampos, V. (2016). Predicting judicial decisions of the European Court of Human Rights: A natural language processing perspective. *PeerJ Computer Science, 2*, e93. https://doi.org/10.7717/peerj-cs.93

European Parliament. (2017). *European Parliament resolution of 16 February 2017 with recommendations to the Commission on Civil Law Rules on Robotics (2015/2103(INL))*. Retrieved from https://www.europarl.europa.eu/doceo/document/TA-8-2017-0051_EN.html

European Commission. (2019a). *Communication: Building Trust in Human Centric Artificial Intelligence*. Retrieved from https://digital-strategy.ec.europa.eu/en/library/communication-building-trust-human-centric-artificial-intelligence

European Commission. (2019b). *Ethics guidelines for trustworthy AI. European Commission, the High-Level Expert Group on AI. 8 April 2019*. Retrieved from https://ec.europa.eu/digital-single-market/en/news/ethics-guidelines-trustworthy-ai

European Commission. (2021a). *A European approach to artificial intelligence*. Retrieved from https://digital-strategy.ec.europa.eu/en/policies/european-approach-artificial-intelligence

European Commission. (2021b). *High-level expert group on artificial intelligence*. Retrieved from https://digital-strategy.ec.europa.eu/en/policies/expert-group-ai

European Commission. (2020). *White paper on artificial intelligence*. Retrieved from https://ec.europa.eu/info/sites/default/files/commission-white-paper-artificial-intelligence-feb2020_en.pdf.

Bundesamt für Justiz. (2021). *Straßenverkehrsgesetz (StVG)*. Retrieved from https://www.gesetze-im-internet.de/stvg/BJNR004370909.html

Globalpolicy. (2021). *Internet Governance Forum (IGF) 2021, 9 December 2021*. Retrieved from https://globalpolicy.ai/en/events-on-globalpolicy-ai/

Innovative Approach to Starting of Outsourcing Company

Vladimir Aleksandrovich Kunin and Nelly Lugert

Introduction

In the modern uncertain and instable economy, companies with fractal business processes (fractals) seem to be more sustainable. The structure of fractals is similar to the structure of the company (having a common concept); while, the degree of freedom in decision making is higher than in a usual business procedure (Warnecke, 1996).

For large corporations, fractal business processes can be spun off into outsourcing subsidiaries, independent in terms of financing and dependent in terms of contracts for production and supply of goods and/or services. However, such spin-off in instable economy requires preparation of risks reducing measures, optimization of the spun-off business process and establishment of economic relations among the outsourcing company and the parent company. In available research papers outsourcing has an ambiguous interpretation. This article relies upon the idea that an outsourcing company is a subsidiary of the customer. Outsourcing is actively used in business, and a financial role of such cooperation can be assessed by D. S. Kurbatov's method (Kurbatov, 2020). As a rule, a suitable business partner and/or contractor is selected by multiple criteria, considering not only a financial benefit, but also a possibility of doing the work under the required standards and before a certain deadline. The customer has a range of requirements to the outsourcing company in terms of business protocol, management, etc. Large companies have their own standards, such as the IWAY protocol for IKEA procurement, or the international quality management system ISO 9001, etc. A large company is interested in cooperating with a reliable business partner that, on the one hand, can

V. A. Kunin (✉) · N. Lugert
St. Petersburg University of Management Technologies and Economics, St. Petersburg, Russian Federation

A. Rumyantseva et al. (eds.), *Challenges and Solutions in the Digital Economy and Finance*, Springer Proceedings in Business and Economics,
https://doi.org/10.1007/978-3-031-14410-3_5

deliver high-quality products and/or provide high-quality services on time, and on the other hand, will work sustainably under the increased external risks. If the outsourcing company is not a startup, there may be certain difficulties in meeting requirements of the customer; and cooperation with a startup bears risks of delayed delivery, poor quality of services/products, etc. Meanwhile, the outsourcing company spun-off from the parent company for a fractal business has necessary competencies due to previous experience and developments. Thus, in fact, a new outsourcing company can be treated as a subsidiary startup, but with accumulated practical experience.

To speed up and ensure efficiency of creation of an outsourcing company (spinning-off a fractal business process into a separate small business), the authors propose that the parent company should create a digital twin of the spun-off fractal business process. Application of a digital twin is aimed at ensuring quality and increasing economic efficiency of the supplied goods/services; it will contribute to a rapid launch of a outsourcing company, increase reliability of forecasting and improve management of the created startup (Gault, 2019). Incorporation of a BPO company will trigger innovative development of small businesses and improve internal business processes of large businesses (Khmelnitsky, 2008), and will become an additional source of income. In particular, expanded access to innovative technologies of a large business will increase innovative potential and competitiveness of a small subsidiary. As a result, it will improve sustainability of small businesses, ensure innovative transformation of the economy and improve socio-economic conditions of the public (Bukh & Heeks, 2018). Also, it may reduce corruption, since there will be no room for political intra-corporate conflicts in the parent company. Basic information about business processes is displayed in such information systems as ERP, CRM, BPMS, for instance, details about orders, logistics, transactions, audit results (Lee et al., 2020). CRM system is a database of orders and customers, and stores information about documents accompanying transactions, customer contacts and order statuses. ERP synchronizes work of departments, business units, controls and monitors sales and other company processes. The authors believe that it is imperative to use database of those systems from a spun-off fractal business process in order to form and develop a scenario for processes interaction in the business model of an outsourcing company.

Materials and Methods

The article uses methods of comparative and scenario analysis, synthesis, synergistic and systematic approaches, as well as methods for creating a digital twin for online promotion as described above (Lugert, 2020). For an outsourcing company, it is recommended to make a 3D visualization of a digital twin for clarity and transparency of management (Kupriyanovsky et al., 2020), but it is not necessary. The difference between a digital twin of a company and a digital twin of online promotion is that the former requires combining all existing processes and developing a scenario for interaction between management and production processes (Halenar et al., 2019). It should be pointed out that formalization of decision-making is not possible for all procedures (Garanin, 2018), this paper only considers modification of a business process suitable for formalization.

Results

Business scale is a traditional indicator to increase profits. Centralization of management is a forced measure that ensures business sustainability. Concentration of control at the top of the business process hierarchy hinders quick identification of problems at lower levels and impedes operational flexibility and effective management. Creation of a management chain based on the vertical structure threatens large businesses with high inertia in forecasting and decision-making. Actually, such form of control system obstructs three requirements of the modern market: reduction of decision-making time, reduction of decision-fulfillment time, and quick launch of products to the market (Warnecke, 1996). This drawback of a vertical management system is fueled firstly, by the need to develop and implement a set of rules governing the management process, and secondly, by the need to organize a complex control system.

The situation with unidentified fractal business processes is especially problematic. These may be departments that perform their functions independently of other company clusters but are located inside another business process and remain concealed from the upper levels of management. Fractals are formed due to various factors:

– Delegating part of the workflow.
– Forced differentiation due to the problematic sector.
– Need for an innovation cluster.

Due to widespread innovative technologies, means and tools, as well as current digital transformation of the economy, competitive corporate development is only possible with effective innovation policy (Götz, 2019), aimed at increasing innovative potential. Such policy includes:

– Intellectual property ratio, defined as a share of intangible assets in the total amount of non-current corporate assets.
– Innovative growth ratio, defined as a share of costs for innovative activities in the total cost of the corporation.
– Innovative share products, defined as a share of innovative products in the amount of sales of the corporation.

These indicators characterize innovative corporate activities and can be treated as factors of innovative potential (Kunin & Tarutko, 2018).

Creation and implementation of innovative materials, ensuring quality of materials and components require new laboratories and business branches (Nikulina,

2019). Therefore, in order to keep together intellectual and material resources and concentrate on solving main development objectives, large businesses should delegate the work on related developments to third-party companies.

Decentralization of large business management can be reached due to cooperation with outsourcing companies, which can be either third-parties or spin-offs from internal fractals. International experience shows that major market players can control quality of labor and management of third-party partners. If a company has introduced a certain quality management system, it must check its partners for compliance with the QMS protocol. This makes additional difficulties due to loss of time for verification, increase in the control staff and dissatisfaction of partners who put a lot of effort to go through the process. Therefore, it seems wise to spin off an outsourcing company, and provide a software solution for business processes management and technological work regulation. A fractal has an inherent potential for self-improvement (Olifir, 2021), therefore, an outsourcing company spun-off from the fractal has good prospects for development. The fractal structure can also be used for other processes, that is, it can be applied in other production, if necessary. (Hlushak & Nykyforchyn, 2018). It should be pointed out that digital twins of key business processes have a universal shell that ensures their application for many companies, by setting parameters to adapt to characteristics of a company. Therefore, construction and practical application of such digital twins shall not require large investments and shall pay off due to optimization of business processes mastered on such digital twins.

Modern technologies ensure creation of a digital twin for the whole company. If a company consists of fractals, the authors propose to delegate a fractal business process to a BPO company together with a digital twin, which should contain required management protocols, management quality standards and technologies.

Creation of a digital twin is directly related to a conceptual model of interaction within the company and presents a multi-level architecture (Qamsane et al., 2021).

Development of a conceptual corporate model is a team work (Godager et al., 2021) aimed at building a structured description, the so-called "corporate model", which represents basic principles and relations of corporate business.

Figure 1 shows a general example of a conceptual model of an outsourcing company. Main interactions occur between process resources, control and transformation of input and output data. Based on the transformations, reporting documents are generated, then the data is checked (analyzed), after which feedback and subsequent information and process management is received.

In the proposed model, relationships are classified by priorities; arrows indicate relations between main elements. Ontology of an outsourcing company is based on the data from the CRM system of the previous fractal business process. The input and output data used in the conceptual model are described in details, these data are digitized. The ontology code generation is made by using formal languages OWL, KIF.

The described outsourcing company model contains components integrated from the previous fractal business process, such as: resources, documentation, logistics,

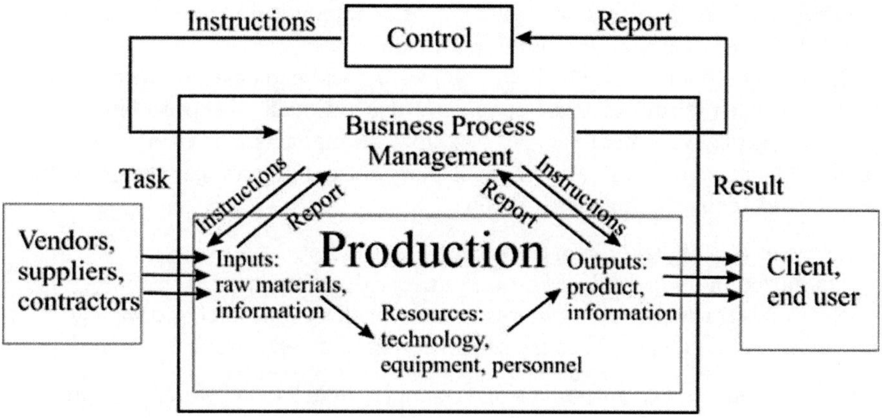

Fig. 1 Example of a conceptual model for an outsourcing company

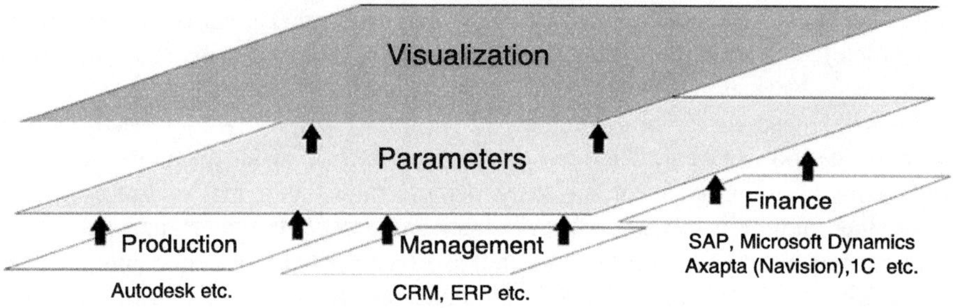

Fig. 2 Information blocks for creating a digital twin

production, etc. A digital twin gives a chance to the outsourcing company owner to gain knowledge and experience.

In order to form a digital twin, it is necessary to digitize blocks of corporate activities and display parameter values either as numerical values or as visualization. Figure 2 shows an example of blocks that contain basic information about corporate business, which may be included in a digital twin. It should be pointed out that each block has auxiliary programs, applications for conducting business that are successfully applied in various industries. For example, using the Autodesk program, you can describe and define technical parameters of produced equipment. The above described CRM, ERP systems help to manage the company. There are also various accounting programs for financial and accounting reporting. Thus, a company that integrated these programs into its will be available create a digital twin both technically and financially.

Creation of a digital twin should be started with programming one block, for instance, the "Production" block. The cost of digital twin development depends on its modification—it can be a 3D rendered object that repeats and interprets parameters of many sensors in production, or an interpretation of numerical parameters obtained by calculations during forecasting.

It should be pointed out that it will be most effective to delegate fractal business processes to BPO subsidiaries—companies that maintain close relations with the parent company even after their spin-off. Since the digital twin is an effective tool to optimize business processes, when a subsidiary outsourcing company is spun-off the digital twin should be applied as follows:

- Creation of a digital twin of the business process.
- Optimization of the business process using a digital twin.
- Spin-off a fractal business process into a subsidiary outsourcing company.
- Transfer of a digital twin to the subsidiary on preferential terms.

Let's consider an example of using a simple digital twin, which can simulate operation of a press with numerical parameters.

We consider financial costs of operating a small vertical press making bales from loose waste paper with and without a digital twin. The manufacturers declare the following characteristics of the press (we use parameters that are directly relevant for this article): temperatures from +40 °C to −40 °C, number of bales produced in 1 hour is 2–4 pieces, weight of a bale depending on raw materials is 250–550 kg, there is no need for commissioning, the equipment is ready for operation, the warranty period is 1 year. The press is operated outside under a canopy, which is an acceptable condition, according to the manufacturer. In the first year of operation in summer-autumn it forms 3 bales per hour on average. In winter, at sub-zero air temperatures, productivity of the press fell to 1 bale per hour, with maximum winter temperature not lower than −10 °C in the first year. In the second year of operation in the warm season, productivity increased again, and in winter with temperature of −15 °C, the press went out of order. It turned out that for uninterrupted operation of the press at sub-zero temperatures, usual oil must be changed to a special one, but productivity still remains 1 bale per hour. In the third year, the concrete slab on which the press was installed collapsed. The cost of the slab is 50,000 rubles, downtime is 1.5 weeks.

Application a digital twin identifies problematic areas in operation of the press in advance; and develops preventive measures to avoid downtime risks or reduce operating costs.

Figure 3 shows that, despite rather high costs of creating a digital twin, its use is rational in terms of finance, as confirmed by statistics of subsequent years of operation. Knowing in advance how the press would behave at low temperatures, the production site should have been placed in a heated room, despite the cost of rent. Production downtime could also have been avoided by prompt oil change and fabrication of the concrete slab in advance. The cost of creation of a digital twin to forecast the work of the press is comparable to the cost of the press itself. Digital twins that do not require lengthy installation and study of various equipment operations are financially affordable for small industries.

Spinning-off a subsidiary outsourcing company from a fractal, on the one hand, is accompanied by a decrease in fixed costs due to reduced employee salaries and

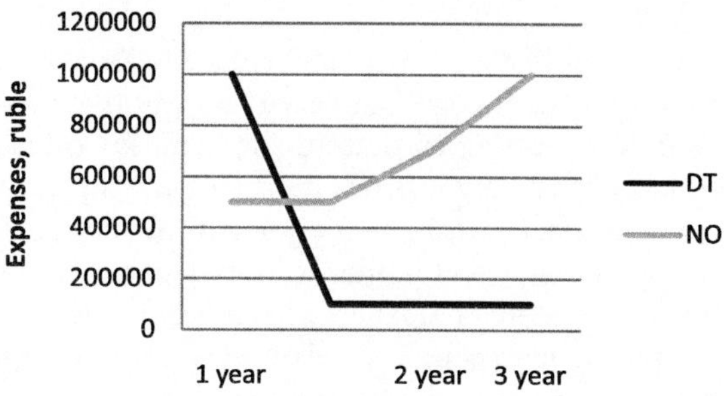

Fig. 3 Graph of press maintenance costs by years with and without the digital twin (DT) and without it (NO)

reduced cost of operating and maintaining equipment, and on the other hand, is accompanied by additional fixed costs associated with paying for services of a BPO company.

Condition for spin-off profitability looks as:

$$|\delta FC_1| > |\delta FC_2|, \quad (1)$$

where.

$|\delta FC_1|$ and $|\delta FC_2|$ respectively, are absolute values of reduced fixed costs and additional fixed costs of the parent company in spinning-off an outsourcing company.

Fulfillment of condition (1) entails an increase in profit from sales and profitability of sales of the parent company, the relative value of the increase is equal in absolute terms to the resulting relative decrease in fixed costs.

$$\Delta_\delta = |\delta FC_1| - |\delta FC_2|$$

The resulting expression should be interpreted as a condition for a cost-effective spinning-off a fractal process into an outsourcing company. By formalizing this condition in the digital twin algorithm, we can automate the decision-making process about spinning-off a fractal into a separate company.

Discussion

The main objective of this research was to develop a concept for a digital twin of a partner outsourcing company, state the reasons of its use and promote its use. The digital twin simplifies creation and ensures high-quality functioning of a startup due to accumulated experience in the fractal business process. An innovative approach to creating an outsourcing company by creating a digital twin, proposed herein, offers a new form of business cooperation through transparency in forecasting, automation of control and interaction with the parent company.

Also, a digital twin can help find the most suitable partner from the market players, for instance, by compliance of management quality with a certain protocol, standard, etc., if there is no need to spin-off a subsidiary outsourcing company.

Before creating an outsourcing company or spinning off a fractal into an independent subsidiary it is recommended to analyze the process of interaction with the parent company. It should be pointed out that creation of an outsourcing company is not an ultimate goal in modernization of a large business. It is necessary to ensure that outsourcing companies do not become monopolists in goods/services, since it can rise an unacceptable level of dependence of large businesses on outsourcing. The digital twin is reconfigurable software that can be used in other related areas. Therefore, the cost of creating a digital twin can be reimbursed by selling or renting the shell to other companies.

Conclusions

The article proved reasons for spinning off a fractal business process into a separate company for further mutually beneficial partner cooperation. The paper also formalized a condition for economic efficiency of spinning off a fractal into a separate outsourcing company. The main processes of an outsourcing company were identified, and a conceptual model of resource interactions was created. The paper proved that creation of an outsourcing company contributes to motivational expansion, that is, new goals for innovative development and implementation of innovative solutions due to high flexibility of a small business start-up. The authors formulated a condition for economic efficiency of spinning off a fractal into a separate outsourcing company; they also showed practical application of a digital twin in order to reduce costs caused by economic risks, and effective use of a digital twin in transformation of business relations when spinning off outsourcing companies.

References

Bukh, R., & Heeks, R. (2018). Definition, concept and measurement of the digital economy. *Bulletin of International Organizations, 13*(2), 143–172. https://doi.org/10.17323/1996-7845-2018-02-07

Garanin, M. A. (2018). The impact of digital twins on the public sector of economy. *Creative Economy, 12*(11), 1733–1758. https://doi.org/10.18334/ce.12.11.39605

Gault, F. (2019). User innovation in the digital economy. *Foresight and STI Governance, 13*(3), 6–12. https://doi.org/10.17323/2500-2597.2019.3.6.12

Godager, B., Onstein, E., & Huang, L. (2021). The concept of enterprise BIM: Current research practice and future trends. *IEEE Access, 9*, 42265–42290. https://doi.org/10.1109/ACCESS.2021.3065116

Götz, M. (2019). The Industry 4.0 induced agility and new skills in clusters. Foresight and STI. *Governance, 13*(2), 72–83. https://doi.org/10.17323/25002597.2019.2.72.83

Halenar, I., Juhas, M., Juhasova, B., & Borkin, D. (2019, May). Virtualization of production using digital twin technology. In *2019 20th International Carpathian Control Conference (ICCC)* (pp. 1–5). IEEE. https://doi.org/10.1109/CarpathianCC.2019.8765940.

Hlushak, I., & Nykyforchyn, O. (2018). Fractal dimensions for inclusion hyperspaces and non-additive measures. *Matematychni Studii, 50*(1), 3–21. https://doi.org/10.15330/ms.50.1.3-21

Khmelnitsky, K. S. (2008). Outsourcing in the development of enterprises and business groups. *Terra Economicus, 6*(4–2), 213–215.

Kunin, V. A., & Tarutko, O. A. (2018). *System of indicators of competitiveness of entrepreneurial structures/Problems of modern economy, 1*(65), 65–68.

Kupriyanovsky, V., Klimov, A., Voropaev, Y., Pokusaev, O., Dobrynin, A., Ponkin, I., & Lysogorsky, A. (2020). Digital twins based on the development of BIM technologies, related ontologies, 5G, IoT, and mixed reality for use in infrastructure projects and IFRABIM. *International Journal of Open Information Technologies, 8*(3), 55–74.

Kurbatov, D. S., & Gao, Lei Fu. (2020). Outsourcing in surface mining operations: decision-making framework with formal algorithm. *Ugol' – Russian Coal Journal, 1*(1126), 26–31. 10.18796/0041-5790-2020-1-26-31.

Lee, S., Choi, I., Kim, H., Lim, J., & Sung, S. (2020). Comprehensive simulation and redesign system for business process and organizational structure. *IEEE Access, 8*, 106322–106333. https://doi.org/10.1109/ACCESS.2020.3000248

Lugert, N. E. (2020, April). Creating a digital twin to the online promotion process for small businesses. In *III International Scientific and Practical Conference "Digital Economy and Finances" (ISPC-DEF 2020)* (pp. 17–21). Atlantis Press.

Nikulina, N. O. (2019). Intellectual decision-making support in the risk analysis of an innovative project. *Design ontology, 9*(3 (33)), 382–397. https://doi.org/10.18287/2223-9537-2019-9-3-382-397

Olifir, D. I. (2021). Fractals in the innovative cluster. *Azimuth of Scientific Research: Economics and Administration*, 10(1 (34)), 240–243. https://doi.org/10.26140/anie-2021-1001-0058.

Qamsane, Y., Moyne, J., Toothman, M., Kovalenko, I., Balta, E. C., Faris, J., et al. (2021). A methodology to develop and implement digital twin solutions for manufacturing systems. *IEEE Access, 9*, 44247–44265. https://doi.org/10.1109/ACCESS.2021.3065971

Warnecke, H.-J. (1996). *Die fraktale Fabrik – Reinbek bei.* Rowohlt.

Implementability of a Process Approach to Strategic Management in the Conditions of Digitalization

Yuliya Ivanovna Rastova, Marina Ivanovna Barabanova, and Mikhail Yurevich Makarov

Introduction

Existence in an era of great challenges determines the need to take advantage of those opportunities for growth, ensuring national security, improving the quality of people's life, developing industries of a new technological paradigm that are opened up by the consistency of priorities and tools for the formation and implementation of strategies at the national, regional, sectoral and corporate levels. And if in Russia at the state level the configuration of the strategic planning system has been formed and experience in implementing the program-target approach in public administration has been gained to this date, then at the corporate level (with the exception of a number of large companies, primarily with public ownership) such requirements for the quality of corporate governance as expanding the planning horizon, clear goal-setting, maintaining a well-thought-out strategic policy, and standardizing the main management processes are yet being comprehended (Miller et al., 2017). At the same time, a significant factor determining global changes in management itself is a sharp increase in the amount of information, the emergence of fundamentally new ways of working with it, changes in the forms of organization, hardware and software tools for developing, adopting, extending and maintaining strategies. It is necessary to structurally—in a business context and guided by the DCED standard requirements—use business intelligence tools (Business Intelligence, BI) and big data

Y. I. Rastova · M. I. Barabanova
St. Petersburg State University of Economics, St. Petersburg, Russian Federation
e-mail: rastova.yu@unecon.ru

M. Y. Makarov (✉)
St. Petersburg University of Management Technologies and Economics, St. Petersburg, Russian Federation

A. Rumyantseva et al. (eds.), *Challenges and Solutions in the Digital Economy and Finance*, Springer Proceedings in Business and Economics,
https://doi.org/10.1007/978-3-031-14410-3_6

analytics (Big Data Analytics, BDA) so that the use of digital technologies is not limited to collecting more data, based on their availability.

One of the important trends in the business environment of the twenty-first century is standardization. When it comes to strategic management, the assessment of its maturity in a company from the standpoint of various standards necessarily includes the following requirements:

- identification, establishment, maintenance, control, and improvement of strategic management processes.
- conformity with the needs and expectations of interested parties to increase their satisfaction and overall perception.

The verified information used in the generation, development, revision and updating of strategies should be based on measurements, research, creative search, and the strategy itself should provide an exhaustive presentation of the aspects and factors that determine the key processes of the organization.

It is fair to say that the strategy as a template (pattern) was studied by H. Mintzberg (1978), as well as R. R. Nelson and S. G. Winter (1982).

The impetus for the process approach uses by the strategic management theorists was the consideration of strategy in conjunction with the organizational actions of the company by D. Schendel and C. W. Hofer (Schendel & Hofer, 1979). At the same time, as N. Khan and N. Korac-Kakabadse (2013) emphasize, the need to consider strategy not as a "decision", but as an "action" was most consistently considered by H. Mintzberg (Mintzberg & Waters, 1985), and A. M. Pettigrew (1992, 1997).

An independent direction of research—"strategy-as-practice" (SAP) has been spreading since the beginning of the 2000s. V. L. Tambovtsev characterizes the essence of the SAP in the following way, ". . .to be practically applicable, research on strategies should focus mainly on specific practices in the development and implementation of strategies" (Tambovtsev, 2011, p. 56).

However, the papers known to the authors are of a theoretical nature, they do not provide for the use of process modeling by software using standard notations, as well as such tools as reference models and frameworks. The question of criteria and indicators of the process maturity of strategic management in the organization is not raised, and, as a result, tools for assessing the sensitivity of target strategic indicators, methods for determining the controllable limits of variation of key performance indicators and reasons for variation are not proposed, and the procedure for implementing rational real corrective actions is not considered.

Materials and Methods

The configuration of corporate strategic management systems from the standpoint of the process approach can be determined by decomposing strategic management processes based on frameworks of varying degrees of detail (conceptual pyramids, cascading to the level of key events, performers, actions).

The function of strategic choice frameworks can be performed by matrices with "product/market" coordinates by Igor Ansoff (1957), "sphere of competition/competitive advantage" by Michael Porter (1980), "stages of the industry life cycle/competitive positions of business units of an organization" ADL/LC, McKinsey's business screen "attractiveness of the economic sector/strengths of the company". At the stage of extending, maintenance and control of strategies, these are strategic maps, key performance indicators and cascading methods (Rastova & Rastov, 2018).

However, the attribute of assessing the strategic management process maturity is, in our opinion, the degree of mastery of the organization management of the tools of the system dynamics theory (SD)—cause-and-effect and flow diagrams, systems of finite difference equations with given initial values levels (Forrester, 1958). Models of system dynamics, along with models of statistical process control and reliability management, are known to be tools for process efficiency. J. D. Sterman (2000), K. D. Warren (2008), H. Rahmandad (2015), M. Barabanova et al. (2018) write about the integration of strategic management concepts with dynamic modeling, about the possibility of searching for best practices in this way.

Simulation dynamic modeling, focused on the use of the positive feedback loops effects, makes it possible to overcome the extreme complexity of the processes of making and implementing strategic decisions in the context of the transformation of the organization environment, the values and expectations of stakeholders, and future needs for resources and technologies.

Results

First, the process approach is the systematic definition and management of processes.

In strategic management, the key processes at the stage of strategy generation are:

- definition of the context of the organization: understanding the environment of the organization and identifying interested parties, establishing the processes necessary to meet their needs and expectations.
- definition of the distinctive features of the organization—its mission, vision, values.
- definition of long-term prospects.

Understanding the external environment of the organization involves the systematization of objective information in terms of legislative and regulatory requirements (agreements), including those specific to the industry; competition; globalization; social, economic, political and cultural factors; innovation and advances in technology; environment. Internal environment includes the types and complexity of activities, relevant processes, resources, levels of competence and knowledge of the organization, management maturity, innovation.

It should be said that in the world practice indicators that characterize the mood of stakeholders: the consumer sentiment index (University of Michigan's Consumer Sentiment Index), the index of leading macroeconomic indicators (Conference Board Leading Index), Consumer Confidence Index, Institute for Supply Management (USA), IHS Markit (Europe), HSBC (China), business activity indices (PMI Manufacturing, PMI Services/PMI non-manufacturing) are widely used.

However, today the level of understanding, assessment of the benefits, risks and opportunities of permanent relationships with consumers and partners is insufficient. Communication processes that meet requirements of interested parties are still being established.

Today, the reporting of organizations is focused on "the triple bottom line" principle and implies two more "bottom lines" in addition to the financial result: social and environmental (Jeurissen, 2000). Information concerning respect for human rights, the fight against corruption, the composition of the board of directors is disclosed (European Union, 2013).

In our opinion, it is high time to talk about "the fivefold bottom line", including indicators of loyalty of interested parties, organizational culture, and maturity of business processes. The best practices to establish the processes necessary to meet the needs and expectations of interested parties can be identified in this way.

At the same time, it should be recognized that studies of the behavior of consumers and other interested parties, based on primary data of individual companies, quickly become outdated, show "gaps" between causes and effects, and imbalances in the decisions made on their basis. Efforts should be made to rapidly develop and spread digital service models in different market segments to collect and transform stakeholder loyalty and satisfaction data using BI and BDA tools into information that is sensitive for making strategic decisions. The volatility of indicators of loyalty (satisfaction) of stakeholders using BI and BDA tools is a reliable sign of a change in the position of organizations and even industries in a competitive environment, their impact on the economic, environmental, and social environment, as well as involvement in social processes.

Changes in the environment require a revision of the mission, vision and values of the organization. Today social responsibility, environmental and cultural factors have become part of the organization environment, and management needs to determine its attitude to the current ESG agenda (Environmental, Social, Governance), as well as to the challenges of the BANI world, with its fragility (Brittle), anxiety (Anxious), non-linearity (Nonlinear), incomprehensibility (Incomprehensible). However, expansionist thinking should remain among the values and corporate ethics of the company.

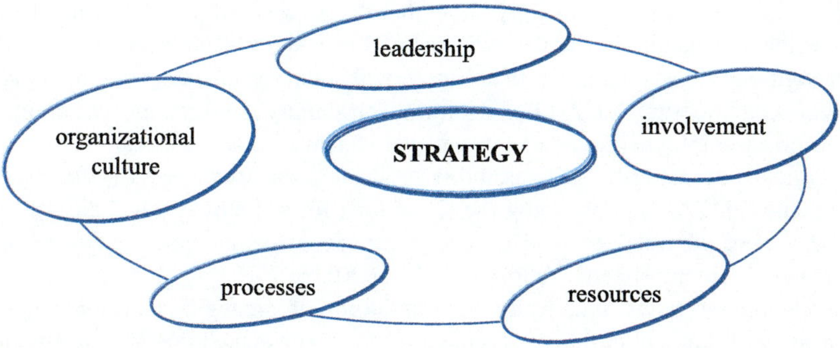

Fig. 1 Eco-environment of developing, adopting, expanding, and maintaining strategies. Source: Developed by the authors

The methodological imperative is the thesis according to which the exclusive competence of the board of directors includes the following matters:

- value creation through mergers and acquisitions.
- use of existing skills and resources to create a competitive advantage by expanding the market.
- development of key competencies such as business processes, the level of staff qualifications, the strength of their motivation, the priority of teamwork, the service provided to customers.

Other issues of strategic management—planning, distribution of duties and powers, systematic measurement and analysis of performance and efficiency, continuous improvement and optimization of procedures are the powers of the company management.

The success of strategic management is determined by a number of conditions that can be called the eco-environment of developing, adopting, expanding and maintaining strategies (Fig. 1).

In order for the strategy to be fully used for business management, it is necessary to demonstrate leadership on the part of managers at all levels. The role of individuals involved in making strategic decisions and organizing their execution is the subject of many papers in the field of strategic management (Pentland, 1999; Khanna et al., 2000; Pettigrew et al., 2001). Leadership as a factor in the strategic management eco-environment is personal participation in all strategic management processes, direct contacts with interested parties, and involvement of staff in continuous improvement.

The involvement of staff is manifested in the understanding of each person of their contribution to the results of the organization activities, in the desire to eliminate obstacles in the work, to increase their competence, knowledge and experience, in the willingness to share their knowledge and experience with colleagues.

Numerous studies convincingly show that the level of employee involvement is one of the key factors in the performance and efficiency of companies, as well as their high market adaptability and ability to withstand the challenges of the modern world (Onuchin et al., 2012). The scientific community is searching for an optimal approach to employee involvement management; methods for measuring involvement are systematized; the possibility of using the key tool of the process approach—PDCA—in describing the algorithms for measuring this indicator, analyzing the results obtained, and choosing directions for increasing engagement is convincingly demonstrated (Kolobov et al., 2020).

It should be noted that from the standpoint of strategic management, it is appropriate to talk about the involvement of other key stakeholders—shareholders, suppliers, banks, guarantors, etc.

The involvement of stakeholders in strategic processes is determined not only by the efforts of leaders, but also by the culture of the organization, consistent with its mission, vision, and values. The board of directors must have at its disposal the results of objective monitoring of key aspects of the culture success, as well as threats to its identity.

When it comes to resources, of course, their availability and quality determine the prospects that are laid down in the organization strategy. Through benchmarking, the organization must gain confidence through continuous improvement that resources are optimally allocated and that all processes are used efficiently and effectively.

A key resource for strategic management is own experienced corporate finance team, able to strategically justify performance, test risks, and secure sound deals on the activities envisaged by the strategic plan. Both these and other key competencies of the company that can become less valuable, including due to external events, should be the subject of regular objective monitoring.

Just as in the case of the use of resources, the efficiency of processes is also compared with leading organizations. The achievement of strategic goals directly depends on how much the outputs of key processes exceed the average indicators for the given sector, the necessary knowledge and skills are determined, and employees are motivated to prevent errors and are involved in work to improve processes and procedures.

Based on the model shown in Fig. 1, we have a finite ordered set of factors—links of the ecological environment (LEE):

$$LEE(p) = \left\{ \begin{array}{l} leadership \\ involvement \\ resources \\ processes \\ organizational\ culture \end{array} \right\},$$

where p is a number of links.

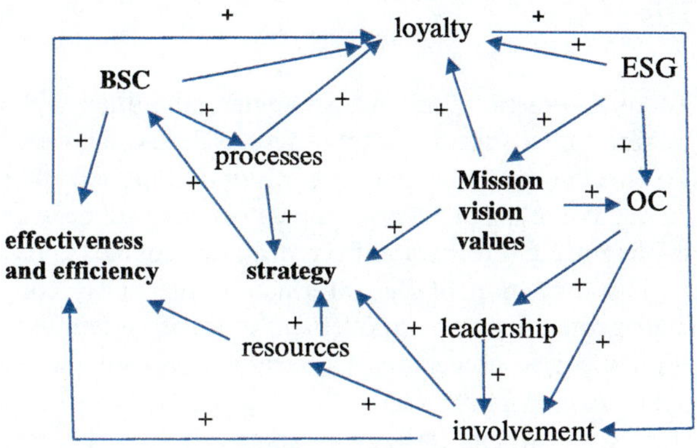

Fig. 2 Cause-and-effect diagram of the strategic management process. Source: Developed by the authors

In the categories of system dynamics, the presented links of the strategic management eco-environment are accelerators that ensure its success. The relationship of the links of the eco-environment with the process of developing, adopting, expanding, and maintaining strategies in the form of a cause-and-effect diagram is shown in Fig. 2.

The task of the right balance of control actions on the links of the eco-environment of strategic management that determine the pace (flow) of their change belongs to the category of semi-structured problems with a multi-criteria solution and the use of positive feedback. In this case, simulation dynamic modeling suggests using vector optimization in analogy with:

$$P = \{p_1^i, p_2^i, p_3^i, p_4^i, p_5^i\},$$

where p^i is the "ideal" value of each of the model parameters.

In order to develop a mathematical model in the form of a dynamic system of simultaneous equations and conduct a simulation on it to search for the ideal vector of changes in the links of the eco-environment that ensure the relevance of the strategy, the effectiveness and efficiency of its implementation, it is necessary to collect a large array of historical and statistical data, determine the values of the converters that determine the nature and degree of dependence of the model parameters on each other. To do this, a consensus must be reached in the business community on the use of unified indicators for assessing model parameters and readiness to disclose such information, including through a specially created digital platform.

Discussion

Today, in practice, “servant” strategies, simplified strategic maps built according to a unidirectional scheme, without feedback and their evaluation are often developed. To develop key performance indicators, simplified approaches are used, methods for alternative selection of target indicators have not been developed, manual calculation is often used, information from disparate sources is used.

The formation of the information-logical structure of the company strategic management using the model and tools of system dynamics will be the reason for formalizing its procedures and results, providing the prerequisites for efficiency and continuous improvement, achieving a new level of competitiveness and relationships with owners, staff, and market.

Models of system dynamics will make it possible to evaluate, analyze and predict the effectiveness of managerial influences on the links of the strategic management eco-environment, and will make it possible to justify managerial decisions on the use of positive feedback effects inherent in it.

The distance that separates us today from the introduction of the described model into business practice is determined by the poor mastery of the tools for assessing the loyalty of consumers and partners, the maturity of processes, the level of organizational culture, and the involvement of key stakeholders.

An attempt to start a discussion about the redundancy of the requirement to master strategic management patterns focused on using the capabilities of system dynamics can be stopped by the argument, “assess the risks in case of refusal to use them deliberately”.

The gradual expansion of the practice of using system dynamics models in the management of strategic management processes will make it possible to single out homogeneous groups of companies in order to identify other latent factors, identify and spread the best practices.

Conclusion

Standardization of strategic management processes in a modern organization on the basis of simulation system-dynamic modeling provides overcoming the complexity of making strategic decisions by assessing the sensitivity of target strategic indicators to changes in one or more parameters of the system-dynamic model, determining the controllable limits of variation of key performance indicators and reasons for variation, as well as the implementation of rational corrective actions, which ultimately ensures an increase in the level of process efficiency in the development and implementation of strategies.

The approach proposed by the authors will make it possible to create development methods and models of digital services for collecting and analyzing data significant for making strategic decisions using BI and BDA tools, as well as for configuring the strategic management process and assessing the process maturity of strategic management in domestic companies in the non-financial sector in accordance with the requirements of the "integrated ability maturity model" using specialized software of BPM systems.

References

Ansoff, I. (1957, Sep–Oct). Strategies for diversification. *Harvard Business Review*, 113–124.

Barabanova, M., Lebedeva, L., Rastova, Y., & Uvarov, S. (2018). Use of system dynamics tools in value-oriented approach in management. *Economic Annals-XXIthis, 173*(9–10), 32–37. https://doi.org/10.21003/ea.V173-05

Forrester, D. (1958). Industrial dynamics – A major breakthrough for decision makers. *Harvard Business Review, 36*(4), 37–66.

Jeurissen, R. Y. E. (2000). Cannibals with forks: The triple bottom line of 21st century business. *Journal of Business Ethics, 23*(2), 229–231. https://doi.org/10.1023/A:1006129603978

Khan, N., & Korac-Kakabadse, N. (2013). Collaborative Corporate Strategy Research Programmes (C.C.S.R.P.) a conceptual integrative strategic framework for a practical research agenda. *Open Journal of Business and Management, 1*, 91–108. https://doi.org/10.4236/ojbm.2013.13011

Khanna, T., Gulati, R., & Nohria, N. (2000). The economic modeling of strategy process: "Clean models" and "dirty hands". *Strategic Management Journal, 21*(7), 781–790. https://doi.org/10.1002/1097-0266(200007)21:7<781::AID-SMJ119>3.0.CO;2-2

Kolobov, A. V., Igumnov, E. M., & Naumov, D. N. (2020). Ensuring personnel involvement in companies' sustainable development. *Strategicheskiye resheniya i risk-menedzhment (Strategic Decisions and Risk Management), 11*(3), 262–271. https://doi.org/10.17747/2618-947X-2020-3-262-271

Miller, A. E., Kuznetsov, S. V., & Rastova, Y. I. (2017). Instruments to assess efficiency of implementation of the strategy of scientific and technological development of the Russian Federation. *Espacios, 38*(33), 16.

Mintzberg, H. (1978). Patterns in strategy formation. *Management Science, 24*(9), 934–948. https://doi.org/10.1287/mnsc.24.9.934

Mintzberg, H., & Waters, J. A. (1985). Of strategies, deliberate and emergent. *Strategic Management Journal, 6*(3), 257–272. https://doi.org/10.1002/smj.4250060306

Nelson, R., & Winter, S. G. (1982). *An evolutionary theory of economic change*. MA Harvard University Press.

Onuchin, A. N., Lutskina, V. V., & Rozin, M. V. (2012). Personnel engagement management. *HRTimes, 20*, 37–40 (in Russ.)

Pentland, B. T. (1999). Building process theory with narrative: From description to explanation. *Academy of Management Review, 24*(4), 711–724. https://doi.org/10.5465/amr.1999.2553249

Pettigrew, A. (1992). The character and significance of strategy process research. *Strategic Management Journal, 13*(2), 5–16. https://doi.org/10.1002/smj.4250130903

Pettigrew, A. M. (1997). What is a processual analysis? *Scandinavian Journal of Management, 13*(4), 337–348. https://doi.org/10.1016/S0956-5221(97)00020-1

Pettigrew, A. M., Woodman, R. W., & Cameron, K. S. (2001). Studying organizational change and development: Challenges for future research. *Academy of Management Journal, 44*(4), 697–714. https://doi.org/10.5465/3069411

Porter, M. E. (1980). *Competitive strategy: Techniques for analysing industries and competitors*. Free Press.

Rahmandad, H. (2015). Connecting strategy and system dynamics: An example and lessons learned. *System Dynamics Review, 31*(3), 149–172. https://doi.org/10.1002/sdr.1541

Rastova, Y. I., & Rastov, M. A. (2018). Strategic management of a modern organization: Concept synergy effect. *Upravlencheskiye nauki (Management Sciencethis), 8*(3), 20–31. https://doi.org/10.26794/2404-022X-2018-8-3-20-31

Schendel, D. E., & Hofer, C. (1979). *Strategic management: A new view of business policy and planning*. Little, Brown, and Co.: Boston.

Sterman, J. D. (2000). *Business dynamics: Systems thinking and modeling for a complex world.* Irwin/McGraw-Hill.

Tambovtsev, V. L. (2011). Research program "strategy as practice" in the study of strategic management. *Rossijskij zhurnal menedzhmenta (Russian Management Journal), 9*(4), 51–62.

The European Parliament & The Council of the European Union. (2013). Directive 2013/34/EU of the European Parliament and of the Council of 26 June 2013 on the annual financial statements. *Official Journal of the European Communities*. Luxembourg: European Union.

Warren, K. D. (2008). *Strategic management dynamics*. Wiley.

Research on the Effects of "DeepFake" Technology for the Modern Digital Space

T. Afanasyeva and I. Yumasheva

Introduction

In recent years, fake news has become a serious menace to society. Fake news refers to content in the style of fictional news, fabricated to deceive the public. The dissemination of false information is carried out primarily through social networks, which makes it possible to achieve a negative impact on millions of users in a short time (Afanasyeva et al., 2019).

Currently, every fifth Internet user receives news through the online platform YouTube, followed by Telegram, Facebook, VKontakte and others. The growing popularity of video platforms underscores the need to create and develop tools to confirm the authenticity of news content. Given the ease of obtaining and spreading disinformation through social networks, online and video platforms, it is becoming increasingly difficult for users to determine the authenticity of incoming information content, which leads to negative consequences of the development of modern digital society and problems in the media (Mirzajonov, 2021).

Modern online users live in an era called the "post-truth era", characterized by digital disinformation, information wars, false information campaigns to manipulate public opinion (Borchard, 2018). Recent advances in information technology have provided conditions for the emergence of the DeepFake phenomenon, as a result of which hyperrealistic videos are created with the exchange of persons leaving a small number of traces of manipulation, completely invisible to users of video platforms and social networks. DeepFake is a product of artificial intelligence applications that

T. Afanasyeva (✉) · I. Yumasheva
St. Petersburg University of Management Technologies and Economics, St. Petersburg, Russian Federation
e-mail: i.yumasheva@spbacu.ru

A. Rumyantseva et al. (eds.), *Challenges and Solutions in the Digital Economy and Finance*, Springer Proceedings in Business and Economics,
https://doi.org/10.1007/978-3-031-14410-3_7

allow combining, replacing, and overlapping images and video clips to create fake videos that look authentic (Kambur, 2021).

DeepFake technology can generate, for example, a humorous, pornographic, or political video of a person when his image and voice are fully used. The scale of the spread of DeepFake and the availability of technology indicates that almost anyone with the skills of a confident user of computer technology can fabricate fake videos that are practically indistinguishable from the original media. Early examples of DeepFake focused on political leaders, actresses, comedians, and other media personalities. At the present stage of development, DeepFake technologies are used not only to attract attention and revenge, but also, in court as evidence for trial, political sabotage, terrorist propaganda, blackmail, etc. (Yumasheva & Grigorieva, 2021).

In order to understand the very essence of the creation technology, it is important to note that DeepFake relies on neural information networks that analyze large sets of data samples, while completely imitating a person's facial expression, mannerisms, voice and intonation. The process of creating DeepFake involves uploading videos of two people into a digital "deep learning" algorithm to teach it to change faces. In other words, DeepFake uses 3D face mapping technology, which is an audio visualization and 3D projection on a physical object of the environment, considering its geometry and location in space. 3D-mapping technologies and artificial intelligence allow you to change a person's face in a video to another person's face, as a result, a complete substitution of the personality as the object of the video is made (Bondarchuk & Kotliar, 2020).

The purpose of the study is to develop recommendations on methods to combat DeepFake technologies. Before formulating effective methods to combat DeepFake, it is necessary to understand the reasons for their existence and the technologies underlying them, therefore, within the framework of the study, the following tasks were set to achieve the goal:

- To analyze the technologies of creation and application of DeepFake technology.
- To identify the advantages and threats that DeepFake technology brings to the media and society.
- To analyze the perception of fake media by modern online users.

Economic-mathematical, structural-logical research methods were used in the work, analytical processing of various documents, regulations, decrees, and laws was carried out. The use of survey methods made it possible to study and conduct a sociological study of the perception of information content by online users.

Materials and Methods

DeepFake became known to the public in 2017, when videos were posted on Reddit by users in which celebrities are compromised by false sexual situations (Oluwole et al., 2021). At the same time, it is important to note that DeepFakes are difficult to

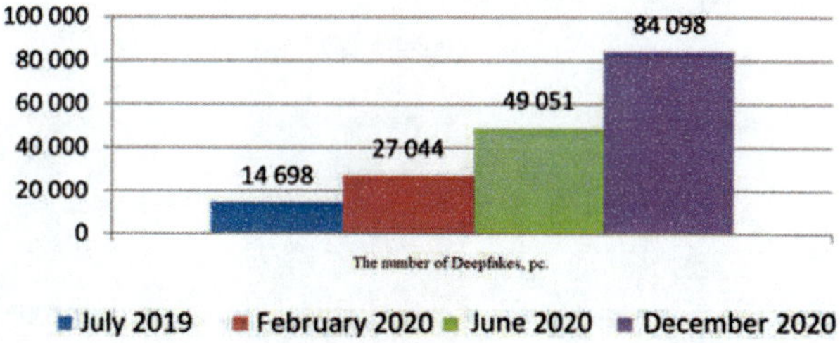

Fig. 1 The number of DeepFakes for the period 2019–2020 years

detect, since they use real frames, can have authentic sound, and are also optimized for high-speed distribution on social networks and video platforms, where conspiracies, rumors and misinformation are easily spread, since users tend to "go with the crowd".

As a result of the combination of multidirectional digital capabilities, most viewers assume that the video being viewed is authentic. At the same time, the ongoing "infopocalypse" encourages people to think that they cannot trust information if it does not come from the social networks they use. At the same time, it is interesting to note that the opinion of family members, close friends or relatives receives a lower degree of trust than social networks, bloggers, online video platforms.

Today, fake videos are common in all spheres of modern society and digital communications, this is since inexpensive equipment and affordable technologies for processing video objects are used in their creation (Vizoso et al., 2021).

In addition, technologies for creating realistic DeepFakes for disinformation purposes are becoming increasingly available, which allows users with initial technical skills to edit videos, change faces and synthesize speech. DeepFake technologies are a symbiosis of generative adversarial networks (GANS), when two artificial neural networks work together to create realistic media.

When creating DeepFake, artificial neural networks are used, which are called "generator" and "discriminator", they simulate the same set of data that are presented in the image, video data and audio data. The "generator" creates new imitated good images, and the "discriminator" works to determine how real the objects and videos of the newly created media are. GAN can view thousands of photos of a person and create a completely new portrait. Soon, generative adversarial networks will be "trained" in the presence of less information to "replace" not only the object's face, but also the entire body and voice.

Even though DeepFake usually requires a large number of images to create realistic sub-cases, researchers have already developed a method for creating fake videos by providing generative adversarial networks with only one photo, for example a selfie. The results of the analysis of the conducted studies showed that for the period 2019–2020 the number of DeepFakes has increased by 84% (Fig. 1), which indicates an exponential increase in their number.

According to the results of the study, it was revealed that, according to experts, the introduction of a legislative ban on the creation and distribution of DeepFake will only worsen the problem. As an example, confirming the validity of this statement,

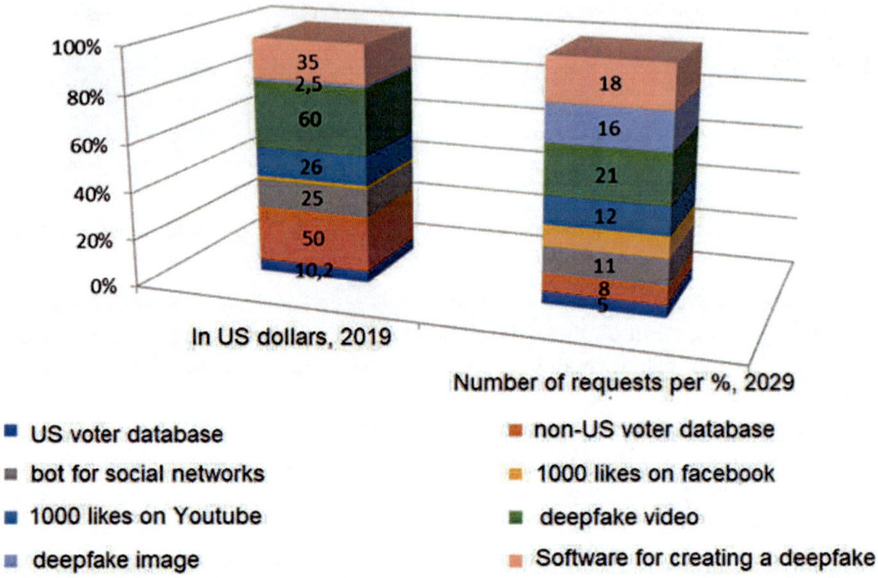

Fig. 2 The level of demand for "illegal data and information products" on the "black" market, 2019 year

experts point to China, where in 2019 fake videos were declared illegal, which allowed the Chinese government to attribute any "inconvenient photo or video" to the category of fakes. According to TrendMicro, a global leader in corporate data protection and cybersecurity solutions, in 2019 DeepFake was in the greatest demand on the "black market", as evidenced by the prices for these products (Fig. 2).

It is important to note that scientists from the SAND Lab at the University of Chicago, after analyzing statistical data, claim that the total monetary losses of global business due to DeepFake approached the $250 million mark in 2019, whereas in 2018 such losses amounted to $176 million.

Consider the positive aspects of the application of DeepFake technology, which have a positive practice of application in many industries, including cinematography, educational media and digital communications, games and entertainment, social networks, healthcare, materials science, various business areas such as fashion and e-commerce. The film industry can benefit from DeepFake technology in many ways. For example, DeepFake technology in cinematography allows you to create digital voices for actors who have lost their voice due to illness or provides updating of the film material instead of re-shooting it. Cinematographers, using DeepFake, can recreate classic movie scenes, create new films with the participation of long-dead actors, using special effects, advanced face editing in post-production and improving amateur video in professional quality.

Deepfake technology allows you to produce voice-over of different types of video objects automatically and efficiently in any language, which allows audiences from different countries to freely watch foreign films and use foreign educational media.

In 2019, as part of an information campaign about malaria with the participation of David Beckham, the issue of language barriers for the audience was resolved by placing educational ads that used visual and voice-changing technologies to obtain the "multilingualism" function. Also, an example of a positive application of DeepFake technology is overcoming language barriers at international video conferences with simultaneous translation of speech and simultaneous changes in facial and mouth movements to improve the quality of eye contact and obtain the effect of participants "speaking the same language".

The technology underlying deepfakes allows you to play multiplayer games and create virtual chat worlds with increased telepresence, creating a natural sound and types of "smart digital" assistants and digital doubles of people, which allows you to qualitatively improve human relations and interactive interaction at large distances. DeepFake technologies can have positive applications in social and medical fields. Deepfakes can help people cope with the loss of loved ones by digitally visualizing the deceased and thereby potentially help grieving people overcome the pain of loss and gradually adapt to the situation. In addition, this technology can digitally recreate a disabled limb or even help people with Alzheimer's disease.

In the context of the global development of digital technologies, researchers are studying the possibility of using GAN to detect anomalies in X-rays and their potential to create virtual chemical molecules for acceleration. Commercial global companies are interested in the potential of DeepFake technology in relation to brands, since it can significantly and qualitatively change e-commerce and advertising (Bendas, 2021).

At the present stage of digitalization of society, four main types of DeepFake producers are distinguished: communities of DeepFake lovers, political players such as foreign governments, and various activists, detractors, and criminal communities such as fraudsters, legitimate entities such as television, media companies, etc.

In general, fans of video content tend to view videos and products created by artificial intelligence as a new form of online humor and a contribution to the development of such technologies as solving an intellectual puzzle, and not to deceive or intimidate people. In this case, the audience considers entertaining, humorous, satirical, and political DeepFakes. Some of the audience is looking for more specific personal benefits, such as raising awareness about the potential of DeepFake to get a high-paying job, for example, related to music videos or TV shows. Thus, both amateurs and legitimate participants, such as TV companies, can cooperate with each other in the process of using DeepFake for various purposes.

While "meme-like" DeepFakes from amateurs can entertain Internet users, "intruders" are also involved in the process of using this technology. Various political players, including political agitators, activists, terrorists, can use DeepFake in disinformation campaigns to manipulate public opinion, to undermine confidence in a particular institution of the country, etc. Therefore, now, in addition to its positive application, DeepFake acts as a weapon of disinformation aimed at interfering in elections, inciting civil unrest, national discord, and other negative global phenomena. At the same time, DeepFake is increasingly being used by fraudsters to carry out market and exchange manipulations, as well as other serious financial crimes. Now, there are already examples when criminals have already used fake audio recordings created by artificial intelligence to impersonate a manager by phone with a request for an urgent money transfer. At the same time, DeepFake video calls for criminal purposes (fraud, financial crimes, etc.) will be available soon, which can be made in real time. So, now, DeepFake technology can already use visual and audio impersonation of managers, for example, from videos available on TED Talk on YouTube.

Results

As part of the study, a survey was conducted of more than 300 specialists in the field of PR, marketing, advertising, and journalism. During the study, it was found out that the majority of respondents—41%—receive information through online media and information sites, social networks were the second most popular—37%. More than half of the respondents, 58%, trust information content. About 43% of respondents doubt the information presented in the videos, they are looking for additional facts and primary sources of news. Thus, it turns out that among Russian online users, more and more people are beginning to think about the reliability of information content. The videos that cause the most doubt among the online audience include videos with the following signs: the absence of an official source, a dubious or absurd title and a little-known source. The results of research are presented in Fig. 3.

In addition to the above criteria, information that promotes the imposition of a certain point of view and the presence of questionable facts, as well as pressure on negative emotions and dramatization is questioned.

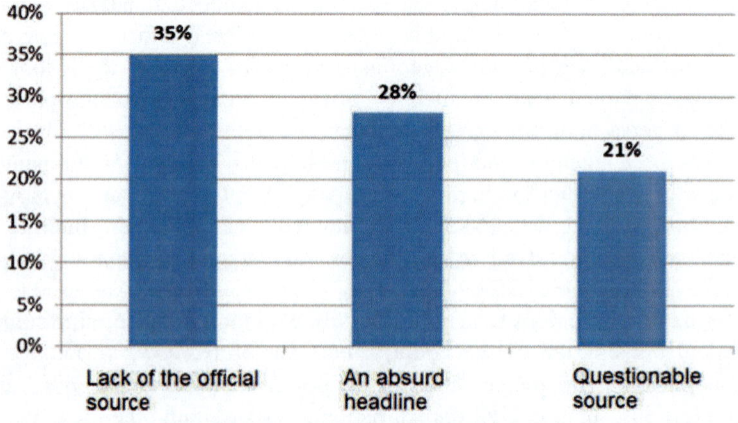

Fig. 3 The main signs of unreliability of information clips according to the online audience in 2021, according to the authors' research

Discussion

An analysis of the practice of legal methods to combat the negative effects and consequences of DeepFake indicates that now four main ones have been formed: legislative regulation, corporate policy and voluntary actions of commercial and non-profit organizations in countering negative DeepFake, digital education and training, improving digital literacy of society, technology for detecting and preventing the spread of DeepFake, authentication of content. Legislation and regulation are obvious means of preventing negative DeepFake.

Currently, DeepFake is not always considered in civil or criminal legislation, although experts in the field of world law have suggested adapting existing laws to cover libel, disinformation, identity fraud or impersonating government officials with the help of fakes. For example, the law of the US state of Virginia against revenge porn recently legislated responsibility for the distribution of "falsely created" images and videos and recognized them as an offense.

Nevertheless, the dynamic development and increasing level of complexity of artificial intelligence technologies requires new types of laws and regulatory framework with the sphere of regulation of DeepFake practice. For example, fakes raise concerns about privacy and copyright, since the visual images of people on DeepFake videos are not exact copies of any existing material, since it can rather be interpreted as new images created by artificial intelligence. Thus, regulators must navigate the complex legal landscape around freedom of speech and property laws in order to properly regulate the use of DeepFake technology.

On the other hand, a proper legal solution to the problem of spreading negative deepfakes should not be a complete ban on this technology. Today, firms working in social networks are widely using tools for using the content posted by users on their websites. One of the options for legislative regulation may be the abolition of the legal immunity of social media companies in relation to the content posted by their users, thus making not only users, but also platforms more responsible for the posted materials. At the same time, it is important to note that modern law currently has at least insufficient influence on "malicious actors", such as terrorists, who can conduct mass disinformation campaigns against other states, nationalities and nations, human values on social networks and online video platforms.

Corporate policy and voluntary actions of commercial and non-commercial organizations in countering negative DeepFake can become a more effective tool against "fakes". For example, politicians and top managers of global business giants can commit themselves not to use illegal digital communication campaigns and commercial and non-profit organizations in countering negative DeepFake tactics in election, information, market campaigns.

From the point of view of working with social media, commercial companies need to ensure compliance with ethical standards and refuse to place at the top of the content those video materials that cause global controversy in the global market (Nazaykin, 2020). Now, few social networks have a policy in relation to countering negative DeepFakes. But, in our opinion, it is social networks that should cooperate to prevent the use of commercial and non-profit organizations in countering negative DeepFake as a weapon for disinformation, and actively introduce transparent, common political ideas to block and remove DeepFake.

Currently, many companies do not remove controversial content, but simply lower its rating to make it harder to find, making it less noticeable in users' news feeds. The increase in the amount of hostility, fake news and disinformation polluting digital platforms has forced some companies to take more active actions, such as suspending user accounts and investing in rapid detection technologies. For example, Facebook cuts off any content that is recognized as false or misleading by third-party fact-checking organizations, does not allow commercial and non-profit organizations to place ads in opposition to negative DeepFake. Instagram's algorithms will not recommend people to view content that is marked as "false" by fact-checking organizations. Among news media companies, the Wall Street Journal and Reuters have formed corporate teams to help and train their reporters to identify fake content.

Conclusion

To create an eco-friendly media space based on improving the interaction of current digital media technologies and the needs of modern society, it is necessary to do the following points:

1. To analyze technologies for creating and applying fake news that threatens people's mental and physical health.
2. To identify the main types of DeepFake and groups of their producers, analyze their goals, objectives, and applications for further neutralization.
3. To search for effective and useful ways of positive application of DeepFake technologies in modern media and society.
4. To analyze online users of digital media behavior, improve the information security of digital content.
5. To implement methods of combating DeepFake technologies in digital media, such as: legislative regulation, development of technologies for DeepFake detection, content authentication and digital methods to prevent the spread of DeepFake.

References

Afanasyeva, T., Iljina, I., & Fetisova, G. (2019). Publicity of non-profit projects as an element of sustainable territorial development. *Conference: MTSDT 2019 – Modern Tools for Sustainable Development of Territories*. Special Topic: Project Management in the Regions of Russia. https://doi.org/10.15405/epsbs.2019.12.05.70.

Bendas, D. (2021). Modern media resources in the communication system of an accordionist. *National Academy of Managerial Staff of Culture and Arts Herald*. https://doi.org/10.32461/2226-3209.4.2021.250247.

Bondarchuk, B., & Kotliar, S. (2020). News in modern media discourse. *Bulletin of Kyiv National University of Culture and Arts Series in Audiovisual Art and Production, 3*(1), 29–37. https://doi.org/10.31866/2617-2674.3.1.2020.202652

Borchard, G. (2018). *The Press and the making of modern media*. A Narrative History of the American Press (pp. 224–240). https://doi.org/10.4324/9781315658667-15.

Kambur, E. (2021). *Emotional intelligence or artificial intelligence: Emotional artificial intelligence*. https://doi.org/10.17932/IAU.FCPE.2015.010/fcpe_v07i2004.

Mirzajonov, A. (2021). The importance of discourse and media text in modern media. *ACADEMICIA an International Multidisciplinary Research Journal, 11*(2), 24–26. https://doi.org/10.5958/2249-7137.2021.00317.7

Nazaykin, A. (2020). Online resources in modern media relations. *Theoretical and Practical Issues of Journalism, 9*(4), 660–672. https://doi.org/10.17150/2308-6203.2020.9(4).660-672

Oluwole, D., Collence, C., & Tinyiko, V. (2021). *Flattening the curve of fake news in the epoch of infodemic: An epistemic challenge*. Deep Fakes, Fake News, and Misinformation in Online Teaching and Learning Technologies (pp. 143–161). https://doi.org/10.4018/978-1-7998-6474-5.ch007.

Vizoso, A., Vaz-alvarez, M., & Garcia, X. (2021). Fighting Deepfakes: Media and internet giants' converging and diverging strategies against hi-tech misinformation. *Media and Communication, 9*(1), 291–300. https://doi.org/10.17645/mac.v9i1.3494

Yumasheva, I., & Grigorieva, O. (2021). Approaches to the classification of media theories. *UNESCO World Heritage in the hands of young people* (pp 258–262).

Innovative Capital as Difficult-to-Identify Factor of Production

Galina Stepanovna Merzlikina

Introduction

Under research there are the concept of 'innovative capital' of a company as a new factor of production, the rationale for the structure of innovative capital, the study of the possibility to identify, account for and analyze the size and dynamics of innovative capital.

The tasks set to achieve the goal of the study were to analyze the definitions of the concept of 'innovative capital' of a company and determine its structure, to look into innovative capital as a factor of production, to identify problems of accounting, evaluation and management of the innovative capital of a company.

The author employs various economic entities as an object of this research. Among them there are enterprises, corporations, small and medium-sized businesses that carry out innovative activities, generate and use innovative capital.

The subject of this study covers the processes of determining and clarifying the structure of innovative capital, evaluation, accounting for and management of it.

The implementation of production processes has now drastically changed. We have seen new, radically new processes, new, radically new products (goods), new tools and mechanisms appear; new business models are being formed and implemented. The industry is evolving from entire automation, computerization to digital transformation of processes, mechanisms and tools. Cutting-edge production is an active, forced and preferable use of Internet technologies, the Internet of things, cloud technologies, and artificial intelligence. The so-called 'smart (intellectual) production', more flexible and more efficient, is being created based on the active and rapid use of all types of innovations (manufacturing, technological, managerial,

G. S. Merzlikina (✉)
Volgograd State Technical University, Volgograd, Russian Federation
e-mail: merzlikina@vstu.ru

A. Rumyantseva et al. (eds.), *Challenges and Solutions in the Digital Economy and Finance*, Springer Proceedings in Business and Economics,
https://doi.org/10.1007/978-3-031-14410-3_8

marketing, financial, and others). There is intellectualization of manufacturing processes under way and, accordingly, known production assets are being transformed into new intellectual objects possessing new characteristics and requiring new methods of accounting, analysis and management.

One would think traditional factors of production (from Latin, *factor* means 'making', 'producing'), such as Land, Labor, Capital, and, added later, Entrepreneurship and Information, will remain active 'doers' of manufactures forever. The traditional factors of production have not gone but have merely become secondary, and such new factors of production as intellectual, innovative, human, digital capital, are coming to the fore. The time has come to interpret theoretically those new factors of production and their role in the process of production.

Industry, a priori, requires innovations. Without innovation, there is no talk about the effectiveness of activities or achieving and maintaining competitiveness. That is why it is essential to measure and analyze the innovation activity of companies. However, currently statistical methods of observation only imply some indicators that characterize either the costs of innovation or the results of innovation activity, and they are completely different in time (Federal State Statistics Service, 2022; Gokhberg et al., 2021; Isaksen & Trippl, 2017; Andrew et al., 2009). The subject of scientific research is *innovative development*, *innovative activity*, *innovative viability* and other nouns following the adjective *innovative* (Zemtsov et al., 2016; Machikita & Ueki, 2015; Rullani et al., 2016; Akhtyamov et al., 2016; Levchenko & Karpenko, 2020; McDowell et al., 2018). In most cases, all the above-mentioned innovation achievements are evaluated using known indicators of innovation activity monitored statistically. However, those concepts do differ in terms of goals, tools, mechanisms, and finally, management methods and, thus, assessment indicators.

Given the necessity to form a company's (region's, country's) innovativeness, one can assume the emergence of a new concept of *innovative capital* and consider it as a new factor of production. The concept of *innovation capital* was first discussed in the research works of L.A Joia (2000) and M. E. Van Buren (1999), later M. Khalique et al.'s (2011) and Gomezelj Omerzel's & Smolčić Jurdana's (2016), Russian scientists also look into the problems of innovative capital (Akhtyamov et al., 2016; Levchenko & Karpenko, 2020; Arenkov & Yaburova, 2018; Ustinova & Alekseyeva, 2020). But, the analysis of the scientific literature on this and related topics showed that it focuses more on the so-called 'intellectual capital' comprising of human, client and structural capitals, social, technological, and spiritual (M. Khalique et al., 2011; Gomezelj Omerzel & Smolčić Jurdana 2016) capitals. *Innovative capital* is more often considered as one of the elements of the intellectual capital structure, which is not clarified; scientific papers cover various points of view. And certainly, innovative capital has not been considered as a factor of production.

The study of the innovative capital of a company, clarifying the essence of this concept, determining its structure, analyzing the possibility of innovative capital to be measured and evaluated will make it possible to arrange innovation activities more reasonably, assess the results achieved and manage the process of innovation activity.

Materials and Methods

This study is based on the main provisions of the theory of economy and company management, the theory of innovation-driven economy, balanced development, the scientific foundations of strategic management, the theory of evaluation and management of business value, methods of statistical observation, analysis and measurement, the principles of company capital formation, presented in scientific publications of foreign and Russian scientists. The potential of using this new concept of *innovative capital* of a company is substantiated herein as the value advanced in certain innovation-oriented resources contributing to successful implementation of innovation activities; its structure is clarified (human, intellectual, patent, digital capital). This paper reveals that the existing methods of accounting and analysis of innovation activities do not make it possible to assess either the value of innovative capital or its transformation, particularly in terms of its constituent elements, which makes it difficult to develop and rationalize innovation activities, does not allow qualifying innovative capital as a new factor of production.

Results and Discussion

The scientific novelty of the study consists in the development of theoretical provisions and recommendations on the possible ways to assess the innovative capital of a company. This paper clarifies the content of innovative capital (human, intellectual, patent and digital capital) and presents a comparative analysis of methods for assessing the effectiveness of innovation activity. The author has revealed major challenges of identifying innovative capital and its subsequent assessment, particularly in terms of unidentifiable intangible assets such as professional knowledge of employees, their creative and digital capabilities required to be assessed using an interdisciplinary approach.

Innovation Activity of a Production Company: Meaning

Innovations have always been essential for industrial manufactures, but in the present context they mostly determine success and development. Researchers point out the critical change that has taken place in the study of corporate production strategies: it was erroneously assumed that the development of digital technologies is exponential. One should agree with (Mainzer, 2020) that scientific and technological progress on the current stage has changed the role of innovations making them develop under to the laws of biological evolution and play a new role of 'mutations', and only after that the market can select products (goods). It is the evidence of the new importance of innovation in modern economic development.

The problems of today's industry (Industry 4.0) are clearly defined as the importance and need for a significant increase in costs to accelerate and localize production processes in order to improve product quality (the modern principle of competitiveness) (Götz & Yankovska, 2020). One of the sources of superiority from the resource-based viewpoint (RBV) should be a unique set of resources (VIRO: valuable, inimitable, rare and organized), which should also include innovative capital. High significance is attached to knowledge, competences that are "'embedded' in an organization and are not easy to transfer" and can hardly be measured and evaluated.

Exploring the Fourth Industrial Revolution, there are also 'dark corners' (unforeseen consequences) observed that require improved management in a digital economy in favor of structural changes, including the capital structure of a company. It has been revealed that new information technologies are revolutionizing products and new 'smart, connectable products', in their turn, launch a new era of competition eager for quick innovations and restrained 'long-running' innovations (Porter & Heppelmann, 2015).

At present, the success of any company is ensured primarily by intangible assets, new technologies, licenses, and innovative capabilities.

Innovative Capital: Concept

The concept of 'capital' has organically entered the modern economic vocabulary. The present-day science is involved in an active study of the issues of (any) capital formation and management, assessment of the value of a business (own capital of an enterprise), capitalization measurement. The cost of capital (business) has become one of the main performance indicators of production activities as well. However, if fixed and floating capitals, human capital (labor resources), then entrepreneurship and information are officially recognized as factors of production (relying on such attributes as terminological clarity and universal recognition, accounting, evaluation and measurement system), new types of capital, including innovative (intellectual) capital, are so far under discussion with their contents, structure, indicators of measurement and evaluation under scientific search and clarification.

The concept of *innovative capital* appears for the first time in the works of Joia L.A. (2000) and Van Buren M. E. (1999), but only as part of intellectual capital. Intellectual capital is discussed in numerous scientific papers (for instance, Tripathy et al., 2015; Jordão & Novas, 2017; Hussinki et al., 2017; Ndou et al., 2018). Innovative capital is considered from the standpoint of strategic management as a strategic innovative capital (Kovacs, 2018). Some research works qualify innovative capital as aggregate costs for innovative projects (Tripathy et al., 2015), or as expenses on innovative resources (Jordão & Novas, 2017). Often innovative capital is viewed as an element of intellectual capital (Hussinki et al., 2017; Ndou et al., 2018).

The author suggested her own definition of the innovative capital of a company as the value advanced in certain innovation-oriented resources contributing to successful implementation of innovations (Babkin & Merzlikina, 2021). It is essential to clarify that the above definition implies an assessment of innovation-oriented resources which makes it possible to attract all available resources for innovation (for instance, equipment, production sites, various types of raw and consumable materials, relevant technological projects and solutions, patterns, employee skills).

Innovative Capital: Structure

The content of the *innovative capital* concept is mostly determined by its structure. However, in the same way that there are multiple points of view on the content of this concept, the structure of innovative capital is also viewed in many different ways. Let us try to consider the evolution of the innovative capital structure based on the well-known and better studied concept of 'intellectual capital'. The concept of 'intellectual capital' was first proposed by Stewart (1997). Now intellectual capital is understood as a set of knowledge and skills of the staff (McDowell et al., 2018; Gomezelj Omerzel & Smolčić Jurdana, 2016), as a factor in the company value increment and a significant resource of the company (Edvinsson & Malone, 1997). The intellectual capital includes human, client, social, technological, spiritual capital (McDowell et al., 2018; Khalique et al., 2011; Gomezelj Omerzel & Smolčić Jurdana, 2016), structural and relational types of capital (Jordão & Novas, 2017). The listed types of intellectual capital are certainly important and necessary for innovation activity as well, but they cannot determine the structure of innovative capital.

The author comes forward with the following structure of a company's innovative capital: human, intellectual, patent and digital capital (Merzlikina, 2020). Perhaps, further development of the innovative capital measuring and evaluating methods will give rise to new components, but so far, the structure is proposed as such. The specified types of capital as components of innovative capital constitute the innovative essence of the total capital.

The human capital of a company has been sufficiently studied (starting from the works of Schultz, 1971) and consists in personal professional knowledge, skills and competences of the employees of a company forming the total human capital (Khalique et al., 2011).

A great number of research papers mentioned above are devoted to intellectual capital, but all of them regard intellectual capital essentially as a modern concept of human capital (higher requirements to employees), without assuming special qualities of employees for innovative activities. However, it should be pointed out that even professional knowledge and skills do not imply special innovative, creative capabilities, that is why the author differentiates human capital and intellectual capital (part of human capital) with the latter distinguished by a set of creative abilities, sometimes referred to as the 'innovation gene', of certain employees of a company who are able to identify novelty, new methods, techniques, processes ('novelty seeking'). Thus, intellectual capital (innovative employees) in essence becomes a catalyst for new ideas, innovations.

Patent capital consists in clearly identifiable intangible assets 'secured' by intellectual property title documents that allow generation of income through using a patent, either from the sale of a license or from the sale of a patent. The author has found no scientific research on patent capital yet. With all the clarity of the content of this concept (capital, everything that generates income, if it can be defined), there is a problem of forming patent capital as such. In order to become the owner of a patent (the owner of patent capital), one requires either significant financial resources, or a strong research base at the enterprise, which is hardly available for everyone.

The most sophisticated and currently demanded component of innovative capital is digital capital, which is generally understood as some digital resources for the digital economy (which seems to be quite a vague notion). Digital capital also includes digital technologies, the process of collecting and processing data, and analytical Big Data management (Krutikov & Gerayeva, 2018; Bughin & Manyika, 2013; Park, 2017; Ragnedda et al., 2020). One particular problem is not so much the proclamation of the necessity and importance of digital capital, but the ability to measure and evaluate it in order to manage it. When evaluating digital capital, it is essential to single out digital capital as access to digital technologies (software, communication facilities, data collection and processing, and digital competences of personnel), without which digital capital will not 'work'. Digital capital is considered (Park, 2017) as a set of conditions that predetermine accessibility and usability of digital services for people. Digital capital is thought to be an accumulation of digital competences and digital technologies that can be isolated, singled out as an independent aspect and measured. Summarizing the above, we would refer to digital capital as a set of tangible and intangible assets allowing the use of digital technologies (Ragnedda et al., 2020).

Innovative Capital as Difficult-to-Identify Factor of Production

There is a problem of accounting for capital in integrated financial reporting (Malinovskaya, 2018). At present, there are six categories of capital suggested (financial, industrial, intellectual, human, social, connected, natural capital), and intellectual capital is considered as a combination of intellectual property and organizational capital. However, the forms of capital types in integrated reporting can be expanded so far as all the six capitals listed above are specified 'for reference'.

Let us try to define innovative capital as a factor of production based on innovation activity indicators. In Russia, they are recorded in *Innovation Activity of Organization* statistical data form No 4 (Federal State Statistics Service, 2022). Based on the analysis of data provided by companies in terms of their innovation activities it has been found that the aspects accounted for include types of innovations (technological, marketing, organizational, environmental) and the total volume of innovative products. Besides, small enterprises engaged in innovation activities are singled out. Thus, it may be concluded based on official information that innovation capital cannot be measured, and therefore it is impossible to conduct a comparative analysis (for enterprises, regions). Due to the lack of accounting for the innovation performance (obtaining additional revenue (its allocation, accounting, tracking) it is impossible to assess the cost of innovative capital. It will only be possible to consider control of the innovative capital cost and balanced management of innovative capital when the operational ('field') data of a particular company/ enterprise are used, when personalized accounting of current and capital costs is carried out and effectiveness of innovative capital and its constituent parts are analyzed. Focus is placed on the indicators of innovation activity and innovative development of a company and a region (Gokhberg et al., 2021). Among them there are indicators characterizing the conditions of innovation activity and innovation potential. But it should be noted that all the indicators under review overlap to a certain extent (indicators of innovative development of a company and a region). And the so-called *process approach* prevails here fixing the implementation of a process (basically, 'today's' costs) and the result of the process implementation of (release of innovative products, advanced technologies used, today's results).

The European Innovation Scoreboard (Isaksen & Trippl, 2017) uses human resource involvement indices and some company performance indicators (investment, performance, efficiency). In the United States human capital ratings are used along with general indicators of economic dynamics and labor productivity, innovation costs and the set of the indicators constitute 'innovative behavior' (a kind of statement of 'input' and 'output') (Andrew et al., 2009).

All the reviewed indicators for evaluating innovation activity do not allow an objective assessment of its effectiveness as long as the indicators compared differ in time (costs incurred today are juxtaposed with the results of already implemented innovations at yesterday's costs). Besides, all the methods offer a certain generalizing index (as an integral indicator) without justifying the weight of either individual indicators or their groups, which distorts the result. So, it is hardly possible to determine the value and dynamics of innovative capital using these indicators. Research work (Merzlikina, 2020) overviews the results of a study of various methods for assessing the level and indicators of innovative development, mainly territories/regions, as information on individual enterprises is not always provided in statistical surveys.

The author of this publication makes an attempt to identify innovative capital. The review of scientific literature and the results of statistical observations showed that there is no generally accepted understanding of this concept, its content, and structural components. Every scholar offers his own interpretation of the concept. This echoes the ancient Indian parable on the elephant and the blind sages. That is, we see a problem of assessing innovative capital arise: how can we evaluate something that everyone defines in their own way? Therefore, before offering methods for assessing innovative capital, it has to be defined. The identification procedure (from Latin, *identifico* means 'to identify') implies that an object is established on the basis of attributes with the most important attribute being its structure, as an inherent feature of the whole. The author proposes the structure of innovative capital to include human, intellectual, patent, and digital capital. Other structural components of innovative capital may be considered just as well, perhaps, but we will focus on the proposed ones.

Difficulties in identifying innovative capital are explained by the complicated determination of its structure and innovative features of seemingly known structural components, the complexity of the subsequent assessment. Currently we suggest the identification of innovative capital as a whole and that of its structural components.

Further research involves identifying approaches to assessing innovative capital. At that, the author, as a professional appraiser, does not see problems in the evaluation of identifiable intangible assets of patent and digital capital (in terms of tangible assets). For this purpose, there are excellent, eternal and universally recognized methods for estimating the value of a business (or rather, estimating intellectual property). One can use all the approaches: cost, comparative, income based. Two methods of the income-based approach are more commonly used, namely, direct capitalization and cash flow discounting. The problems here may only be in a correct and reasonable formation of cash flows and the determination of the discount rate or capitalization ratio.

However, the assessment of unidentifiable intangible assets will require at least an interdisciplinary approach. How to assess professional knowledge (human capital)? How to assess intellectual capital as part of innovation ('novelty seeking' mentioned earlier)? How to assess the creative abilities and digital competencies of employees? One can assume that most likely these will be expert assessments, but who is going to specify the assessment model and a list of characteristic features?

Thus, the author has come forward with her version of the identification of innovative capital (clarifying its structural components) as a factor of production and indicated significant current difficulties in identifying innovative capital such as the fuzziness of the structure and the lack of special methods for assessing unidentifiable intangible components of innovative capital.

Conclusion

Based on the research conducted the author draws the following conclusions.

1. Innovation development is, a priori, inherent in cutting-edge industrial production. The study of production processes changes the approach to assessing the factors of production; they transform both quantitatively and qualitatively.
2. The content of the *innovative capital* concept has been clarified to denote the value advanced in certain innovation-oriented resources that contribute to the successful implementation of innovative activities.
3. The structure of innovative capital is proposed to include human, intellectual (special creative competences of personnel), patent, digital (both tangible and intangible assets) capital.

4. It is concluded that innovative capital is difficult to be identified as long as information (statistically observed) is usually different in time. Integrated reporting can only take into account identifiable tangible and intangible assets. A considerable part of the unidentifiable intangible assets that are part of the innovative capital 'drops off the radar.
5. Challenging issues of identifying innovative capital are determined as the complexity of its structure and subsequent evaluation. It has been found that known methods for assessing intellectual property can be used to assess the identified intangible components of innovative capital, and the assessment of unidentifiable intangible components of innovative capital (professional and creative abilities and digital competences of personnel) requires an interdisciplinary approach and more complex further research.

This study may be continued to design a methodology for assessing innovative capital, to determine indicators for assessing innovative capital, to look into management of the innovative capital cost.

References

Akhtyamov, M. K., Gonchar, E. A., & Tikhonova, N. V. (2016). Evaluation of intellectual capital of organization as element of intrinsic value of enterprise. *Creative Economy, 10*(8), 945–960. https://doi.org/10.18334/ce.10.8.35851

Andrew, J. P., DeRocco, E. S., & Taylor, A. (2009). *The innovation imperative in manufacturing: How the United States Can Restore Its Edge*. The Boston Consulting Group.

Arenkov, I. A., & Yaburova, D. V. (2018). Improved management of intellectual capital as factor in development. *Business Strategies, 5*(49), 33–35.

Babkin, A. V., & Merzlikina, G. S. (2021). Justification of interrelation of the innovative capital of the enterprise and smart manufacturing. *π-Economy, 14*(3), 86–102. https://doi.org/10.18721/JE.14307

Bughin, J., & Manyika, J. (2013). Measuring the full impact of digital capital McKins. *Quart, 4*, 88.

Edvinsson, L., & Malone, M. (1997). *Intellectual capital*. Harper Business.

Federal State Statistics Service. (2022). *Costs of innovative activity of organizations (since 2010)*. Retrieved from https://www.gks.ru/folder/14477

Gokhberg, L. M., Ditkovsky, K. A., Kuznetsov, I. A. ... & Fursov, K. S. (2021). *Indicators of innovation activity: 2021*. Moscow, Higher School of Economics National Research University. https://doi.org/10.17323/978-5-7598-2375-9.

Gomezelj Omerzel, D., & Smolčić Jurdana, D. (2016). The influence of intellectual capital on innovativeness and growth in tourism SMEs: Empirical evidence from Slovenia and Croatia. *Economic Research-Ekonomska istraživanja, 29*(1), 1075–1090. https://doi.org/10.1080/1331677X.2016.1211946

Götz, M., & Yankovska, B. (2020). Industry 4.0 as factor of company competitiveness in post-transitional economy. Foresight and STI. *Governance, 14*(4), 61–78.

Hussinki, H., Ritala, P., Vanhala, M., & Kianto, A. (2017). Intellectual capital, knowledge management practices and firm performance. *Journal of Intellectual Capital, 18*(4), 904–922. https://doi.org/10.1108/JIC-11-2016-0116

Isaksen, A., & Trippl, M. (2017). Innovation in space: The Mosaic of regional innovation patterns. *Oxford Review of Economic Policy, 33*(1), 122–140.

Joia, L. A. (2000). Measuring intangible corporate assets: Linking business strategy with intellectual capital. *Journal of Intellectual capital?, 1*(1), 68–84.

Jordão, R. V. D., & Novas, J. C. (2017). Knowledge management and intellectual capital in networks of small-and medium-sized enterprises. *Journal of Intellectual Capital*. https://doi.org/10.1108/JIC-11-2016-0120.

Khalique, M., Nassir Shaari, J. A., & Isa, A. H. B. M. (2011). Intellectual capital and its major components. *International Journal of Current Research, 3*(6), 343.

Kovacs, O. (2018). The dark corners of industry 4.0–Grounding economic governance 2.0. *Technology in Society, 55*, 140–145.

Krutikov, V. K., & Gerayeva, E. V. (2018). Digital capital and innovation entrepreneurial activity. *Economy and Entrepreneurship, 3*, 703–706.

Levchenko, L. V., & Karpenko, O. A. (2020). Innovative capital: Institutional acceleration of formation and development. *Economic Sciences, 188*, 13–17. https://doi.org/10.14451/1.188.13

Machikita, T., & Ueki, Y. (2015). Measuring and explaining innovative capability: Evidence from Southeast Asia. *Asian Economic Policy Review, 10*(1), 152–173. https://doi.org/10.1111/aepr.12093

Mainzer, K. (2020). Technology foresight and sustainable innovation development in the complex dynamical systems view. Foresight and STI. *Governance, 14*(4 (eng)), 10–19.

Malinovskaya, N. V. (2018). The concept of multiple capitals in integrated reporting. *International Accounting, 21*(6 (444)), 700–713. https://doi.org/10.24891/ia.21.6.700

McDowell, W. C., Peake, W. O., Coder, L., & Harris, M. L. (2018). Building small firm performance through intellectual capital development: Exploring innovation as the "black box". *Journal of Business Research, 88*, 321–327. https://doi.org/10.1016/j.jbusres.2018.01.025

Merzlikina, G. S. (2020). Innovation development of region: Essential architecture of indicators. *π-Economy, 13*(5), 55–64. https://doi.org/10.18721/JE.13504

Ndou, V., Secundo, G., Dumay, J., & Gjevori, E. (2018). Understanding intellectual capital disclosure in online media Big Data: An exploratory case study in a university. *Meditari Accountancy Research, 26*(3), 499–530. https://doi.org/10.1108/MEDAR-03-2018-0302

Park, S. (2017). *Digital capital*. Palgrave Macmillan. https://doi.org/10.1057/978-1-137-59332-0

Porter, M. E., & Heppelmann, J. E. (2015). How smart, connected products are transforming companies. *Harvard Business Review, 93*(10), 96–114.

Ragnedda, M., Ruiu, M. L., & Addeo, F. (2020). Measuring digital capital: An empirical investigation. *New Media & Society, 22*(5), 793–816. https://doi.org/10.1177/1461444819869604

Rullani, E., Cozza, C., & Zanfei, A. (2016). Lost in transition: Systemic innovations and the new role of the state in industrial policy. *Economia e Politica Industriale, 43*(3), 345–353.

Schultz, T. W. (1971). *Investment in human capital. The role of education and of research*. The Free Press, xii, 272 pp.

Stewart, T. A. (1997). *Intellectual capital: The new wealth of organizations*. Doubleday/Currency.

Tripathy, T., Gil-Alana, L. A., & Sahoo, D. (2015). The effect of intellectual capital on firms' financial performance: An empirical investigation in India. *International Journal of Learning and Intellectual Capital, 12*(4), 342–371.

Ustinova, L. N., & Alekseyeva, N. S. (2020). Structuring definitions of terminological apparatus in intellectual capital research. *Scientific and Technical Bulletin of SPbSPU. Economic Sciences, 13*(3), 41–56. https://doi.org/10.18721/JE.13303

Van Buren, M. E. (1999). A yardstick for knowledge management. *Training & Development, 53*(5), 71–78.

Zemtsov, S., Muradov, A., Wade, I., & Barinova, V. (2016). Factors of innovation activity of Russian regions: What is more important – Human or capital? *Foresight, 2*, 29–42. https://doi.org/10.17323/1995-459X.2016.2.29.42

Challenges in Professional Translation: Implications for Linguistic Education and the Modern Digital Economy

Anna-Maria Arias, Maria Fedyukovskaya, and Alexander Fedyukovsky

Introduction

In the Strategy for the Information Society Development, approved in Russia in May 2017, the concept of the digital economy is defined as business activities where the key production factor is digital data; the processing of large volumes and the use of analysis results of these data, compared with traditional forms of business, makes it possible to significantly increase the efficiency of various types of production, technologies, equipment, storage, sale, delivery of goods and services (Bershadskaya et al., 2014).

In the Program "Digital Economy of the Russian Federation"(Vinokurov, 2021), developed for the period until 2024, one of the five basic areas of development is identified as *personnel* with technical background and digital literacy, and *education* (Zagloel et al., 2021).

Digital literacy is recognized by the European Union as one of the eight key competencies necessary for full life and successful professional activities (Hockly, 2021). The ability to work with digital technologies is becoming a constant and prerequisite for most specialties. The uniqueness of digital technologies is that citizens can acquire the competence in many other areas more effectively.

Based on main characteristics of the innovative economy, digital technologies contribute to the translation of updated knowledge and increased amount of information. That is, the modern practice of using digital technologies generates the

A.-M. Arias · M. Fedyukovskaya
St. Petersburg University of Management Technologies and Economics, St. Petersburg, Russian Federation

A. Fedyukovsky (✉)
Peter the Great St. Petersburg Polytechnic University (SPbPU), St. Petersburg, Russian Federation

A. Rumyantseva et al. (eds.), *Challenges and Solutions in the Digital Economy and Finance*, Springer Proceedings in Business and Economics,
https://doi.org/10.1007/978-3-031-14410-3_9

demand for the skills necessary for processing, analyzing, interpreting large amounts of data, including working with artificial intelligence. At the same time, the quality of digital education is determined not only by forming in-depth digital skills, which involve working with big data, but also by forming the culture of perceiving modern digital content, readiness for changes because of the economy digitalization. In addition, the very renewal of digital technologies requires the transfer to continuous education and constant training of specialists in many fields and industries (Kolbachev & Pahomova, 2019).

The needs of the digital economy significantly affect the contemporary linguist image and force to review the meaningful content of the labor functions performed by a specialist who speaks a foreign language, to adapt to new professional challenges (Netolicky, 2019). That is why the professional standard "Specialist in the Field of Translation", approved in 2021, directs the vector of translation training towards the synthesis of translation theory and practice with computer technologies and organizational and managerial skills. The influence of digitalization on the translator profession is seen from the fact that in all generalized labor functions of the professional standard there is knowledge and skills to use computer technologies for translation. As a result of modern technological processes, the language from the specialty is becoming the language for the specialty. In the Atlas of New Professions, the profession of "Digital Linguist" is listed, aimed at solving the problems of information processing and creating sense using computer technologies, as well as the search for technical interaction between the person and the computer (Bago et al., 2017). Often, a contemporary translator acts as an editor, whose tasks include either choosing a translation option that is correlated with the original sense, or editing the text translated in accordance with the communicative purpose of the original text. Here is the end of the new professions' list for a linguist-translator, who does not speak in favor of developing this type of professional activities.

In this regard, the question arises: what are the priorities in modern linguistic education: obtaining a high level of language training or mastering the competencies that allow working in the digital environment?

Material and Methods

The authors use the following methods which allow them to consider the contemporary issues and to formulate the research conclusion:

Solid sampling method—a method of context analysis based on the material of Susan Collins's novel "The Hunger Games" and the translation of its 100 sentences into Russian, conducted by the *Yandex. Translator* machine translation system.

Students' questionnaire method. There were 101 undergraduate students at St. Petersburg University of Management Technologies and Economics.

Method of statistical analysis based on the results of the studying the subject "Practical Course of Translation". The period of mastering the subject: three terms (from the fifth term of the third year to the seventh term of the fourth year).

Results

Digital technologies have always been the most popular in linguistic education, since they provide a large amount of information and a variety of opportunities to work with audio and video materials that significantly identify the process of teaching foreign language(s) and increase the motivation of students and their willingness to master foreign language(s) for professional activities (Gavrilenko, 2020; Tolmach, 2021).

Foreign language education is usually based on the creation of an artificial foreign language environment. The use of audiovisual means in the process of linguistic education allows to "blur" the boundaries of an artificial language environment and make different communicative situations, during foreign language or translation lessons, closer to the real ones.

Digital technologies are designed to affect the sound and visual modality of the person. Psychologists have paid much attention to the psychological features of training and the role of audiovisuals in foreign language classes (Ovsyanko & Kyrychenko, 2021). Audiovisuals turn to the sensory-perceptual organization of people, which is the central channel of their information perception. At the same time, there is a simultaneous reliance on the work of several receptors, which leads to the activation of mental processes: memory, attention and, as a result, the effectiveness of students' perception of information is significantly increased.

As practice shows, digital technologies not only contribute to the qualitative transformation of linguistic education, but they also made it possible to translate the process of mastering foreign language(s) into a remote format (Nugumanova et al., 2020).

In case of distance linguistic education, it is advisable to assert the need for electronic pedagogical support, which is considered as an optimally organized interactive electronic educational environment (Vaganova et al., 2020).

The electronic educational environment involves feedback from teachers, allows to identify and analyze the pedagogical needs of students in the learning process, to take measures to overcome difficulties, namely, the adequate and advisable use of electronic resources, thereby increasing the level of motivation and providing the students with self-control opportunities. The electronic educational environment properly organized increases the potential professional success of future specialists in the conditions of the economy digitalization.

According to the analysis of the students' performance, the result of mastering professional translation competencies was not significantly influenced using distant educational technologies. The level of absolute achievement in the subject "Practical Course of Translation" in mixed-format training is 99%, during the training in the traditional format −97%. Thus, it can be assumed that the training format has an insignificant impact on the quality of translation training.

The other challenge is how ready the students, when graduating, will be to demonstrate their digital skills to solve professional problems. In this regard, the curricula for translation specialty are to necessarily include the subjects which teach machine and computer-aided translation. The other prerequisite for training linguist-translators is the skill of critical analysis and adjustment of the quality of machine and computer-aided translation. According to the results of the survey, future linguists-translators actively use software products to solve translation, communication or search problems. Among the 101 students, 64 preferred *Yandex.Translator* to other online services, which is 64%. The other 36% preferred either *Google* (28%) or *PROMT* (4%) or *DeepL* (4%).

The analysis of language material translated from English into Russian with *Yandex.Translator* demonstrated the presence of different errors: lexicosemantic, grammatical (morphological and syntactic), spelling, punctuation ones. There were stylistic errors which manifest in violation of the traditional use of language units in speech. At the same time, some errors can be simultaneously ranked as errors of different levels, e.g., lexicosemantic, grammatical, and stylistic ones.

For example, the sentence "*I'm guessing the second*" was translated as "*Я угадываю секунду*". "*Second*" is not only a noun "секунда". The word is preceded by a definite article "*the*", which indicates an ordinal numeral ("второй"), but *Yandex.Translator* failed to identify the value of the ordinal numeral.

There are also some discrepancies when translating proper names, which indicates a low degree of their adaptation when translating with the service.

Here is another example where the online translator was not able to find the necessary equivalent to express the value of "draw lots": the sentence "*It's time for the drawing. Effie Trinket says as she always does, "Ladies first!" and crosses to the glass ball with the girls' names*" was translated as "*Пришло время для рисования. Эффи Тринкет, как всегда, говорит: "Дамы вперед!"—и подходит к стеклянному шару с именами девочек*".

The word "*drawing*" here is not *painting (рисование)* according to the machine translator.

Grammatical errors which include morphological and syntactic errors also occurred in the translation performed by the popular online translator. These are errors in the structure of the language unit, non-compliance with the norms of word formation, the norms of syntactic connection between the words in the phrase and sentence. In the material studied there are a lot of grammatical errors associated with the misrepresentation of the verb mood, aspect and tense, the transmission of elliptical syntactic structures which are typical of the English language.

The machine translation system analyzed also makes spelling and punctuation errors.

Due to the errors at lexical, spelling, and punctuation levels in computer-aided translation there are errors distorting the meaning: "*And the protocol has become rusty*" (*Протокол стал ржавым*). Or: "*The Career pack sets off at a run just as dawn begins to break, and birdsong fills the air*" (Карьерная стая пускается в бег, как только начинает светать, и птичье пение наполняет воздух).

The occurrence of such errors is explained by literal translation performed by the service. The service translates all language units without performing translation transformations, without considering the differences between the languages' systems.

The result of the analysis of translation performed by *Yandex.Translator* is as follows: among 100 sentences there are 18 translated without distorting the meaning, even though the sentences were taken out of the context, in 82 sentences there are different errors. Therefore, post-editing of machine translation is necessary, since the translation service does not consider the complexity of translating a number of structures, polysemy, borrowing, linguistic-cultural and collocation errors, which entails semantic distortions, as well as the need to perform translation transformations to eliminate the semantic "conflict" while translating. The most convincing imperfection of machine translation is manifested in translating fiction, which requires a deep analysis of the original text.

Discussion

In forming the whole complex of knowledge, abilities and skills acquired in obtaining the linguistic education, several areas of implementation of the interdisciplinary approach can be distinguished:

1. Formation of the humane and tolerant personality capable of intercultural dialog and polylog. Students are prepared to become participants in multicultural dialog, able to know someone else's culture, critically perceive their own and integrate "their own" and "someone else's" in the process of professional communication. Multicultural formation and personal development occur within the framework of the development of such disciplines as intercultural communication, linguistic studies, and linguo-cultural studies.
2. Formation of the system of encyclopedic knowledge of language, the principles of its functioning, about communication of language and thinking in general and formation of the system of knowledge of the language(s) learned, and development of speech skills and discourse strategic skills during studying practical disciplines of foreign language(s) and translation.
3. Connection of linguistics and computer technologies. Modern requirements for the linguist-translator include the possession of software used in solving professional translation tasks. Due to the integration of theoretical linguistic and computer disciplines in this block, it is implemented at the final stage of the development of linguistic educational programs of higher education. Undoubtedly, the interdisciplinary approach allows the transfer of research methods from one scientific discipline to another, organizing binary (double) interdisciplinary disciplines on this principle, examples of which are machine and automated translation, corpus linguistics, and the study of parallel texts. The mastery of computer competencies also occurs in a mixed training mode when remote educational technologies are implemented.

The digitalization of education, because of the economy digitalization, is a set of processes aimed at creating the electronic educational environment in the university and including students and teachers in using the content of this environment (Godin

& Terekhova, 2021). The processes of digitalization of education involve the training of such specialists who are fluent in mobile and Internet technologies and are ready for continuous training (i.e., advanced training) using electronic means, as a result of which they are guaranteed to be in demand in the modern labor market.

Requirements imposed by digitalization on participants in the educational process are:

- Computer competence of teachers and students.
- Readiness of teachers and students to transform their activities in terms of the educational process digitalization.
- Advanced skills in self-organization and self-training of students.

In relation to the teacher, computer technologies require additional competencies, the main of which are:

- Owning e-learning software: conducting classes and creating e-learning materials.
- Organizational competence.

Characteristics of educational process in digitalization terms are:

- Personalization, i.e., implementation of a personal-oriented approach in training.
- High rate of provision of educational material.
- One-time provision of information to many people.
- Level of information perception and mastery of substantive aspects.

Thus, on the one hand, the application of computer technologies makes education continuous, accessible, and personal-oriented (Kameneva, 2021).

On the other hand, the use of digital technologies, being introduced into all spheres of life, including the educational process, does not guarantee the effectiveness of education for contemporary students belonging to the digital generation. Dissonance is manifested in the fact that students, being "digital natives" (Springer Fachmedien Wiesbaden, 2018), experience difficulties with the use of digital technologies, precisely in the process of their own education.

The possible reasons for this "conflict" can be considered as the non-formation of the educational process participants' skills in using digital resources, the lack of "live" contact with teachers and fellow students, a large amount of educational material offered for independent study. Therefore, in the educational process digitalization, there are new challenges for both teachers and students, and the need to constantly increase their own computer competence for a qualitative existence in the modern educational space.

Conclusion

In the contemporary period of the pandemic and the actual communication constraints, digital technologies remove the space-time restrictions imposed on the process of teaching foreign language(s) and translating with the traditional format of education. They make it possible to involve native speakers or members of the professional community in the teaching process, which is ultimately designed to increase the level of proficiency in the foreign language and the quality of professional training of linguistics graduates.

Digitalization is to become an integral part of linguistic education and to help acquire and improve the abilities and skills of critical and responsible using digital technologies in real professional activities and the contemporary information society.

Training translators for their professional activity, in addition to the formation, development and realization of their digital skills, represents formation of integral characteristics of the specialist training quality, which are related to the presence of a whole and meaningful complex of knowledge and skills in certain cross-disciplinary issues, which is due to the need to cover several different subjects and implies the presence of a certain level of reference skills which may not be digital ones. In other words, the subjects, using an interdisciplinary approach in linguistic education, retain their independence, enriched by new research principles and are designed to bring the specialist of foreign language(s) to the qualitative level which is to meet modern professional requirements.

References

Bago, P., Preradovic, N. M., Boras, D., & Ljubešić, N. (2017). Educating digital linguists for the digital transformation of EU business and society. In *INFuture2017: Integrating ICT in Society* (pp. 71–78). https://doi.org/10.17234/INFUTURE.2017.8.

Bershadskaya, L., Chugunov, A., & Trutnev, D. (2014, November). Information society development in Russia: Measuring progress and gaps. In *Proceedings of the 2014 Conference on Electronic Governance and Open Society: Challenges in Eurasia* (pp. 7–13). ACM https://doi.org/10.1145/2729104.2729122.

Gavrilenko, N. N. (2020). Formation of the digital competence of the translator. In V. I. Karasik (Ed.), *Topical issues of linguistics and teaching methods in business and professional communication, vol. 97. European Proceedings of Social and Behavioural Sciences* (pp. 548–553). European Publisher. https://doi.org/10.15405/epsbs.2020.12.02.73

Godin, V. V., & Terekhova, A. (2021). Digitalization of education: Models and methods. *International Journal of Technology, 12*(7), 1518–1528. https://doi.org/10.14716/ijtech.v12i7.5343

Hockly, N. (2021). *Digital literacies*. Springer Nature. https://doi.org/10.1007/978-3-030-79143-8_110

Kameneva, I. (2021). The actual questions of the realization of the personality-oriented educational paradigm in the context of digitalization. *E3S Web of Conferences, 273*, 12060. https://doi.org/10.1051/e3sconf/202127312060

Kolbachev, E., & Pahomova, A. (2019). The development of digital education in the Russian Federation: Risks and threats. *Bulletin of the South-Russian State technical University (NPI) Series Socio-Economic Sciences, 1*, 4–11. https://doi.org/10.17213/2075-2067-2019-1-4-11

Netolicky, D. M. (2019). *The role of professional standards in professional learning. Transformational professional learning: Making a difference in schools* (1st ed.). Routledge. https://doi.org/10.4324/9780429324277

Nugumanova, L. N., Shaukhutdinova, G. A., & Jakovenko, T. V. (2020, May). Some aspects of digitalization processes in education. In *2nd International Scientific and Practical Conference "Modern Management Trends and the Digital Economy: from Regional Development to Global Economic Growth" (MTDE 2020)* (pp. 197–200). Atlantis Press. https://doi.org/10.2991/aebmr.k.200502.032.

Ovsyanko, G., & Kyrychenko, S. (2021). Psychological features of studying a foreign language for professional purposes. *International Scientific Journal of Universities and Leadership, 11*, 158–167. https://doi.org/10.31874/2520-6702-2021-11-1-158-167

Springer Fachmedien Wiesbaden. (2018). We are digital natives. *ATZelektronik worldwide., 13*, 22–25. https://doi.org/10.1007/s38314-018-0054-5

Tolmach, M. (2021). Digital Technologies in Education: Possibilities and trends of application. *Digital Platform: Information Technologies in Sociocultural Sphere, 4*(2), 159–171. https://doi.org/10.31866/2617-796X.4.2.2021.247474

Vaganova, O. I., Zhidkov, A. A., Chelnokova, E. A., & Barabina, I. E. (2020). Innovative activities in the electronic educational environment. *Karelian Scientific Journal, 9*(4 (33)), 60–63. https://doi.org/10.26140/knz4-2020-0904-0016

Vinokurov, I. (2021). Digital economy in the Russian Federation. Problems and prospects for development. *Public Administration, 23*, 60–66. https://doi.org/10.22394/2070-8378-2021-23-5-60-66

Zagloel, T. Y. M., Surjandari, I., Berawi, M. A., Asvial, M., Harwahyu, R., Suryanegara, M., Setiawan, E. A., Suwartha, N., & Maknun, I. J. (2021). Digital economy and technology development in the Russian Federation. *International Journal of Technology, 12*(7), 1323–1327. https://doi.org/10.14716/ijtech.v12i7.5439

The Quality of the Digital Ecosystem in the Financial Sector

Elena Ivleva, Elena Sintsova, and Nina Shashina

Introduction

Let us pay attention to some features of modern approaches to the definition of the digitalization factor as an accelerator of economic growth, hastening transformation processes in the real and financial sectors of the economy.

There are specific approaches to assessing the role of digital technologies as a factor accelerating economic development. From the complete denial of this role in the studies of Porokhovskiy (2020). Before the limitations of the scope of various kinds of institutional traps and barriers to entry into the industry in the works of Stepanov and Kovalchuk (2020). The first approach is based on statistics of an annual decline in the growth rate of gross national product and a decrease in the growth rate of labor productivity of the world's leading economies. In any case, the digital economy is associated with phenomena: the platform economy and the ecosystem economy.

Moreover, researchers are interested not so much in the quantitative assessment of economic growth as in assessing changes in its quality. The use of digital technologies makes it possible to identify a fundamentally new quality of financial and non-financial services consumption. In all economies of the world entrepreneurship in the consumer services sector dominates in the structure of activities. In the economy of innovation-oriented and efficiency-oriented countries of the world, the sector of services to business is growing. This is an obvious and significant

E. Ivleva (✉) · E. Sintsova
St. Petersburg University of Management Technologies and Economics, St. Petersburg, Russian Federation

N. Shashina
St. Petersburg Institute of Economics and Management, St. Petersburg, Russian Federation
e-mail: n.shashina@spbacu.ru

A. Rumyantseva et al. (eds.), *Challenges and Solutions in the Digital Economy and Finance*, Springer Proceedings in Business and Economics,
https://doi.org/10.1007/978-3-031-14410-3_10

transformation within the economic structure of national economies, affecting the construction industry and the consumer services sector, along with the mining industry sector. In Russia the share of entrepreneurs providing commercial, informational, and financial services to businesses is significantly lower than in innovation-oriented countries. However, the national reports of the Russian Federation on global monitoring of entrepreneurship show the growth of this economy sector both within the framework of established entrepreneurs—7.6%, 6.4%, 23.6% in 2018, 2019 and 2020, respectively, and within the framework of early-stage entrepreneurs—11.8%, 9.4% and 17.1%, respectively (GEM, 2021). In parallel with the quantitative growth of the share of the sector of services to business, their quality is changing, as they are provided within the framework of digital platforms and ecosystems. It is noted by Radkovskaya et al. (2021). The quality of banking services is changing: green methods, technologies, green financial instruments are being developed. Banks use green ecosystem tools, develop green offices, provide reports on sustainable development, examine investment projects from the viewpoint of sustainable development.

The digital transformation of the Russian real and financial sectors of the economy makes it possible to identify a number of institutional traps associated with these processes. The theory of institutional traps of V. M. Polterovich (1999) is also applicable to the institutional organization of the digital economy, since it is associated with the peculiarities of the implementation of rules and standards, overcoming institutional voids. Voids and traps can be associated with the action of destabilizing factors of development, insufficiently effective financial and insurance support for participants in the economic process, the presence of administrative barriers, property problems, etc. The directions of digital transformation cannot be limited only to the development of digital infrastructure, healthcare, and digitalization of public services. Changes are also recorded in the financial sector of the economy. The institutional contours of digitalization traps are constantly changing and are the subject of study in many scientific papers. In the papers of Stepanov and Kovalchuk (2020) direct and indirect economic traps of digitalization are discussed most often. These are pricing traps, traps of asset dematerialization, non-market distribution of digital income, and others. A particularly discussed management approach to overcome such traps is the approach from the viewpoint of sustainable development (SD) and ESG—factors of growth. It is interesting to identify the financial traps of ESG—growth associated with emergency management decisions at the level of macro and mega economic systems that are currently demonstrated by the USA, Russia, the European Union, etc. These are traps in the perception of the process of energy transition from traditional energy sources to renewable and environmentally friendly ones. The energy transition will be long. It cannot be interpreted as one-time, including financial decisions of the European Union. Hence the economic and political pressure on the modern energy market. Modern researchers explore the issues of sustainable development of communities in the digital economy (Okrepilov et al., 2019; Kuzmina et al., 2019). Ritman et al. (2017), Vlasova et al. (2016) describe the tools for sustainable development of enterprises and securing the key advantages of entrepreneurial structures. We will provide a link to our attempts to determine the approaches to such managerial choice (Burgonov et al., 2019; Burgonov et al., 2020; Ivleva et al., 2016; Ivleva & Rumyantseva, 2021). The contours of both exchange and over-the-counter trading are changing. A new green sector is being formed on the securities exchange. Banks are planning and implementing debut green bond issues.

Special opportunities for the growth of entrepreneurial activity have been identified as a result of the implementation of digitalization technologies during the coronavirus pandemic. The digital technologies choice stimulates the innovative activity of all forms of entrepreneurship, and also slows down the growth of statism in the Russian economy. The promotion of remote technologies, platform and ecosystem services has being actively implemented during the coronavirus pandemic. Moreover, we should note the parallel growth in the expansion rate of the domestic national product and the digitalization of the banking sector of the economy, expanding the range of platform financial and non-financial services in Russia. Digitalization as a factor and technology of economic growth makes it possible to make a choice, including emergency managerial decisions that are the result of overcoming specific managerial traps. The depth and duration of coronavirus cycles, the directions of corrective financing and the tools for emergency management decisions are changing during periods determined by the virus recurrences. It once again determines the need for not single, but systemic managerial decisions.

However, along with the issues of determining the connection between the digitalization factor and the acceleration of economic development, identifying direct and indirect traps of digitalization and making decisions to overcome them, there are growing problems associated with the deep web. In the structure of indicators of the non-observed economy, this sector of financial flows sets new and increasingly serious tasks of economic security (Ivleva et al., 2016; Pimenova et al., 2020) related to the management of risks and threats of the digitalization of the economy. These are the problems of certification of new digital products and services, the so-called Info-Gypsies in the Internet space, etc.

Materials and Methods

As part of the content analysis, the list of sources used does not include in full the publications of the authors of the article. At the same time, publications of specialized economic journals containing different approaches within the framework of the research problem are included. Applied approaches of structural, systemic, and comparative analysis were used in the study of the digital transformation of the financial sector of the economy. The approach used has been developed due to the fact that:

1. Transformation processes in the digital economy lead to the formation of digital platforms, platform companies, platform business models, and ecosystems. The

research approaches of the authors of scientific papers differ both in determining the quantitative and qualitative assessments of the noted changes.
2. The ecosystem quality of financial and non-financial services is growing as part of these changes, which is considered by us in the focus of ESG changes. An example is banking services, which include the reduction of waste activities, waste recycling, energy savings, ESG loans, ESG risks, ESG ratings. Within the securities market, this is the formation of the green finance sector, ESG bonds, exchange-traded funds that comply with ESG principles.

Results

The study of the financial services quality in the digital economy has a number of features:

- Analysis of modern periodical papers allows us to identify controversial approaches in determining the role of the digitalization factor itself in accelerating economic development and its quality increasing.
- The development of digital technologies in the financial sector of the economy entails an increase in the ecosystem quality of financial services, both for business and the population. Ecosystem services are convenient and accessible. They are institutionally and infrastructurally circumspect. They consider world methods, technologies and financial instruments. Financial and non-financial companies within the ecosystem are evolving, constantly attracting new digital platforms, ensuring a change in the range of financial services and an increase in their quality.
- The requirements for financial institutions in the context of digital transformation are changing very quickly. Today these are not only distance selling opportunities, but also regulatory requirements related to green finance and green growth promotion.
- The institutional traps associated with the digital transformation of the real and financial markets include the financial traps of ESG growth, including the problems of implementing financial instruments for the energy transition from traditional energy sources to green technologies. Digital technologies do not change the strategy of the relationship between production and the environment, but they can significantly change growth technologies and expand the range of financial instruments for green finance within the framework of green growth. In the long term, financial flows in the financial market are redistributed in favor of companies with an ESG strategy.

Discussion

The study allows us to formulate three controversial problems:

- Some studies of economic development are based on the assertion that digitalization is an accelerator of economic growth, others refute this approach in economics, emphasizing that the growth rate of gross domestic product and the growth rate of labor productivity of the world's leading economies are declining.
- Infrequent studies are devoted to the formation of a new ecosystem quality of the financial sector of the digital economy, and the ecosystem quality of financial services. Ecosystems of financial institutions are focused on quality growth, since they combine partner services focused on finding quality growth and using world experience in achieving competitiveness in the sustainable development outlines. Very few researchers raise this problem.
- The theory of traps and voids of the digital transformation of the financial and non-financial sectors of the economy has not been formed, which explains the differences in research approaches and interpretations.

Conclusion

Summing up, we define several conclusions:

Modern economies are developing under the influence of new factors such as digitalization and the coronavirus pandemic. Expanding the scope of the study, taking into account these non-economic factors, is possible within the framework of institutional analysis. The experience of stimulating the development of financial institutions is represented by world methods, technologies and financial instruments (Ivleva & Rumyantseva, 2021; Polterovich, 1999).

The digital transformation of the economy, including its financial sector, should be assessed based on ESG—ecosystem growth criteria. It cannot be limited to the development of digital infrastructure, digital public services, digital healthcare technologies, etc. In the Russian economy, the budget for the implementation of the Digital Economy of the Russian Federation national project is estimated as the budget of annually missing investments in the Russian economy that emphasizes the presence of institutional traps in the implementation of all identified approaches.

Economic growth should be seen as ecosystem growth. In this regard, the definition of a new quality of growth as an ecosystem quality of financial and non-financial services consumption by all economic entities seems to be justified (Aleksandrovich et al., 2020; Barajas et al., 2017; Vlasova et al., 2016; Ivleva & Rumyantseva, 2021).

In the statistics of the non-observed economy, the sector of the so-called deep web is changing, with which new and increasingly serious tasks of managing the risks and threats of the digitalization of the economy are associated.

References

Aleksandrovich, B. E., Aubakirovna, E. L., Andreevich, B. P., & Nikolaevna, K. S. (2020, March). Organizational and economic mechanism for managing the development of the cultural sphere of the Russian regions in the digital economy. In *Proceedings of the III International Scientific and Practical Conference* (pp. 1–4). https://doi.org/10.1145/3388984.3390627.

Barajas, A., Shakina, E., & Fernández-Jardón, C. (2017). Acceleration effect of intangibles in the recovery of corporate performance after-crisis. *Research in International Business and Finance, 42*, 1115–1122.

Burgonov O., Golubetskaya N., Ushakova E., & Chirkova T. (2020). The architecture of strategic management of innovative development of companies in the digital economy. In *E3S web of conferences. Key trends in transportation innovation, KTTI 2019* (Vol. 157, p. 04017). https://doi.org/10.1051/e3sconf/202015704017.

Burgonov, O., Kruglov, D., & Ivleva, E. (2019, October). Features of industrial management in the field of ecology. In *IOP Conference Series: Materials Science and Engineering* (Vol. 618, No. 1, p. 012073). IOP Publishing.

GEM. (2021). *Global entrepreneurship monitoring*. Russia 2020/2021. Retrieved from https://gsom.spbu.ru/research/research_statistics/gem/

Ivleva, E., Pak, K. S., & Nagornaya, K. (2016). Regional indicators of the safety development of the entrepreneurship economy. In *MATEC Web of Conferences* (Vol. 73, p. 07019). EDP Sciences. https://doi.org/10.1051/Matecsconf/20167307019.

Ivleva, E. S., & Rumyantseva, A. Y. (2021). The coronavirus pandemic as a factor in corrective financing. *Economics and Management., 27*(11), 910–915. https://doi.org/10.35854/1998-1627-2021-11-910-915

Kuzmina, S., Morozova, N., Pimenova, A., & Agafonova, A. (2019). Sustainable architecture: Trends in higher professional architectures' education. In *MATEC Web of Conferences* (Vol. 265, p. 07017). EDP Sciences. https://doi.org/10.1051/matecconf/201926507017.

Okrepilov, V., Kuzmina, S., & Kuznetsov, S. (2019, March). Tools of quality economics: Sustainable development of a 'smart city' under conditions of digital transformation of the economy. In *IOP Conference Series: Materials Science and Engineering* (Vol. 497, no. 1, p. 012134). IOP Publishing. https://doi.org/10.1088/1757-899x/497/1/012134.

Pimenova, A. L., Vaganova, V. A., Yahvarov, E. K., & Kuzmina, S. (2020, March). Digitalization tools in ensuring the efficiency of the enterprise activity. In *Proceedings of the III International Scientific and Practical Conference* (pp. 1–4). https://doi.org/10.1145/3388984.3390641.

Polterovich, V. (1999). Institutional traps and economic reforms. *Economics and Mathematical Methods, 35*(2), 3–20.

Porokhovskiy, A. F. (2020). Digitalization and artificial intelligence: Prospects and challenges. *Economics. Taxes. Law., 2*, 84–90.

Radkovskaya, N. P., Klochkova, E. N., & Lvova, Y. N. (2021). Digital transformation of the Russian financial market as a key strategy in the post-COVID period. *News of the St. Petersburg State University of Economics, 3*(129), 48–53.

Ritman, N., Pimenova, A., Kuzmina, S., & Baranova, A. (2017, October). Strategic factors of development of the enterprises. In *IOP Conference Series: Earth and Environmental Science* (Vol. 90, no. 1, p. 012092). IOP Publishing. https://doi.org/10.1088/1755-1315/90/1/012092.

Stepanov, I. M., & Kovalchuk, Y. A. (2020). Economic traps of the artificial intelligence introduction. *Economics. Taxes. Law., 13*(2), 92–101.

Vlasova, M., Pimenova, A., Kuzmina, S., & Morozova, N. (2016). Tools for company's sustainable economic growth. *Procedia Engineering, 165*, 1118–1124. https://doi.org/10.1016/j.proeng.2016.11.828

Human Potential in the Sustainable Development of Regions

Natalia Voronina

Introduction

The concept of sustainable development implies a special approach to the management of regions, in which the use of the potential available to a particular region is carried out to preserve and expand the opportunities available to people (Nesterov, 2018). The concept of sustainable development was first formulated at the UN Conference in 1992 as meeting the needs of present and future generations. The content of the term determines the existence of two fundamental factors: environmental sustainability; socio-economic sustainability (Gruznevich, 2017).

The main goal of the sustainable development of the region is to provide conditions for the formation, accumulation, development, rational use and conservation of human potential and the preservation of the biosphere.

The concept of sustainable development includes three components: economic development, environmental conservation, and social progress.

The economic prerequisites for sustainable development are associated with a reduction in natural resource potential and imbalances in economic development.

Environmental prerequisites are caused by the widespread degradation of the surrounding environment.

Social progress is manifested in the strengthening of social contribution to economic activity. An active social policy in the regions aimed at reducing the effects of social risks and improving the quality of life of the population ensures an increase in the quality of human potential, which leads to the sustainability of socio-economic processes (Leontieva & Ignatova, 2020).

The new format of the study of the possibility of sustainable development of regions is productive, assuming the following key components: the state, the person,

N. Voronina (✉)
Moscow Aviation Institute (National Research University), Moscow, Russian Federation

A. Rumyantseva et al. (eds.), *Challenges and Solutions in the Digital Economy and Finance*, Springer Proceedings in Business and Economics,
https://doi.org/10.1007/978-3-031-14410-3_11

society, the business environment. The relevance of the study of this problem is primarily since today conceptual and methodological approaches to the formation of sustainable development of the regions of the Russian Federation have not been fully developed.

The purpose of the study is to clarify the interdependence of the processes of realizing human potential and the formation of sustainable regional development.

Materials and Methods

The concept of sustainable development inherently proceeds from the balance of human activity in the environment, economy and social sphere, and the social component of sustainable development implies that humanity acts as the central or main member of the Eco socioecological system and a person is not an object, but a subject of development. This means that a person should participate in the processes that form the sphere of his life, facilitate the adoption and implementation of decisions, monitor their execution.

The specificity of sustainable development is precisely that it is influenced by the magnitude, qualitative characteristics and standard of living of human potential, which is understood as the combination of physical and spiritual forces of its inhabitants (Legotina & Chumakova, 2018), which can be used to achieve individual and social goals.: both instrumental, related to providing the necessary conditions for life-activity, and existential, including the expansion of the human abilities themselves and the possibilities of his self-realization.

In the modern world, human potential acts as the most important factor of economic growth since the use of all other development resources depends on its condition. Among all the components of the national wealth of any state, it is the human potential that plays a leading role. According to the World Bank, in developed countries, human capital as an economic expression of a country's human potential makes up from 68% to 76% of the total national wealth. That is, the main share of the national wealth lies in people (Mudrecov, 2016).

However, at the level of the constituent entities of the Russian Federation, the formation of regional social policy still rarely considers the indicators of the level of development of the human potential of the region and the quality of life. Until now, when developing a strategy for the sustainable development of regions, emphasis is placed on economic development to the detriment of human development goals, which ultimately has a negative impact on the socio-economic development of both territories and the country. The problem of reproduction and improvement of human potential remains extremely relevant at all levels of government: global, state, regional and municipal.

Results

Since the 1990s, regional development programs have been developed in the Russian Federation with regional concretization and orientation on the use of existing prerequisites for stabilizing and improving the socio-economic situation of the respective territories, a standard for the development of competition in Russian regions has been introduced (Analytical Center under the Government of the Russian Federation, 2022). The space of Russia's regional development is significantly differentiated, both from the point of view of the selected development trajectories, and the initial state, and the standard of living, and the policy of the regional government. The most striking manifestation of inequality is the ratio of indicators of sustainable development within its space (Table 1).

The formation of a rating of the quality of regional financial management allowed us to assess the management activities of the regions and to conduct a gradation: high, medium, low level of management activity.

Regions with the highest rating (high): The Chuvash Republic, the Orenburg region, the Vladimir region—a total of 24 regions (24 regions in 2018, 27 regions in 2017).

Regions with an average rating (average)—a total of 43 regions (47 regions in 2018, 47 regions in 2017).

The region with the minimum rating (low): The Jewish Autonomous Region in total—18 regions (14 regions in 2018, 11 regions in 2017).

The main results of the assessment:

1. Not reducing the coefficient of covering expenditures of the budgets of the subjects of the Russian Federation with their own funds without attracting borrowed funds.
2. Improving the quality of budget planning in the subjects of the Russian Federation, expressed in reducing the number of amendments to the laws on budgets of the subjects of the Russian Federation during the financial year.
3. Reduction of the average value of the indicator for the Russian Federation, reflecting the ratio of overdue accounts payable of the subjects of the Russian Federation and budgetary and autonomous institutions of the subjects of the Russian Federation to expenses.
4. Reduction of the average level of debt burden of the budgets of the subjects of the Russian Federation.
5. Absence of cases of late submission of reports to the Federal Treasury

The main shortcomings in the management of regional finances:

1. An increase in the number of subjects of the Russian Federation with poor quality of regional finance management.
2. In some regions during the year there was an increase in expenses that were not provided with an increase in their own revenue sources.
3. In some subjects of the Russian Federation, incentive tax benefits for property tax and corporate income tax are recognized as ineffective.

Table 1 Rating of Russian regions on various grounds

No	Region	Rating of Russian regions by quality of life (results of 2020) (Ria, 2020)	Rating of Russian regions by quality of life (results of 2020) (Ria, 2020) Regional Finance Management Quality Rating (Ministry of Finance of the Russian Federation, 2020) B-High C-Average N-Low	ESG Rating of Russian regions (as of 2020) (Sustainable Development Platform, 2020)
1	Moscow	1	B	2
2	St. Petersburg	2	B	23
3	Moscow oblast	3	C	20
4	Republic of Tatarstan	4	C	1
5	Belgorod region	5	B	42
6	Krasnodarskiy kray	6	B	45
7	Leningrad region	8	B	8
8	Voronezh region	7	B	12
9	Khanty-Mansi autonomous Okrug.	10	B	4
10	Kaliningrad region	9	B	54
11	Yamalo-Nenets autonomous Okrug.	12	B	19
12	Lipetsk region	11	C	3
13	Nizhny Novgorod region	15	C	49
14	Tyumen region	14	B	6
15	Sverdlovsk region	13	C	7
16	Samara region	16	C	26
17	Rostov region	19	C	34
18	Kursk region	20	B	5
19	Sevastopol	17	C	–
20	Tula region	18	C	22
21	Kaluga region	21	B	10
22	Novosibirsk region	22	C	30

(continued)

Table 1 (continued)

No	Region	Rating of Russian regions by quality of life (results of 2020) (Ria, 2020)	Rating of Russian regions by quality of life (results of 2020) (Ria, 2020) Regional Finance Management Quality Rating (Ministry of Finance of the Russian Federation, 2020) B-High C-Average N-Low	ESG Rating of Russian regions (as of 2020) (Sustainable Development Platform, 2020)
23	Stavropol territory	23	C	48
24	Republic of Adygea	27	B	25
25	Chelyabinsk region	24	C	27
26	Sakhalin region	34	B	15
27	Kamchatka Krai	30	C	74
28	Ryazan region	26	C	16
29	Republic of Bashkortostan	25	C	46
30	Khabarovsk territory	31	N	33
31	Yaroslavl region	28	C	18
32	Vladimir region	33	B	50
33	Ulyanovsk region	29	C	17
34	Orenburg region	32	B	55
35	Oryol region	40	N	63
36	Penza region	37	C	44
37	Smolensk region	35	C	38
38	Tambov region	43	N	61
39	Chuvash Republic	44	B	13
40	Bryansk region	41	B	40
41	Primorsky Krai	46	B	57
42	Republic of Crimea	47	C	
43	Murmansk region	36	C	53

(continued)

Table 1 (continued)

No	Region	Rating of Russian regions by quality of life (results of 2020) (Ria, 2020)	Rating of Russian regions by quality of life (results of 2020) (Ria, 2020) Regional Finance Management Quality Rating (Ministry of Finance of the Russian Federation, 2020) B-High C-Average N-Low	ESG Rating of Russian regions (as of 2020) (Sustainable Development Platform, 2020)
44	Volgograd region	42	C	59
45	Ivanovo region	52	C	66
46	Krasnoyarsk territory	38	C	28
47	Udmurt Republic	39	C	58
48	Perm region	49	C	60
49	Saratov region	48	C	9
50	Magadan region	50	N	79
51	Tomsk region	45	N	51
52	Republic of Mordovia	51	N	36
53	Tver region	58	C	14
54	Kemerovo region	56	N	68
55	Irkutsk region	63	C	32
56	Republic of Dagestan	64	N	31
57	Pskov region	55	N	80
58	Astrakhan region	54	C	69
59	Vologda region	59	B	29
60	Omsk region	53	C	43
61	Chukotka auth. District	68	N	41
62	Republic of Mari El	61	C	37
63	Kirov region	60	C	52
64	Republic of Khakassia	57	N	70
65	Republic of Sakha (Yakutia)	70	C	35

(continued)

Table 1 (continued)

No	Region	Rating of Russian regions by quality of life (results of 2020) (Ria, 2020)	Rating of Russian regions by quality of life (results of 2020) (Ria, 2020) Regional Finance Management Quality Rating (Ministry of Finance of the Russian Federation, 2020) B-High C-Average N-Low	ESG Rating of Russian regions (as of 2020) (Sustainable Development Platform, 2020)
66	Novgorod region	65	B	30
67	Kostroma region	62	N	24
68	Altai territory	72	B	64
69	Amur region	67	C	56
70	Republic of North Ossetia—Alania	76	N	47
71	Komi Republic	69	B	78
72	Republic of Karelia	73	C	76
73	Nenets autonomous Okrug	66	C	21
74	Chechen Republic	71	C	75
75	Arkhangelsk region	74	C	71
76	Republic of Kalmykia	78	N	83
77	Kabardino-Balkarian rep.	75	B	81
78	Republic of Ingushetia	79	N	72
79	Kurgan region	81	N	73
80	Republic of Altai	80	C	64
81	Republic of Buryatia	77	C	62
82	Jewish autonomous region	83	N	77
83	Karachay-Cherkess Republic	84	C	67
84	Trans-Baikal territory	82	C	65
85	Republic of Tyva	85	N	82

4. In some regions, the indicators of the “road maps” have not been met.
5. Eight subjects of the Russian Federation violated the terms of agreements on the provision of budget loans.
6. 23 subjects of the Russian Federation violated the terms of agreements on measures for socio-economic development and improvement of public finances for 2019.

The development of the region largely depends on its attractiveness for investment activities. Today, an increasing number of investors are choosing regions for investments, considering their compliance with the principles of sustainable development. ESG ratings are an assessment tool from the point of view of regional sustainability. In 2006, the UN formulated the principles of responsible investment. The ESG assessment is based on the analysis of three key factors affecting the sustainability and ethics of investments (Principles of Responsible Investments—PRI): Environmental and social risks (Social) and the quality of corporate Governance (Governance). Adherence to the principles of sustainable development of the region is assessed using ESG factors.

The European rating agency RAEX-Europe compiles an ESG rating of Russian regions, which allows assessing environmental and social risks, as well as the quality of public administration, within the same scale. In accordance with the best international practices, the Agency assesses not only the level of environmental, social and management risks, but also how effectively the regions cope with these risks. ESG data is mainly qualitative: it is primarily non-financial data that is difficult to quantify in monetary terms.

В основе большинства ESG рейтингов, лежит принцип совмещения уровня подверженности риску и оценки эффективности его нивелирования.

The crisis phenomena of 2020 caused by the pandemic did not bypass the investment sphere. The volume of investments in fixed assets in Russia, according to Rosstat, decreased by 1.4% against an increase of 2.1% in 2019. The decline was caused by the results of the second and third quarters, when restrictive measures were introduced, and many business processes simply stopped.

The leaders in investments in fixed assets per capita are the northern regions: Nenets Autonomous District (2046 thousand rubles), Yamalo-Nenets Autonomous District (1975 thousand rubles) and Chukotka Autonomous District (629 thousand rubles). At the same time, if the Nenets Autonomous District and Chukotka occupy high positions with rather modest investment volumes, mainly due to the small population, then the Yamalo-Nenets Autonomous District is characterized by a high absolute investment volume, which amounts to 1.075 trillion rubles, and is second only to Moscow, which is the leader in this indicator.

The lowest indicators of investments in fixed assets per capita by the end of 2020 were noted in the Karachay-Cherkess Republic (40.1 thousand rubles), Kostroma Region (41.2 thousand rubles), the Republic of Ingushetia (42.8 thousand rubles).

The positive dynamics of investments in 2020 was shown by 34 subjects of the rus-sian Federation, in one region the volume of investments did not change, in 50 there was a decline. At the same time, 12 regions showed an increase of more than 10%, and 22 regions showed a decrease of more than 10%.

The most severe deterioration in dynamics took place in the Republic of Sakha (Yakutia), where the volume of investments in fixed assets decreased by 49.1% compared to last year, investments in Sevastopol decreased by 40.9% and by 39% in the Republic of Altai (RIA, 2021).

The statistics of various departments available in Russia allows us to build a system of indicators to assess the exposure of regions to risks and the degree of their leveling.

The Republic of Tatarstan, Moscow, and the Lipetsk Region with a high and medium level of management remain the most attractive for investment.

The least attractive for investment are the Republic of Tyva, the Kabardino-Balkarian Republic, and the Republic of Kalmykia.

At the same time, there is a specific situation in the assessment of regions. If, for the most part, the assessment of regions in three positions: quality of life, investment attractiveness, quality of management coincides in degree, then in a number of regions this is not the case. For example, the Kabardino-Balkarian Republic ranks 75th in terms of quality of life, 81st in terms of investment attractiveness, while at the same time high quality of management is noted. The regions in which the discrepancy in the estimates is revealed include Belgorod Region, Krasnodar Territory, Kaliningrad Region, Vladimir Region, Chuvash Republic, Bryansk Region, Primorsky Territory, Saratov Region, Vologda Region, Novgorod Region, Kostroma Region, Altai Territory, Komi Republic.

The differentiation of regional development is since the emergence of technological, social and other innovations occurs unequally. The modern stage of Russian modernization imposes new requirements for the formation and realization of the human potential of regions. The introduction of digitalization contributes to the reduction of regional inequality in the country, as it allows to develop infrastructure, develop, and actively implement digital technologies in various spheres of life, as well as to form the necessary regulatory and legislative framework for their application.

The 20th Survey of the heads of the world's largest companies in 79 countries showed that the priority business areas for development.

An integral part of human potential is human capital, as a set of acquired knowledge, skills, skills necessary to meet the needs of a person and society as a whole. As can be seen from the data in Table 2, the least attention is paid to the human factor in Russia. The human potential of the region is formed because of the

Table 2 Priority business areas

Directions	All countries (%)	Russia (%)
Innovations	23	13
Human capital	15	7
Digital and technological resources	15	20
Customer service quality	10	15
Competitive advantage	10	13

Source: (Mudretsov, 2016)

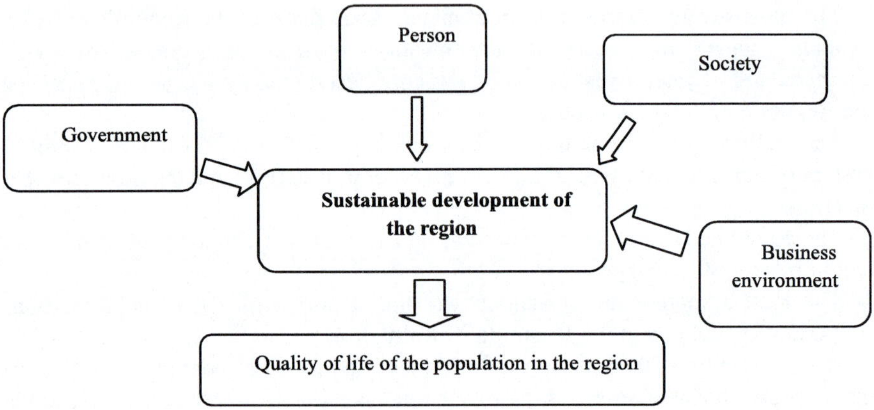

Fig. 1 Factors influencing the sustainable development of the region (compiled by the author)

interaction of the potentials of various socio-demographic and socio-professional communities living and operating in a particular region. The human potential of the region can be considered as a combination of the needs and abilities of the population of the region, on the one hand, and its readiness, on the other, to realize these capabilities by performing socially significant functions and roles.

Discussion

The main prerequisites for the sustainable development of Russia are a large territory with preserved non-renewable natural resources and natural ecosystems, human potential, and economic resources. To achieve sustainable development, it is necessary to direct economic resources to the development of human potential (Shelekhova, 2002).

The quality of human potential that lags the objective needs of the development of socio-economic relations significantly hinders social development as a whole (Eremina, 2012).

The formation of processes aimed at the sustainable development of regions relates to the following components: state activity, people, society, business environment, which is shown in Fig. 1.

The state is considered from the point of view of an element whose activity is connected with the formation and provision of:

- The direction of development of the region, taking into account its geographical position, resource potential, economic potential.
- Regulatory and legislative support of activities.
- Balanced development of the biosphere.
- Development strategies of the Russian Federation.

A person is considered from the point of view of an element whose activity is related to ensuring the implementation and implementation of the development programs of the region on the one hand as a labor resource, on the other hand as the main consumer of the result of sustainable development.

Society is viewed from the point of view of an element whose activity is associated with the formation of the necessary social opinion and tendencies aimed at the desire to live in better conditions and strive for their formation and maintenance.

The business environment is viewed from the point of view of an element whose activities are related to economic and social development in accordance with the region's development strategy and goals. The formation of the business environment depends on the number and structure of the available workforce, the level of education, professional and qualification composition, demographic trends. Thus, the role of social factors of economic development increases.

Achieving sustainable development of the region is possible only with the interaction of all these elements, whose activities are characterized by the desire to achieve a single goal.

Conclusion

The transition to sustainable development of the regions should ensure a balanced solution to the problems of socio-economic development and preservation of a favorable environment and natural resource potential, meeting the needs of present and future generations of people (President of the Russian Federation, 1996).

The directions of transition to regional sustainable development should include:

- Development of a strategy for the development of the territories of the Russian Federation, which is based on the principle of balanced development of regions.
- Creation of a legal framework.
- Development of a system to stimulate the social sulfur of the regions' activities aimed at forming the involvement of the population in achieving the Sustainable Development Goals.
- Assessment of the economic capacity of the region's ecosystems to increase the investment attractiveness and quality of life in the region.
- Formation of an effective system for promoting the ideas of sustainable development and the creation of an appropriate system of education and training.

The transition to sustainable development will require coordinated actions in all spheres of society, an adequate reorientation of the social, economic, and environmental institutions of the State, whose regulatory role in such transformations is fundamental.

To manage the process of transition to sustainable development and evaluate the effectiveness of the funds used, it is necessary to set targets and restrictions with the provision of a procedure for monitoring their achievement. The targets can be expressed in indicators characterizing the quality of life, the level of economic development and environmental well-being, which allow achieving a balanced development as one of the factors of sustainable development.

The problems of achieving sustainable development of regions require further research and the development of appropriate methodological tools for assessing the real state and progress in this area.

References

Analytical Center under the Government of the Russian Federation. (2022). *Implementation of the Competition development standard in Russian regions*. Retrieved from https://ac.gov.ru/projects/done/project/vnedrenie-standarta-razvitia-konkurencii-v-rossijskih-regionah-17.

Eremina, E. V. (2012). The importance of human potential in the development of the region. Еремина, Е. В. (2012). Значение человеческого потенциала в развитии региона. *Russian Journal of Education and Psychology, 5*, 1.

Gruznevich, E. S. (2017). *Development, sustainable development and balanced sustainable development, and their interrelation at the regional level: Terminological aspects*. Tendencii i perspektivy razvitiya.

Legotina, I. A., & Chumakova, E. A. (2018). Systematization and analysis of approaches to the definition of human potential in economic theory. *Scientific Bulletin of the Volgograd Branch of RANEPA Series: Economics, 2*, 100–104.

Leontieva, A. G., & Ignatova, I. V. (2020). The quality of human potential as a factor of entrepreneurship development in the region. *Bulletin of the Altai Academy of Economics and Law, 3–2*, 220–227.

Ministry of Finance of the Russian Federation. (2020). *The results of the assessment of the quality of regional finance management for 2019*. Retrieved from https://www.minfin.ru/common/upload/library/2020/11/main/Rezultaty_otsenki_za_2019_god.pdf.

Mudrecov, A. F. (2016). The human potential of sustainable development. *Regional Problems of Economic Transformation, 8*(70), 98–103.

Mudretsov, A. P. (2016). Human potential of sustainable development. *Regional Problems of Economic Transformation, 8 (70*, 98–103.

Nesterov, A. K. (2018). *Sustainable development of regions*. Encyclopedia of the Nesterovs. Retrieved from http://odiplom.ru/lab/ustoichivoe-razvitie-regionov.html.

President of the Russian Federation. (1996). *The concept of the transition of the Russian Federation to sustainable development*. Retrieved from https://docs.cntd.ru/document/9017665.

Ria. (2020). *Rating of Russian regions by quality of life*. Retrieved from https://ria.ru/20200217/1564483827.html.

RIA. (2021). *RIA rating. Russia today*. Retrieved from https://riarating.ru/regions/20210305/630195904.html.

Shelekhova, A. M. (2002). *The main provisions of the strategy of sustainable development of Russia*. Retrieved from http://www.nsc.ru/win/sbras/bef/strat.html.

Sustainable Development Platform. (2020). *ESG rating of Russian regions, 2020*. Retrieved from https://sdpl.ru/761-esg-reyting-rossiyskih-regionov-2020.html.

Labour Resources use in the National Economy: G7 and Russia Comparative Analysis

Svetlana Pshenichnikova, Elena Kolesnik, and Vladimir Plotnikov

Introduction

In the past two years, the global economy and national economies have been developing under the powerful influence of the Covid-19 pandemic (Borio, 2020; Nadezhina et al., 2021; Pirogova et al., 2021; Vertakova & Feoktistova, 2020; Zhang et al., 2020). This led, in most cases, to a slowdown in economic growth or even to its termination. Prospects for further development are closely linked to the success in the fight against the pandemic. At the same time, economic dynamics is governed not only by random factors (for example, a pandemic), but also by fundamental factors. The main fundamental factor of economic growth is the number of resources involved in the economic turnover (extensive factor) (Senhadji, 2000; Shabbir et al., 2020), as well as their quality (intensive factor) (Pogodaeva & Senchenko, 2017; Saleem et al., 2019).

It is the resources that determine the long-term socio-economic dynamics (Akaev & Hirooka, 2009; Pradhan et al., 2019). Therefore, their analysis and assessment of their impact on economic growth should be priority issues for study. Within the framework of this article, an analysis was made of the use of one of the types of resources—labour resources. The research was carried out on empirical materials from the G7 countries (G7: Canada, France, Germany, Great Britain, Italy, Japan, USA) and Russia. When referring to these countries together, the article uses the term "G8". The purpose of the article is to analyse labour resources using in the national economies of G8 countries.

S. Pshenichnikova (✉) · E. Kolesnik · V. Plotnikov
St. Petersburg State University of Economics, St. Petersburg, Russian Federation

A. Rumyantseva et al. (eds.), *Challenges and Solutions in the Digital Economy and Finance*, Springer Proceedings in Business and Economics,
https://doi.org/10.1007/978-3-031-14410-3_12

Materials and Methods

The research used data from international (World Bank, OECD, etc.) and Russian (Federal Agency of State Statistics of Russia) statistics, as well as quantitative and qualitative assessments and developments cited in the literature and business media. The analysis of the use of labour resources in the G8 was carried out by examining data on employment. The analysis used a binary classification of types of employment—flexible and rigid:

1. Flexible employment is associated with high labour market volatility. With flexible employment, the market reacts to changes with small losses in time, effort, costs, and efficiency. Flexible employment is provided by two factors: flexibility of the workplace; flexibility of payment for work. Employment flexibility means weakening the value of the standard work contract. Temporary, seasonal contracts, work on schedule, self-employment, outsourcing, etc. are being developed (Vertakova & Kurbanov, 2016).
2. Hard employment is associated with the execution of traditional labour contracts. On the one hand, it gives rise to high inertia and weak adaptability of the labour market. On the other hand, tough employment is the stabilizer of the economy, allowing it to be more resilient in the face of crises.

The concept of flexible and rigid employment was developed by Pshenichnikova in relation to the macroeconomic level of the economic system (Pshenichnikova, 2017). Pshenichnikova's concept is based on econometric modelling. Flexible employment implies a unidirectional, synchronous change in the dynamics of the number of employed and the dynamics of the volume of GDP. The coefficient of employment elasticity in terms of GDP in the long run is equal to or close to this value (unit elasticity). Rigid employment is due to either multidirectional or asynchronous changes in the studied indicators. In this case, the coefficient of elasticity takes values in the range from zero to one, or more than one. Then, in general terms, the dependence of employment on the volume of GDP can be interpreted as inelastic or elastic.

Rigid employment is heterogeneous in essence and form of manifestation. It may differ not only in the value of the elasticity coefficient, but also in the peculiarities of the relationship between the dynamics of employment and GDP. There are several typical situations:

1. Percentage fluctuations in GDP are more significant than fluctuations in the size of the employed labour force. This is "straight" rigidity.
2. The fall in GDP is more catastrophic compared to employment. This is "reverse" rigidity, which implies the impossibility of reducing employment. In this case, employment is maintained within a certain range, naturally, there may be an inefficient use of labour. At the same time, unemployment does not increase significantly, and, consequently, social tension does not reach the "boiling point". At the same time, the decline in production, while maintaining the number of

employees and the absolute size of the wage fund, reduces the amount of income of other factors of production.

In an earlier study (Pshenichnikova & Pshenichnikov, 2016; Romanyuk et al., 2019) A classification of the G7 countries and Russia was proposed by the type of labour strength. As a result, two groups of countries were identified:

1. Germany, Russia, France, Japan. From the point of view of identifying the types of macroeconomic employment in these countries, at the beginning of the study period, the inverse rigidity of employment is observed, which then turns into direct rigidity of employment. That is, the type of employment in these countries is "mixed rigid". The transition from reverse to direct employment rigidity in Germany took place in 2003, in Russia in 2005, in France in 2004, and in Japan in 2010. The inverse employment rigidity determines the unstable dynamics of economic growth and labour productivity. This transition indicates positive processes: there is a positive growth rate of labour productivity, stabilization of the situation in the traditional sector of the economy and the creation of conditions for the further development of the innovation sector.
2. Great Britain, Italy, Canada, USA. The largest GDP growth in 2018 compared to 1995 is achieved in Canada (301.9%) and the United States (292.1%). In this group of countries, the type of employment is direct, rigid, which indicates an efficient sectoral structure of production, as well as an effective structure of the labour force consumed. Constant positive growth rates of labour productivity arising from the direct rigidity of employment indicate a favourable economic situation in both the traditional and innovative sectors of the economy.

Results and Discussion

A study was carried out on the demographic situation, economic activity of the population, gender and age composition of the population, unemployment rate, level of education, state of fundamental science, growth of labour productivity, level of wages. Based on this analysis, it is possible to single out the features of the quantitative and qualitative characteristics of the reproduced labour force in the G7 countries and Russia.

1. The number of employed in different countries differs significantly (Table 1). For example, in Russia, the number of employed is more than two times less than in the American economy: in the Russian Federation in 1990–75.3 million people, in 2014–75.7 million people, in 2017–72.1 million people; in the USA in 1990—151.0 million people, in 2017–153.3 million people. The maximum number of employed among the group of countries studied was reached in the USA in 2013–163.1 million people.
2. After the 2008 crisis, the G8 countries have seen a drop in GDP and a slight increase in employment, except for Japan, where the number of employed fell from 67.4 million people in 1997 to 65.3 million people in 2013. In 2009, labor

Table 1 The number of employed labour force in the G7 and Russia, 1990–2017, thousand people

	1990	1991	1992	1993	1994	1995	1996	1997	1998	1999	2000	2001
Russia	75,325	75,421	75,580	73,289	71,003	71,160	70,322	68,837	67,877	72,978	73,524	72,311
G8 share, %	18.8	18.6	18.6	18.1	17.5	17.4	17.1	16.7	16.4	17.2	17.3	17
Great Britain	29,340	29,231	28,933	28,729	28,733	28,720	28,847	29,006	29,032	29,353	29,589	29,574
G8 share, %	7.3	7.2	7.1	7.1	7.1	7.0	7.0	7.0	7.0	6.9	6.9	6.9
Germany	37,155	39,773	39,693	39,756	39,800	39,931	40,203	40,405	40,645	40,491	40,445	40,481
G8 share, %	9.3	9.8	9.8	9.8	9.8	9.8	9.8	9.8	9.8	9.6	9.5	9.5
Italy	23,830	24,006	23,255	23,134	22,964	22,826	22,925	22,947	23,078	23,239	23,311	23,519
G8 share, %	5.6	5.9	5.7	5.7	5.7	5.6	5.6	5.6	5.6	5.5	5.5	5.5
Canada	14,669	14,746	14,720	14,807	14,931	15,023	15,151	15,386	15,647	15,949	16,206	16,465
G8 share, %	3.7	3.6	3.6	3.7	3.7	3.7	3.7	3.7	3.8	3.8	3.8	3.9
USA	131,154	131,437	133,025	134,067	136,192	137,995	139,994	142,528	144,665	146,800	148,723	149,967
G8 share, %	32.8	32.5	32.7	33.1	33.6	33.8	34.1	34.5	34.8	34.7	34.9	35.2
France	25,716	25,696	25,887	26,017	26,102	26,258	26,595	26,587	26,775	27,060	27,220	27,283
G8 share, %	6.4	6.4	6.4	6.4	6.4	6.4	6.5	6.4	6.4	6.4	6.4	6.4
Japan	63,257	64,403	65,335	65,755	66,082	66,372	66,872	67,447	67,427	67,194	67,006	66,861
G8 share, %	15.8	15.9	16.1	16.2	16.3	16.3	16.3	16.3	16.2	15.9	15.7	15.7
G8 total	400,446	404,713	406,428	405,554	405,807	408,285	410,909	413,143	415,146	423,064	426,024	426,461
	2002	2003	2004	2005	2006	2007	2008	2009	2010	2011	2013	2017
Russia	73,020	72,208	72,999	73,773	74,661	75,788	76,421	76,350	76,185	76,213	75,928	72,100
G8 share, %	17.0	16.8	16.8	16.8	16.9	17.0	17.0	17.0	16.9	16.9	16.7	16.7
Great Britain	29,866	30,169	30,394	30,662	31,108	31,205	31,573	31,690	31,783	31,965	32,381	32,000
G8 share, %	7.0	7.0	7.0	7.0	7.0	7.0	7.0	7.0	7.1	7.1	7.1	7.4
Germany	40,430	40,287	40,868	41,394	41,849	42,190	42,267	42,411	42,465	42,390	42,144	41,700
G8 share, %	9.4	9.4	9.4	9.4	9.5	9.5	9.4	9.4	9.4	9.4	9.3	9.6
Italy	23,874	24,247	24,738	24,674	24,864	24,923	25,289	25,133	25,151	25,282	25,472	23,000
G8 share, %	5.6	5.6	5.7	5.6	5.6	5.6	5.6	5.6	5.6	5.6	5.6	5.3

Canada	16,936	17,353	17,583	17,738	17,964	18,342	18,656	18,751	18,930	19,123	19,488	18,400
G8 share, %	3.9	4.0	4.1	4.0	4.1	4.1	4.2	4.2	4.2	4.2	4.3	4.3
USA	151,083	152,030	153,226	155,191	156,990	158,168	159,731	159,692	159,300	160,635	163,166	153,300
G8 share, %	35.2	35.3	35.3	35.4	35.5	35.5	35.5	35.5	35.4	35.6	35.9	35.4
France	27,609	28,246	28,443	28,717	28,863	29,100	29,332	29,645	29,752	29,847	30,029	26,900
G8 share, %	6.4	6.6	6.6	6.6	6.5	6.5	6.5	6.6	6.6	6.6	6.6	6.2
Japan	66,380	66,122	65,855	65,899	65,937	66,259	66,142	65,917	66,190	65,952	65,383	65,300
G8 share, %	15.5	15.4	15.2	15.0	14.9	14.9	14.7	14.7	14.7	14.6	14.4	15.1
G8 total	429,198	430,662	434,106	438,048	442,236	445,975	449,411	449,589	449,756	451,407	453,993	432,700

Source: UNCTAD, World Bank

Table 2 Unemployment rate in the G8, 2019

Country	Unemployment rate, %
Italy	10.0
France	8.4
Canada	5.7
Russia	4.5
United States	3.7
Great Britain	3.7
Germany	3.1
Japan	2.4

Source: International Labor Organization

force employment increased in the UK—by 0.3%, Germany—by 0.3%, Canada—by 0.4%, France—by 1.2%. In the USA, labor force employment increased significantly over the period under study—by 32 million people, namely from 131.1 million people in 1990 to 163.1 million people in 2013, but then decreased by ten million people.

As can be seen from these indicators, crises in the global economy have had a significant impact on fluctuations in the number of employees and the level of employment. The level of economic activity (the ratio of the economically active population to the total population) in the Russian economy has grown in recent decades. It reached the maximum value among the studied countries and amounted to 53%. The minimum value of the indicator was observed in 2012 in Italy (42%), and the maximum value (52%)—in Germany. Thus, the level of economic activity in Russia is higher than in the G7, which indicates a more intensive use of labor in the Russian economy.

According to data for 2019 (as in previous years), the lowest unemployment rate in the G8 is observed in Japan (2.4%), and the highest in Italy (10.0%) (Table 2). Russia occupies, as is the case with the level of employment, the average position in terms of unemployment, behind Japan, Germany, Great Britain, and the United States.

Labor productivity growth is observed in all G8 countries. This is because in these countries a lot of attention is paid to technological progress and innovation (Bodrunov & Plotnikov, 2017). The Russian Federation has the lowest labor productivity during the period 1990–2013. In the G8 countries, on average, labor productivity per hour worked increased by 38.82% over 1990–2013. In the United States, this parameter is the highest, which indicates the presence of a highly skilled workforce and the widespread use of modern technology. According to the data for 2019, the leader among the G8 countries in terms of production per employee (in purchasing power parity) is the United States (USD 116,384). The lowest values of the indicator are in Japan and Russia (Table 3).The average monthly nominal wages in the G8 countries, except for Japan (decrease by 1.02 times), are growing steadily. In Canada, during the study period, it increased 1.9 times, in France—1.8 times, in Italy—2 times, in Great Britain—2.4 times. The minimum wage in 2013 was recorded in Russia, and the maximum—in Canada.

Table 3 The volume of production per 1 employee (in 2011 prices, at PPP, 2019)

Country	Volume of production, USD
USA	116,384
France	96,446
Italy	92,296
Germany	90,492
Canada	85,726
Great Britain	81,370
Japan	75,384
Russia	75,384

Source: International Labor Organization

Conclusion

An analysis of the materials presented allows us to assert that most of the quantitative parameters characterizing the labor force in the Russian economy correspond to the level of the G7 countries. These indicators include the level of official unemployment, the share of the economically active population. However, in terms of the qualitative characteristics of the total labor force, Russia lags far behind the economically developed countries of the world. This is evidenced by such indicators as the low share of spending on education and science in the structure of GDP; low proportion of specialists with secondary vocational education; low costs for training one student and one scientist; ineffective system of training and retraining of personnel; unfavorable demographic situation; low wages; low level of labor productivity.

The unfavorable reproductive characteristics of the total labor force in Russia are holding back economic growth. It is necessary to minimize them through the implementation of a well-thought-out public policy. To achieve this goal, it is necessary to solve a set of tasks, the most important of which should be the following: targeted training of specialists for the innovation sector; allocation of additional funds to finance fundamental science; implementation of various social projects and programs to increase the number of research and teaching staff; taking measures to reduce unemployment.

Analysis of data on the reproduction of labor in the Russian economy allows us to conclude about a general decrease in the efficiency of this process. During structural transformations for the transition of the economy to a market type of management, the Russian economy, which in the Soviet period of development occupied a leading position in the quantity and quality of the labor force, began to reduce efficiency. The observed inverse rigidity of employment testifies to this: the characteristics of the reproduced labor force reduced the efficiency of the functioning of the national economy. Structural transformations, together with the introduced market institutions, have reduced the overall efficiency of the sectors of the economy.

The G8 countries can be grouped into two groups in terms of labor productivity. The leading positions are occupied by the USA, France, and Canada, in which there has been a constant and steady growth in labor productivity over the period

1990–2013. The rest of the countries can be included in the second group. The socio-economic successes of the developed countries of the world are due to the effective use of production factors, namely, the interaction of a highly qualified workforce and an active purposeful investment process. The efficiency of interaction of factors of production is expressed in the growth of labor productivity.

The qualitative and quantitative characteristics of the reproducible labor force, when interacting with investments in the economic system, can provide a change in the type of employment of the labor force; change the ratio of the scale and volume of the traditional and innovative sectors. As a result, the efficiency of the functioning of the entire economic system may change, since a competitive innovation sector is able to ensure the progressiveness of economic development, and in the long term, affect the type of economic system.

References

Akaev, A. A., & Hirooka, M. (2009, April). A mathematical model for long-term forecasting of the dynamics of innovative economic activity. In *Doklady Mathematics* (Vol. 79, No. 2, pp. 275–279). SP MAIK Nauka/Interperiodica.

Bodrunov, S., & Plotnikov, V. (2017). Institutional structures influence on the technological development of the economic system. In *Proceedings of the 30th International Business Information Management Association Conference, IBIMA 2017-Vision 2020: Sustainable Economic development, Innovation Management, and Global Growth* (pp. 2658–2665).

Borio, C. (2020). The Covid-19 economic crisis: Dangerously unique. *Business Economics, 55*(4), 181–190.

Nadezhina, O., Zaretskaya, V., Vertakova, Y., Plotnikov, V., & Burkaltseva, D. (2021). European integration risks in the context of the COVID-19 pandemic. *International Journal of Technology, 12*(7), 1546–1556. https://doi.org/10.14716/ijtech.v12i7.5396

Pirogova, O., Temnova, N., & Markova, E. (2021). Online education in Russia: Status and development trends. In *E3S Web of Conferences (Vol. 258). EDP Sciences*. https://doi.org/10.1051/e3sconf/202125810020.

Pogodaeva, T., & Senchenko, N. (2017). Institutions and economic growth in the resources intensive use economies. In *Country Experiences in Economic Development, Management and Entrepreneurship* (pp. 105–123). Springer.

Pradhan, R. P., Arvin, M. B., Nair, M., Bennett, S. E., & Bahmani, S. (2019). Short-term and long-term dynamics of venture capital and economic growth in a digital economy: A study of European countries. *Technology in Society, 57*, 125–134.

Pshenichnikova, S. N. (2017). *The structure of the economic system: The role of the labor force*. Kultinform-Press.

Pshenichnikova, S. N., & Pshenichnikov, R. S. (2016). The impact of employment and investment on economic growth in the G8 countries. *Proceedings of the St. Petersburg State University of Economics, 3*(99), 7–16.

Romanyuk, I., Pshenichnikova, S., & Plotnikov, V. (2019). The relationship of employment, economic growth and economic structure (The example of Russia). *Revista ESPACIOS, 40*(32), 30.

Saleem, H., Shahzad, M., Khan, M. B., & Khilji, B. A. (2019). Innovation, total factor productivity and economic growth in Pakistan: A policy perspective. *Journal of Economic Structures, 8*(1), 1–18. https://doi.org/10.1186/s40008-019-0134-6

Senhadji, A. (2000). Sources of economic growth: An extensive growth accounting exercise. *IMF Staff Papers, 47*(1), 129–157.

Shabbir, A., Kousar, S., & Kousar, F. (2020). The role of natural resources in economic growth: New evidence from Pakistan. *Journal of Economics, Finance and Administrative Science.* https://doi.org/10.1108/JEFAS-03-2019-0044

Vertakova, Y. V., & Feoktistova, T. V. (2020). Implementation of anti-crisis measures for the population and business in the context of the COVID-19 pandemic. *Economics and Management, 26*(5), 444–454. https://doi.org/10.35854/1998-1627-2020-5-444-454

Vertakova, Y., & Kurbanov, A. (2016). Outsourcing as a form of disintegration process administration in the economy. In *Vision 2020: Innovation Management, Development Sustainability, and Competitive Economic Growth* (pp. 2158–2167).

Zhang, D., Hu, M., & Ji, Q. (2020). Financial markets under the global pandemic of COVID-19. *Finance Research Letters, 36*, 101528.

Formation of a Comprehensive "Tourism and Travel" Digital Platform in the Russian Federation

Viktor Valerievich Lavrov

Introduction

The development of digital technologies is a modern trend in the economy in almost all industries and in most countries of the world. Numerous digital platforms are being created in education, healthcare, trade, and many other industries and spheres of activity. In the context of stagnation of business processes due to the influence of external factors, such as the COVID-19 pandemic, the decline in economic growth and restrictions on movement around the world, the development of digital platforms is of great importance for the tourism sector. According to V. A. Plotnikov, a well-known Russian researcher, "Digitalization is a modern stage in the development of informatization, which is characterized by the predominant use of digital technologies for generating, processing, transmitting, storing and visualizing information due to the emergence and spread of new hardware tools and software solutions" (Plotnikov, 2018, p. 17). Many researchers around the world are studying the processes of adapting tourism to new forms of development in the digital age. For instance, an article by Tokuro Matsuo, Masaki Samejima and Kiyoto Hashimoto states: "As the Internet develops, tourism information systems are increasingly developing for B2C use. Online user preferences are analyzed through data mining, and the system makes a trip plan based on user preferences" (Matsuo et al., 2013). The active development of new technologies and the use of various databases and payment systems contributes to the growth of new directions in tourism, the fact, which is explored in the article by Weidong Wu, who states: "Concepts such as digital tourism, virtual tourism and smart tourism have emerged in the tourism industry. In the process of the development of information tourism in the future, tourists should be taken as the center to integrate tourism information resources"

V. V. Lavrov (✉)
St. Petersburg State Institute of Film and Television, St. Petersburg, Russian Federation

A. Rumyantseva et al. (eds.), *Challenges and Solutions in the Digital Economy and Finance*, Springer Proceedings in Business and Economics,
https://doi.org/10.1007/978-3-031-14410-3_13

(Wu, 2020). New technologies for the use of information resources are implemented not only in the interests of the tourism sector of a country as a whole, but also to increase the profitability of individual facilities, e. g., zoos. In this area, the experience of researchers Asri Mulyani, Dide Kurniadi and others is interesting; they note the importance of information technologies in tourism for the development of the zoo system: "This research aims to plan a strategy for implementing tourism information systems to increase the number of visitors, with a case study on tourism in Cikembulan animal parks in Garut district, West Java, Indonesia" (Mulyani et al., 2020). Research on tourism shows that the development of the industry is currently characterized by the introduction and use of a variety of innovative digital technologies and interfaces to optimize interaction with tourists and reinforce competitive advantages and influence in this area.

Methodology

In this article, the author used the following methods of scientific research. An important role is given to empirical research methods, in particular, the study and analysis of various sources of information related to the subject of the study. Theoretical research methods were actively used, in particular, classification and systematization of individual elements of tourist activity, analysis, for the study of specific elements of the Tourism and Travel digital platform, and synthesis, as a generalization of individual elements into a single interconnected system.

Results and Discussion

In many countries of the world, various digital platforms are being actively developed, combining a large number of information resources and data. In the Russian Federation, work is underway to develop digital resources for the tourism sector. "Important directions of the development of digital technologies in the field of tourism include the transfer of all public services related to the implementation of tourist activities into electronic form; the creation of conditions for the formation of a tourist ecosystem; the introduction and development of multilingual tourist assistance services, including information services, navigation services and self-service; the development and implementation of an electronic visitor map and a similar mobile application in cities and regions of the Russian Federation" (Government of the Russian Federation, 2019, Sect. 13).

"At the present stage, one can note the trends of "The digitalization of business" in the field of tourism in the Russian Federation, which is quite promising when business is enlarged, the application of new technological solutions, the use of a real-time bundling system, new booking services, online services for direct sales of certain services to tourists and tour packages with the possibility of self-construction of the tour" (Bogomazova et al., 2019, p. 44).

"In Russia over the past decade, several platforms for tourism have been developed and presented. The StavTravel platform is a mobile application (online map). The Altai-Today mobile guide contains information about the sights of Altai and popular routes for tourists. The Metis platform serves as an assistant for tour operators to study customer feedback" (Kumova, 2021, pp. 25–26).

The application is currently working in test mode and only in Moscow. Reinforcing the position of tourism in each country is an opportunity to activate economic processes, which will help to accelerate the growth of economies and overcome the severe crisis that many companies and corporations, especially those in the field of the tourism industry, find themselves in. In this area, it is interesting to note the experience of scholars as Taric Dogru and Umit Bulut, who in their research intended "to analyze the casual link between tourism development and economic growth in seven European countries <...> suggesting that economic growth and tourism development are interdependent" (Dogru & Bulut, 2018). The study and analysis of various areas of improvement of the tourism sector, with due account for the potential of modern technology, prove the relevance of the topic of the research and the need to study the development and implementation of digital technologies in the modern tourism system. These days, tourists can browse the websites of tour operators to get information on the name of tours, countries, names of hotels, trip duration, the price of the tour they are interested in, and the booking procedure. Hotel websites provide a certain range of information for tourists: booking options, length of stay, types of room, room price, catering types, and additional services and terms and conditions. The existing international booking systems, such as trivago, etc., provide a great opportunity for trip simulation: you can choose the country, the dates of arrival and departure, room category, the name of the hotel, transfer, booking, and so on. Digital payment and transport information systems are rapidly developing. Interactive maps with the names of hotels, cultural heritage sites, modes of transport, and navigators have also been developed. Therefore, a tourist who decides to arrange his or her trip independently or choose the types and destinations of travel, has to do research on large number of separate information resources or contact sales managers who offer tourist products. Based on the conducted research, the author proposes the formation of Tourism and Travel comprehensive digital platform, which combines many resources necessary for travel in a single information space. In Fig. 1, the author presents a block diagram of tourism as a system of interrelated elements.

Depending on its various characteristics, the corresponding classifications of types of tourism are formed. Figure 2 shows some of the elements of tourism included in the tourism scheme presented above.

The structural scheme of tourism developed by the author as a system of interrelated elements includes almost all categories of items, such as types of tourism, types of recreational activities, organization of a tourist trip, accommodation facilities, means of transport, types of food, country, region of stay, duration of a trip, prices for tours or trips, type of visa. The structural scheme of tourism represents

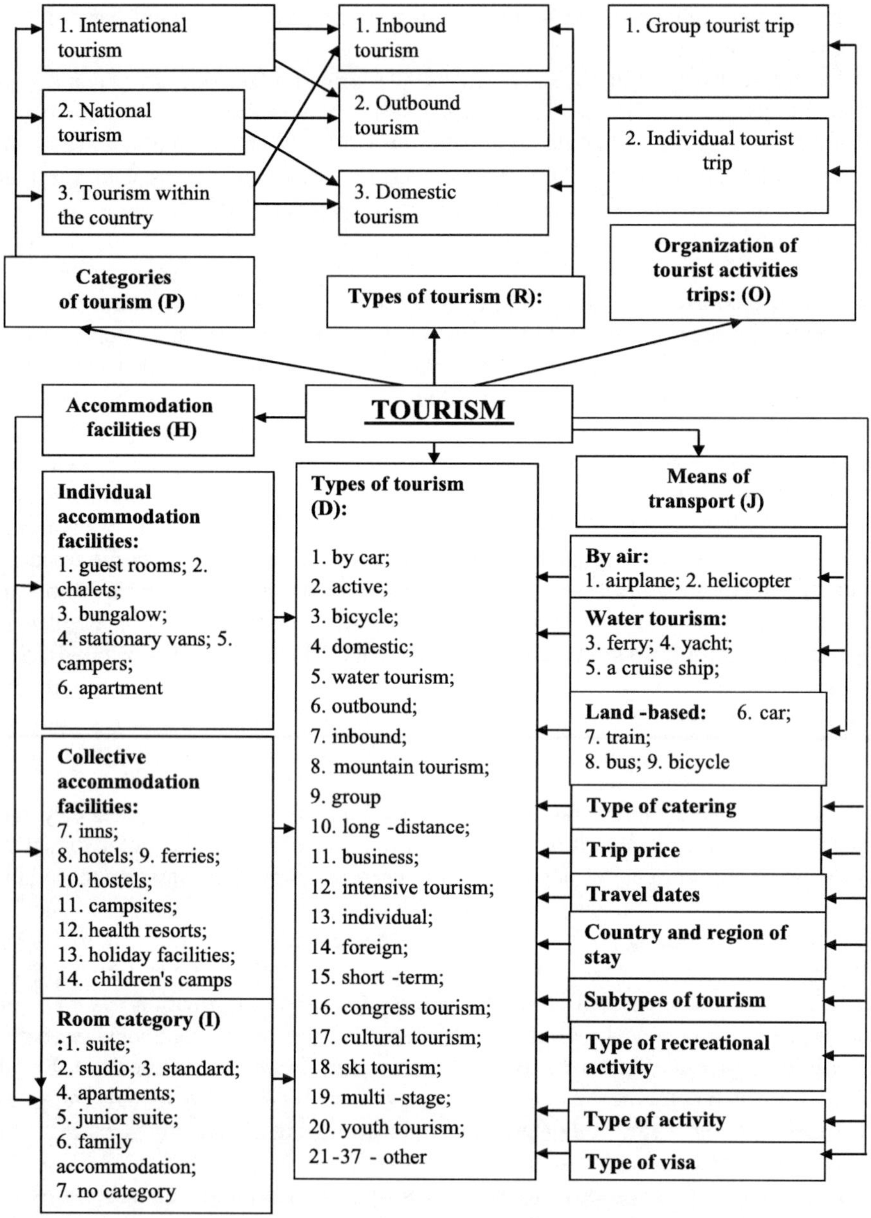

Fig. 1 A block diagram of tourism as a system of interrelated elements

the most important direction of industrial and regional development of tourism in the Russian Federation, as, according to the author, it is the basis for the formation of Tourism and Travel comprehensive digital platform. To attract more tourists to

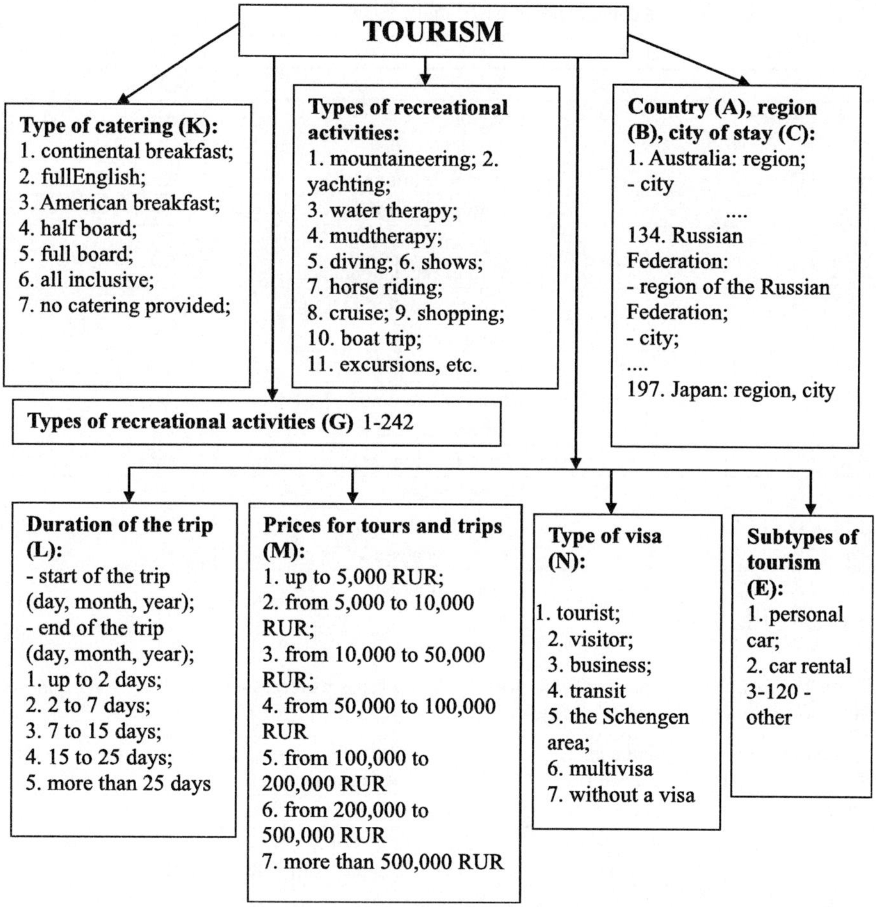

Fig. 2 A block diagram of tourism as a system of interrelated elements (continued)

specific recreational areas both globally and in the Russian Federation, it is necessary to create a model for the formation of a tourist route based on personal choice. This model can function based on the digital platform "Tourism and Travel" proposed by the author, which allows to coordinate all the elements of tourism selected by a potential tourist, expanding the range of available options and speeding up the search for a tour or route with the fulfillment of all the conditions presented. The model of a tourist route creation based on personal choice is presented in Table 1.

The book by A. S. Kuskov et al. "Recreational geography" includes, on pages 278–281, a table that lists 37 main types and 120 subtypes of tourism. On pages 492–494, the classification of 67 types of recreational activities and 242 types of recreational pursuits is presented. The numbering of these categories was provided by the author. Table 2 provides examples of some of the characteristics of tourism used as the basis for its classification.

Table 1 Model of tourist route creation based on personal choice

#	Designation of the indicator	Name of Indicator	Number of parameters
1	A	Country	A_1–A_{197}
2	B	Regions	B_1–B_n
3	C	Cities, localities	C_1–C_m
4	D	Type of tourism	D_1–D_{37}
5	E	Subtype of tourism	E_1–E_{120}
6	F	Type of recreational activity	F_1–F_{67}
7	G	Kind of recreational activities	G_1–G_{242}
8	H	Accommodation facilities	H_1–H_{14}
9	I	Room category	I_1–I_7
10	J	Means of transport	J_1–J_9
11	K	Type of catering	K_1–K_7
12	L	Duration of the trip	L_1–L_5
13	M	Price of a tourist trip	M_1–M_7
14	N	Type of visa	N_1–N_7
15	O	Organization of a tourist trip	O_1–O_2
16	P	Tourism category	P_1–P_3
17	R	Type of tourism	R_1–R_3

Produced by the author

Table 2 Examples of types and subtypes of tourism, types of recreational activities and types of recreational pursuits

#	Designation of the indicator	Name of indicator	Example of an indicator
1	D_8	Type of tourism	Mountain tourism
2	E_{28}, E_{29}, E_{30}, E_{31}	Subtype of tourism	- hiking (28), skiing (29), cycling (30) routes in mountainous areas; mountain climbing (31);
	F_7	Type of recreational activity	Ski slope (7)
4	G_{26}, G_{27}, G_{28}	Kind of recreational activities	Slalom (26), downhill race (27), snowboard (28)

Performed by the author on the basis of data of (Shpilko, 2012, p. 278, 492)

To studying the functioning of this model, Table 3 provides an example of a potential tourist's personal choice of a specific tourist route in the Russian Federation.

A special feature of Tourism and Travel digital platform proposed by the author is that the model can be applied in any country of the world, taking into account the available tourist and recreational resources and the specifics of tourist activities there.

"For tourism and recreation activities, an important role is played by the information infrastructure that supports the formation, functioning, and development of the single information environment and the means for information interaction at all levels of tourism development. Basically, all tourism sector enterprises use

Table 3 Example of forming a tourist route based on a personal choice

Item No.	Name of Indicator	Selection parameter	Individual set of tourism elements
1	Country	A_{134}	Russian Federation
2	Regions of the Russian Federation	B_{23}	Altai Krai
3	Cities within the region of the Russian Federation	C_2	Barnaul
4	Type of tourism	D_1	Car tourism
5	Subtype of tourism	E_2	Rented car
6	Type of recreational activity	F_{22}, F_{53}, F_{55}	Landscape observations (22), trekking (53), hiking (55)
7	Kind of recreational activities	G_{87}, G_{192}, G_{201}	Seeing unique landscapes (87), hiking (192), driving and hiking (201)
8	Accommodation facilities	H_{11}	Camping
9	Room category	I_7	No category
10	Means of transport	J_6	Car
11	Type of catering	K_5	Full board
12	Duration of the trip	L_3	7–15 days
13	Price of a tourist trip	M_4	From 50,000 to 100,000 RUR
14	Type of visa	N_7	Without a visa
15	Organization of a tourist trip	O_1	Group tourist trip
16	Tourism category	P_2	National tourism
17	Type of tourism	R_3	Domestic tourism

Produced by the author

information and communication technologies to some extent, since their main activities are based on information processing, including various booking systems, search engines, means of communication with partners, end users, etc." (Lavrov, 2021, p. 146).

An interesting study focused on the processes of using video and photo materials for positioning tourist resources. Thus, the article by Ricardo Kleinlein et al. examined the significance of photos and videos from various holiday destinations that were found on various websites. "Each user was presented with 50 randomly selected images, each along with the question of whether it was suitable or not, in terms of aesthetics, for the touristic promotion of the town. A total of 232 people participated" (Kleinlein et al., 2019). In this way, a database of photos of interesting places in various countries is formed to promote tourist products. Also, on the basis of video files obtained from different network resources, it is possible to create virtual tours and develop video content on trips to different countries of the world. Digital services in the field of tourism development are helpful not only for group tours organisation, but also for individual trips. In his article Craig Leith focuses on the creation of comfortable tourist trips for people who want to spend their holiday alone. "Solo tourism as a concept will continue to fragment. The individual motivations, behaviour and experience of the solo tourist will depend on a range of internal and external factors." (Leith, 2020). In addition to the development of additional types and directions of tourism, it is necessary to reform the tourism industry system across the country. "To ensure the effective development of the system of state regulation and support of tourist activities in Russia, the author recommends the implementation of the following measures: participation in major Russian and international exhibitions of the tourism industry; the creation of a common database on tourism, which would contain information about tours and excursions, a numbered fund and accommodation facilities; attraction of investments for reconstruction and construction of tourist facilities" (Lavrova & Plotnikov, 2018).

Undoubtedly, ongoing modernization of the hotel infrastructure, the introduction of new innovative technical means and additional services is of great importance for the development of tourism. "Tourists and city dwellers could always choose a suitable option for themselves, depending on their preferences and income level. The hospitality industry is not only about a human approach to service, but also providing coordinated technical issues in customer service (Wi-Fi connectivity, high quality phone calls, alerting about customer desires, technologies for fast processing and ordering, and much more)." (Pirogova et al., 2021). From the standpoint of a systematic approach, the development of digital technologies in tourism is part of a comprehensive process of technological renewal of the country's economy.

"Technological development is seen as a factor in ensuring national economic security. A promising economic model is built on the ability to generate innovations, develop and introduce new technologies into production. These abilities must be realized within the country" (Bodrunov et al., 2017). Therefore, the ability to use information technologies in tourism to meet the needs of customers based on the study of their interests, personality traits, purchasing behavior, analysis of photos of past trips in social networks, etc. is an important element in increasing the purchasing activity of tourists.

Conclusions

In the context of increasing competition between sales markets and, on a national scale, between competing companies and organizations of the tourism industry, the diversification of tourist activities and development of their structural features both in the country as a whole and in individual regions is the most important task in the tourism industry optimization, which in turn leads to an increase in the profitability of the industry and an increase in the level of competitiveness of tourism, which is a very important direction for the development of regions and the country as a whole, in the context of modern economy.

The concept of development of the comprehensive Tourism and Travel digital platform proposed by the author makes it possible to combine the data related to tourism that are already available in the information interfaces into a more detailed and interconnected system that can be used for selecting types of tourism, types of recreational activities, and all the relevant characteristics necessary for forming a route and choosing an individual or group tour. The model examined here can be applied to any country in the world, and probably digital platforms created for the development of tourism in separate countries will merge in the near future, giving rise to an international digital platform for the development of tourism based on personal choice, with all the necessary services and interfaces.

References

Bodrunov, S., Plotnikov, V., & Vertakova, Y. (2017). Technological development as a factor of ensuring the national security. In *Vision 2020: Sustainable Economic development, Innovation Management, and Global Growth* (pp. 2666–2674).

Bogomazova, I. V., Anoprieva, E. V., & Klimova, T. B. (2019). Digital economy in the tourism and hospitality industry: Trends and prospects. *Service in Russia and Abroad, 13*(3 (85)), 34–47.

Dogru, T., & Bulut, U. (2018). Is tourism an engine for economic recovery? Theory and empirical evidence. *Tourism Management, 67*, 425–434. https://doi.org/10.1016/j.tourman.2017.06.01

Government of the Russian Federation. (2019). *Tourism development strategy in the Russian Federation for the period up to 2035*. Retrieved from http://government.ru/docs/37906/

Kleinlein, R., García-Faura, Á., Luna Jiménez, C., Montero, J. M., Díaz-de-María, F., & Fernández-Martínez, F. (2019). Predicting image aesthetics for intelligent tourism information systems. *Electronics, 8*(6), 671. https://doi.org/10.3390/electronics8060671

Kumova, D. M. (2021). The use of platforms based on artificial intelligence in the field of tourism. *Service in Russia and Abroad, 15*(3 (95)), 18–26.

Lavrov, V. V. (2021). *The evolution of tourism development in the Russian Federation: From the tourism industry to the digital ecosystem in tourism*. Publishing House of SPbSEU.

Lavrova, T., & Plotnikov, V. (2018). The development of tourism: The experience of Russia. In *MATEC Web of Conferences* (Vol. 170, p. 01028). EDP Sciences. https://doi.org/10.1051/matecconf/201817001028.

Leith, C. (2020). Tourism trends: Lifestyle developments and the links to solo tourism. *Journal of Tourism Futures, 6*(3), 251–255. https://doi.org/10.1108/JTF-11-2019-0126

Matsuo, T., Samejima, M., & Hashimoto, K. (2013). Development of tourism information systems. *IEEJ Transactions on Electronics, Information and Systems, 133*(3), 562–566. https://doi.org/10.1541/ieejeiss.133.562

Mulyani, A., Kurniadi, D., Septiana, Y., & Wahyono, T. (2020). Strategic planning for the implementation of tourism information systems. *International Journal of Scientific & Technology Research, 9*(2), 4714–4717.

Pirogova, O., Shanygin, S., Lavrova, T., & Plotnikov, V. (2021). The hospitality industry development factors (case of St. Petersburg). In *E3S Web of Conferences* (Vol. 244, p. 10038). EDP Sciences. https://doi.org/10.1051/e3sconf/202124410038.

Plotnikov, V. A. (2018). Digitalization of production: Theoretical essence and development prospects in the Russian economy. *Izvestiya Sankt-Peterburgskogo gosudarstvennogo ekonomicheskogo universiteta, 4*, 16–24.

Shpilko, S. P. (2012). Classification of types of tourism: From theory to practice. *Bulletin of the National Academy of Tourism, 1*, 9–15.

Wu, W. (2020, February). Analysis of digital tourism, virtual tourism and wisdom tourism. In *The International Conference on Cyber Security Intelligence and Analytics* (pp. 18–25). Springer. https://doi.org/10.1007/978-3-030-43309-3_3.

Promoting Film Tourism as a Marketing Promotion of Audiovisual Products in the Russian Media Market

Viktor Valerievich Lavrov, Nina Aleksandrovna Seliverstova, and Svetlana Anatolyevna Fatova

Introduction

In modern economic conditions, the Russian media market faces the crucial task of identifying and implementing effective and acceptable strategies for promoting audiovisual products. The audiovisual works include motion pictures, television films, video films, slide shows, and other cinematic works. Filming locations are increasingly becoming destinations for tourists, which contributes to the formation of new tourist destinations. On the basis of this direction, film tourism was formed as a new direction of tourism. On the one hand, film tourism helps to attract the attention of tourists to those countries, cities where significant film events took place and where famous actors were filmed. On the other hand, the development of this area of tourism contributes to the promotion of film products, as the number of views increases, lists and ratings of films are created, the shooting of which took place in cities and areas of specific countries. Currently, online travel is being created as a new type of video-visual content, the basis of which is that the traveler makes daily reports on the passage of a specific route. It can be hiking trips to various hard-to-reach places or travel to different countries of the world. This film production is very popular among tourists who want to repeat this route, which attracts more people who want to join the journey, combining it with watching a video report. Consequently, the evolutionary processes of combining the film industry and tourism as an example of mutually beneficial cooperation aimed at promoting both film production and the development of new tourist destinations for travel seems to be the most important and modern trend of our time.

The audiovisual products are aimed at meeting a person's spiritual needs. "In this regard, there is no doubt that the culture of film consumption closely interacts with

V. V. Lavrov (✉) · N. A. Seliverstova · S. A. Fatova
St. Petersburg State Institute of Film and Television, St. Petersburg, Russian Federation

A. Rumyantseva et al. (eds.), *Challenges and Solutions in the Digital Economy and Finance*, Springer Proceedings in Business and Economics,
https://doi.org/10.1007/978-3-031-14410-3_14

the general culture of modern Russian society, which reflects the interests of various social groups. As it emerges in the sphere of consumption, the film is, on the one hand, a product created by its producers, and, on the other, the film itself produces the preferences of the audience, and, therefore, the need itself" (Sapelko & Ovsyankina, 2017).

Film production is both a cultural phenomenon and an economic sector. From the point of view of culture, making a film is a creative process. It is aimed at transmitting certain artistic information to the viewer and forming their cultural values. From an economic point of view, the film can be seen as a product that requires competent promotion, like any other product.

These days, in the context of modern innovations that provide new ways of information transmission, and the possibility to show films on television, in cinemas, and online, it is becoming increasingly more important to promote films based on the use of creative ways to attract the attention of the audience to this type of media production.

The dynamic development of the Russian media market has led to tougher competition among the media. Each media outlet is active in four markets: advertising, consumer, content, and media assets. On each of them, the publication must act in a special way, must choose its own marketing strategy (Mehrabyan & Balashova, 2020).

World statistics show that today films as a mass consumption product can no longer exist without promotion. In the media business, there is a firm belief that creating a film is only half the job, as the most important thing is to find its audience and compete for it. For instance, the promotion of The Last Exorcism included an original method: a girl appeared in a popular video dating chat, flirted with guys, and then morphed into a demon. Soon the reaction of the guys appeared online. The film, made on $two million budget, ended up earning $ 68 million (Rosenberg, 2010).

The Russian media market has developed a rather peculiar commercial and corporate model for the formation of a film distribution strategy, which is a specific sequence of stages of influencing the consumer: "awareness—attitude—preference—purchase." The preparation of each marketing decision, in this technological scheme, is characterized by the goals that have to be achieved within a more or less specific period of time.

The process of forming a film promotion complex, as well as its composition, is largely determined by the strategy that the organization uses to ensure sales, i.e., do they act indirectly, advertising to the film distribution company, or directly, creating demand from the end consumer. Based on the above, the following strategies are traditionally outlined, with the help of which each participant in the audiovisual market expects to solve their marketing tasks:

The push strategy is to motivate intermediary organizations to bring the product from the producer to the end user. In the film business, such intermediaries are film distribution organizations that follow the producer immediately through their distribution channels. Thus, the product is "pushed" to the market along the chain until it reaches its final consumer. Based on this strategy, special events are regularly organized to support and motivate intermediaries and to convince them that the product will sell well.

The pull strategy is aimed at creating an active advertising and information impact immediately on the target consumer, in this case, on the audience. Thus, the interest caused by advertising among consumers makes the latter "pull" the product from the sellers. The task of the seller of the service (cinemas, TV channels) is to provide the end user with comprehensive information about the film so that by the time of the release the consumer (viewer) has formed their attitude to this film. The tools of this strategy include the entire range of well-known advertising techniques and technologies for promoting film products that allow for attracting a potential audience.

The problem of choosing a strategy for promoting an audiovisual product is more complex than it may seem at first glance. This is due to the fact that a strategic choice always has to be not only appropriate, but also feasible. The forces operating in the marketing environment are uncontrolled factors that need to be identified and properly responded to. It should also be considered that the standardization of approaches to the development of strategies reduces the cost of promotion, but at the same time makes the promotion less effective.

According to Bolan and Williams (2008), an important role in the promotion of tourist services is played by the image of the place as a filming location, which is created by film tourism. "Consumers can be very sensitive to images which are important determinants of what a service customer purchases. In relation to image, the medium of film can have a very influential impact on its audience. Consumers may make purchase choices in a range of areas as a result of what they have seen in the movies.".

Therefore, the formation and development of various ways to promote audiovisual products in the media market is the most important direction in attracting attention to new film products, which helps to increase mass viewing and box office of the film.

The main ways to promote film products based on the use of communication tools are presented in Table 1.

Performed by the authors based on data from Koshetarova and Krikunova (2017a, 2017b).

Detailed information on methods for promoting film products is presented in the following list (Koshetarova & Krikunova, 2017a, 2017b):

- Organization of crowd scenes shooting with the participation of residents of the city and tourists who are in the region. For instance, Peter Jackson decided to make the information about his Hobbit film as open and accessible as possible. Jackson became one of the first directors to video blog from the set of his film. In his 10–15-minute videos, the director provided detailed descriptions of the filming process, provided virtual tours around the film set, shared his impressions, secrets of production and mentioned some events from the plot.
- Creation of official groups of the film in social networks, with detailed information about the actors, the specifics of the film, and the shooting locations, which

Table 1 The main ways to promote film products based on the use of communication tools

No	The name of the method of promoting film products	Examples of using methods
	Organization of mass filming with the participation of city residents and tourists	"The hobbit"—video blog from the set of the film
	Creation of official movie groups in social networks	"Super 8"—the distribution of encrypted information on social networks to attract the attention of users
	Development of an action plan to promote the film	"The last weekend"—meetings with the director, theatrical performance with the participation of leading artists, photo shoots with artists, press conferences for journalists
	Distribution of press releases, search for investors, announcement of a competition for the best movie poster	Active activities to inform potential consumers of film products, covered by the media
	Holding photo exhibitions with the involvement of everyone	This direction of film promotion is aimed at personal interest and further distribution of photos on social networks.
	Development and implementation of advertising products	"Ted": TV advertising, social media content advertising, distribution of merchandise advertising the movie, information about the movie in the press, information about the movie's ratings and broadcast locations
	Trailer release	One of the main ways to promote the film, the first acquaintance with the basics of the plot and the artists
	Use for landscape filming of famous tourist destinations	Many films feature well-known tourist destinations, which draws attention to the viewing of the film and creates nostalgic memories of travel.

attract a growing interested audience. For instance, Steven Spielberg and J. J. Abrams, the creators of "Super 8," understood that the advertising campaign of the film should be impressive and memorable. Before filming began, they encrypted a link to the website in an ad for Iron Man film. All the information, including numbers, names, and dates, was carefully encrypted, and people took time to solve the puzzles. As a result, after a year of games, the film's revenue amounted to 206 million against 50 million spent. The shooting of the film was disguised by the production of two fictional films; finally, the premiere turned out to be stunning.

– The development of an action plan for the promotion of a film (meetings with the director, a theatrical performance with the participation of leading actors, photo sessions with the cast, press conferences). An interesting idea was implemented by the creators of the Last Weekend: www.ruskino.ru website featured the portal "Expedition after the film." For example, the website could be used to express one's opinion about the script, the replicas of the main characters, or their costumes. Some scenes were rewritten based on the opinion of potential viewers.

- Distribution of press releases, search for investors, announcement of a competition for the best film poster, in which anyone can participate through the social network system.
- Holding photo exhibitions with open participation (photos related to the film).
- Development and implementation of advertising products (TV commercials, content advertising in social networks, distribution of film-related merchandise, information about the film in the press, information about its ratings and places where the film is on). So, for the release of Ted movie, the creators registered pages for the teddy bear in all social networks, and TalkingTed app was created for smartphones.
- Trailer release. The success of the premiere is determined not only by the right choice of location and format: the advertisement of "the first film of 2006" ("Day Watch"), instead of the traditional "Happy New Year", successfully used the phrase "Happy New Watch";
- Use of well-known tourist destinations for outdoor shooting or the creation of new tourist attraction based on information about the filming locations. The set of the film "Burnt by the Sun 2. Exodus" was visited by Vladimir Putin and Dmitry Medvedev, which attracted huge attention to the film.

In the competitive context of the media content market, modern types of promotion of audiovisual products and the search for potential consumers of the services are of great importance. For example, the British Tourism Authority actively promotes various tourist destinations in the UK based on the rapid development of film tourism. "By publishing a Potter map, the aim was to create a UK-wide trail featuring principal sites from the book and film. The third edition of the Map, produced in June 1999, included 67 films and TV programmes, 110 locations on a grid-referenced map and many more locations mentioned in the text. The Map was distributed to 300,000 overseas visitors in North America, Singapore, Australia, Japan and Europe during 1999. The Southwest of England has provided the location for many films and television dramas in recent years. Historically, it is an area associated with famous authors and their books. For example, with the author, Agatha Christie, who set her novel Evil Under the Sun on Burgh Island" (Busby et al., 2003).

The consumer audience of a film is the audience targeted by the marketing efforts of the company and many people of various professions (director, screenwriter, composer, etc.) who created the motion picture. The consumer value (or perceived value) of such a film product is determined only by its audience. In the end, it is the number of viewers who came to the cinema that is the criterion for the success and commercial appeal of a film project. The key to success is a clear understanding of the needs of the audience and the ability to use this knowledge effectively (Budilov & Leszko, 2018).

Escher and Zimmermann (2001), in their article Geography meets Hollywood, note that in order to form the tourist image of a territory, it is necessary to study the impact of outdoor shooting on the perception of the viewer and the potential tourist. "Feature films considerably shape people's behaviour and their everyday perception of landscape. Landscape should be a part of the movie's geography since the roles of landscapes in feature films are diverse and lasting. Thus, the authors can isolate the following functions of landscapes in movies: landscape as setting, landscape as guarantor for credibility and authenticity, landscape as metaphor or symbol, landscape as myth, landscape as actor, landscape as location, and landscape as destination of location tourism".

In recent decades, there has been an increase in the export of films outside the country of origin. Since the early 2000s, French companies have been actively investing in the marketing and distribution of national films in a large number of foreign markets, including Russia. The need to promote film production determines the need for its high-quality translation into different languages. Acquaintance with the film begins with its title, defined as a film name (Anisimov et al., 2019).

To a large extent, another very important group is involved in the promotion of a film premiere: the media, bloggers, film critics, and film scholars. For example, after the premiere of the Russian film "Paragraph 78», influences started their targeted work online to form a "positive information field around the film, » with specially trained people spreading positive reviews about the film online and counteracted any negative opinions. Croy and Walker (2003), in their article Rural Tourism and Film, examine the important issues of reviving the economy in rural areas by attracting visitors through the development of film tourism. "In search of diversification, rural regions can target tourism and cinema. In particular, for film tourism, it is necessary to predict the number of visitors and assess the potential negative consequences. Broken Hill, located in the heartland of Australia, has been used to explore the complementary integration of cinema into regional development and tourism strategies.".

Therefore, the more efforts are made to promote a new type of film by the participants of the film project, the more the audience will be aware of the premiere and interested in it, which in turn will increase the rating of the film after its theatrical release. For example, the campaign to promote Psycho, developed personally by Alfred Hitchcock, was supposed to surround the film in an aura of mystery: he sent his assistant out to buy up as many copies of Bloch's novel as possible, and chose not to show Psycho to critics and not to arrange traditional screenings for the chosen ones.

Methods

The article used empirical research methods, such as observation, which is the study of the processes of promoting film products in modern conditions and drawing conclusions about the importance of this type of consumer attraction to this type of film content and comparison, which is an opportunity to compare various actions to promote films and identify the main, most promising areas for promoting film production in modern conditions.

Particular attention was paid to the method of a systematic literature review on the topic of film promotion in the modern media market.

On the basis of the studied material, the classification of film tourism directions is developed, and an example of the authors' modeling is provided that shows the development of a scheme for promoting audiovisual products based on the use of tourist destinations.

Results

The results of marketing activities to promote film products are:

- Firstly, increasing the level of popularity of the film.
- Secondly, increasing the potential audience.
- Thirdly, the release of the film on the media market of the Russian Federation, and then on the international media market.
- Fourthly, the formation of a valuable image for the film.
- Fifthly, obtaining additional funding and increasing fees from film rentals.

To expand film audiences and increase the popularity of the film products, it is necessary to engage in the active development of film tourism, to attract tourists to places of pilgrimage related to specific films and to combine the interest of tourists in tourist and recreational destinations with interest in the shooting locations or places where the main events of the films unfold. Film tourism, which combines the interest in travel and interest in shooting locations of famous films, has been gaining popularity recently. Hudson and Ritchie (2006), in the article Promoting Destinations via Film Tourism, examine the most important issues of the development of film tourism and its impact on increasing interest in tourist destinations. "Film tourism is a growing phenomenon worldwide, fueled by both the growth of the entertainment industry and the increase in international travel. It identifies the optimum marketing factors that encourage film tourists to visit destinations that appear (or are depicted) in the movies. Factor analysis reveals four types of marketing activities in which destinations can engage to promote film tourism: proactive efforts to encourage producers and studios to film at the location, efforts to generate media publicity around the film and its location, marketing activities that promote the film location after production, and peripheral marketing activities that leverage film tourism potential".

Film tourism is a type of cultural tourism based on the films that visitors have watched, including TV series or animation, and may include visiting the places where the action is set in; places where filming took place (including film studios), which may not coincide with the location where the story is set in, or theme parks that recreate the setting of the films. Film tourism also includes visits to film festivals. It can be noted that the film industry and tourism are interconnected, which contributes to the development of the regional economy.

It was found that the tourist market in the Russian Federation is more represented by film tours dedicated not to one film, but to a group of films shot in a certain area. So, the Perm company "Green Wind" organizes a rafting tour to the filming locations

Table 2 The most popular film tourism destinations

Item No	Film title	Name of the city, locality	Country name
1	Roman holiday	Rome	Italy
2	Sideways, USA	Napa Valley, Santa Barbara County, California	USA
3	Vicky Cristina Barcelona	Barcelona	Spain
4	The avengers, spider-man	Streets of New York	USA
5	Sherlock Holmes and Dr. Watson	London	Great Britain
6	The Lord of the rings	Queenstown, shire	New Zealand
7	Harry potter, highlander	Glencoe Valley,	Scotland
8	Cast away	Monuriki Island	Fiji
9	Perfume: The story of a murderer	Ponte Vecchio bridge	Florence
10	Troy	Malta	Malta

Source: completed by the authors based on data from Belozerova (2017)

of the film "The Geographer Drank His Globe Away" along the Usva River. The jeep tour "Prisoner of the Caucasus" was created in 2016 for the 50th anniversary of the film by the Jeeping Tour sports club. (Garbuz & Anokhin, 2020, pp. 51–52).

According to the Tourism Competitive Intelligence study, more than 40 million tourists choose to visit the countries that have become the main locations of popular films; Table 2 shows the most popular world destinations of film tourism.

One of the examples is the experience of New Zealand, where the world-famous Peter Jackson trilogy the Lord of the Rings was filmed. The state conducted a large-scale advertising campaign in order to identify the country with Middle earth. As a result, by 2004 the number of tourists in New Zealand increased from 1.7 to 2.4 million. According to experts, only the first part of the Lord of the Rings trilogy generated $41 million profit. "Following the success of prominent feature films shot on location, including Tolkien's wildly popular The Lord of the Rings, New Zealand boasts an impressive film tourism industry. New Zealand is a vast expanse of natural beauty and a magical world of fantasy on screen, and its tourism imagery, including the ways in which savvy local tourism boards have in recent decades used the country's film representations to sell New Zealand as a premiere travel destination." (Leotta, 2011).

After the release of the Game of Thrones episodes filmed in Dubrovnik (Croatia), the annual number of visitors there increased by 10%. The Croatian government had to limit visits to the city to 4000 people daily, and five tourist routes related to the filming locations of the series were created. The Harry Potter film series helped increase the attendance of all locations involved in the shooting in the UK by 50%. The number of visitors to the Sydney National Park, which was the shooting location in Mission Impossible-2, increased by 200%. Troy, a historical blockbuster, brought the Turkish Çanakkale Province a 73% increase in tourist traffic, and the number of visitors to Miami went up by 150% thanks to CSI Miami.

Table 3 Popular directions of Russian film tourism

Item No	Film title	Name of the city, locality
1	Love and pigeons	Medvezhyegorsk, Karelia
2	Ivan Vasilievich changes profession (IvanVasilievich: Back to the future)	Rostov
3	Kidnapping, Caucasian style	Alushta
4	The diamond arm	Novorossiysk, the Black Sea coast of the Krasnodar Krai
5	The dawns here are quiet	Karelia
6	How I ended this summer	Valkarkai, a polar station on the coast of the east Siberian Sea
7	The geographer drank his globe away	Usve River, perm Krai
8	Leviathan	Teriberka
9	Peculiarities of the National Hunt	Protochnyi settlement, Priozersky District, Leningrad region
10	The horde	Selitrennoe village, astrakhan region
11	The Crimean bridge. Made with love!	Crimea
12	Chernobyl: Zone of exclusion	Kolchugino, Moscow region.

Source: completed by the authors

The main trends in the development of modern world film tourism were identified. Among tourists, a long-term spectator interest in films around the world is formed, as evidenced by the effect of the popularity of films that have gathered a mass audience and large box office. Another example of the influence of the film industry on tourism is the creation of fan clubs—communities of fans of a particular film history. Fan meetings are held with film crews of films, for example, the ComicCon festival, which was held in Moscow in 2018, was attended by more than 170 thousand people (Udeneeva & Saiej, 2019).

Table 3 presents the Russian film tourism destinations. The table provided here shows some shooting locations or places where stories of films are set. Very often, these territories are used to attract tourists, and new tourist destinations gradually develop on the basis of famous films. On the other hand, directors choose filming locations that are well-known as established tourist destinations, which contributes to even greater interest in the film.

We conclude that the media sphere is not only a condition for the development of tourism, but also a part of the service. In modern hotels, for example, the availability of a wide range of international television channels is a necessary part of high-quality service. Film and tourism play an increasingly important role in the development of the economy of many countries. This is noted by many businesspeople, politicians, and public figures of the world, including Taleb Rifai, the Secretary-General of the United Nations World Tourism Organization (UNWTO).

Experts say that a successful combination of two industries (the film industry and the tourism industry) can generate particularly impressive results, which is confirmed by the spectacular example of Pembrokeshire (England). Thanks to Harry

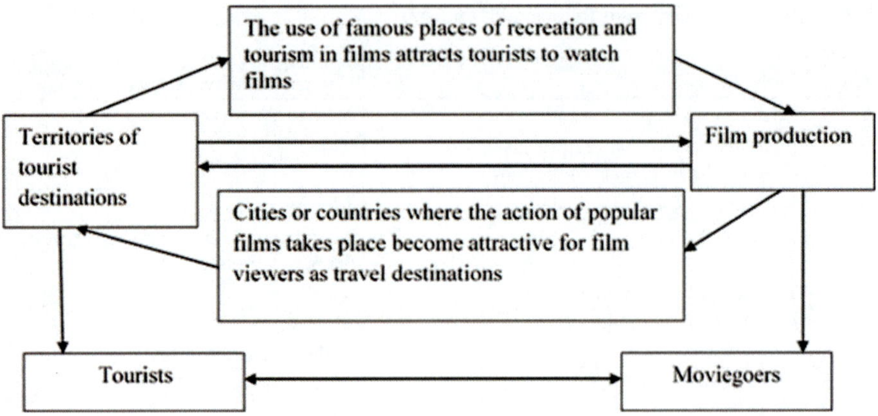

Fig. 1 The scheme of audio-visual products promotion based on the use of tourist destinations

Potter films and Ridley Scott's Robin Hood, this county is annually visited by four million tourists, who leave 600 million euros in the local budget. A deeper analysis of the points of contact between tourism and media arts is possible, which will increase the efficiency of using tools and resources for the development of both areas (Koshetarova & Krikunova, 2017a, 2017b).

The Association of Film Commissioners International (AFCI Cineposium) held a symposium in St. Petersburg in September 2019, which was attended by representatives of the film industry from more than 50 countries, members of national film commissions and producer organizations, companies such as Marvel Studios, The Walt Disney Studios, Netflix, HBO, Lionsgate, Warner Brothers, Youtube, etc. We would like to mention that the previous symposia have been held in cities with rich film background, such as Los Angeles, Atlanta, Barcelona, and New York.

St. Petersburg can take its place in the film tourism market, as it has many historical and thematic locations. In fact, St. Petersburg already has some experience as a destination of international film tourism: the BBC series "War and Peace", which was shown in many countries around the world two years ago, made the city on the Neva a place of pilgrimage for fans of the film. Despite the fact that a significant part of the shooting of this film took place in Lithuania, the Northern Capital created a detailed route to the locations of the screen saga. The tour includes a visit to the Winter Palace and Yusupovsky Palace, the State Russian Museum, the Mikhailovsky Castle, and the famous suburbs of St. Petersburg.

Based on the presented data on the definite relationship between the tourism industry and the film industry, the authors developed a scheme for promoting audiovisual products based on the use of the tourist destinations. Figure 1 shows a diagram of the mutual influence and interrelation between media interests and tourist interests of potential moviegoers and potential tourists. The merging of the tourism industry and the film industry based on the use of tourist destinations in films creates a joint potential audience of viewers and tourists; this increases both the profitability of the tourism industry and the rating of films, which, in turn, contributes to an increase in revenue of regional budgets.

It can be concluded that on the one hand, the development of the film industry makes regional tourist destinations more popular, especially in little-known regions, while on the other hand, popular tourist territories are the basis for the development of storylines in films. Thus, film tourism is formed as a synthetic direction bringing together the segment of moviegoers and the segment of tourists and combining them into a single whole.

Discussion

One of the most important aspects of the promotion of films is to attract the attention of the legislative and executive bodies of the state. The main centers of the film industry in Russia are Moscow and St. Petersburg. These cities have film production at global level. In other Russian regions, film production and the promotion of film products do not reach their full potential, which creates the need for more intensive measures that involve active use of media communications.

According to experts, in order to increase the rating of feature films and box office revenues, it is necessary to attract budgets as much as possible, create favorable conditions for regional film producers, and use such a financial incentive as rebate. It is necessary to provide information support for regional projects with the help of major mass media. Regional film studios should conduct independent continuous work on the implementation of various forms of Internet marketing and hold special events at the regional level (regional and city competitions, quizzes, thematic cultural and leisure programs, etc.), which help attract the attention of the local audience to the work of directors" (Koshetarova & Krikunova, 2017a, 2017b).

It should be noted that Russia is currently moving slowly into the international film tourism market, but there are already some serious developments in the organization of such activities. In our country, we have the Russian Office of Film Commissions; in addition, local film commissions are organized in 12 regions of the Russian Federation. The functions of film commissions include selecting "content at the place of shooting" for filmmakers, assistance in organizing the shooting at the location, and solving administrative issues. "The Kaliningrad Region is the most active in this area: the budget of this region even includes a special line of expenses related to the activities of the film commission."

Films are the soft power that develops the image of the country. On the other hand, this generates business, because when a film group comes to a region, it brings money there. They live in a hotel, rent equipment, use a range of services. It should be noted that the Ministry of Economic Development of the Russian Federation has prepared a roadmap for the development of the tourism industry for the next six years, where the goal is stated, for the first time ever, of the promotion of tourism to Russia through film and animation, as the most relevant and efficient tools that have not been used in our country before.

According to experts, this new strategic vision will help to hold Cineposium at a high level and show foreign film producers all the potential of the film making process in Russia. St. Petersburg is already developing promotional tours for participants and guests of this industry event; the business program, with its workshops and working sessions, will also be intense.

Conclusion

Thus, film tourism as the most important factor in the promotion of audiovisual products in the Russian media market will help to enhance the potential of tourist destinations in the development of film tourism.

Many scientists and practitioners in the field of the film industry and the tourism industry support the direction of combining the capabilities of the film industry and the tourism industry in order to mutually beneficial cooperation and joint promotion of both film industry products and the development of new tourist destinations and areas of increased tourist interest.

Cooperation and parallel development of these two industries will contribute to the formation of modern marketing thinking, the accumulation of positive experience in the organization of advertising activities and will systematize practical competencies in the use of communication tools that allow for creative development and effective implementation of strategies for promoting audiovisual products in the Russian media market.

References

Anisimov, V. E., Borisova, A. S., & Konson, G. R. (2019). Linguistic and cultural localization of film titles. *Russian Journal of Linguistics, 23*(2), 435–459.

Belozerova, Y. M. (2017). The role of media arts in the development of tourism and the economy of the regions. *The Science of Television, 13*(3), 43–56.

Bolan, P., & Williams, L. (2008). The role of image in service promotion: Focusing on the influence of film on consumer choice within tourism. *International Journal of Consumer Studies, 32*(4), 382–390.

Budilov, V. M., & Leszko, V. V. (2018). Trends and prospects of film industry development in Russia and in the world. In *Topical Issues and Prospects of Economic Development in the Context of Modernization: Collection of Articles of the International Research and Practice Conference* (pp. 16–21).

Busby, G., Brunt, P., & Lund, J. (2003). In Agatha Christie country: Resident perception of special interest tourism. *Tourism, 51*(3), 287–300.

Croy, W. G., & Walker, R. D. (2003). *Rural tourism and film-issues for strategic regional development. New directions in rural tourism* (pp. 115–133). Ashgate.

Escher, A., & Zimmermann, S. (2001). Geography meets Hollywood die Rolle der Landschaftim Spielfilm. *GeographischeZeitschrift, 89*(4), 227–236.

Garbuz, D. A., & Anokhin, A. Yu. (2020). Cinema tourism as a technology to increase the tourist attractiveness of the region. In *Tourist and recreational potential and features of tourism and service development* (pp. 49–58).

Hudson, S., & Ritchie, J. B. (2006). Promoting destinations via film tourism: An empirical identification of supporting marketing initiatives. *Journal of Travel Research, 44*(4), 387–396.

Koshetarova, L. N., & Krikunova, E. A. (2017a). Specifics of the promotion of full-length feature films by means of socio-cultural activities. *Bulletin of the Tambov university. Series: Social Sciences, 4*(12), 67–76.

Koshetarova, L. N., & Krikunova, E. A. (2017b). Features of the promotion of full-length feature films by means of socio-cultural activities. *Bulletin of the Tambov university. Series: Social Sciences, 4*(12), 67–76.

Leotta, A. (2011). *Touring the screen: Tourism and New Zealand film geographies*. Intellect Books.

Mehrabyan, L. G., & Balashova, S. P. (2020). Features of the development of a marketing strategy in the media sphere. *International Journal of Humanities and Natural Sciences, 11*(3), 176–179.

Rosenberg, A. (2010). *The last exorcism' chatroulette prank gets the word out (NSFW)*. Retrieved from http://www.mtv.com/news/2437395/the-last-exorcism-chatroulette-prank-gets-the-word-out-nsfw

Sapelko, S. N., & Ovsyankina, A. V. (2017). Modern approaches in the promotion of the film. In *Bulletin of Scientific Conferences* (no. 1–5, pp. 160–161). LLC Consulting Company Yukom.

Udeneeva, N. V., & Saiej, A. I. (2019). Some aspects of the development of world film tourism. *Scientific Bulletin of MGIIT, 5*, 24–32.

Law in the Era of Digitalization

Vitaly D. Sattarov, Roman A. Romashov, and Vladislava V. Kosareva

Introduction

The era of digitalization or the digital era is usually called the stage of development of human society that comes after the transition from industrialism to new, more advanced forms of economic and political structure based on the priority of knowledge, information and technology in people's lives, the activities of organizations and public authorities.

A large-scale transformation of public relations, caused by the widespread spread of various technical achievements, such as cryptocurrencies, cyberspace, robotics and many others into everyday reality, the practice of most citizens, seems characteristic of modernity.

It is natural that these conditions give rise to challenges for the law, due to the need for effective ordering of such interactions of subjects that did not exist before or were characterized by an insufficient degree of relevance. In other words, the law should strive to reflect the legal reality in the most appropriate way.

The problems of adapting legal regulation to the requirements of the time today acquire the character of a promising area of scientific research, in the development of which both domestic and foreign scientists took an active part. Among the Russian authors, such as A.A. Vasiliev (Vasiliev et al., 2020), A.A. Kartskhiya (2019), V. Ya. Lyubashits (Lyubashits & Osinsky, 2019), T.N. Mikheev (2019), etc. should

V. D. Sattarov
Perm Institute of the Federal Penal Service, Perm, Russian Federation

R. A. Romashov
Murmansk Arctic State University, Murmansk, Russian Federation

V. V. Kosareva (✉)
St. Petersburg University of Management Technologies and Economics, St. Petersburg, Russian Federation

A. Rumyantseva et al. (eds.), *Challenges and Solutions in the Digital Economy and Finance*, Springer Proceedings in Business and Economics,
https://doi.org/10.1007/978-3-031-14410-3_15

be especially singled out. Among foreign researchers, a significant contribution to the disclosure of the topic was made by A. de Franceschi, R. Schulze (2019), D. Banisar (2006) and other scientists.

At the same time, it should be noted that the legal literature currently reveals in detail only a small part of the issues that inevitably escalate in the context of changes taking place in society. Many aspects of the legal regulation of digital technologies continue to require the close attention of the scientific community.

Based on the above prerequisites, this paper will attempt to investigate several the most controversial trends in the development of law and legislation related to the introduction of technical innovations into all spheres of human activity.

Materials and Methods

A wide range of methods of scientific cognition was used in the preparation of the article, including such fundamental general scientific methods as analysis and synthesis, induction and deduction, formal logic techniques.

The comparison method made it possible to correlate various technical means and features of their legal regulation, to obtain conclusions from this basis that are significant for the work. The historical method was used in the study of the process of the emergence of the concept of "artificial intelligence" and, in general, in the analysis of the development of information and digital technologies. The system method has become the basis for a comprehensive assessment of neural networks, artificial intelligence, and robotics as interrelated objects of legal regulation. The formal legal method was used in the process of processing, analyzing the current regulatory framework, identifying its inherent disadvantages and advantages.

We believe that the combination of these methods makes it possible to obtain reasonable and correct conclusions on the designated topic.

Results

The transition to the digital age is clearly reflected in many aspects. First, it is characterized by the widespread dissemination of individual phenomena and technical innovations, which a few decades ago the overwhelming majority of the public considered only utopia or fiction. For example, such phenomena should include artificial intelligence, the emergence of humanoid robots, neural networks, distance education, virtual reality, etc.

Even though the use of these scientific achievements brings undoubted benefits, it also creates certain risks, including those caused by threats of software failures, hacker attacks, loss of privacy. These circumstances entail the need for a legal understanding of the essence and features of the listed phenomena, the addition of the regulatory legal framework with rules and regulations that will be able to regulate the social relations associated with them with a high degree of effectiveness.

As a matter of priority, such an object of legal regulation as a neural network should be considered. In science, this concept is usually understood as a set of artificially created neurons that allow simulating individual intellectual functions, including generalization and analysis of data, learning, and even the ability to find original solutions to problem situations. So, science knows the precedents in which neural networks rediscovered the laws of nature discovered by scientists (Yasnitsky, 2015).

The order of functioning of these technical means copies the work of the brain, but on a much smaller scale, since modern computers are still not able to provide a volume of neural connections that would be commensurate with human. Nevertheless, even in the existing limited conditions, the characterized technical means has advantages over other software tools, which consist in the ability to adapt to the environment and the tasks set, to adapt to the environment. Such the peculiarity was the reason for the active introduction of technology in the field of business, medicine, advertising, law, physics, etc.

In addition, the widespread use of neural networks is directly influenced by the fact that the technology in question is a basic component of the facial recognition system, computer programs, software of the latest military technologies, voice assistants and search services.

Recognizing and accepting the prospects that the active use of the scientific achievement under study opens for humanity, it should not be overlooked that in the absence of detailed legislative regulation, it will also produce the formation of certain risks and dangers. In particular, the inviolability of information about a person's private life, personal, family secrets, personal data, and other confidential information may be put at risk.

In accordance with the first section of Article 24 of the Constitution of the Russian Federation, the collection, storage, use or dissemination of data on a person's private life without his consent is not allowed. Nevertheless, neural networks of search services regularly monitor, and record completed queries, adjusting search results to a specific individual, collecting information for advertising purposes, promoting products and brands.

Despite the prohibition of the use of personal data without the consent of the subject, endless offers of services, loans, loans, regular spam, and other unwanted notifications have become commonplace. According to K.S. Gulyaev, problems in the field of personal data protection are manifested not only in targeted advertising, but also in the work of social networks. Since such services accumulate a large amount of personal information, they act as a convenient tool for marketers, insurers and loan organizations (Gulyaev, 2018). These facts confirm that there are significant shortcomings and problems in the current legislation.

In general, the concept of "neural network" is rarely found in Russian legal acts and for the most part only in narrow, technical aspects. For example, the mention is present in several national standards, methodological recommendations, strategies, development concepts adopted within the framework of the activities of various departments. For this reason, we consider it expedient to develop and adopt a regulatory legal act that regulates in detail the procedure and conditions for the use of neural networks, the purposes of their application, and will limit the possibilities of collecting information about citizens.

The problem of insufficient degree of ordering of the analyzed technical means is aggravated by the fact that it acts as the foundation of most of the developed variants of artificial intelligence. The latter is most often understood as a system of technical devices capable not only of learning, but also of imitating human behavior.

If you briefly delve into the history of this topic, you can find that the idea of artificial intelligence in science originated in the 50s of the XX century. It was at this time that a largely philosophical question was raised for the first time—will a computer be able to communicate with a person so well as to convince him that he is also a person? We believe that the answer to the voiced question today will not be so unambiguous that a few decades ago, when only a few of the most progressive authors could respond positively to it.

It should be noted that one of the key features of artificial intelligence is traditionally considered to be the ability to think creatively and adjust one's own behavior when external conditions change. The potential to go beyond the usual equation distinguishes the analyzed engineering innovation from traditional computing machines, which further increases its value and expands its application options. So, from the words of A.V. Chechkin's specificity of smart systems is manifested in the fact that they can independently "update, develop, modify, adapt and, if necessary, can partially transform their own theater of action, their problem area" (Chechkin, 2021), which makes them an extremely flexible and useful tool.

It should be noted that the listed features of the technology in question are not exhaustive since there is no unified approach to the essential characteristics of artificial intelligence in science. Such a topic remains a highly relevant, debatable area of research. For example, V.A. Shestak and A.G. Volevodz said that it is impossible to formulate an exact definition of the characterized technology for the reasons of the existence of various robotic systems (Shestak & Volevodz, 2019).

The legal support of artificial intelligence is also characterized by a gap. For example, in Russian law, in the absence of legislative regulation of this issue, subordinate regulation is being actively improved. The most significant Russian legal act in this area is the National Strategy for the Development of Artificial Intelligence for the period up to 2030 (hereinafter referred to as the Strategy) adopted by Decree of the President of the Russian Federation No. 490 of October 10, 2019.

It is noteworthy that this by-law offers its own, legal definition of the desired concept. Thus, in accordance with the Strategy, artificial intelligence is a complex of technological solutions that allows simulating human cognitive functions (including self-learning and finding solutions without a predetermined algorithm) and obtaining results comparable to at least the results of human intellectual activity when performing specific tasks.

At the same time, in our opinion, the disadvantages of the above concept exceed its advantages, if only because the term "cognitive functions" is too broad a definition, which is ambiguously revealed in psychology and cannot be used to denote the phenomenon of law.

Since the above strategy does not disclose which, cognitive functions are in question in the definition, perhaps it should be simplified and stated as follows: artificial intelligence makes up a set of technical solutions that determine the system's ability to learn, simulate individual human behavior traits and make non-standard decisions without a predetermined algorithm.

Additionally, it should be noted that the Decree of the Government of the Russian Federation dated August 19, 2020 No. 2129-r approved the Concept of the development of regulation of relations in the field of artificial intelligence and robotics technologies until 2024, which directly indicates that the improvement of technologies poses serious challenges to the legal system and requires the development and implementation of special legislative regulation, taking into account the specifics of the widespread use of technologies. We believe that such a statement correctly characterizes the prevailing conditions. It may also be advisable to include it in the text of the Strategy, for example, in section "Conclusion', to make the current subordinate regulatory framework more consistent.

In continuation of the topic, special emphasis should be placed on the fact that the development of the issue of the consequences of placing artificial intelligence in a robotic body has a high scientific value. Since the resulting creature because of such manipulations can become very similar to a person over time, it is inevitable to raise the question of the need to grant him rights, even if they will have some restrictions.

In modern legal theory, the existence of rights and freedoms is usually attributed to the exclusive properties of only those subjects that belong to the species Homo sapiens. This fact is explained by the fact that scientists traditionally understand a person as a special socio-biological being of the highest order. Professor V.V. Orlom said that people, from the standpoint of scientific materialism, are not a local and random episode of evolution, but a necessary result of the infinite development of matter, its "highest color" arising "with iron necessity" inherent "in the very nature of matter." This is a microcosm and a unique being of its kind, which arose because of the continuous creation of itself (Orlov, 1998).

Considering the stated position and without denying its scientific value, we believe it is possible to look at the essence of the problem from a different position, reduced to identifying the most important signs of a person as a unique being. Thus, the distinctive features of a person should include the presence of a developed second signaling system (speech), logical and creative thinking, the ability to act in the interests of a social group, sensuality, and self-awareness.

If robots can imitate some of these features already today, then at least the last three are still inaccessible to them. Based on the opinion of the American scientist L. Solum, the electronic systems currently existing remain only programs with logistical support attached to it. Having no free consciousness, intentionality, any personal feelings, and emotions, they cannot be characterized in any other way (Solum, 1992). In view of the above, it seems correct to conclude that it would be premature to endow them with legal personality (Gabov & Khavanova, 2018).

Under these conditions, it is natural to assume that an artificial intelligence placed in a robotic body will be able to obtain rights that will be like human ones only if it meets the basic characteristics of the Homo sapiens species. In other words, a robot with rights is a being that has developed self-awareness, a desire for self-preservation, is distinguished by the ability to understand and share people's feelings, can act altruistically, etc.

Until robotics reaches such a high level of development, we believe that it will be inappropriate to talk about the legal personality of artificial intelligence. Let us add that the above conclusion is also confirmed by the insufficient development of the regulatory framework, which regulates the sphere of robotics and its application in practice.

In addition to the legal acts mentioned in the work, most of which are directly or indirectly related to the analyzed object, the concept of robot is found only in a number of national standards. In this matter, the Russian Federation lags significantly behind the leading states in the field of technological progress (Japan, South Korea, China, the USA), which currently have broad, progressive legislation in the analyzed area (Shestak & Volevodz, 2019).

We believe that this circumstance appears to be a direct consequence of the imperfection of the Russian regulatory legal framework in the field of regulating information relations and its inconsistency with international standards set out in the works of well-known foreign lawyers and human rights defenders (De Franceschi & Schulze, 2019; Banisar, 2006).

Discussion

The results of this study demonstrate an interesting pattern associated with the fact that the degree of scientific elaboration of legal problems in the era of digitalization increases in proportion to the complexity of the characterized object. If neural networks rarely fall into the field of professional interests of legal scientists, have practically no legislative basis, then the topic of artificial intelligence is already significantly more often reflected in legal works, has a certain base, which consists mainly of subordinate regulatory legal acts.

At the same time, the problem of the legal personality of robots in the absence of a detailed legislative foundation attracts the closest attention of the scientific community. To confirm this circumstance, we will give examples of the most common approaches to this issue.

Thus, V.Ya. Lyubashitsa and A.S. Osinsky should be considered among the opponents of granting robots rights, according to whom a robot is not a replacement for a person, but only an obedient executor of his will. He must take over the performance of routine work, freeing a person for creativity (Lyubashits & Osinsky, 2019). Such a belief is traditionally inherent in the most conservative scientific circles and is currently often criticized.

A significant number of scientists take the opposite position, which is to recognize the possibility of granting rights to the subjects in question soon. For example, according to researchers such as P.P. Sergun and A.M. Bobrov, robots have already become social beings and actively participate in public relations, since people can no longer do without them. Under these conditions, it is natural that artificial intelligence will soon be endowed with a status that will allow it to take part in legal relations (Sergun & Bobrov, 2019).

Considering the above approaches, it is impossible not to pay attention to the proposal to recognize smart robots as special status as "electronic persons" or similar legal structures ("mechanical persons", "biomechanical persons", "quasi-subjects", etc.). The essence of these ideas, in many presented in the scientific literature (Baranov, 2018; Gadzhiev & Voynikanis, 2018), always consists in endowing robotic intelligence with such a volume of rights that will be less in comparison with human.

Considering the above, the concept presented in this paper, related to the moment of granting robots rights, can act as one of the solutions to emerging problems.

In addition, the legal characteristics of neural networks as an object of legal regulation are of value. We believe that the social relations associated with this object deserve careful study by the scientific community and can become a promising area of scientific work.

Conclusion

Thus, the results of this study indicate that the development of public relations in the era of digitalization will require significant changes to the current legislation. Special attention should be paid to the legal regulation of such important achievements of technology and science as neural networks, artificial intelligence, robots.

In relation to neural networks and artificial intelligence, it seems necessary to develop and adopt fundamental regulatory legal acts that will regulate the specifics of their operation.

The most controversial issue is the regulation of relations related to the determination of the legal status of artificial intelligence placed in a robotic body. The endowment of smart robots with rights and freedoms like human ones will be justified only on the condition that they begin to correspond to the distinctive features that distinguish the species Homo sapiens in comparison with other creatures. That is, to obtain legal personality, they must have developed self-awareness, feelings, and emotions, be able to share the feelings of other participants in legal relations, be capable of altruistic actions.

We believe that this topic is a highly promising area of scientific research that reveals current legal problems related to maintaining order in public relations of the digital era.

References

Banisar, D. (2006). *The right to information in the age of information. Human rights in the global information society* (pp. 77–89). The MIT Press.

Baranov, P. P. (2018). Legal regulation of robotics and artificial intelligence in Russia: Some approaches to solving the problem. *Severo-kavkazskii yuridicheskii vestnik= North Caucasus Legal Vestnik, 1*, 39–45.

Chechkin, A. V. E. (2021). The thesis about the presence of artificial intelligence. *Intelligent Systems. Theory and Applications, 25*(1), 29–49.

De Franceschi, A., & Schulze, R. (Eds.). (2019). *Digital revolution-new challenges for law: Data protection, artificial intelligence, smart products, blockchain technology and virtual currencies*. CH Beck.

Gabov, A. V., & Khavanova, I. A. (2018). The evolution of robots and the law of the XXI century. *Bulletin of Tomsk State University, 435*, 215–233. https://doi.org/10.17223/15617793/435/28

Gadzhiev, G. A., & Voynikanis, E. A. (2018). Can a robot be a subject of law (search for legal norms to regulate the digital economy)? Right. *Journal of the Higher School of Economics, 4*, 24–48. https://doi.org/10.17323/2072-8166.2018.4.24.48

Gulyaev, K. S. (2018). Human right to the internet, rights on the internet and when using internet things: New trends. *Precedents of the European Court of Human Rights, 1*, 29–37.

Kartskhiya, A. A. (2019). Digital transformation of law. *Monitoring of law enforcement, 30*(1), 25–29. https://doi.org/10.21681/2226-0692-2019-1

Lyubashits, V. Y., & Osinsky, A. S. (2019). Digital law and information technologies in the era of globalization: Problems of theory. *Bulletin of the Moscow University of the Ministry of Internal Affairs of Russia, 7*, 19–23. https://doi.org/10.24411/2073-0454-2019-10368

Mikheeva, T. N. (2019). On the question of the legal foundations of digitalization in the Russian Federation. *Bulletin of the O.E. Kutafin University (MGUA), 9*, 114–122. https://doi.org/10.17803/2311-5998.2019.61.9.114-122

Orlov, V. V. (1998). *The history of human intelligence*. Perm University Press. Ch. 1,2. Prehistory - myth - religion - enlightenment - Kant - Hegel.

Sergun, P. P., & Bobrov, A. M. (2019). Subject-object education as a structural element of administrative legal relations. *Perm Legal Almanac, 2*, 138–145.

Shestak, V. A., & Volevodz, A. G. (2019). Modern needs of legal support of artificial intelligence: A view from Russia. *All-Russian Journal of Criminology, 13*(2), 197–206. https://doi.org/10.17150/2500-4255.2019.13(2).197-206

Solum, L. B. (1992). Legal personhood for artificial intelligences. *North Carolina Law Review, 70*(4), 1231–1287.

Vasiliev, A. A., et al. (2020). *The transformation of law in the digital age*. Altai University Press.

Yasnitsky, L. N. (2015). Neural networks are a tool for obtaining new knowledge: Successes, problems, prospects. *Neurocomputers: Development, Application, 5*, 48–56.

Teacher's Identity in Online ESP and Music Education. Teacher's Primus Inter Pares Factor

Sergey Sayarovich Ivanov, Marina Vianorovna Zolotova, and Tatiana Borisovna Sidneva

Introduction

Massive compulsory shift to e-learning due to COVID-19 pandemic in the last few years has challenged teachers' identity in many aspects. In this paper these aspects are described and ways to cope with this kind of stress and recover a stable teacher's identity are suggested.

There is plenty of evidence that the promptly accomplished once in a lifetime shift from face-to-face classroom studies to online education was a major cause of stress, as it usually happens when a longstanding traditional way of life is suddenly and abruptly changed. Apart from psychological side, physiological impact was also great.

Teachers as representatives of one of the most stressful professions appeared in the front line trying to cope with the unexpected changes. Some of them experienced emotional distress which led to nervous breakdowns as well as physical manifestations of stress, undermining their health. In this respect, the findings of the paper and the solutions proposed are more than ever acute today.

As soon as the focus of this article is on professional, psychological and social impacts, undermining the professional identity in question, many other effects (health, economic, schedule and time-management, etc.) are beyond the scope of our research.

A lot of researchers provide facts that this change resulted in teachers' "identity struggles" (Tao & Gao, 2018) as a necessity to build, or even "rebuild their

S. S. Ivanov (✉) · M. V. Zolotova
National Research Lobachevsky State University of Nizhny Novgorod, Nizhny Novgorod, Russian Federation

T. B. Sidneva
Glinka Nizhny Novgorod State Conservatoire, Nizhny Novgorod, Russian Federation

A. Rumyantseva et al. (eds.), *Challenges and Solutions in the Digital Economy and Finance*, Springer Proceedings in Business and Economics,
https://doi.org/10.1007/978-3-031-14410-3_16

[professional] identities" (Kanno & Stuart, 2011). What was needed was not simply to adjust the methodology to online teaching, but to "reevaluate fundamental beliefs and values about what it means to be a teacher, both practically and theoretically (Simon, 2012, p. 3). For many teachers the most incomprehensible thing was being pushed "to make a shift in their educational priorities" (Rebenko, 2020, p. 201) and (especially for the veterans) the most painful and bitter experience was suddenly losing the leading and active role in classroom teaching and being put off to the peripheral roles of the profession, from being "the sage on the stage to the guide on the side" (King, 1993, p. 30).

All the above observations describe challenges faced by foreign language teachers. The same refer to musicians with the only difference that music teachers were affected in a more severe way as the flair of a performer on the stage, emanated also in the classroom and inspiring the disciples, appeared to fade away in the online environment.

This brings about the following chapter in which materials and methods of the research are provided.

Materials and Methods

This paper deals with foreign language and music teachers' identities, the ways to reconsider some of its integral parts and attempt to reach a balance between a stable off-line and online identities. We compare these two types of professionals, as it will help to make more useful, mutually enriching, and clear-cut conclusions than being confined only to one specific sphere. The specialists' experience, challenges and solutions are considered in Russia's higher educational establishments and are mostly the result of our collaboration with colleagues who are experts in the two subjects in question. We also use other published materials.

Herewith, relevant analysis methods have been selected.

Observation method was applied to find proper examples of teachers' identity dissonance and related problems in the fields concerned, listed in published papers and feedback provided by professionals.

Chronological analysis has also been employed in the paper, singling out synchronous and asynchronous aspects of working in both professional fields. This also helped to highlight dynamic characteristics of the teaching process in ESP and music.

Such a crucial factor as humanitarian component missing or minimalized in e-learning and methods to humanize the education process in depersonalized online environment are proposed.

The article is structured basing on comparative analysis principle which helped to track common challenges and reach applicable solutions.

To reach a deeper understanding of the problem and bring about the necessary results the above methods were incorporated in holistic approach to music and language education respectively which is described further on.

Results

To provide terminological clarity it is necessary to state that "identity" and "role" are different notions. Any person can play different roles, change them of their own free will; moreover, teachers are not professional actors and acting is not the core activity of the career. A definition which gives a general understanding of what constitutes identity was provided by Alsup (2006) stating that, "Success as a teacher is attached to a sense of professional identity that integrates the intellectual, emotional and the physical aspects of a teacher's life. It means being able to combine the core identity or personal beliefs and sense of self with a professional identity" (p. 36). In simpler terms, roles are variable, as any teacher can take on different roles at different times, whereas identity is a constant complex notion.

Some researchers (Tao & Gao, 2018; Rebenko, 2020; Qi et al., 2021) operate with the same model of ESP teachers' identity as given below (Fig. 1).

Even though teachers are not professional actors, what they do in the classroom is a kind of performance, but what they bring in is not only the role but also their identity. Bearing in mind that every face-to-face class or a rehearsal is a performance, the above model of teacher's identity can be complemented by three main

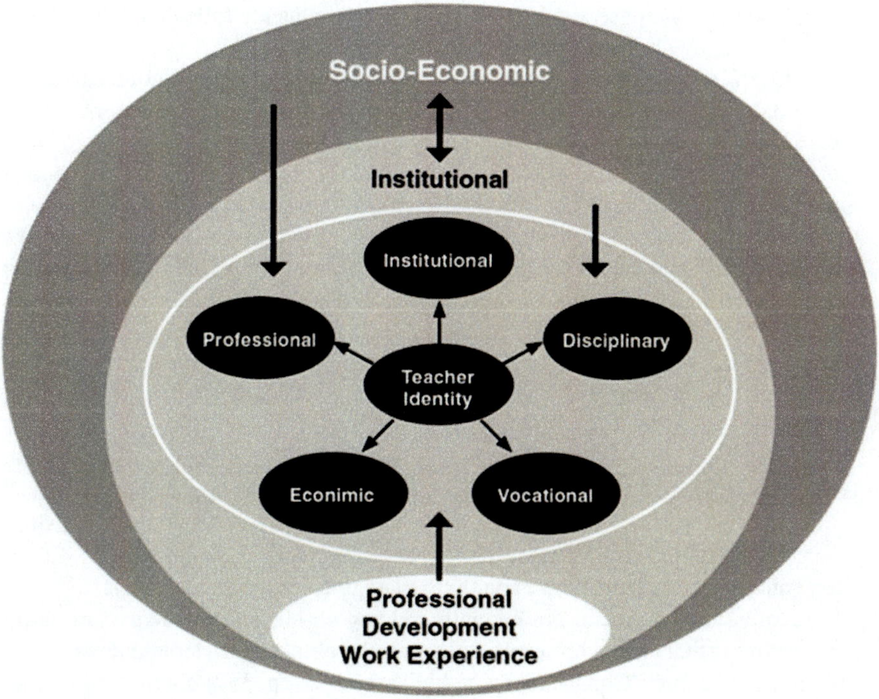

Fig. 1 A refined model of ESP teacher identity

elements of a performance: (1) setting, (2) the appearance and (3) the manner (Goffman, 1959).

The setting is the location together with its properties where the performance takes place. Its geometry includes a hall or a classroom, where the acting starts and where it finishes; each participant in this location knows where and when to sit or stand, when to engage and when to keep still and silent and what constitutes appropriate behavior.

The appearance is about the dress code, and the manner is the style and emotional colouring in which a performance is conducted.

All the above elements are absent or lacking in e-learning process. Education is a two-way traffic and teacher's identity is undermined and defragmented on either side of it: from inside (because of inner conflict) and from outside (how the teacher's image and identity is perceived by other people). It was determined that, "A teaching and learning experience deprived of face-to-face interactions is a major source of dissonance" (Simon, 2012, p. 3).

Absence or lack of all the described above segments of traditional teacher's identity online contribute to its defragmentation and destruction from inside. To be more specific, looking at Fig. 1, we can see that *every aspect* of teacher's identity is challenged in the online environment:

1. The instructional segment. As different from real-time classes, giving instructions and overseeing how they are carried out is desynchronized online. The instructor cannot see at what pace and how well the instructions are followed, which has a destabilizing effect.
2. The disciplinary segment. As different from real-time classes, no one can discipline listeners online. There are technical means allowing the conference organizer to switch off somebody's microphone, to delete a user from the session, but it can be disruptive to the whole process and in the absence of emotional contact interpreted by users to be not a method of punishment but as a technical failure or malfunction. From the students' side, it is very easy to avoid doing certain tasks as no one can force them to do something and control their activity real-time. As it is characteristic of all forms of online work and studies—the disciplinary focus was shifted from the supervisor to self-discipline of everybody involved. In our case, students had to accept some of the responsibilities traditionally assumed by teachers. It also happened because there are "no norms for their behavior in virtual classrooms" (Polin, 2004, p. 43).
3. The vocational segment. "Online teaching further threatens teachers' identity by challenging some of the very reasons why they became teachers in the first place" (Simon, 2012, p. 47). Many aspirations, self-image as a teacher faced disillusionment in the online-environment, deprived of geometrical configurations, dress-code and norms of behavior. Though, the last two are debatable points.
4. The professional segment. Traditionally treating on-line students like extra-mural or part time students, "who were not among the elite and . . . looked down on as inferior education" (Gunawardena & McIsaac, 2004, p. 357) continues pressing upon teachers' perception. This leads to setting lower requirements to these

"underdogs", lower expectations and getting weaker feedback and poorer results in return. Such prejudiced mindset undermines professional fulfillment and brings about a sense of dissatisfaction and inefficiency.
5. Economic segment (or to be more specific, material or income). Online teaching has traditionally been viewed as some extra source of income and a part time job, only additional to real-time educational activity. Those who were involved in it naturally benefited from it materially but have only been viewed by their peers as doing something extra, not very important and only for the sake of money.

The above five factors have been undermining and ruining teachers' identity from the inside, creating the inner source of dissonance. Simultaneously, the professionals who were trying to cling to something analogous to traditional aspects, hurriedly reevaluating their beliefs and views to preserve their identity, were hurt from the outside. Students stopped perceiving them as teachers in their traditional sense; some students appeared to be more versatile using modern technologies and did not hesitate to demonstrate it, some showed that they didn't need to be disciplined any longer and indifferently disobeyed. With such characteristics as charisma, sense of humour, acting skills devalued, depersonalized, and blocked in the online environment, teachers failed to impress and rule the audience and had to leave the central part of the stage without applause. Education process abruptly turned from being teacher-centered to learner-centered.

For their part, learners also faced numerous challenges and encountered some difficulties. Not willing and not being ready to take on some roles and responsibilities previously held by teachers was only one source of disturbance. Having to take them on filled a lot of learners with omnipotence; that they could choose to be guided by anything they were able to find on the Internet. With so many resources, most of which impossible to check for their trustworthiness and not knowing how to check them due to the absence of critical thinking in this sphere, a student "is like a man in interstellar space, who has no way of knowing which direction is up and which is down" (Orwell, 1949, p. 198). Herewith, it is necessary to provide evidence in both language and musical sphere.

In language sphere the most troublesome are cases with pronunciation and translation. Picking up some new words, students turned to on-line resources helping them to pronounce them correctly. Not knowing where to search and what sources to trust, a lot of them learned to pronounce the words incorrectly. It is common knowledge that in most cases it is easier to learn something anew than to correct the deeply rooted mistakenly learned variants. It makes the situation even worse when a learner is absolutely convinced that their variant is correct and refers to some sites or links which appear to be fake or amateurish. Dealing with translation issues, a lot of learners don't bother to critically assess what text is being translated and collate some of its parts with the context, completely and blindly relying on e-translators. As a result, many get illogical and ridiculous outcomes and misunderstanding. An illustrative example here is provided by a student reading an article about a debate in the House of Lords and translating the word "lord" as a synonym to the word "god", because a dictionary says so. All this happened due to affordance of final solutions and not knowing the path to walk to reach these solutions. The teacher should play the role of a guide here, providing trustworthy sources, references, and standards.

In musical sphere the situation appeared to be the same and sometimes even more outrageous. Having found standard samples of playing a musical piece, young performers tried to imitate these. Weak and poor copies (if any) came out as a result. It happened because the music students disregarded some of basic physiological aspects, which include correct breathing for wind instruments, physical environment guaranteeing the correct breathing and sounding like posture when standing and sitting, arm position etc.; it may seem weird, but some professionals bring their own chair to the concerts (Playing technique, 2021; Barilo, 2020). Individual physiological traits must also be taken into consideration and dealt with by a professional in each individual context. Some of young performers were unaware that no matter how good the microphone is, the recorded or the on-line transmitted sound can be distorted (Mechetina, 2020); disregarded the quality of the recordings and on-line broadcasts in which the distorted sound appeared not worth copying. What is more, as soon as instrumental performance is determined by both physiological and physical factors, it is noteworthy, that incorrect breathing techniques can cause health disturbances for winders and can be even traumatic for young players stringing their instruments (Barilo, 2020). All this hindered or blocked the ability of young performers to copy the original. The same as with language education, it happened due to affordance of final solutions and not knowing the path to walk to reach these solutions.

The sudden and compulsory shift from traditional classroom studies to e-learning stressed both teachers and learners, causing their identity dissonance. Like in any stressful situation the bodily response is directed to restore the initial state (Selye, 1982) and "human being strives to create holistic identities" (Simon, 2012, p. 48). There is plenty of evidence provided by scholars and psychologists that a teacher with disrupted or unstable identity cannot function efficiently, cannot be satisfied by the learning outcomes (Simon, 2012). Teachers' psychological stress, overload, lack of professional fulfillment and feeling inefficient online "lead to depression, irritability, lower their self-esteem and cause psychosomatic symptoms and addictions (Schonfeld et al., 2017; Petrakova et al., 2021). The inability to function normally under stress and the desire to obtain a stable holistic identity call for prompt solutions, some of them suggested in the following chapter.

Discussion

Having found themselves in "interstellar space" both in psychological and practical terms, most teachers began to adjust to the on-line environment by transferring their classroom practices online. As a result, modifications made to their existing courses appeared to be minor and their online courses were mostly classroom-based and teacher-centered. Thus, they tried to make on-line format fit their face-to-face identities, instead of adapting their professional identities to the on-line format. Such attempts were generally not very successful because, as it was mentioned in the previous chapter, the audience cannot be affected by teachers' personalities online and some of the practices became asynchronous (as different from face-to-face format). Moreover, certain standard traditional practices had to be abandoned either because they were not adjusted to the online format or some teachers were unwilling to demonstrate them online, i.e., to go public, because online activities are recorded and archived and can be viewed by many, as opposed to the rigid privacy of the classroom.

The solutions described below are based on two behavior strategies to manage stress: "emotionally-oriented (e.g., positive reassessment of the situation) and problem-oriented (e.g., problem solving strategy)" (Lazarus & Folkman, 1984).

As soon as teachers' identity does not exist without beliefs, attitudes, sense of achievement and efficiency, we suggest applying the first emotionally oriented strategy to positively reassess the role of the teacher in the online environment which in the end will help to build up holistic online teacher's identity. The starting point here is to grasp that being shifted to the online environment does not abolish the role of the teacher and "the virtual presence of an instructor does not diminish the central role of teaching" (Simon, 2012, p. 11). Skepticism undermines teachers' identity, so to acquire holistic online identity teachers need to honestly believe that online education *is effective*. They also need to abandon any hope that e-learning is something temporal and after a short period of time the pandemic will be over, and things will get back to their traditional way. Such hopes create doubts, and doubts are disruptive in acquiring a holistic identity. Unfortunately for the unbelievers, online education is here to stay and what happens today is forever.

Many researchers point to the fact that in the online environment teachers are instructors and facilitators (Simon, 2012; Rebenko, 2020; Qi et al., 2021). We suggest viewing this factor positively and calling it *primus inter pares*. As it was mentioned before, in "interstellar space" the role of instructor must be recognized as indispensable both by teachers and students. Learners must get accustomed to the fact that guidance and the ability to critically assess myriads of online materials and samples can be provided by the teacher alone. For their part, having accepted *primus inter pares* role, teachers need not only demand the final product or demonstrate it. Of course, the final product is important, but it shows what is done rather than how it is done. This brings us to problem-solving strategies which require a humanitarian component to bridge the digital divide.

In music education, showing how to play a piece or part of it certainly takes place in a real classroom, but it should not be limited to that in the online environment (during online sessions via Skype, Zoom etc.). A lot of students download, and use backing tracks to rehearse and study a music piece individually. These (even good ones) do not take into consideration individual physiological and psychological peculiarities, let alone the agogics—tiny changes or deviations introduced in the tempo during artistic performance. As a result, it becomes very difficult, especially for learners, to physically perform a music piece using an instrumental, and it blocks any ability to agogically interpret a music piece which is characteristic of high-level performance skill. The role of a “music online instructor” here is to provide indispensable guidance not only by showing how to play, but, ideally, recording the accompaniment adjusted to the learner’s individual features. This recording and discussion will also serve as a guidance in further steps, when searching for readily available backing tracks. It will help to critically assess them and choose the one, that is most suitable for the player; as well as to avoid ridiculous but, unfortunately, real situations when a performer negligently downloads an instrumental and starts performing but when the key in the recording is suddenly changed—sits silently in a shock and does not know what to do and what to play. “With the universal algorithmization and virtualization of education, all innovations should in no case capture even a small “patch“ of the territory of consciousness responsible for personal growth, the ability to think critically and form opinion” (Sidneva et al., 2021, p. 685). Consequently, the solution for teachers can be formulated as follows: do not just show the final product that you are able to produce as a soloist but rather take on the status of facilitator or *primus inter pares*, inserting a humanitarian component in online classes and show how to walk the path. As for physiological component, emerging technologies are being tested today, promising future possibilities to teach hands-on techniques online.

For language education the solution suggested seems very similar to the music sphere described above. *Primus inter pares* factor implies designing student-centered online courses where the teacher plays the role of coordinator and instructor rather that the soloist. Just like in music education, the teacher’s role is indispensable when presenting standards and guidance of pronunciation, intonation, and other prosodic elements. “The online instructor” should also encourage critical thinking in translating and interpreting various texts, collating the final product with its context before presenting it. Seemingly primitive but practical solution here for the language teacher is the same as for the music one—to show how to do the task, e.g., when asking students to make a summary—do it yourself during the online session, play the role of *primus inter pares* and show them how to walk the path.

Using the behavior strategies to manage stress will help to overcome difficulties and challenges, and what is more important—build up a stable healthy online teacher-student identities, without which the whole process is bound to be less effective and deprived of professional and emotional fulfillment.

Conclusion

Abrupt and sudden shift to online education caused both teachers’ and learners’ stress resulting in teachers’ identity dissonance. Those teachers who had stable real-time identities must develop healthy online identities in present day situation. Faced with difficulty applying the face-to-face teaching approaches with which they were familiar to e-learning, prevented them from being fully satisfied with their online teaching. It happens because teacher’s identity comprises not only practices, but is an evolving complex notion consisting of beliefs, expectations, perception of self on the one hand, and affordances and technological versatility—on the other hand. For the educational process to be effective and satisfying for everybody involved, teachers must have healthy stable identities. Solutions to build up holistic online identity suggested in the paper are demonstrated by the examples of foreign language and music teachers and are generally applicable to both. They include:

1. Recognizing the *primus inter pares* factor and developing positive attitude to it.
2. Learning to live in the online environment as this is not a temporary challenging period—this is an integral part of our life.
3. Synchronizing where possible teacher-student interaction and inserting human component in online classes to bridge the digital divide.
4. Designing student-centered rather than teacher-centered online courses

The proposed solutions open discussion area for further methodological and practical research and disputes in the fields concerned. Moreover, the findings of the paper could be generalized and applied by professionals teaching other subjects with such limitations as lack of technologies to teach hands-on techniques online and only emerging or developing comprehensive online courses. As it is stated in the article, a teacher working in today's realities is more than a teacher; *primus inter pares* factor appears to be broader than simply a teacher's role: it implies tracking and keeping up with technological changes, collaborating with various specialists, selecting adequate content and designing online courses and above all—trying to humanize the e-learning process. Consequently, *primus inter pares* factor can also be applicable to teaching in such spheres as medicine, acting, social work to name a few.

References

Alsup, J. (2006). *Teacher identity discourses: Negotiating personal and professional spaces*. Routledge.

Barilo, T. (2020). *Pluses and minuses of distance education at a music school*. Retrieved from https://monch-dmsh.murm.muzkult.ru/media/2020/11/23/1245112155/Doklad_Barilo_noyabr_2020.pdf.

Goffman, E. (1959). *The presentation of self in everyday life*. Bantam Doubleday Dell Publishing Group.

Gunawardena, C. N., & McIsaac, M. S. (2004). Distance education. In D. H. Jonassen, *Handbook of research for educational communications and technology* (2. Bs) (pp. 355–395).

Kanno, Y., & Stuart, C. (2011). Learning to become a second language teacher: Identities in practice. *The Modern Language Journal, 95*(2), 236–252. https://doi.org/10.1111/j.1540-4781.2011.01178.x

King, A. (1993). From sage on the stage to guide on the side. *College Teaching, 41*(1), 30–35.

Lazarus, R. S., & Folkman, S. (1984). *Stress, appraisal, and coping*. Springer Publishing Company.

Mechetina, E. (2020). *Distance learning in professional piano lessons*. Retrieved from https://www.classicalmusicnews.ru/articles/distancionnoe-obuchenie-mechetina/

Orwell, G. (1949). *Nineteen eighty four*. Novel. Signet classic.

Petrakova, A. V., Kanonire, T. N., Kulikova, A. A., & Orel, E. A. (2021). Characteristics of teacher stress during distance learning imposed by the COVID-19 pandemic. *Education Issues, 1*, 93–114. https://doi.org/10.17323/1814-9545-2021-1-93-114

Playing technique. (2021). Retrieved from http://the-clarinets.net/english/clarinet-how-to-play.html.

Polin, L. (2004). *Learning in dialogue with a practicing community. Learner-centered theory and practice in distance education: Cases from higher education* (pp. 17–48). Lawrence Erlbaum.

Qi, W., Sorokina, N., & Liu, Y. (2021). The construction of teacher identity in education for sustainable development: The case of Chinese ESP teachers. *International Journal of Higher Education, 10*(2), 284–298. https://doi.org/10.5430/ijhe.v10n2p284

Rebenko, M. (2020). Modelling ESP teacher identity in Ukrainian tertiary education. *Journal of Teaching English for Specific and Academic Purposes, 8*(3), 201–213. https://doi.org/10.22190/JTESAP2003201R

Schonfeld, I. S., Bianchi, R., & Luehring-Jones, P. (2017). Consequences of job stress for the mental health of teachers. In *Educator stress* (pp. 55–75). Springer, . https://doi.org/10.1007/978-3-319-53053-6_3.

Selye, H. (1982). *Stress without distress*. PROGRESS.

Sidneva, T., Musichenko, V., & Gudkov, A. (2021, January). Creative universities in the digital age: Transformation of academic traditions and new strategies. In *Second Conference on Sustainable Development: Industrial Future of Territories (IFT 2021)* (pp. 681–686). Atlantis Press. https://doi.org/10.2991/aebmr.k.211118.120.

Simon, E. (2012). *The impact of online teaching on higher education faculty's professional identity and the role of technology: The coming of age of the virtual teacher* (Doctoral dissertation, University of Colorado at Boulder).

Tao, J. T., & Gao, X. A. (2018). Identity constructions of ESP teachers in a Chinese university. *English for Specific Purposes, 49*, 1–13. https://doi.org/10.1016/j.esp.2017.09.003

Economic Mechanism for Assessing the Efficiency of the Department as a Business Unit of a Higher Educational Institution

Natalya Leonidovna Ketoeva, Maria Andreevna Kiseleva, and Ksenia Sergeevna Volodina

Introduction

At the present stage of development of society, education is increasingly referred to as a service sector, therefore, a higher educational institution, in this regard, is considered as an enterprise that provides not only educational services, but also provides research and innovation activities (Ketoeva et al., 2020b). Practical experience and the requirements of management science show that the success of a higher educational institution is achieved by the interested participation of each link, including each employee. A special place in the system of higher education belongs to the department (Andreeva & Mizova, 2018). Due to the fact that at the level of the department there is a main relationship with students and graduate students, young scientists and researchers and a direct educational, cultural, social impact on them (Ketoeva et al., 2020a). The effectiveness of the university directly depends on the activities of the department and its head, this is especially important in today's rapidly changing world, aimed at the development of information technology (Tumbas et al., 2019). With the help of process detection, it is possible to determine and orient: the activities of the university—to business processes; university management system—on the management of each business process separately, and all business processes in general; the quality system of the university—on the quality of software technology for the implementation of business processes.

N. L. Ketoeva (✉) · M. A. Kiseleva · K. S. Volodina
National Research University "Moscow Power Engineering Institute", Moscow, Russian Federation
e-mail: ketoyevanl@mpei.ru; VolodinaXS@mpei.ru

A. Rumyantseva et al. (eds.), *Challenges and Solutions in the Digital Economy and Finance*, Springer Proceedings in Business and Economics,
https://doi.org/10.1007/978-3-031-14410-3_17

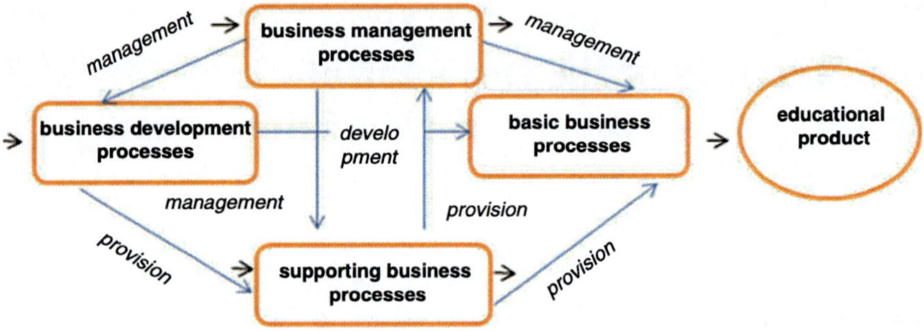

Fig. 1 The relationship of business processes in an educational organization (Ketoeva et al., 2021b)

The authors define business processes as systems of various interrelated and regulated activities in which the influencing influences and features of resources are controlled (Prokhorova et al., 2017).

Methods for describing business processes for higher educational institutions are similar to any enterprise. However, the specifics of the activities of higher educational institutions (Universities) present their own requirements when highlighting business processes. In universities, educational, research (Ketoeva et al., 2021b) and innovation activities (Neprokina, 2019) are distinguished, which include certain functions and sets of business processes aimed at achieving certain goals. The optimal interconnection of all business processes within an organization forms a system level, or business system (Fig. 1).

Thus, all business processes can be classified according to the level of management. Strategic speed includes business processes of development and management (rector's office, vice-rector for educational activities, vice-rector for scientific activities, vice-rector for economics). It is here that a new quality of business processes arises, which is absent at other stages—a qualitative concept, including organization, the main strategic goal. Statistical decisions made at the highest level set the target orientation, the orientation of the entire set of business processes of the organization, formulating priorities and orienting the requirements for achieving results.

The level of tactical management includes supporting business processes (training management, training department, directors of institutes). And the tactical level is the main business processes, i.e. direct activities of the departments (head of the department, faculty).

Based on the foregoing, it is possible to define the department as part of the university system, a kind of strategic business unit in the educational services market. The department is the main structural subdivision of the university, carrying out educational, methodological and research activities in one or more areas, while realizing its main task—the training of highly qualified personnel.

Each of the activities eventually creates a unified management strategy, the result of which is the formation of a general rating of departments. In order to correctly evaluate and calculate this, a new economic mechanism is needed that evaluates the entire complex activity of departments as business units.

Thus, the purpose of the study is to develop an economic mechanism for evaluating the effectiveness of the department as a business unit of a higher educational institution.

The results of the scientific research are maximally aimed at improving the methods and means of managing the complex activities of the university's business units (Ketoeva et al., 2021d). The economic mechanism presented in this article is part of the software product algorithm developed by the authors for monitoring the complex activities of departments. The finished software product will be a practical tool for use in the activities of higher educational institutions, in order to increase their competitiveness and attractiveness in the global and domestic markets.

The results of the scientific research affect the implementation of the national project "Digital Economy", the implementation of tasks in the field of innovative development, such as the transition to integrated efficiency of business processes and automation of management systems. The successful solution of the set goal and the receipt of the planned results is ensured by correct formulations of tasks arising from the positive experience of previous studies.

The continuation of research on this scientific topic will be reflected in the dissertation work on the topic: "Regulation of financial processes of management of complex university activities in the digital economy."

Materials and Methods

The methodological base of the study was made up of methods: dialectical scientific knowledge and private scientific (analysis, synthesis, comparison, logical and system-structural analysis, formalization, analysis of legal documents), modeling.

There are many works by Russian and foreign researchers devoted to assessing the economic efficiency of the university's business units. In the context of this scientific topic, it is necessary to focus on three key aspects of modern research (legal, educational and scientific), which will form the basis for the development of an economic mechanism:

1. The normative legal documents regulating the activities of higher educational institutions are: Federal Law "On Education in the Russian Federation" dated December 29, 2012. No 273-FZ (last edition), ORDER No 475 dated March 24, 2020 "On Approval of Performance Indicators of Federal Budgetary and Autonomous Educational Institutions of Higher Education Subordinate to the Ministry of Science and Higher Education of the Russian Federation, and the Work of Their Managers, based on the results of which incentive payments are established to the heads of such institutions.

2. Works related to research in the field of developing an economic mechanism for evaluating the effectiveness of departments (Grigorash, 2018; Vorobiov & Murzaeva, 2017).
3. Works devoted to the problems of regulating business processes in higher education (Nedosekin, 2002; Trachenko, 2012; Karsanov & Kutsuri, 2017; Kutsuri et al., 2018), Johnston & Shroff-Meta, 2000). Despite the fact that a number of studies have already been conducted on this topic, this area requires additional study. It should also be noted that the existing regulations and the bulk of publications on this issue are devoted to methods, models of financial management at universities, and not to the creation of a practical model or software product that would be effectively applied in practice. At the same time, in our opinion, in the conditions of the digital economy, it is the assessment of the effectiveness of business process management in universities that requires special attention.

Results

Economic activities, together with relationships and interactions, represent a network of organizational processes. Based on the process approach, the activities of the department can be divided into business processes (BP) of management, basic and supporting business processes, business development processes (Bolshakova et al., 2020).

The management business processes include the management of budgetary and non-budgetary funds of the department, personnel management and risk management (Rumyantseva et al., 2019).

For the main business processes of the department, all processes that add value are characteristic, such as: educational activities, research and innovation activities, marketing activities management and interaction with related markets.

Supporting business processes are information resources, administrative and economic activities, logistics.

To the BP of development, we include the development of a system of indicators for evaluating the effectiveness of the department's activities.

In general terms, the activities of the department, as a business unit of an educational institution, the authors presented in Fig. 2.

The economic mechanism of the functioning of the organization should be focused on achieving strategic development goals, such as ensuring sustainable development and strengthening competitive positions in the market (Kamirova, 2017).

Based on this goal, the mechanism of the functioning of the organization should contribute to the solution of the following tasks:

1. Formation of conditions for the most complete satisfaction of consumer demand for services (product) within the educational services market.

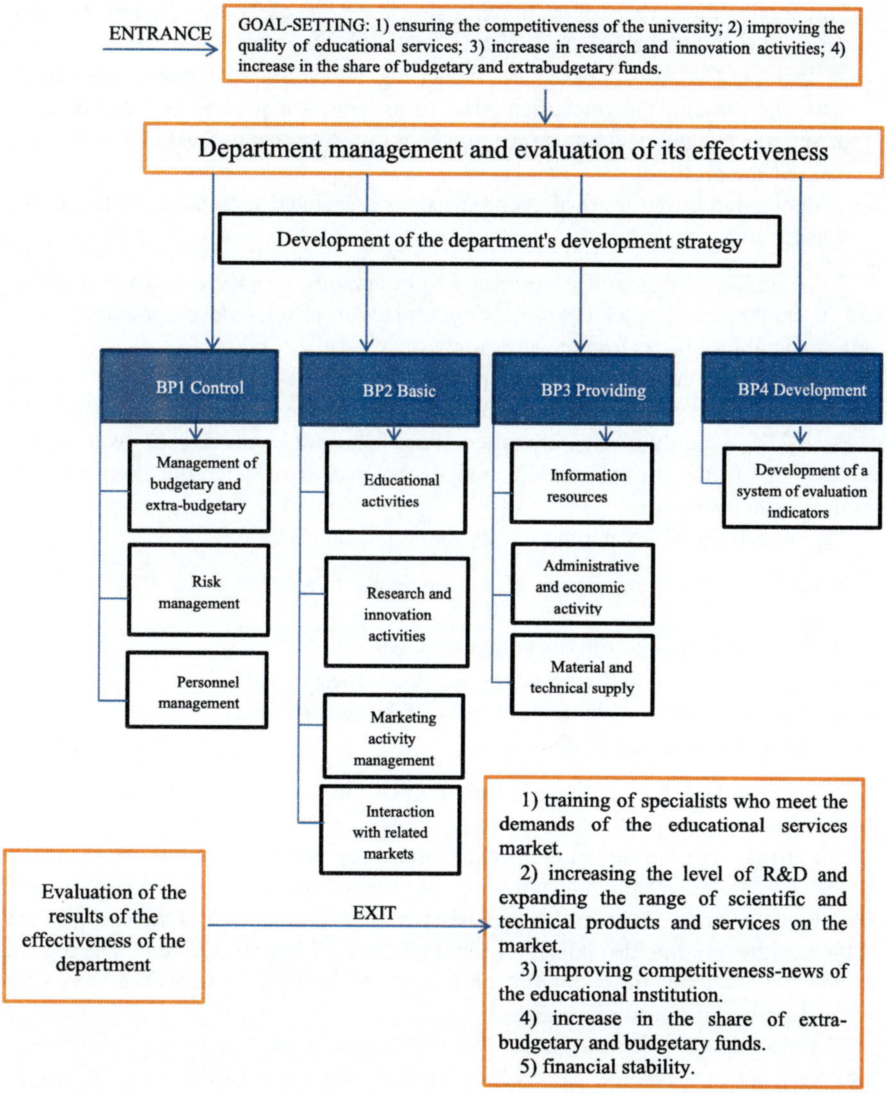

Fig. 2 Scheme of the activities of the department as a business unit of a higher educational institution (Ketoeva et al., 2021c)

To solve this problem, it is necessary to study the market, develop an effective commodity and pricing policy.

2. Ensuring the economy of economic and financial activities.

This problem is solved on the basis of minimizing the cost of labor, material, financial resources for the organization of production processes, for the implementation of commercial operations, and the performance of organizational management functions.

3. Maximizing the amount of profit remaining at the disposal of the organization and ensuring its effective use.

   This task is solved based on increasing the volume of the organization's activities through the implementation of an effective pricing and depreciation policy, tax planning, optimization of the ratio of the capitalized and consumed parts of the profit.
4. Minimization of the level of economic risks associated with the activities of the organization.

To solve this problem, it is necessary to effectively manage commercial, financial, investment and other risks. Risk minimization ensures development stability and predictability of performance results.

The economic mechanism developed by the authors for assessing the level of efficiency of the department's activities will increase the competitiveness (Ketoeva et al., 2021c) not only of the department, but of the entire educational institution in the market of educational services and in the market of scientific and technical products and services.

The objectives of developing an economic mechanism are:

1. Ensuring the competitiveness of the university (Bagiev et al., 2021; Erznkyan, 2021).
2. Improving the quality of educational services.
3. Increasing research and innovation activities (Joseph, 2017).
4. Increasing the share of budgetary and extrabudgetary funds.
5. Ensuring financial sustainability.

To achieve these goals, it is necessary to formulate a strategy clearly and correctly for the development of the department, as a vector for further work.

The structure of the economic mechanism is proposed to be built in the form of a sequence of activities aimed at assessing the level of efficiency of each department. The implementation of these activities will not only form an idea of the current state of the department, but also adjust its activities for further improvement. The logical scheme for evaluating the economic mechanism for the effectiveness of the activities of the departments is shown in Fig. 3.

The evaluation of the criteria for the activities of each department is calculated separately for each type of activity: educational (educational) (Bagiev et al., 2021) and research and innovation activities (Bogomolova et al., 2015). The formulas are compiled by the authors and within the framework of this article are proposed as an algorithm for evaluating criteria.

As an example, we will evaluate the criteria of educational (training) activities. The effect of changes in the number of students on the amount of financial receipts to extra—budgetary funds is calculated by the formula:

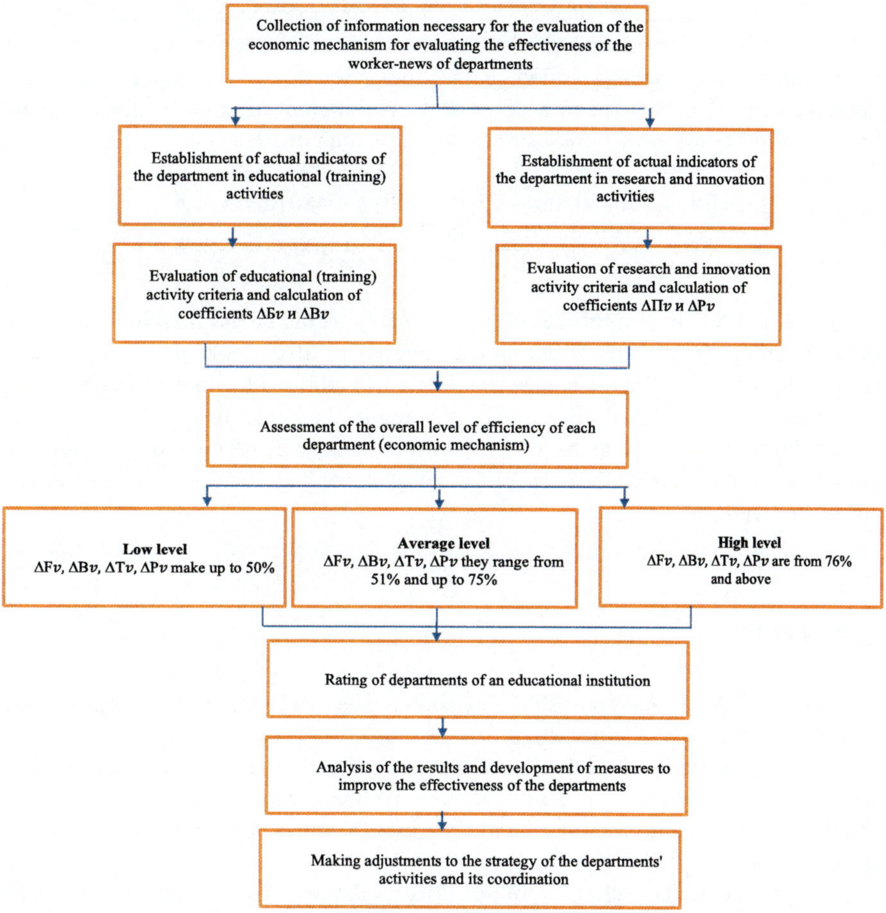

Fig. 3 The logical scheme for evaluating the economic mechanism of the effectiveness of the activities of the departments

$$\Delta \mathrm{Bv} = \sum_{i=1}^{n} \frac{V_{rep} d_{irep} Ц_{irep} - V_f d_{if} Ц_{if}}{V_f d_{if} Ц_{if}} \times 100\%$$

Where, ΔBv—change in extrabudgetary income from student education, which occurred due to the number of students.

n—number of specialties.

Vrep, Vf—the total number of contract students in the reporting and base periods, respectively.

di—share of students studying on a paid basis in i specialty.

Цi—the cost of education for one student of the i specialty.

The data for the calculation were taken from one of the leading energy universities of the Russian Federation, from the reports of indicators for the last 3 years.

According to the results of calculations, we saw a direct dependence on the change in the number of enrolled students on the amount of extra-budgetary funds. Since the total number of students enrolled in the university in 2020, compared to 2019, decreased by 21%, this led to a decrease in the amount of extra-budgetary funds by 48,070,000 rubles. This indicator is 20% less than in 2019.

$$\mathrm{Bv} = 190960000 - 239030000 = -48070000 \ \text{rub.}$$
$$\Delta \mathrm{Bv} = \frac{190960000 - 239030000}{239030000} \times 100\% = -20\%$$

One of the reasons for this decline, presumably, could be the situation caused by the pandemic, because the admission campaign in 2020 took place under strict restrictions and significant changes in the order of admission and enrollment.

Based on the results obtained, a score rating of the departments (Ketoeva et al., 2021a, 2021d) from the highest to the lowest value will be compiled, where the departments with the highest scores themselves will be additionally motivated, stimulated and encouraged.

Discussion

Having studied various approaches of economists (Hurwitz, 1891; Blaug, 1994; Izmalkov et al., 2008; Kleiner, 2020; Joseph, 2017) to the definition of the concept of the category "economic mechanism", we can conclude that the economic mechanism of an organization is a set of economic methods, methods, forms, tools, levers of influence on economic relations and processes (Ketoeva et al., 2021a) occurring in the organization.

At the moment, the authors (Mollick, 2014; Shkunova & Pleshanov, 2017) are in the process of implementing the developed mechanism in the university, in order to monitor the complex activities of departments. According to the forecast data, after the implementation, it is expected:

- an increase in the financial indicators of the university's business units to 5%.
- an increase in the competitiveness of the university in the educational services market.
- identification and reduction of inefficient business units of the university by 2%.
- an increase in the number of applicants by 10–15%.
- an increase in the indicators of research and innovation activities by an average of 3%.

The achievement of these goals affects the implementation of tasks in the field of innovative development, such as the transition to integrated efficiency of business processes and automation of management systems. The economic mechanism developed by the authors can be applied in the economic, social and educational spheres.

Conclusion

The economic mechanism developed by the authors for evaluating the activities of the department, from the point of view of educational (educational) activities, will allow to assess the demand for specialties and identify inefficient business units of the university, which in the complex will contribute to the growth of the level of training of students, the level of qualification of teachers and financial support of the department.

From the point of view of research and innovation activities, it makes it possible to evaluate and, if necessary, influence the methods of improving indicators in this area.

Successful implementation of all set goals will contribute to the growth of competitiveness of universities and attractiveness in the global and domestic markets.

The continuation of research on this scientific topic will be reflected in the dissertation work on the topic: "Regulation of financial processes of management of complex university activities in the digital economy."

Acknowledgements The investigation was carried out within the framework of the project "The mechanism for monitoring the integrated activities of the departments of the National Research University "MPEI" to ensure the competitiveness of the business unit of the educational institution" with the support of a grant from NRU "MPEI" for implementation of scientific research programs "Energy", "Electronics, Radio Engineering and IT", and "Industry 4.0, Technologies for Industry and Robotics in 2020–2022.

References

Andreeva, A. N., & Mizova, E. M. (2018). Digital economy: New business opportunities. *Economy and Business, 4*, 19–21.

Bagiev, G., Omarov, M., & Afanasiyeva, E. S. (2021). Marketing – mastering, rethinking. *Prospects of the Problem of Modern Economy, 2*(78), 110–114.

Blaug, M. (1994). *A simple lesson in economic methodology* (doctoral dissertation, thesis).

Bogomolova, I. K., Zadorozhnyaya, E. K., & Shevchenko, I. K. (2015). Budgetary efficiency of higher schools and financial management monitoring. *Modern Problems of Science and Education, 1–1*, 486–492.

Bolshakova, A. V., Shakurov, A. A., & Slanov, V. P. (2020, September). Modern university marketing management: Specific nature. In IOP Conference Series: Materials Science and Engineering (Vol. 9401, 012147). IOP Publishing.

Erznkyan, B. G. (2021). Problems of development of institutional economics. *Economy of modern Russia, 3*, 40–48. https://doi.org/10.33293/1609-1442-2021-3(94)-40-48

Grigorash, O. V. (2018). *Modern technologies for assessing the performance effectiveness of the technical university department*. Sreda. https://doi.org/10.31483/r-21565

Hurwitz, A. (1891). Über Riemann'sche Flächen mit gegebenen Verzweigungspunkten. *Mathematische Annalen, 39*(1), 1–60.

Izmalkov, S., Sonin, K., & Yudkevich, M. (2008). The theory of economic mechanisms (Nobel prize in economics 2007). *Voprosy economiki, 1*, 148–157. https://doi.org/10.32609/0042-8736-2008-1-4-26

Johnston, B., & Shroff-Meta, P. (2000). *Financing of higher education and its accessibility: An international comparative analysis of the cost of higher education and the policy pursued for financial support of students*. University Management: Practice and Analysis, New York.

Joseph, E. (2017). *Aoun robot-proof: Higher education in the age of artificial intelligence*. MIT Press.

Kamirova, A. N. (2017). Strategic planning: Orientation of the university to sustainable development and competitive advantages. Management consulting, (1 (97), 166–171.

Karsanov, T. K., & Kutsuri, G. N. (2017). The problem of money development in the course of financing budgetary and autonomous institutions. *Business and Society, 3*, 8–9.

Ketoeva, N. L., Lisin, E. M., Kiseleva, M. A., Korkin, V. S., & Zargaryan, M. T. (2020a). Development and structural analysis of the organizational management model of information interaction of subjects of the scientific and educational process. *Economics and Entrepreneurship, 12*, 957–961. https://doi.org/10.34925/EIP.2021.125.12.190

Ketoeva, N. L., Kiseleva, M. A., & Kotelnaya, V. K. (2020b). Methods of assessing the man-agement of research activities of universities in Russian and world practice. *Bulletin of the South-Russian State Technical University (NPI) Series Socio-economic Sciences, 13*(4), 37–45. https://doi.org/10.17213/2075-2067-2020-4-37-45

Ketoeva, N., Kiseleva, M., & Dranitsyna, V. (2021a). Development of a mechanism for the Management of University Research Activities Based on the principles of sustainable development. In *E3S Web of Conferences* (Vol. 295, p. 05007). EDP Sciences. https://doi.org/10.1051/e3sconf/202129505007.

Ketoeva, N. L., Zargaryan, M. T., & Volodina, K. S. (2021b). An interaction model Devel-opment between the university and the related markets elements. In *Proceedings of the International Scientific and Practical Conference on Sustainable Development of Regional Infrastructure* (pp. 707–712). https://doi.org/10.5220/0010596307070712

Ketoeva, N. L., Kiseleva, M. A., & Volodina, K. S. (2021c). The economic mechanism for evaluating the effectiveness of the university chair as a business unit of a higher educa-tional institution. *Kreativnaya ekonomika, 15*(12), 4571–4584. https://doi.org/10.18334/ce.15.12.113851

Ketoeva, N., Kiseleva, M., & Sysoeva, E. (2021d). Digital Management of Department as a business-unit of educational institution. In *IV International Scientific and Practical Conference* (pp. 1–6).

Kleiner, G. (2020). Intellectual economy of the digital age. Digital age: The steps of evolution. *Ekonomika i matematicheskie metody, 56*(1), 18–33. https://doi.org/10.31857/S042473880008562-7

Kutsuri, G. N., Shanin, S. A., Frumina, S. V., Gardapkhadze, T., & Ivanova, E. V. (2018). Russian practice of identifying and assessing budget risks. *Journal of Applied Economic Sciences, 13*(3), 711–719.

Mollick, E. (2014). The dynamics of crowdfunding: An exploratory study. *Journal of Business Venturing, 29*(1), 1–16.

Nedosekin, A. O. (2002). *Management of accumulative constituent funds of the obligatory retirement insurance using fuzzy sets approach*. Finite-2002 report.-Minsk.

Neprokina, I. V. (2019, May). Functional model of the head of the third-generation university department. In Institute of Scientific Communications Conference (pp. 407–414). Springer

Prokhorova, V., Chernikova, V., Novoselova, N., Hendon, A., & Seraya, N. (2017). Clustering of modern industrial enterprises as a criterion of successful activity. *International Journal of Applied Business and Economic Research, 15*(23), 393–402.

Rumyantseva, I. A., Krotenko, T. Y., & Zhernakova, M. B. (2019, November). Digital competencies: Requirements for information Technologies in the Framework "Management

University-Industry-Science-Market". In The International Scientific and Practical Forum "Industry. Science. Competence. Integration" (pp. 754–762). Springer

Shkunova, A.A., & Pleshanov, K. A. (2017). The organization of the project activities of university students: The results of the scientific research and development prospects. Bulletin of Mininsky University, (4 (21), 4. https://doi.org/10.26795/2307-1281-2017-4-4

Trachenko, M. B. (2012). Problems of adaptation in Russia of the world experience of implementation of result-oriented budgeting. Bulletin of the Moscow university. Ser. 6. *Economics, 1*, 47–56.

Tumbas, P., Sakai, M., Pavlicevic, V., & Rakovic, L. (2019, March). Digital competencies in business informatics curriculum innovation. In *Proceedings of the 13th International Technology, Education and Development Conference (INTED)*, Valencia, Spain (pp. 11–13).

Vorobiov A. E., & Murzaeva A. K. (2017). Technique of the assessment of efficiency of activity of chair. World of science. *Pedagogy and Psychology, 5*(5). Retrieved from https://mir-nauki.com/PDF/37PDMN517.pdf.

Digital Business Models in the Manufacturing Sector

Vitaly A. Mordovets, Yulia V. Meleshko, and Olga Dmitrovna Ugolnikova

Introduction

Digital transformation is an inevitable process for any enterprise wishing to adapt to the new conditions of the digital economy. Digital technologies of the fourth industrial revolution have already found wide application in banking, retail, telecoms, logistics, public administration, education and health care. "Most researchers see digital industrialization as a process that affects progress (change) in five areas: engineering and technology, production, management, institutions and society", noted earlier (Makarova et al., 2021, pp. 151–152). However, the potential for using new technologies in industrial production has not been fully disclosed. The digital transformation of an industry requires more time, more resources, and a fundamental reform of the company's culture. At the same time, the growing demands for the individualization of goods, for its service, as well as the complication of the production of high-tech products oblige manufacturers to develop and introduce new technologies on an ongoing basis.

Despite the high rates of digital technologies development and an increase in their importance in the competitiveness of the industry, to this day there is no consistent methodological approach to determining the digitalization of industry, the breadth of its scope has not been determined, all possible consequences for business models

V. A. Mordovets (✉)
St. Petersburg University of Management Technologies and Economics, St. Petersburg, Russian Federation

Y. V. Meleshko
Belarusian National Technical University, Minsk, Republic of Belarus

O. D. Ugolnikova
St. Petersburg State Economic University, St. Petersburg, Russian Federation

A. Rumyantseva et al. (eds.), *Challenges and Solutions in the Digital Economy and Finance*, Springer Proceedings in Business and Economics,
https://doi.org/10.1007/978-3-031-14410-3_18

have not been analyzed, which makes it difficult to develop a competitive strategy industrial enterprise.

The authors of this article proceed from the assumption that the digital transformation of an industrial enterprise goes beyond the automation of individual business processes (for example, production, logistics or service). As we already wrote "The development of digital production and digital services entails the digitalization of business models. The result is flexible, adaptive, individualized industrial production through the triad of digital production, digital services and digital business models" (Mialeshka, 2021, p. 358). The main advantage of the technologies of the fourth industrial revolution is the creation of new business models that change the traditional perception of interaction with customers and the creation of customer value proposition.

The scientific idea of the authors is to reject the need to develop digital technologies as an end in itself for industrial enterprises. The authors proceed from the fact that a balanced approach to the selection of new technologies is necessary. Often, the use of expensive technology becomes a tribute to fashion and the main positive result of its implementation is to improve the image of the enterprise. This effect is amplified in conditions of technological uncertainty arising during the next industrial revolution.

Most modern research on the problems of digital transformation of an industrial enterprise is limited to studying the obvious consequences of digitalization (operation optimization, employee productivity, safety and security, supply chain management etc.). The digitalization of separate business processes is also studied in some detail, including considering specifics of different industries (mining, machine manufacturing, energy, agricultural, and others). At the same time, a systematic understanding of the impact of digitalization on the activities of an industrial enterprise, linking digital production, digital services and digital business models, has not yet been developed; there is no unified approach to the meaningful description of digital business models of industrial enterprises; new principles of behavior of industrial enterprises in the digital network economy have not been developed.

The purpose of this work is to identify the main directions of business models transformation of industrial enterprises influenced by digital technologies and based on this, to determine ways to improve the competitive strategy of an industrial enterprise in a networked digital economy.

Materials and Methods

The study is based on the scientific works devoted to the problems of digital transformation of the economy, primarily in the manufacturing sector. The methodological framework of the research is represented by the theoretical provisions of post-market and neo-industrial concepts, Industry 4.0, risk economics and network economy, technological uncertainty. General scientific methods of analysis and synthesis, philosophy-economic and institutional approaches were used.

Results

Koch and Windsperger (2017) rightly note that "there are more implications of digital technology then mere increases in efficiency. Changes concern not only enhancements in production technology but also challenge established business models, environmental conceptions, and sources of competitive advantage" (p. 25). Although recently in various areas of research, increased attention has been paid to the organizational consequences of digitalization of production and services, there is still no systematic study of the consequences of the impact of digital technologies on the business models of an industrial enterprise and its competitive strategy.

If earlier "IT strategy has been separated from business strategy since it has been considered to be a functional strategy aligned with and subordinated to a chosen business strategy" (Koch & Windsperger, 2017, p. 24), then in Industry 4.0 the IT strategy is integrated into the business strategy. It is reasonable to say that the competitive strategy of an industrial enterprise lies in a digital business strategy aimed at deliberately using the unique capabilities of digital technologies embedded in industrial production to gain a competitive advantage.

Many researchers consider the digitalization of business models as the final stage of the digital transformation of an enterprise, following the "creation of a digital layer", that is, the technical digitization of production elements and processes. In particular, Aagaard et al. (2019), Ross et al. (2017), Kraus et al. (2018), Ushakova et al. (2021) agree that the first step of digital transformation is carried out at the operational level and includes the standardization of business processes and optimization of operations through the implementation of technologies and software, and the second uses purely digital technologies to formulate, target and personalize alternative proposals with the purpose of forming a new value proposition.

We must agree with Johnson et al. (2008), who argues, that a digital platform or digital solution can open up a new era of transformational growth, but when you look under the hood of a company and see what really drives it, its business model is the engine of transformation. Successful business models are distinguished by an exceptionally strong consumer value proposition and a stable, scalable system in which all elements are easily combined, complementing each other. The creation of industrial products that take into account the individual preferences of consumers (consumer value propositions) is based on the use of data obtained from the Industrial Internet of things (sensors with which industrial products are equipped) and directly by the end consumers themselves (through feedback, in primarily through the Internet, for example, through online stores).

"The availability of bulk data allows development of services, which were not possible so far, like navigation systems with user-driven traffic information. Availability of bulk data allows a variety of new business models. In combination with third party services as weather, calendar, payment services, geolocation or historical data, new levels of organization and scheduling are possible. The possibilities are endless"—Drath and Horch (2014, p. 58) are optimistic. The limits of the development of business models based on data seems to be the topic of a separate study, in connection with which, within the framework of this work, we will not delve into the study of the indicated problems, we will only note the controversy of the thesis about its limitlessness. However, Drath and Horch are right about the fact that it is thanks to big data that a large number of new business models and ways to ensure the competitiveness of an enterprise appear.

Brettel et al. (2014) write: "For many manufacturing companies, value added services provide an appropriate opportunity to differentiate themselves in addition to high product quality in order to ensure a strong competitive position. On top, long term service contracts can help to leverage risks of high demand volatility, as the actual product serves as a platform for further service sales over the time of utilization <. . .>. Embedded Systems of smart products and machines will enable entirely new remote maintenance concepts <. . .>" (p. 40). In this context, services and applications (digital services) should be viewed not as a functional component of industrial production, but as a new way of creating value proposition for an industrial enterprise and a source of its competitive advantages. In this regard, the demand for business service providers from among developers with knowledge of the subject area for operational development of applications is already growing in the labor market.

The described patterns of development of digital business models only partially reflect the changing approaches to ensuring the competitiveness of the industrial complex, since the transition to a customer-oriented model (implemented, among other things, through digitalization) is already a consequence (external manifestation). In order to reveal the mechanism of the impact of business models digitalization on the competitiveness of industrial enterprises, it is necessary to turn to the organizational and managerial features of Industry 4.0.

The widespread and ever deeper integration of digital technologies into the enterprise economic activities leads to the formation of a digital ecosystem characterized by a high degree of dynamism. The turbulence of the digital ecosystem is predetermined by several reasons. First, the specificity of digital technologies themselves, which are developing at a very high rate. New technologies constantly emerging are closely intertwined with each other and with those already in use, thereby constantly changing the digital infrastructure of production, services and business models. Industrial enterprises objectively (due to technological uncertainty) do not have sufficient information to give preference to one technology or another, which makes it difficult to develop a digital transformation strategy. At the same time, in an effort to gain the advantages of a technology leader, enterprises (and countries) are instigating a technology race.

The situation is further complicated by the fact that "against the background of the next change in technical and technological structures and the development of global speculative finance, the role of social and functional innovations is increasing, which are used to coerce actors into a given behavior through the targeted presentation of information on the Internet and traditional media" (Solodovnikov, 2020, p. 22). To promote products, enterprises are increasingly using not traditional marketing strategies, but social and functional innovations that allow "using information concentrates of various designs" to "effectively redistribute (redistribute) material resources (change ownership relations) without damaging the property itself" (Solodovnikov, 2018, p. 40). These post-market methods of competition are proving to be extremely effective, further exacerbating technological uncertainty.

Secondly, the reason for the high degree of uncertainty in the digital ecosystem also lies in the network nature of the interaction of participants in the production process, which prevails in modern digital industrial production. The need for interaction between different-level economic entities (the state, large and small enterprises) in Industry 4.0 is predetermined by the fact that the competencies and capacities of individual entities are not enough to create new products in a short time in high-tech, science-intensive and customer-oriented industrial production.

The digital platform, first of all, is designed to solve technical problems of coordinating various production systems and processes, thereby removing barriers to interaction between independent business entities. By lowering operating costs and improving communication and engagement with all stakeholders in the ecosystem, including customers, partners, suppliers and distributors, digital technology opens up new opportunities in terms of skills exchange, open innovation, co-creation and partnerships between companies (Coviello et al., 2017). Thus, having solved the problem of technical "alignment", the digital platform opens up opportunities to unite, recombine and complement the resources of various organizations, for example, production facilities, labor resources, personnel competencies and qualifications, intellectual property, and finance at the inter-company and cross-sectoral levels. The functionality of each individual firm increases with the deepening of its specialization in its core competencies.

The specifics of creating use value in digital networks (generative value created by several enterprises, which has a pronounced service dominant, inter-firm and inter-industry character) predetermines dynamic interaction between participants in the production process. The combination and recombination of the resources of various enterprises, including those related to various industries and types of activity, allows creating new unique customer values, but assumes constantly renewing interfirm and intersectoral manufacturing ties. Coviello et al. (2017) characterize the markets emerging around digital platforms as current, instant and transaction-specific. Manufacturing ties between network actors are established to solve a specific problem, that is, situationally, and then disintegrate. Industry boundaries are blurring, relatively stable industry structures are disappearing, and businesses are abandoning linear value chains in favor of dynamic networks.

In these conditions, value creation occurs in unforeseen ways and competition for the same resources that can be used to produce various products (and not only manufacturing ones) increases. Along with it, competition in the finished industrial product market is intensifying, becoming more and more individualized through industrial services and digital business models. The digital ecosystem is becoming a hyper-competitive environment where industrial enterprises compete for digital technologies, the resources of their counterparties and the end consumer. As a result, industries never reach equilibrium (Smith et al., 2001).

Koch and Windsperger (2017) came to the logical conclusion that "the higher the degree of digitization, the more the firm creates competitive advantage through value co-creation of the interconnected firms" (p. 23). At the same time Selander et al. (2013) is absolutely right, arguing that participation in ecosystems is not an option for individual firms, but a necessity. Thus, digital business models, being a prerequisite for ensuring the competitiveness of industrial enterprises in Industry 4.0, involve enterprises in a dynamic, turbulent and highly competitive digital ecosystem, which in turn requires a change in the enterprise's approaches to ensuring competitiveness.

Discussion

In a dynamic, turbulent and highly competitive digital ecosystem, traditional sources of sustained competitive advantage, implying a relative stability of the environment, are described, for example, by Poter (2014) or Barney (1995) are losing their relevance. More and more scientists reject the concept of sustained competitive advantages in favor of temporary advantages due to the impossibility of preserving them in the conditions of globalization and hypercompetition (D'Aveni et al., 2010; Tanriverdi et al., 2010). "At present, in economic science, the study of cooperation between economic entities as the basis for increasing their competitiveness is becoming increasingly important. At the same time, the emphasis is on the fact that the Smithian views on the market situation as a field of competition between all and all are hopelessly outdated today", – emphasizes Solodovnikov and Ivanova (2016, p. 188).

Within the framework of traditional theories of creating competitive advantage, the strategy of a firm in a changing environment can be aimed either at trying to predict future changes better than competitors, or as quickly as possible to adapt to changing circumstances. However, none of these strategic models can be fully implemented in the digital ecosystem, since the probability of predicting in conditions of technological uncertainty and network interaction tends to zero, and the pace of change is increasing and enterprises do not have time to adapt.

Based on the advantages of the digital platform, Eisenhardt and Martin (2000) formulate the concept of "dynamic capabilities", assuming that a permanent unstable state contributes to increasing the flexibility and adaptability of industrial enterprises and thereby becomes a source of sustained competitive advantage. Koch and Windsperger single out "interorganizational network structure" as "the only remaining relevant source of sustained competitive advantage" (p. 18), which provides enterprises with dynamic capabilities. "Firms are embedded in dynamic and turbulent ecosystems characterized by high degrees of uncertainty, value is co-created with several other firms and strategy-making is based on actively shaping the environment. In the digital environment firms participate in intertwined networks—the one remaining constant under these circumstances—to create value and compete"—consider these authors (Eisenhardt & Martin, 2000, p. 20). Miles et al.

(2009) drawing on the concept of dynamic capabilities, concludes that in order to support large vertically integrated firms, organizations create multi-firm networks and community-based structures to rely on a strategy of persistent exploration of an expanding set of complementary markets whose participants continuously adapt technologies to new uses (p. 65).

While agreeing that such "dynamic capabilities" can provide an enterprise with additional competitive advantages, we note that this concept should not be taken as an absolute. In this case, the principle that we stated earlier with regard to popular theory of the ratio of production factors by Heckscher and Ohlin: "we are not talking about the fact that all these intellectual constructions are theoretically inconsistent and do not have a certain epistemological potential, but that they are very abstract and, in their practical application, require serious real-ontological concretization" (Solodovnikov et al., 2021, pp. 020006–1).

Firstly, as rightly pointed out by D'Aveni, Dagnino and Smith, there is apparently a lack of evidence that this applies to different contexts and time periods (2010, p. 1372). Secondly, the situational nature of production relations has a downside. Hervé et al. emphasize: "This makes it more difficult to conclude long-term relationships with actors integrated at the time into any network in question. Because digital technologies are increasing the number of instantaneous, brief and interrelated interactions, the pace of these encounters is also accelerated" (Hervé et al., 2020, pp. 33–34). It is assumed that the enterprise must be in constant search for partners, constantly changing and adapting to new market requirements, in such conditions time gaps between projects are inevitable. During this period, the enterprise will be able to continue functioning at the expense of accumulated reserves (or attraction of borrowed funds), which, as a rule, are small (or absent) for SME. If the company fails to find new partners quickly enough, it will be forced to leave the market. In an effort to increase their competitiveness, SME can acquire a narrow specialization, which in the future, however, will complicate the search for new partners, the more they fill a narrower niche.

Thus, a dynamic production network becomes a source of sustained competitive advantage only for enterprises that are able to constantly change and experience periods of "inactivity", and therefore have sufficient resources for this. Most small and medium-sized enterprises will not be able to meet such criteria, and therefore the problem of economic stability will be especially acute for them. In general, this does not reduce the effectiveness of the functioning of the most dynamic production network, which is open to new participants coming to replace it.

In a dynamic production network built on the basis of the platform, consisting of many partners, the problem of developing a common concept for creating a new value proposition inevitably arises. Baldassarre et al. (2017) comment that conceptualizing a strong value proposition becomes even more complex as it requires understanding and managing multiple needs and goals in a multi-stakeholder network in order to create shared value as a result.

Another theoretical response to the challenges of a dynamic digital ecosystem was the concept of environmental impact through its networks (Fjeldstad et al., 2012; Snow, 2015; Baldwin & von Hippel, 2011; Bøllingtoft et al., 2012;

Chesbrough, 2003). Read et al. (2009) write that "to the extent we can predict the future, we can control it" (p. 2), therefore, under the influence of a network of partners, the environment becomes endogenous. "Thus, firms operating in digital ecosystems make strategic decisions and choose competitive actions by effectuating their environment"—Koch and Windsperger explain (2017, p. 11). According to these authors, "Firms are capable to efficiently effectuate ecosystems by attracting heterogeneous actors to co-create value on their digital platforms and controlling these platforms through technical boundary resources (such as APIs and SDKs)" (Koch & Windsperger, 2017, pp. 21–22).

In accordance with the concept of environmental impact, an enterprise functions simultaneously in several interconnected dynamic network that predetermine the directions and methods of its economic activity, and after that the competitive strategy of the enterprise. At the same time, global economic and technological uncertainty complicates not only the choice of technologies and the range of products that enterprises intend to produce, but also the choice of networks and partners. In this part, the theories described above need to be improved. Solving the problem of the uncertainty of the environment by actively influencing it through the formation of networks, they do not answer the question, but in accordance with what principle such networks are created.

A more promising approach is experimental. The development of digital technologies (primarily virtual modeling and forecasting), and network forms of interaction that provide an opportunity to quickly find the right partner, contribute to the implementation of an experimental approach. This approach provides for a series of experiments (which are faster and cheaper) before the start of the implementation of any project, which make it possible to understand the prospects of this project and put into practice already proven solutions. The transition to this kind of strategic planning, based on constant experimental activity and decision-making based on the results of the experiments, requires a change in the logic of management thinking and the development of a corporate culture of "experiment".

Conclusion

In Industry 4.0, an enterprise's competitive strategy is a digital business strategy that aims to deliberately exploit the unique capabilities of digital technologies embedded in industrial production and products to gain a competitive advantage. Big data creates new services and business models based on data, which are characterized by a high degree of customer focus and individualization. Services and applications (digital services) become a new way to create value for an industrial enterprise and a source of its competitive advantages.

At the same time, digital business models, being a prerequisite for sustained competitive advantages in Industry 4.0, involve manufacturing enterprises in a dynamic, turbulent and ultra-competitive digital ecosystem. This, in turn, requires a change in the company's approaches to competitiveness. The reasons for these specific properties of the digital ecosystem lie, firstly, in the global technological uncertainty and high rates of development of digital technologies themselves, and secondly, in the network nature of the interaction of participants in the production. As a result, manufacturing enterprises face not only the need for rapid product development and flexible manufacturing, but also the need to operate in a complex digital ecosystem.

The relevance of traditional sources of competitive advantage (combination of resources, enterprise positioning, etc.) in the digital network economy is decreasing. Some scientists consider the digital ecosystem as a source of dynamic capabilities, which are not sufficient for sustained competitive advantage of manufacturing enterprises. In conditions when customer values are the result of cooperation between several economic entities, and it is no longer possible to predict development or adapt to an uncontrolled and destructively changing environment, manufacturing enterprises can integrate or create their own digital intersectoral network, thereby gaining the ability to control or influence evolution of business environment. This causes the transition from relatively static industries to dynamic digital network, and from the point of view of a strategy for competitive advantages—to the active formation of the environment. At the same time, global economic and technological uncertainty complicates not only the choice of technologies and the range of products that enterprises intend to produce, but also the choice of networks and partners. A promising approach to solving this problem for manufacturing enterprises can be strategic planning based on constant experimental activity and decision-making based on the results of the experiments.

References

Aagaard, A., Aagaard, A., & Harrison. (2019). *Digital business models*. Springer.

Baldassarre, B., Calabretta, G., Bocken, N. M. P., & Jaskiewicz, T. (2017). Bridging sustainable business model innovation and user-driven innovation: A process for sustainable value proposition design. *Journal of Cleaner Production, 147*, 175–186. https://doi.org/10.1016/j.jclepro.2017.01.081

Baldwin, C., & von Hippel, E. (2011). Modeling a paradigm shift: From producer innovation to user and open collaborative innovation. *Organization Science, 22*(6), 1399–1417. https://doi.org/10.1287/orsc.1100.0618

Barney, J. B. (1995). Looking inside for competitive advantage. *Academy of Management Perspectives, 9*(4), 49–61.

Bøllingtoft, A., Donaldson, L., Huber, G. P., Håkonsson, D. D., & Snow, C. C. (2012). *Collaborative communities of firms: Purpose, process, and design*. Springer.

Brettel, M., Friederichsen, N., Keller, M., & Rosenberg, M. (2014). How virtualization, decentralization and network building change the manufacturing landscape: An industry 4.0 perspective. *International Journal of Information and Communication Engineering, 8*(1), 37–44. https://doi.org/10.5281/zenodo.1336426

Chesbrough, H. W. (2003). *Open innovation: The new imperative for creating and profiting from technology*. Harvard Business Press.

Coviello, N., Kano, L., & Liesch, P. W. (2017). Adapting the Uppsala model to a modern world: Macro-context and microfoundations. *Journal of International Business Studies, 48*(9), 1151–1164. https://doi.org/10.1057/s41267-017-0120-x

D'Aveni, R. A., Dagnino, G. B., & Smith, K. G. (2010). The age of temporary advantage. *Strategic Management Journal, 31*(13), 1371–1385. https://doi.org/10.1002/smj.897

Drath, R., & Horch, A. (2014). Industrie 4.0: Hit or hype? [industry forum]. *IEEE Industrial Electronics Magazine, 8*(2), 56–58. https://doi.org/10.1109/MIE.2014.2312079

Eisenhardt, K. M., & Martin, J. (2000). Dynamic capabilities: What are they? *Strategic Management Journal, 21*(10/11), 1105–1121. https://doi.org/10.1002/1097-0266(200010/11)21:10/113.0.CO;2-E

Fjeldstad, Ø. D., Snow, C. C., Miles, R. E., & Lettl, C. (2012). The architecture of collaboration. *Strategic Management Journal, 33*(6), 734–750.

Hervé, A., Schmitt, C., & Baldegger, R. (2020). Internationalization and digitalization: Applying digital technologies to the internationalization process of small and medium-sized enterprises. *Technology Innovation Management Review, 10*(7), 28–40.

Johnson, M. W., Christensen, C. M., & Kagermann, H. (2008). Reinventing your business model. *Harvard Business Review, 86*(12), 57–68.

Koch, T., & Windsperger, J. (2017). Seeing through the network: Competitive advantage in the digital economy. *Journal of Organization Design, 6*(1), 1–30. https://doi.org/10.1186/s41469-017-0016-z

Kraus, S., Palmer, C., Kailer, N., Kallinger, F. L., & Spitzer, J. (2018). Digital entrepreneurship: A research agenda on new business models for the twenty-first century. *International Journal of Entrepreneurial Behavior & Research, 25*(2), 353–375. https://doi.org/10.1108/IJEBR-06-2018-0425

Makarova, I. V., Lepesh, G. V., Ugolnikova, O. D., & Meleshko, J. V. (2021). Analysis of directive and policy documents on digital industrialization of the Russian Federation and the Republic of Belarus. *Voprosy gosudarstvennogo i municipal'nogo upravlenija, 1*, 150–172. (in Russian).

Mialeshka, Y. V. (2021). New industrialization and trends in modernization of Belarusian industry. *Nauka i tehnika (Science and Technique), 20*(4), 357–364. https://doi.org/10.21122/2227-1031-2021-20-4-357-364. (in Russian).

Miles, R. E., Miles, G., Snow, C. C., Blomqvist, K., & Rocha, H. (2009). The I-form organization. *California Management Review, 51*(4), 61–76.

Poter, M. E. (2014). *Competitive strategy: Techniques for analyzing industries and competitors*. Free Press.

Read, S., Dew, N., Sarasvathy, S. D., Song, M., & Wiltbank, R. (2009). Marketing under uncertainty: The logic of an effectual approach. *Journal of Marketing, 73*(3), 1–18. https://doi.org/10.1509/jmkg.73.3.001

Ross, J., Beath, C., & Sebastian, I. (2017). *Digitized ≠ digital* (Vol. 17). MIT Center for Information Systems Research Research Brief.

Selander, L., Henfridsson, O., & Svahn, F. (2013). Capability search and redeem across digital ecosystems. *Journal of Information Technology, 28*(3), 183–197. https://doi.org/10.1057/jit.2013.14

Smith, K. G., Ferrier, W. J., & Ndofor, H. (2001). Competitive dynamics research: Critique and future directions. *Handbook of strategic management, 315*, 361.

Snow, C. C. (2015). Organizing in the age of competition, cooperation, and collaboration. *Journal of Leadership & Organizational Studies, 22*(4), 433–442. https://doi.org/10.1177/1548051815585852

Solodovnikov, S. Y. (2018). Risk economy. *Jekonomicheskaja nauka segodnja (Economic Science Today), 8*, 16–55. https://doi.org/10.21122/2309-6667-2018-8-16-55. (in Russian).

Solodovnikov, S. Y. (2020). Planning engineer training in economic security in the mineral mining sector: Theory and practice. *Gornyj zhurnal (Mining Journal), 11*(2280), 20–25. https://doi.org/10.17580/gzh.2020.11.01. (in Russian).

Solodovnikov, S. Y., & Ivanova, T. (2016). Impact of the expansion of the customs union on the economic competitiveness of Belarus and Armenia. *Jekonomicheskaja nauka segodnja (Economic Science Today), 4*, 188–195. https://doi.org/10.21122/2309-6667-2016-4-188-195. (in Russian).

Solodovnikov, S. Y., Serhiyevich, T. V., & Meleshko, Y. V. (2021, September). The relative advantages of modern cross-country technological cooperation in the context of the transition to industry 4.0. *AIP Conference Proceedings, 2389*(1), 020006. https://doi.org/10.1063/5.0064421

Tanriverdi, H., Rai, A., & Venkatraman, V. (2010). Research commentary – Reframing the dominant quests of information systems strategy research for complex adaptive business systems. *Information Systems Research, 21*(4), 822–834. https://doi.org/10.1287/isre.1100.0317

Ushakova, E., Voronina, E., Fugalevich, E., & Mikhaylova, M. (2021). Digitalization of the economy as a development trend in Russia. *Jekonomika i upravlenie (Economics and Management), 27*(3), 175–182. https://doi.org/10.35854/1998-1627-2021-3-175-182. (in Russian).

Increasing of the Technological Cooperation Efficiency in the Context of Digitalization

Elena V. Ushakova, Sergey Yu Solodovnikov, and Tatsiana V. Serhiyevich

Introduction

The problem of technological cooperation is the subject of research by many scientists. The research on the cooperation in the commercialization of innovations is considered in the papers of Aarikka-Stenroos et al. (2014), Abdul Hamid and Abd Rahman (2014), Kim et al. (2022). Many scientists have achieved success in international technology transfer research (Soares et al., 2020; Palaco et al., 2022). Many scientists are studying the international and interfirm cooperation in new conditions. Impact of the digitalization on the cooperation and innovation is revealed by Alcácer et al. (2016), Amit and Han (2017), Gault (2019), Bessonova and Battalov (2021). We also explored the problems of technological cooperation in the new conditions. In previous studies we have identified the factors and risks that arise by the expansion of international technological cooperation. We revealed the factors of international technological cooperation expanding: change in the dynamics of foreign trade as a result of consumption growth in China and other developing countries of produced by them goods as a result of increased interior demand; trend of declining share of intermediate goods and services in international trade; the growing influence of new and emerging technologies on world trade; the necessity of transition to Industry 4.0; an unprecedented degree of concentration in several countries of the most important raw materials required for the current industrialization stage; unprecedented rise of global social inequality (Solodovnikov, 2021,

E. V. Ushakova (✉)
St. Petersburg University of Management Technologies and Economics, St. Petersburg, Russian Federation

S. Y. Solodovnikov · T. V. Serhiyevich
Belarusian National Technical University, Minsk, Republic of Belarus
e-mail: serhiyevich@bntu.by

A. Rumyantseva et al. (eds.), *Challenges and Solutions in the Digital Economy and Finance*, Springer Proceedings in Business and Economics,
https://doi.org/10.1007/978-3-031-14410-3_19

p. 124). We also revealed the risks arising from the international technological cooperation expansion in current conditions: the emergence of a new phenomenon—the economy of risks; institutional international, civilizational and cultural differences; increased economic diversity, accompanied by the complication and exacerbation of political and economic relations; overestimated needs expectations of individuals; a decrease in the trust level in society due to a decline of social capital at the society and an increase in interclass and intraclass contradictions; weakening of the state role in the management of social, economic, technical and technological processes in society; the complexity of adaptation of institutions of partnership society-state-business in the context of quick digitalization. These risks largely determine the current forms, mechanisms, and tools of interfirm technological cooperation. At the same time, the dominant factor in this process is the consistent development of a digital society. It is revolutionizing not only the economic system of society (in which the digital economy is developing of outstripping rates), but also almost all socio-cultural relations. In connection with the emergence of new factors and risks of international and interfirm technological cooperation in the context of the digitalization of the economy, the features and effectiveness of technological cooperation are changing. It requires research of new tools for improving of the technological cooperation efficiency in the context of digitalization, which is the purpose of this paper.

Materials and Methods

The subject of the study is the technological cooperation efficiency in the context of digitalization. The purpose of this paper is the design of new technological cooperation efficiency increasing tools in the context of digitalization. The authors, being adherents of the scientific school in the field of studying the modernization of the economy, are united by the methodology of the subject-activity approach. Research works of domestic and foreign scientists-economists, scientific reports, data of the research reports carried out by the authors were used to write this article. The general scientific methods such as system approach, unity of historical and logical, institutional approach were used during the research. The institutional approach let us consider the dialectical contradiction between traditional institutions ensuring the interaction of the object under study and its traditional external environment and the need for the evolution of these institutions in order to adapt to new conditions.

Results

The most gnosiologically useful and methodology correct scientific publications devoted to the phenomenon of digitalization is the article by T.N. Yudina «"Surveillance capitalism" as "digital econom"» and/or "digital society"» (Yudina, 2018).

Let us immediately make a clause. We have chosen this article not because it is the only philosophical and economic work of T.N. Yudina on this topic. On the contrary, we have chosen this article, since this work, first of all, is a quintessence of a whole series of science-intensive dedicated to digitalization publications by T.N. Yudina (Yudina, 2016, 2017, 2019; Geliskhanov et al., 2018; Yudina & Kupchishina, 2019; Yudina & Balashov, 2020). Secondly, this article is the most methodologically suitable for the study of the increasing of technological cooperation efficiency in the context of digitalization.

T. N. Yudina could show the general and specific in the development of modern economics. In essence, this author could develop the concept of «surveillance capitalism» introduced by S. Zuboff (2015) and describe it as «the very essence of the "digital economy" and/or "digital society"» (Yudina, 2018, p. 14). At the same time in our opinion T.N. Yudina discovered the phenomenological nature of modern society and/or the digital economy, including revealing its essence as a noumenon. The noumenon is a philosophical concept that was introduced in Neoplatonism for denoting the world of intelligible entities. I. Kant, proceeding from his transcendental idealism, used the concept of noumenon to designate a sphere that transcends sensory experience. The traditional relationship of essence and phenomenon is replaced by Kant's rigid opposition of the sphere of everyday facts and scientific experience (phenomenon) to the unknowable world of "thing-in-itself". Today a phenomenon is understood as a set of properties of a system and/or a phenomenon that is a result its individual development. According to this methodological approach, the phenomenon is a combination of the general (development principle) and the special (unique).

Let us consider from this point of view the statement of T.N. Yudina: «However, many compatriots, including scientists, don't see the phenomenon of the "digital economy" in general, taking it for a noumenon. The phenomenon of "CE" *(digital economy—authors' note)* contains leaps towards the quantitative development of the economy and, at the same time, essential threats affecting individuals and society as a whole» (Yudina, 2018). If we consider the noumenon in the neo-platonic sense (since it is unlikely that T.N. Yudina believes that many of her compatriots, including scientists, adhere to the methodological and/or ideological positions of Kantian transcendental idealism), then the named noumenon (as a principle of development) doesn't contradict the phenomenological approach, in which all real objects are considered as a combination of general (development principle) and special (unique). The named noumenon is a part of phenomenological approach. At the same time, this methodological clarification doesn't refute the fact that many scientists considering the phenomenon of the digital society and/or the digital economy exaggerate the importance of the external similarity of many economic processes before and after the digital revolution. It is in the content analysis and synthesis of the latter that the scientific contribution of this Russian scientist lies. The scientific contribution of the Russian scientist T.N. Yudina lies precisely in the meaningful analysis and synthesis of these economic processes.

The beginning of the twenty-first century is characterized by the emergence of a new type of economic rent, namely digital rent. Digital rent is appropriated primarily by Big Tech—Alphabet, Amazon, Apple, Meta as well as Microsoft, «whom belong a new type of capital—big data—in fact, a source of digital rent» (Yudina, 2018, p. 14). But big data isn't the only source of digital rent in current economy. K. Birch and D.T. Cochrane say that «new forms of specifically digital rentiership are emerging as the result of Big Tech. The ecosystems that Big Tech create and control represent a heterogenous assemblage of technical devices and platforms, as well as users and developers, legal contracts and rights, collective standards, etc. These digital ecosystems enable Big Tech to make economic rents in new ways that reinforce their techno-economic power, while undermining the political, social, and economic capacity of others to shape the future» (Birch & Cochrane, 2021, p. 11). T.N. Yudina rightly notes that «the analog, non-digital economy represented and represents real relations between people in the process of production, distribution, exchange and consumption of real goods and services and institutions. The "digital economy" leads to a virtual artificial world, it is "watching capitalism" or "surveillance capitalism" with its institutions and organizations» (Yudina, 2018, pp. 14–15). Considering the problem of interaction between the virtual and analogue economies in the context of design of new technological cooperation efficiency increasing tools in the context of digitalization, it is necessary to understand that the analogue economy provides all the material goods consumed by households. No matter how we develop the digital economy, a person can never do without food, water, clothing and many things in the material world.

In the same way, the digital transformation of industry, initially focusing on the maximum possible digitalization of the industrial complex (according to the criteria of economic feasibility and/or environmental safety and/or humanization and safety of production processes), has generated a new digital reality—a parallel virtual display of production processes. This virtual display allows not only to improve the processes of direct production, including through the widespread use of robots (material and virtual), but also to develop industrial services very quickly. It led to a new industrial revolution and significantly changed the business models in industry. «A firm's business model defines how it delivers (supply model) and profits from (revenue model) the customer value it creates through the effective satisfaction of needs» (Cachon, 2020, p. 15). In the context of digitalization, fundamentally new revenue models are emerging, implying, for example, "payment" for a service by the user with access to his personal data (social networks, video hosting). At the same time with these processes, the accumulation and capitalization of BIG DATA is actively accelerating and, accordingly, the increase in digital rent. All this is changing the main tools for increasing of the technological cooperation efficiency in the context of digitalization, when traditional tools are not only supplemented, but also largely replaced by IT tools, including through the formation of digital platforms business. At the same time, in order to provide a competitive national industrial complex (as a basis for technological cooperation, too) it is necessary to maintenance and develop non-digital (analog) modern technologies in industry.

As a result of the development of the digitalization processes investors became an alternative by investing in digital technologies in industry: to invest in the development of the industry itself and industrial services or in the virtual shell of a digitized industry (expecting obtaining more revenue through digital rent or for doing business for information security). While «the state policy of the analyzed countries *(Russia and Belarus—author's note)* in the field of digital industrialization <...> is characterized by the continuing self-determination of states in the new economic conditions» (Makarova et al., 2021, p. 164), it seems promising for enterprises of the Republic of Belarus and the Russian Federation to develop technological cooperation not only in the direction of product cooperation and industrial services, but in the direction of accumulation and capitalization of big data generated by the ongoing digitalization of national industrial complexes in order to obtain digital rent. The feasibility of such a strategy is confirmed by the technical and economic nature of the new industrial production based on «cyber-physical systems, which involve the digitization of equipment and industrial products, the creation of their "digital twins" and control of production processes in virtual space in real time using algorithms for automated decision-making» (Mialeshka, 2021). The transition to a new type of industrial production under the influence of digitalization is impossible without robotization proceeding.

The necessity of the robotization of the national industrial complex of the Republic of Belarus and the Russian Federation is today supported by many economists. Current robotization is one of the most promising areas of technical and technological modernization of the national economy, leading to the transformation of social and labor relations. The expansion of the production and use of robots in the economies of the Republic of Belarus and the Russian Federation as a component of modernization will make it possible to make a technological breakthrough, which will entail changes in value chains, in producer-consumer relations, and an increase in the competitiveness of the domestic economy. It should be noted that the potential for a rapid increase in the use of disembodied (intangible) robots in the financial, trade, commercial and law spheres of the economies of the Republic of Belarus and the Russian Federation has largely been exhausted. Accordingly, public and private business begins to move more actively towards the development and use of material (including industrial) robots. All these processes will affect the change in labor relations and the labor market (Bogatyreva et al., 2021, p. 9). The decline of the number of employees engaged in medium and low-skilled mental work in the spheres of the national economy that have undergone accelerated robotization will inevitably have an impact on the labor market and labor relations. The impact of robotization on employment will appear itself in fundamental changes in its structure. At the macro and micro levels, the transformation of the employment structure will be accompanied by social instability and changes in income distribution.

New industrialization in the context of digitalization requires finding long and cheap money both to finance the domestic industry and to develop international technological cooperation. The most important factor in the expansion of international current technological cooperation is «the strengthening of international political-economic competition. The competition is manifested in the increasing frequency of trade conflicts, increased sanctions pressure on individual producers and entire states, and the introduction of new measures of tariff and non-tariff foreign trade regulation» (Solodovnikov et al., 2021). And even if these instruments don't affect directly belarusian production or export of goods and services, but «the redistribution of markets as a result of the strengthening protectionist measures made by individual countries creates risks of displacement Belarusian products» (Solodovnikov & Sergievich, 2020, p. 67). In addition, as the practice of recent years has shown, restrictive and sanctioning instruments of political and economic competition are often used unilaterally by countries with large economies, while «countries with a small open export-oriented economy, such as Belarus, are usually very limited in the free use of protectionist measures of international trade» (Serhiyevich, 2020, p. 140). That is why the task of strengthening interfirm technological cooperation is being actualized, first of all, in the Union State of Belarus and Russia. The Republic of Belarus and the Russian Federation must have strong state-owned banks. This will provide long and cheap money for financing the development of industry and increasing the technological cooperation efficiency. For example, Romanian scientists analyze the lessons of industrial policy in their country in recent years: «Learning from the experience of the past 27 years, which shows that we cannot expect capital finance and the funding of local industry as long as the Romanian state-owned banks currently hold only 5% of the total banking system assets» (Chivu et al., 2017, p. 168). They also add, that «creating a Romanian-capital bank and/or a Sovereign Investment Fund, for the economic development of Romania, a step that is not only imperative, but it is also a matter of common sense» (Chivu et al., 2017, p. 168). The strong state-owned banks focused on supporting national industry are necessary because of that «digitalization has made the financial market global» (Vardomatskya et al., 2021, p. 6). The globalization of financial markets increases the risks of the technological cooperation efficiency. Thus, the creation of a financing mechanism for industrial modernization is a tool for increasing technological cooperation in the context of digitalization. Thus, the creation of a financing mechanism for industrial modernization is another tool for the technological cooperation efficiency increasing in the context of digitalization. With the development of the digital economy and the emergence of economy of risks, the costs of ensuring the safety of its operation increase. Recall that «external threats, challenges to the country's economic security are often implemented through mechanisms of undermining the competitiveness (economic security) of industrial enterprises» (Solodovnikov, 2020, p. 21), including through the offense of stable relations of technological cooperation. The variety of these mechanisms undermining the competitiveness of industrial enterprises increases with the digitalization of the economy. M. Christen and E. Bangerter describe the fundamental problems that are widely used for explaining why it is seemingly hard to defend IT systems: «asymmetry between defense and offense», «complexity of ICT systems»; «software is inherently insecure today» including software vulnerability; «lack of attribution and consequences for the attacker» (Christen & Bangerter, 2017, pp. 246–247). All these factors, along with «the impact of time pressure on cybersecurity behavior» (Chowdhury et al., 2019), determine the risks of significant vulnerability to the technological and economic security of industrial enterprises. These risks are increasing in the context of technological cooperation, since in the process of inter-firm interaction responsibility for ensuring security is partially blurred. As a result, such an important tool as providing cybersecurity is required in order to increase the technological cooperation efficiency in the context of digitalization.

Discussion

We began our contribution with a review of the results that we obtained in our research of the topic of technological cooperation in the context of digitalization. We first discuss new factors of international technological cooperation expanding in current conditions and of arising herewith risks. Then we showed that in connection with the emergence of new factors and risks of international and interfirm technological cooperation in the context of the digitalization of the economy, the features and effectiveness of technological cooperation are changing. It requires research of new tools for improving of the technological cooperation efficiency in the context of digitalization. Then we paid attention to the methodological foundations of our research, using the theoretical results of the Russian scientist T.N. Yudina. The following three points are of particular importance: the digital economy should be considered both as a phenomenon and as a noumenon; digitalization has led to the emergence of a new type of economic rent—digital rent, which is formed in connection with the emergence of big tech and big data; digitalization creates new business models through fundamentally new revenue models. All of this is cardinally transforming the main tools for increasing the technological cooperation efficiency in the context of digitalization. Our research made it possible to establish the main technological cooperation efficiency increasing tools in the context of digitalization: firstly, digitalization of industrial and business processes in the national industrial complex; secondly, robotization of industrial processes; third, maintenance and development of analog (non-digital) technological processes in industry; fourth, ensuring the development of industry with long and cheap money; fifth, providing cybersecurity.

There are some limitations to this study, which open possibilities for further studies. Our results are limited by the features of the economic policy of a particular country and are relevant, first of all, for the Republic of Belarus and the Russian Federation. «Any country in the world implements state economic policy. But due to the diversity of its goals, the configuration of tools and priorities in the countries of the world are significantly different, due to the different visions of national elites of key national interests in the economic sphere. Therefore, the formation of the state's economic policy takes place on the basis of critical reflection and use of the experience gained in domestic economic theory and business practices, taking into account global trends in the development of economies in the world» (Vertakova et al., 2020, p. 3). In our contribution, we proceeded from the national priorities of the economies of the Republic of Belarus and the Russian Federation to build a super-industrial economy with a strong industrial complex. Our conclusions can be supplemented with new technological cooperation efficiency increasing tools, taking into account new challenges in the context of sanctions pressure on the Republic of Belarus and the Russian Federation.

Conclusion

The paper includes the new tools of the increasing of the technological cooperation efficiency in the context of digitalization: digitalization of industrial and business processes in the national industrial complex; robotization of industrial processes; maintenance and development of analog (non-digital) technological processes in industry; ensuring the development of industry with long and cheap money; providing cybersecurity. Further research can be continued in the direction of finding ways to overcome organizational and economic problems arising by technological cooperation and innovation creation.

References

Aarikka-Stenroos, L., Sandberg, B., & Lehtimäki, T. (2014). Networks for the commercialization of innovations: A review of how divergent network actors contribute. *Industrial Marketing Management, 43*(3), 365–381. https://doi.org/10.1016/j.indmarman.2013.12.005

Abdul Hamid, N., & Abd Rahman, A. (2014). A systematic literature review on the success factor of innovation commercialization performance. In *Paper presented at the ISTMET 2014 – 1st International Symposium on Technology Management and Emerging Technologies, Proceedings* (pp. 199–204). https://doi.org/10.1109/ISTMET.2014.6936506

Alcácer, J., Cantwell, J., & Piscitello, L. (2016). Internationalization in the information age: A new era for places, firms, and international business networks? *Journal of International Business Studies, 47*(5), 499–512. https://doi.org/10.1057/jibs.2016.22

Amit, R., & Han, X. (2017). Value creation through novel resource configurations in a digitally enabled world. *Strategic Entrepreneurship Journal, 11*(3), 228–242. https://doi.org/10.1002/sej.1256

Bessonova, E., & Battalov, R. (2021). Digitalization as a tool for innovative economic development. *Economic Annals-XXI, 186*(11–12), 66–74. https://doi.org/10.21003/EA.V186-08

Birch, K., & Cochrane, D. T. (2021). Big tech: Four emerging forms of digital rentiership. *Science as Culture, 2021*, 1–15. https://doi.org/10.1080/09505431.2021.1932794

Bogatyreva, V. V., Bobrik, M. Y., & Serhiyevich, T. V. (2021). Evolution of labor relations in the Republic of Belarus in the context of digitalization and robotization of the economy. *Economic Science Today, 13*, 6–14. https://doi.org/10.21122/2309-6667-2021-13-6-14

Cachon, G. P. (2020). A research framework for business models: What is common among fast fashion, e-tailing, and ride sharing? *Management Science, 66*(3), 1172–1192. https://doi.org/10.1287/mnsc.2018.3275

Chivu, L., Ciutacu, C., & Georgescu, G. (2017). *Deindustrialization and reindustrialization in Romania*. Palgrave Macmillan. https://doi.org/10.1007/978-3-319-65753-0

Chowdhury, N. H., Adam, M. T., & Skinner, G. (2019). The impact of time pressure on cybersecurity behaviour: A systematic literature review. *Behaviour & Information Technology, 38*(12), 1290–1308. https://doi.org/10.1080/0144929X.2019.1583769

Christen, M., & Bangerter, E. (2017). Is Cyberpeace possible? In *The nature of peace and the morality of armed conflict* (pp. 243–263). Palgrave Macmillan. https://doi.org/10.1007/978-3-319-57123-2_13

Gault, F. (2019). User innovation in the digital economy. *Foresight and STI Governance, 13*(3), 6–12. https://doi.org/10.17323/2500-2597.2019.3.6.12

Geliskhanov, I. Z., Yudina, T. N., & Babkin, A. V. (2018). Digital platforms in economics: Essence, models, development trends. *St. Petersburg State Polytechnical University Journal, 11*(6), 22–36. https://doi.org/10.18721/JE.11602

Kim, J., Paek, B., & Lee, H. (2022). Exploring innovation ecosystem of incumbents in the face of technological discontinuities: Automobile firms. *Sustainability (Switzerland), 14*(3). https://doi.org/10.3390/su14031606

Makarova, I. V., Lepesh, G. V., Ugolnikova, O. D., & Meleshko, J. V. (2021). Analysis of directive and policy documents on digital industrialization of the Russian Federation and the Republic of Belarus. *Public Administration Issues, 1*, 150–172.

Mialeshka, Y. V. (2021). New industrialization and trends in modernization of Belarusian industry. *Science and Technique, 20*(4), 357–364. https://doi.org/10.21122/2227-1031-2021-20-4-357-364

Palaco, I., Kim, S. K., Park, M. J., & Rho, J. J. (2022). Exploring capabilities of international technology transfer intermediaries between emerging and developed countries. *Journal of Technology Transfer, 47*(1), 307–352. https://doi.org/10.1007/s10961-021-09849-2

Serhiyevich, T. V. (2020). Stimulating the production of intensive renewal goods. *Science and Technique, 19*(2), 139–147. https://doi.org/10.21122/2227-1031-2020-19-2-139-147

Soares, J., Romero, F., & Lopes Nunes, M. (2020). Innovation and technology transfer: A framework for clustering intermediation roles. In *Paper presented at the 2020 IEEE International Conference on Technology Management, Operations and Decisions, ICTMOD 2020*. https://doi.org/10.1109/ICTMOD49425.2020.9380600.

Solodovnikov, S. Y. (2020). Planning engineer training in economic security in the mineral mining sector: Theory and practice. *Gornyi Zhurnal, 11*(2280), 20–25. https://doi.org/10.17580/gzh.2020.11.01

Solodovnikov, S. Y. (2021). Factors responsible for expansion of Belarusian-Romanian technological cooperation. *Economic Science Today, 13*, 119–125. https://doi.org/10.21122/2309-6667-2021-13-119-125

Solodovnikov, S. Y., & Sergievich, T. V. (2020). Challenges and prospects for the development of ferrous metallurgy in the Republic of Belarus. *Chernye Metally, 11*(1067), 62–68. https://doi.org/10.17580/chm.2020.11.09

Solodovnikov, S. Y., Serhiyevich, T. V., & Meleshko, Y. V. (2021, September). The relative advantages of modern cross-country technological cooperation in the context of the transition to industry 4.0. *AIP Conference Proceedings, 2389*(1), 020006. https://doi.org/10.1063/5.0064421

Vardomatskya, L., Kuznetsova, V., & Plotnikov, V. (2021). The financial technologies transformation in the digital economy. In *E3S Web of Conferences* (Vol. 244, p. 10046). EDP Sciences.

Vertakova, Y., Plotnikov, V., & Babich, T. (2020). Conceptual framework of state economic policy in the technological and social transformation conditions. In *E3S Web of Conferences* (Vol. 164, p. 11016). EDP Sciences. https://doi.org/10.1051/e3sconf/202016411016

Yudina, T. N. (2016). Perspective of digital economy. *Journal of Northeast Asia Studies, 5*, 20–24.

Yudina, T. N. (2017). Digitalization as a trend of modern economic development of the Russian Federation: Proycontra. *State and Municipal Government., 3*, 139–143. https://doi.org/10.23394/2079-1690-2017-1-3-139-143

Yudina, T. N. (2018). «Surveillance capitalism» as «digital economy» and/or «digital society». *Theoretical Economy, 4*(46), 13–17.

Yudina, T. N. (2019). Digital segment of the real economy: Digital economy in the context of analog economy. *St. Petersburg State Polytechnical University Journal. Economics, 12*(2), 7–18. https://doi.org/10.18721/JE.12201

Yudina, T. N., & Balashov, A. M. (2020). Digital transformations in economy management and state-private partnership role. *E-Journal Public Administration, 80*, 300–320. https://doi.org/10.24411/2070-1381-2020-10075

Yudina, T. N., & Kupchishina, E. V. (2019). «Digital» economy institutional infrastructure formation in the Russian Federation. *St. Petersburg State Polytechnical University Journal. Economics, 12*(4), 9. https://doi.org/10.18721/JE.12401

Zuboff, S. (2015). Big other: Surveillance capitalism and the prospects of an information civilization. *Journal of Information Technology, 30*(1), 75–89. https://doi.org/10.1057/jit.2015.5

Business Models Transformation in Light Industry in the Context of Digitalization

Tatiana A. Borisova, Tatiana V. Serhiyevich, and Kirill V. Kliyev

Introduction

The scientific interest in the study of theoretical and practical aspects of business models in the world economic literature has been increasing since the late 90s twentieth century. This interest's increasing is due to the emergence of fundamentally new forms of management under the influence of digitalization, which is changing the organization of business processes and the mechanisms of interaction with contractors and consumers. A large difference in the efficiency of enterprises operating in similar sociocultural and economic conditions leads to attempts to analyze their successful practices. At the same time, researchers are seeking to find out the universal patterns determining the effectiveness of a particular business model. A unified approach to understanding the essence of a business model in the economic literature doesn't exist. Many researchers pay attention to the nature of interaction with the consumer as one of the main characteristics of the business model (Johnson et al., 2008; Osterwalder & Pigneur, 2010). Most researchers proceed from the assumption that the business model includes the full range of interactions with business partners participating in the value chain (Amit & Zott, 2001, 2015; Chesbrough & Rosenbloom, 2002; Osterwalder & Pigneur, 2010; Teece, 2010; Magretta, 2011). At the same time, we state the insufficiency of the theoretical study of the concept of «business model», since the existing studies are presented in the format of case study of specific business practices or describe the

T. A. Borisova (✉) · K. V. Kliyev
St. Petersburg University of Management Technologies and Economics, St. Petersburg, Russian Federation

T. V. Serhiyevich
Belarusian National Technical University, Minsk, Republic of Belarus
e-mail: serhiyevich@bntu.by

A. Rumyantseva et al. (eds.), *Challenges and Solutions in the Digital Economy and Finance*, Springer Proceedings in Business and Economics,
https://doi.org/10.1007/978-3-031-14410-3_20

patterns of development of business models in special institutional, economic and cultural conditions. The transformation of business models in the context of digitalization has been studied even less, while both the Republic of Belarus and the Russian Federation «set as their strategic goal the adaptation of society, economy and industry to the new realities of digitalization» (Makarova et al., 2021, p. 154). There are no papers devoted to the study of the business models in the light industry specifics in the new conditions. This contribution will fill the theoretical and methodological gap in the study of the transformation of business models in the light industry in the context of digitalization.

In earlier studies, we obtained conclusions regarding the specifics of business models in modern light industry and some trends in their transformations (Serhiyevich, 2020b). In addition, we studied the factors and trends in the value chains transformation in light industry in the context of the robotization of the economy affecting the business models transformation (Serhiyevich, 2021). Digitalization is one of the most important factors in the transformation of the business model of a light industry enterprise. Therefore, the impact of digitalization on the transformation of business models in the light industry requires further detailed research, which is the purpose of this article.

Materials and Methods

The object of our study is business models in light industry in the context of digitalization. The purpose of this paper is the design of the business models transformation in light industry in the context of digitalization.

The methodology of the authors is based on the subject-activity approach. Research works of scientists-economists, data of the research reports carried out by the authors were used to write this article. General scientific methods (analysis, synthesis, unity of historical and logical), system approach, institutional approach was applied during the research.

In the study of the impact of digitalization on the business models transformation in the light industry, we are faced with the methodological task of correctly defining the boundaries of the business model. This affects the consideration of digitalization as an external or internal factor in the transformation of the business model. The formulation of this problem is due to the fact that today the production of goods is carried out by a network of interacting entities that are closely interconnected in production chains. The reality of the network economy shows that the value chain changes the sequence of links and loses its traditional for the industrial era linearity, acquiring network features (Serhiyevich, 2021). At the same time, considering a business model as a network of interacting entities has some applicability limitations in studying the impact of digitalization on the business models transformation in the light industry. These limitations are due to the following factors. Firstly, the participants in the value chain have different (sometimes—opposite) interests' directions. Secondly, enterprises are involved in many different value chains. Thirdly, relationships between participants in the value chain are determined by the action of market mechanisms. However, in our opinion, it is acceptable to consider a network of firms within the same business model in the case when these firms are interdependent to such an extent that relations between them are formed under the influence of not market but regulated by focal firm mechanisms. In this case, the impact of digitalization on the value chains evolution will be considered as an internal, and therefore manageable factor.

When identifying the impact of digitalization on the business models transformation in the light industry, we will proceed from an understanding of transformation as a process of development—«irreversible qualitative changes appearing in three forms: progress, regression and neutral (one-plane) change» (Solodovnikov, 2018, p. 9). Taking into account the above methodological remarks, the task to find out the impact of digitalization on the business models' transformation in the light industry means determining which connections and relationships that have arisen under the influence of digitalization become stable, regularly repeated (they change the structure of the system under consideration while maintaining its system properties).

Results

External factors of business models digital transformation include technical and technological accessibility of digital tools, economic accessibility of digital tools, digitalization of public life, product lifecycles acceleration, growth of the dynamism of the external environment. The internal factors of business models' digital transformation are the type of economic activity and the specifics of production and business processes, the motivation and interest of top management in digital transformation, the dynamics of digital skills of personnel. The ambiguity in evaluation the business model's digital transformation lies in the fact that the introduction of digital tools into business practice is widespread today, and the digitization of single processes does not always indicate a transition strategy to a radical business infrastructure digitization. Some digital elements are being introduced into the system of internal production and business processes, into the system of interaction with partners and consumers, as well as into the goods or services. S. J. Berman researches the digital transformation as a factor in creating new business models and identifies two key complementary areas for this transformation: «reconfiguring the customer value proposition—what is being offered—and reshaping the operating model—how it is delivered» (Berman, 2012, p. 17). The impact of digitalization on the transformation of business models in the light industry is characterized by three components. Firstly, what digital technologies and tools are used by the enterprise (from the creation of websites to the transition to cyber-physical systems). Secondly, what is the degree of implementation of digital technologies along the value chain (digitization of the entire value chain or its individual links—for example, just e-commerce). And thirdly, how the use of digital technologies is transforming the mechanism of reproduction of use value (including the revenue model and the supply model). New business models are emerging in the light industry and the fashion industry, producing not only goods, but new related services (for example, an online stylist) or digital goods (digital apparel collections). At the same time, it should be considered that today the choice of a particular technology can be carried out not because of strategic planning based on a rational evaluation of technologies or a scientifically based forecast, but under the influence of a cross-country or cross-firm technological race (including the digital race). Thus, when studying the digital transformation of business models in the light industry, it is fundamental not only what digital technologies and tools are used by enterprises (although this is certainly important for further understanding of our topic), but how their use affects the recombination of labor, financial, material, information, scientific, technical and other resources for the creation and appropriation of use value, in other words, what new sources of use value can be formed by the introduction of digital technologies and tools.

«World trends require the activation of ways to increase the competitiveness of any country, which objectively implies the need for innovation and digitalization of all spheres and areas of economic activity» (Vertakova et al., 2020). The light industry belongs to the traditional sectors of the economy and at the same time has an increased potential for the convergence of digital and traditional technologies. Digital trends in the field of light industry include blockchain, augmented reality, and artificial intelligence. However, many technical innovations used in light industry are not digital, but their use in the economic activities of enterprises makes it technologically possible and cost-effective to introduce technical and technological systems that transform the business model—additive technologies, robotic systems, radio frequency identification etc.

3D design, virtual product modeling and 3D printing provide many benefits for light industry enterprises. Thanks to the new opportunities for the development of prototypes of shoes and apparel in digital form, the time and material costs of creating collections can be large reduced. Digitalization of the workflow brings time advantages—for example, the time for prototypes design is reduced from several days to several hours. These processes accelerating allows the enterprise to respond more quickly to changes in consumer demand. Using additive technologies, «the process of creating value becomes more compact. Transport costs and risks are reduced since the transportation of material items is replaced by the transfer of digital files. Increasing the role of these files shifts the distribution of value added to the link of the digital design of the value chain, which can large change the configuration of value-added distribution» (Dementiev et al., 2018, p. 69). 3D technologies in the collections design and goods production are used, for example, by such companies as Marc Cain, Adidas, Hugo Boss, Brax, etc. 3D modeling makes possible production customization in the light industry, when the size of the manufactured product is determined taking into account the individual characteristics of the consumer. Moreover, additive technologies make it possible to quickly produce related items (accessories, implements, etc.).

New opportunities for light industry enterprises are provided using robotic systems. Robots in light industry are most often used directly in the production process, in logistics and warehouse services, as well as in the processes of accounting and inventory of goods. Radio-frequency identification technologies allow not only to automate accounting processes and large speed up inventory, but also to receive real-time information about commodity flows throughout the entire product lifecycle. These technologies are able of changing the architecture of value chains. The robotic systems use allows the company to refuse to outsource production operations for the manufacture of fabrics, clothing, and footwear. Digital verticalization and value chains simplification contribute to the transition to a dynamic organization of production and its localization within a country or region. The chain simplification reduces the intellectual property infringement risks and prevents the ideas losing problem. The reshoring trend since the early 2000s confirms it empirically. Many apparel and fashion corporations are facing challenges in quality control, supply chain transparency and the need to speed up supplies. Therefore, they have adopted strategies to reshore their production back to home country, which pushed the development of automation systems in light industry, allowing robots to replace cheaper labor in economically less developed countries. The basing of production in economically developed countries also provides the advantage of using reputational capital. Thus, consumers associate the quality of goods with the country of origin. At the same time, value chains have different trends—simplification and complexity, which indicates an active transformation of their structure (Serhiyevich, 2021). Instead of some chain links (for example, sewing plants), other links appear based on the need for digital products and services (for example, interaction with Microsoft, Adobe and Autodesk). In addition, the use of new digital channels of communication with the consumer (both in terms of collecting information about changes in consumer behavior and in terms of product promotion) entails the formation of new inter-industry relations.

Digital tools are transforming product promotion systems. «Although fashion is more free-spirited creativity and craftsmanship, the fast pace of technology is forcing the fashion industry to maximize profit through E-tailing» (Agarwal, 2019, p. 405). As a rule, E-commerce doesn't replace, but only complements the capabilities of traditional off-line stores, which requires the integration of various promoting goods channels. The buyer combines on-line and off-line interaction tools with the seller. Fittings in traditional stores are complemented by online price comparison and online orders, while fitting clothes in a virtual fitting room encourages the consumer to visit a real store. «The new business models, mainly based on customer engagement, are primarily influenced by a change in the distribution channels oriented to an omni-channel strategy. Along the consumer journey, digital tools enhance different dimensions of the customer experience, particularly informativeness, entertainment, and social presence» (Matarazzo et al., 2021, p. 653). At the same time, in the context of the development of the Internet, logistics and digital marketing, the boundaries of product markets are expanding. The consumer gains access to new offers, often personalized according to the history of his preferences. Consumer loyalty reduces sustainability, and to maintain it, it is not enough to provide just a product—it is important to create a social attitude. The current enterprises in the light industry and the fashion industry make content. «With information about products becoming as important as the products themselves, almost every company is now in the business of creating and delivering "content"—information that is personal, relevant and timely when accessed by the customer» (Berman, 2012, p. 17). In a society characterized by prosumerism, the consumer becomes an active participant in the fashion system, seeking to interact as closely as possible with brands through various media.

Actually, the digitalization impacts on the work of each enterprise, causing a change in management approaches to business processes. At the same time, the digitalization of individual processes entails the need for the digitalization of related functions. For example, the introduction of digital and robotic systems into production is inevitably accompanied by the digitalization of management and accounting processes. This allows the manager to control all stages of the product lifecycle in real time, since digitalization makes it possible to improve the mechanisms of information exchange and reduce transaction costs arising from this exchange. The digitalization also makes it possible to reduce the links of information exchange, increase access to information to more network participants, as well as helps to increase the transparency of information and the speed of its motion (Serhiyevich, 2020a, p. 144). The transition to multi-channel interaction with the consumer also requires new forms of promotion coordination, not only for the integration of several promotion channels (for example, traditional and online stores), but also for the integration of various promotion channels and channels of communication with the consumer (websites, social networks).

The business model digitalization creates a need for new skills and competencies. «Thus, competences in cyber physical systems or analytical data processing know-how are necessary and require changes in workforce qualifications. If manufacturers do not possess these resources, they need to develop a partner network to source them externally» (Rachinger et al., 2018). However, the increasing demand for digital skills is only one of the consequences of the development of the digital economy. The influence of the digital economy extends to the organization of labor, forms of professional communications, structures of horizontal and vertical interactions, professional and qualification characteristics of labor resources. The interaction of new structural and traditional divisions (including the redistribution of functions, responsibilities, and power within the enterprise) is a factor in the evolution of the organizational structure in the context of the business model digitalization. The need for new skills and competencies, due to the business model digitalization, can be realized by creating or expanding the enterprise digitalization department and supplementing the staff functions for digital business transformation (from the implementation of the Internet of things and robotic systems to digital marketing and e-commerce).

At the same time, some researchers of the digital transformation of business models in the light industry and the fashion industry idealize the positive impact of digital technologies. P. Bertola and J. Teunissen believe that «digital transformation, properly driven, could reshape the fashion industry into a more sustainable and truly customer-driven business» (Bertola & Teunissen, 2018). This is an overly optimistic view of industry transformation. Firstly, there is often a quasi-customer focus, when "joint creativity with the consumer" is replaced by its simulation, and "individual design" is only a choice from the alternatives specified by the manufacturer. The functional purpose of such tools is mainly marketing and image-building. Modern digital marketing tools based on digital algorithms form a special information space around the consumer. Secondly, digitalization is a source of not only competitive advantages, but also new risks and threats. Making production more customer-oriented and involving the consumer in the process of creating a product, the enterprise faces new risks. Leveling the effect of some risks, digitalization generates new risks, and its impact on the transformation of both the business model itself and the entire value chain is not unambiguously positive. The digitalization of single elements of value chains creates difficulties in the interaction of its various participants due to the increased diversity of the technological structure, which leads to incompatibility of standards and information loss. The solution to this problem can be «the creation, around the "smart factory", of a network of flexible decentralized manufacturing and service units. Differently from a vertically integrated and centralized model, this modular and distributed system, empowered by real-time data exchange, can better fit quick changes and reduce the impact of critical issues on the entire system, isolating the problem into the inefficient units» (Bertola & Teunissen, 2018). Virtualizing value chains and providing real-time access to product lifecycle data is made possible by new RFID tracking technologies (such as RFID tags) and dedicated product lifecycle management (PLM) software. The potential for business models' transformation in the context of digitalization is limited by the environment of interactions—the formation of a "smart" network is constrained by insufficient digitalization of all subjects of interaction (consumers, competitors, suppliers, government bodies).

Discussion

The present study is another step towards comprehensive research of the digital transformation of business-models, contributing to the development of an effective communication among the different researchers in the digital economy area. This paper demonstrates that transformation of business-models is increasingly attracting the interest of both academics and practitioners, in particular when referring to digitalization. This interest is expected to have a further growth due to the increasing interaction digital tools with traditional economy spheres, as well as due to the increasing impact of digitalization's on human behaviors.

We began our contribution with a review of the results that we obtained in our recent research of the topics of business-models, light industry, and fashion industry in the context of digitalization. We first discuss methodological approaches and restrictions of our research. Then we described external and internal factors of digital transformation of business models.

We have shown limitations in evaluation the impact of digitalization on the business models transformation. The digitization of single processes does not always indicate a transition strategy to a radical business infrastructure digitization. Some digital elements are being introduced into the system of internal production and business processes, into the system of interaction with partners and consumers, as well as into the goods or services. Then we described what components characterize the impact of digitalization on the transformation of business models in the light industry and described in detail the technologies that affect the change in business models in the light industry. We also focused on the consequences of business models digitalization in the light industry: the need for digitalization of related functions; the need for new skills and competencies; idealization of the positive impact of digital technology and the emerging risks of the digital race; increasing of the diversity of the technological structure of value chains.

Nevertheless, the research has some limitations. Our contribution can be supplemented by the study of perspective digital technologies that can be used in the light industry and the fashion industry in the future and influence the business models transformation. Due to the Covid19 pandemic extension and the increasing of cross-country trade contradictions and restrictions on cross-border goods transition, value chains in light industry will increase the potential for digitalization. So, it requires a further increase in research of the industry digital transformation.

Conclusion

We found out in our research that the business model transformation under the influence of digitalization is a process of irreversible essence changes in the mechanism of reproduction of the use value. These changes are due to the introduction of digital technologies and tools in production and business processes, in systems of interaction with other participants in the value chain and consumers as well as in the goods or services. We found that new sources of use value in light industry are emerging under the influence of digitalization as a result of an increase in the speed of design, production and distribution of goods; customization; value chains transformation and dynamic organization of production; the home country reputational capital use; attracting highly qualified personnel to the industry; redistribution of capitalization between individual value creation processes; transition to hybrid value creation models combining virtual and real processes; the domino principle in the digitalization of an enterprise and industry.

References

Agarwal, V. (2019). Technology, sustainability, and consumer expectation—New ways of thinking about future fashion. In *Research into design for a connected world* (pp. 403–411). Springer. https://doi.org/10.1007/978-981-13-5974-3_35

Amit, R., & Zott, C. (2001). Value creation in e-business. *Strategic Management Journal, 22*(6–7), 493–520.

Amit, R., & Zott, C. (2015). Business models. In *Encyclopedia of the social & behavioral sciences* (2nd., pp. 33–36). Springer.

Berman, S. J. (2012). Digital transformation: Opportunities to create new business models. *Strategy & Leadership.* https://doi.org/10.1108/10878571211209314

Bertola, P., & Teunissen, J. (2018). Fashion 4.0. Innovating fashion industry through digital transformation. *Research Journal of Textile and Apparel.* https://doi.org/10.1108/rjta-03-2018-0023

Chesbrough, H., & Rosenbloom, R. S. (2002). The role of the business model in capturing value from innovation: Evidence from Xerox Corporation's technology spin-off companies. *Industrial and Corporate Change, 11*(3), 529–555. https://doi.org/10.1093/icc/11.3.529

Dementiev, V. E., Ustyuzhanina, E. V., & Evsukov, S. G. (2018). Digital transformation of value chains: «smail curve» can become «scowling». *Journal of institutional studies, 10*(4), 58–77. https://doi.org/10.1016/B978-0-08-097086-8.73040-6

Johnson, M. W., Christensen, C. M., & Kagermann, H. (2008). Reinventing your business model. *Harvard Business Review, 86*(12), 57–68.

Magretta, J. (2011). *Understanding Michael Porter: The essential guide to competition and strategy*. Harvard Business Press.

Makarova, I. V., Lepesh, G. V., Ugolnikova, O. D., & Meleshko, J. V. (2021). Analysis of directive and policy documents on digital industrialization of the Russian Federation and the Republic of Belarus. *Public Administration Issues, 1*, 150–172.

Matarazzo, M., Penco, L., Profumo, G., & Quaglia, R. (2021). Digital transformation and customer value creation in made in Italy SMEs: A dynamic capabilities perspective. *Journal of Business Research, 123*, 642–656.

Osterwalder, A., & Pigneur, Y. (2010). *Business model generation: A handbook for visionaries, game changers, and challengers* (Vol. 1). Wiley.

Rachinger, M., Rauter, R., Müller, C., Vorraber, W., & Schirgi, E. (2018). Digitalization and its influence on business model innovation. *Journal of Manufacturing Technology Management.* https://doi.org/10.1108/jmtm-01-2018-0020

Serhiyevich, T. V. (2020a). Stimulating the production of intensive renewal goods. *Science and Technique, 19*(2), 139–147. https://doi.org/10.21122/2227-1031-2020-19-2-139-147

Serhiyevich, T. V. (2020b). Transformation of business models in light industry in the context of robotization. *Economic Science Today, 12*, 73–81. https://doi.org/10.21122/2309-6667-2020-12-73-81

Serhiyevich, T. V. (2021). Factors and trends in the transformation of value chains in the conditions of robotization: The case of light industry. *Newsletter of North-Caucasus Federal University, 3*(84), 120–128. https://doi.org/10.37493/2307-907X.2021.3.15

Solodovnikov, S. Y. (2018). Chaos definition and its role in complicated systems' development. *Economic Science Today, 7*, 5–18. https://doi.org/10.21122/2309-6667-2018-7-5-18

Teece, D. J. (2010). Business models, business strategy and innovation. *Long Range Planning, 43*(2–3), 172–194.

Vertakova, Y., Plotnikov, V., & Babich, T. (2020). Conceptual framework of state economic policy in the technological and social transformation conditions. In *E3S Web of conferences* (Vol. 164, p. 11016). EDP Sciences. https://doi.org/10.1051/e3sconf/202016411016

Constitutional Principles of Forming the Foundations of the Digital Economy in the Russian Federation

Marina Vladimirovna Rybkina and Dariia Andreevna Pavlova

Introduction

Digitalization of social relations in a certain way makes changes in the content and structure of legal regulation. It is impossible not to note the fact that digitalization itself (a conditional concept. This concept includes not only technological processes and production, these processes cover the entire system of social relations both in one, separate state, and in the world community. Simultaneously with digital technologies, the structures and content of the economy and management are changing. Such previously unknown concepts appear, as a digital state and digital economy, in the disclosure of which it is necessary to focus, first, on the standards established by the Constitution and the entire world community. Constitutional principles lay the foundations of the digital economy, approve general rules and restrictions regarding the security of the state, society and citizens. Constitutional principles are further developed in legislation, the purpose of which is to formulate concepts, norms that establish the regulation of the main content of the digital economy.

However, at present in the Russian Federation there is no system of regulatory legal regulation of individual areas that make up the complex concept of the digital economy. The purpose of this article is to determine the place and role of constitutional principles in the formation of both the system of legal regulation of the digital economy and its individual elements.

M. V. Rybkina (✉) · D. A. Pavlova
St. Petersburg State University of Economics, St. Petersburg, Russian Federation

A. Rumyantseva et al. (eds.), *Challenges and Solutions in the Digital Economy and Finance*, Springer Proceedings in Business and Economics,
https://doi.org/10.1007/978-3-031-14410-3_21

Materials and Methods

The research used such general scientific methods of cognition as induction, deduction, analysis, synthesis and a system-structural approach to identify the constitutional foundations of regulation of the digital economy. Among the private law methods, one can single out the formal-logical and comparative-legal method, which were used to develop recommendations and conclusions of the study. In order to conduct research, the scientific literature related to the problems and analysis of the current legislation in the field of legal regulation of the digital economy from 2017 to 2021 was studied. The work touches upon the research of such authors as Sturgeon (2019), Gureev et al. (2020), Bukht and Heeks (2017), Chazhaeva et al. (2020), Mingaleva and Mirskikh (2019), Voskresenskaya et al. (2020).

Particular attention was paid to the influence of constitutional norms and principles on the development of doctrine and legislation in the field of digital economy. Based on the subject of the study, such normative legal acts as the Constitution of the Russian Federation, Presidential Decree of the Russian Federation No 203 of May 6, 2017, Federal Law of the Russian Federation No 259-FZ of July 31, 2020 and Federal Law of the Russian Federation No 259-FZ of August 02, 2019, as well as studies by Horsley (2021), Olwor (2021), Osavelyuk (2020), Plotnikova (2021), Daly (2021), Kombarova et al. (2021), Mazayev (2021), Dennett (2021) and other.

Results

Taking into account the trends in the development of gradual digitalization of modern society, one of the key issues is the formation of an appropriate regulatory framework in the field of regulation of the digital economy, which by its properties should act not only as a regulator of narrow-industry relations, but also their constitutional and legal integrity.

Considering the basis of the economic constitution (taking into account the interpretation of this definition as a set of constitutional provisions related to economic activity), it is permissible to assume that one of the key amendments to the Constitution of the Russian Federation, which sees the legislator's attempt to constitutionally consolidate the continuous policy of digital economy development, giving it a key status, is Article 75.1: "Conditions for sustainable economic growth of the country are created in the Russian Federation". This amendment (supplement) directly agrees with Article 8 of the Constitution "the unity of the economic space is guaranteed in the Russian Federation < ...> freedom of economic activity"; articles 34 "everyone has the right to freely use their abilities and property for entrepreneurial and other economic activities not prohibited by law"; with paragraph (e) of Article 71 "establishing the foundations of federal policy and federal programs in the field of <...> economic <...> development of the Russian Federation" and so on.

Taking into account also part 1 of Article 17 of the Constitution "in the Russian Federation, the rights and freedoms of man and citizen are recognized and guaranteed in accordance with the generally recognized principles and norms of international law and in accordance with the Constitution", as well as the provisions of a number of concepts, strategies and other strategic planning documents, the current legislative norms cannot fully reflect the systemic approach to the regulation of the digital economy and the creation of adequate conditions in conjunction with the above provisions of the Constitution (including the implementation of legal protection for the protection and processing of personal data and digital profile on the Internet, the protection of the rights of subjects in the field of digital rights and digital currency, and so on).

For example, Federal Law of the Russian Federation No 259-FZ of July 31, 2020 and Federal Law of the Russian Federation No 259-FZ of August 02, 2019 do not contain sufficient norms establishing the responsibility of entities carrying out their activities with digital currency, digital financial assets and utilitarian digital rights. It is permissible to conclude that these acts are an attempt by the legislator at the initial level to introduce them into circulation as objects of law and conduct "conditional" regulation of their activities, thereby leaving intentional gaps that will be filled based on the further activities of these entities.

It is permissible to assert that in order to comply with the provisions specified in Articles 8, 17, 34, 71, 75.1 of the Constitution of the Russian Federation, the legislator at the federal level needs to develop appropriate regulations in the field of personal data protection (in terms of the implementation of protection mechanisms), the rights and obligations of citizens engaged in their activities in the digital economy, which will take into account the current international experience and complement the international treaties of the Russian Federation, develop appropriate grounds for bringing persons to responsibility., violating not only the existing legislation, but also provide for sanctions for abuse of the right and the introduction of citizens and legal entities in view of its relative novelty, taking into account the features that are characteristic of the objects of the digital economy.

These acts will contribute to the implementation of Article 46 of the Constitution of the Russian Federation, the provisions of which contain a guarantee to everyone for judicial protection of his rights and freedoms, since at this stage the right to judicial protection, within the framework of these federal laws, is guaranteed only if a number of conditions are met. For example, according to part 6 of Article 14 of Federal Law of the Russian Federation No 259-FZ of July 31, 2020, the claims of a certain circle of subjects are subject to judicial protection only if they inform tax and other authorities about the facts of possession of digital currency and the completion of civil transactions and (or) transactions with digital currency." In this case, the logic of the legislator of controlling the turnover of digital currency is traced, however, taking into account the current practice of the experience of courts of general jurisdiction, other direct cases of judicial protection, under other conditions, are difficult to trace, since the courts, despite the adoption of relevant legal acts on the recognition of digital rights and digital currency as objects of law, do not see the actual basis for the implementation of the protection of the rights and legitimate interests of persons carrying out activities with these objects (for example, resolution of the Ninth Arbitration Court of Appeal of the Russian Federation of May, 15, 2018 No 09AP-16416/2018 in case No A40-124668/2017, decision of the Moscow City Court of the Russian Federation of April, 18, 2018 in case No 7-4313_2018).

Discussion

One of the fundamental acts containing the prerequisites for the formation of a regulatory framework in the field of regulation of the digital economy is the Decree of the President of the Russian Federation dated May 6, 2017 No 203. In this Decree, the digital economy is understood as an economic activity, the peculiarity of which is data in digital form (including technology), as a key factor of production. This means that the key difference from traditional economic activity is the presence of a digital component. However, the term digital economy may be subject to a different interpretation, where the digital economy is considered as an economic model in which the transformation of existing classical business models takes place (Sturgeon, 2019). In this model, there is a direct relationship between the development of the economy, taking into account the achievements and developments of subjects in the field of information technology, microelectronics and telecommunications (Scharf, 2021), the expansion of the introduction of personalized service models and the increasing importance of direct interaction between the manufacturer and the end user (Gureev et al., 2020). On the other hand, it is permissible to understand the digital economy as a certain part of economic activity that directly interacts with a business model based on digital goods and services (Bukht & Heeks, 2017). We propose the following definition, the digital economy is the use of information technologies in the emerging economic activity of participants in economic turnover, the distinctive features of which are innovative technologies, sustainable systems countering cybercrime, protecting personal data and ensuring the protection of the rights of owners. Thus, the following elements should be included in the foundations of the digital economy:

1. Provision of energy resources and digital technologies.
2. Formation of the structure of state and municipal management of the processes of digitalization of the provision of public services and the formation of law enforcement structures in this area.
3. Management of labor resources in the context of changing formats of working conditions and labor functions.
4. Determination of the basic needs of the labor market.
5. Training and retraining population.

It should be noted that information technologies are only a means of solving legal problems. In a certain way, these tasks are solved in various ways due to the almost complete absence of regulatory legal regulation. Under certain conditions, the fundamental constitutional and legal ideas (principles of science) in the field of constitutional law can become norms-principles and thereby lay the basic guiding influence on the regulation of social relations, as well as influence the improvement of the effectiveness of both constitutional and legal regulation and regulation of the digital economy (Osavelyuk, 2020). The fundamental constitutional principles in this regard can be called the principles that enshrine the priority of recognizing, respecting and protecting human rights, which include: the right to privacy, the right to freely seek, receive and disseminate information by any legal means, the right to judicial protection, the right to identification, the right of access to the electronic network, the right to free use of one's abilities and property for business activities and otherwise economic activity not prohibited by law, right of ownership. The enumeration of these constitutional principles makes it possible to turn to a possible classification of constitutional principles, which, of course, must be considered when forming the legal foundations of the digital economy. The proposed classification divides the principles into general social (freedom, respect for human rights, justice, legality, dignity of the individual); general democratic (human rights and freedoms as the highest value, the rule of law, equality before the law and the courts, recognition of private property and economic freedom of the individual); special legal (the highest legal force of the Constitution and laws, the inadmissibility of the retroactivity of the law, the presumption of innocence, the inadmissibility of abuse of the right); principles defined by the Constitutional Court of the Russian Federation (legal certainty, proportionality, balance of private and public interests) (Plotnikova, 2021). Changes in law are pronounced when we turn to digital rights. This new legal category is connected primarily with the constitutional right to freedom of information in its diverse manifestation.

Thus, the digital economy has a number of key features, among which, from a legal point of view, it is permissible to single out that one of the qualitative indicators of this economic detail is the level of development of information technologies and their active introduction into the sphere of public relations (digitalization), which requires appropriate legal regulation.

However, considering the unstable development of the model of the digital economy of the Russian Federation (Chazhaeva et al., 2020), the following legal aspects occupy a special place:

- the need to ensure the information sovereignty of the Russian Federation, information security, protection of state and bank secrecy, taking into account the increase in cases of unfair use of information (Mingaleva & Mirskikh, 2019) (for example, violation of legislation in the field of protection of intellectual property and personal data);
- the need to create an up-to-date regulatory framework in the field of regulation of the digital economy, in which special legal mechanisms can be provided, by their properties having the necessary flexibility, taking into account the pace of development of digital objects and legal relations in the field of the digital economy.

Also, several scientists at this stage of development of the digital economy highlight the need to develop mechanisms for bringing to justice persons who violate the legislation in the field of the digital economy (Voskresenskaya et al., 2020). Understanding that the existing model of formation of the regulatory legal regulation system in Russia cannot cope with the constantly changing trends in digitalization, we suggest paying attention to international experience. The peculiarity of the Constitutions of common law states is not only their form, but also the approach to content. In particular, the most applicable is the functional approach, which defines the basic and additional functions of the Constitution that have an impact on the entire legal system (Horsley, 2021), the definition that the Constitution forms the foundation of public administration of a modern democratic society (Olwor, 2021), and the principles can be formulated not only in the law, but also in the acts of the judiciary (Daly, 2021). There is no definitive list of constitutional rights and values of common law. However, the highest values are: justice, the rule of law, fundamental human rights and democracy (Dennett, 2021). Communication from the Commission to the European Parliament, the Council, the European Economic and Social Committee and the Committee of the Regions on a Single Market for Intellectual Property Rights Boosting creativity and innovation to provide economic growth, high quality jobs and first-class products and services in Europe shows the need to reduce legal barriers to investing in the economies of the European Union. It should be noted that the so-called "mixed" type of legal regulation used in countries with stable liberal economies shows the greatest effectiveness before the application of established regulatory systems. The recent Constitutional reform has shown the possibility of development in this direction, pointing to the possibility of the existence of the basic principles of the digital economy, but without including them in the text of the Constitution itself (Mazayev, 2021). The flexibility of the forms of legal regulations governing the process of digitalization of the economy should become a benchmark soon for the legal space of Russia.

Conclusion

The authors propose the following conclusions:

1. The system of regulatory legal regulation of the digital economy should include both centralized and decentralized elements. The centralized elements include laws adopted in the main areas of operation of the Concept, approved by the Decree of the President of the Russian Federation dated May 6, 2017 No 203, decentralized include the emerging judicial practice, judicial acts adopted by the Constitutional Court of the Russian Federation, as well as agreements adopted by collective participants in economic relations.
2. The significance and role of the Constitution consists, first of all, in the protection of human and civil rights and freedoms and the formation of a stable legal order that ensures the security of the individual of the state in the digital environment. The entire system of legal regulation of the information environment should be built on the basic constitutional principles.

3. For the effectiveness of work on the creation of collective agreements between participants in economic turnover in the context of digitalization of social relations, it is necessary to formulate requirements for these documents for their legal use.

References

Bukht, R. & Heeks, R. (2017). *Defining, conceptualising and measuring the digital economy*. Development Informatics Working Paper, 68. https://doi.org/10.2139/ssrn.3431732

Chazhaeva, M., Serebryakova, A., Tashkulova, G., & Atabekova, N. (2020). Sustainable development of the digital economy on the basis of managing social and technological threats. *Digital Economy: Complexity and Variety vs. Rationality, 87*, 49–56. https://doi.org/10.1007/978-3-030-29586-8_6

Daly, P. (2021). The generation and guardianship of constitutional principles. Revue générale de droit 2021: 1-35 doi: https://doi.org/10.2139/ssrn.3923167.

Dennett, A. (2021). *Constitutional rights and principles. Public Law Directions*. (pp. 1–188). Oxford University Press. https://doi.org/10.1093/he/9780198870579.003.0009

Gureev, P., Degtyareva, V., & Prokhorova, I. (2020). National features of forming a digital economy in Russia. *ISC Conference - Volgograd, 2020*(1100), 13–20. https://doi.org/10.1007/978-3-030-39319-9

Horsley, T. (2021). Constitutional functions and institutional responsibility: A functional analysis of the UK constitution. *Legal Studies*, 1–21. https://doi.org/10.1017/lst.2021.32

Kombarova, E., Kabyshev, V., Zametina, T., & Troitskaya, T. (2021). Constitutional principles of Russian public authority system with regard to constitutional reform. In *Selection and peer-review under responsibility of the Organizing Committee of the conference*, pp. 335–340. https://doi.org/10.15405/epsbs.2022.01.54

Mazayev, V. (2021). The search for new meanings continues (Concerning the Amendments to the Constitution of the Russian Federation). *Lex Russica, 7*(176), 15–31. https://doi.org/10.17803/1729-5920.2021.176.7.015-031

Mingaleva, Z., & Mirskikh, I. (2019). The problems of digital economy development in Russia. *Digital Science*, 48–55. https://doi.org/10.1007/978-3-030-02351-5_7

Olwor, N. (2021). *Constitutional law: Principles and practices*. Elsevier. https://doi.org/10.2139/ssrn.3789181

Osavelyuk, E. (2020). The concept and types of principles of constitutional law of Russia. *Education and Law, 3*, 58–62. https://doi.org/10.24411/2076-1503-2020-10308

Plotnikova, I. (2021). The place and role of constitutional principles in the system of principles of Russian law. *Bulletin of Saratov State Law Academy, 3*(140), 28–37. https://doi.org/10.24412/2227-7315-2021-3-28-37

Scharf, N. (2021). The evolution and consequences of digital rights management in relation to online music streaming. *Legal Studies, 2021*, 1–20. https://doi.org/10.1017/lst.2021.26

Sturgeon, T. (2019). Upgrading strategies for the digital economy. *Special Issue: Digitalization and Global Strategy, 11*, 34–57. https://doi.org/10.1002/gsj.1364

Voskresenskaya, E., Vorona-Slivinskaya, L., & Achba, L. (2020). Digital economy: Theoretical and legal enforcement issues in terms of regional aspect. *E3S Web Conferences, 164*(8), 09016. https://doi.org/10.1051/e3sconf/202016409016

Optimization of Control in the Management of Transport Business Objects Using the Game Theory Methods

Alexander Vyacheslavovich Gudkov and Ruslan Yakovlevich Vakulenko

Introduction

In an article on control in art management (Gudkov, 2022), it was mentioned that the most effective and most resistant method of choosing the object of checking is randomization. That is, among the set of employees, a subject to check, a subset is equally likely to be selected that falls under direct control actions. An example of checking homework by a teacher in a journal was given. If a student from the list is chosen absolutely equally likely, then this will save the teacher from the opposition of "logic and patterns". For example, failure to do homework after you have already been checked yesterday, because "the second time the same student is usually not checked" or, if it is noticed that the teacher calls in alphabetical order, then those whose last name has passed cannot try hard etc.

This article sets the goal of expanding randomization methods by applying elements of the Game theory—the mathematical science of optimal strategies in games (Mazalov, 2010; Kremlev, 2016). One of its main goals is to find a way to prescribe the "rules" or "conditions" of the game in such a way that the "gain" of the compiler of the conditions is maximum. Note that the concept of "game" in mathematics means "conflict" that proceeds according to certain rules.

Conflict, in turn, in mathematics is called any situation of interaction between competing parties with different strategies, interests, points of view, different ranking of factors influencing on their decision making, different set of actions to achieve

A. V. Gudkov (✉)
Glinka Nizhny Novgorod State Conservatoire, Nizhny Novgorod, Russian Federation

R. Y. Vakulenko
Nizhny Novgorod State Linguistic University named after N. A. Dobrolyubov, Nizhny Novgorod, Russian Federation

A. Rumyantseva et al. (eds.), *Challenges and Solutions in the Digital Economy and Finance*, Springer Proceedings in Business and Economics,
https://doi.org/10.1007/978-3-031-14410-3_22

goals, etc. At the same time, each of the parties has the goal of obtaining the greatest possible individual gain from competitors.

One of the notable parts of game theory is contract theory. This is a science that solves the problem of drawing up a contract (a set of rules, regulations, agreements, etc.) in which the benefit of the drafter of the contract (employer, manager) in the process of its implementation would be maximum (Golovan et al., 2005).

Modern economic theory, called "neoclassical economics" (Seligman, 1968) in a number of sources, is characterized by a set of rules, the alleged implementation of which creates a certain hypothesis about the actions of counterparties (participants in the conflict, players)—a statement that somehow approximates the construction of a model or localizes uncertainty in solving a problem. Here are the main ones:

1. All participants in the competitive economic process, all parties to the conflict are rational players, that is, they clearly understand their goals. It is worth noting that for different subjects the criterion for the success of the game can be different and the goals they pursue are also different. For example, in one conflict, one participant considers the best gain to be the maximum saving of a resource, for example, time, and the other—the greatest profit. Profit maximization, as the most common criterion for the success of the gaming process, is called in the modern economy "a narrow understanding of rationality" (Becker, 1976).
2. In the vast majority of conflicts, for lack of an opportunity to describe other conditions for their course, it is assumed that each participant in economic relations is maximally aware of the goals, actions and strategies of the other participants. That is, he perfectly understands what others are doing at any moment, and his assumptions about the further behavior of competitors are also assumed to be known. The depth of planning of all participants in the conflict is assumed to be maximum, if not infinite (Bolotova et al., 2015). That is, by calculating its strategy of behavior, any subject can build arbitrarily long and complex chains of conclusions and assumptions about the actions of competitors and their reaction to them.

   For example, a manager assumes that an employee, when fulfilling a work agreement, has a certain set of goals known to him. Consequently, in a particular situation, he will act in a certain way or choose one of the known set of behaviors. Then, taking into account certain known goals of some third party, he changes the strategy of behavior and takes such and such a different step, and so on.
3. The principle of zugzwang or the principle of "do no harm" (Tsvetaeva & Dzgoeva, 2019). If any next step with any strategy only worsens the situation, this step should not be taken, it is better to leave everything as it is. A natural consequence of this statement is the postulate: «Any poorly functioning control mechanism is usually better than no control at all» (Perevedentsev, 2018).
4. If the collusion of the participants in the conflict is possible and improves (or at least does not reduce) the gain of each of them, then this collusion will always occur.

Alexey Savvateev in his lectures gave an example of a tender for the sale of four exclusive goods on a market where there are 4four large players and many small participants who are not able to resist them in the auction. In this case, in the auction, four monopolists agree in advance and each bids on only one of the four pre-agreed goods. Bidding is easily and profitably won four times, and each monopolist receives its product at the best price, because none of the small participants is able to offer a higher price, and its direct competitors do not participate in the auction due to collusion.

Transport business as an economic-mathematical model can be quite accurately described by the following conditions: firstly, transport is a service sector. As for the delivery of cargo or ensuring a continuous flow of commodity and raw materials interactions of economic entities, and for the transportation of people, then the criteria for the success of the process, such as speed, uninterruptedness and reliability, are added to the conditional indicator of passenger comfort and satisfaction with the services provided. Each of these criteria, both generally and divided into component parts, needs to be controlled.

Secondly, services can be conditionally described as a business that produces kilometers of transported objects from fuel, transport depreciation, financial and other resources. This generalized schematic description makes it possible to understand the evidence of control actions and their applicability to the transport business in the broadest sense of its model. Further in the article, the concept of "employee" implies the widest range of employees of the transport enterprise who need to be controlled, and the concept of "production regulations" (Fayol et al., 1992) implies the set of rules, instructions and technical specifications that are necessary for execution by the employee of the transport business.

Materials and Methods

Development of Control Rules and Sanctions Mechanisms

Control is a necessary, but expensive, resource-intensive event (Gudkov, 2022). On the one hand, each supervisory employee must be paid, on the other hand, each period of time in the manager's schedule devoted to supervisory actions is costly.

Let us consider a situation in an arbitrary conflict, in which the condition is satisfied for each of the participants: any deviation from a certain strategy of game behavior worsens the payoff function for it. That is, no matter how he changes the behavior that is optimal in a certain sense, he will win at least a little, but less. Obviously, in this case, no player is interested in changing his strategy alone. This situation in game theory is called the Nash equilibrium, in honor of the mathematician, economist, Nobel laureate John Nash (Kremlev, 2016).

If in such a situation there is no collusion of the players, in which their payoff will increase, then the situation is called a Strong Nash equilibrium or a collusion-resistant Nash equilibrium (Mazalov, 2010).

Consider an example: at the enterprise, several employees violate the production rules. Supervisory entities are not enough to always detect all violations of all employees. Without loss of generality, we assume that there is only one supervisor. In one iteration of the control process (for example, in 1 h or 1 day), he can check and detect a violation of only one employee. He knows for sure everyone who violates, but he can only choose one for verification. In this example, game theory has proven the existence of two Nash equilibria:

1. Each employee is sure that all his colleagues, fearing checking, will definitely not violate the regulations, and if he violates alone, he will be detected by control with 100% probability. Let the "price" of an iteration of an "honest" workflow, that is, how much an employee will save if he violates and his violation was not revealed, is *P*. Suppose, if his violation did not go unnoticed and control revealed it, then he will pay a "fine" 10*P*. It is clear that in the situation described in the first equilibrium, it is more profitable for him to work honestly, because otherwise, he is "caught" on the violation is guaranteed (with probability 1). This balance is realized if all other employees are also afraid to violate, because each of them is sure that the others fear the same. As noted earlier, the action that players anticipate from others is rational for them too. The player anticipates that the rest of them act in a certain way within their optimal strategies, and that is exactly what they do.
2. Suppose each employee is sure that many of his colleagues will violate the regulations. Without loss of generality, let us assume that all employees of the enterprise will do this. Therefore, he understands that the checking will reveal his violation with the probability $p = 1/N$, where *N* is the number of violators. If assumed by the worker $N > 10$, then on average for one checking iteration at the game price described above, each worker loses $\frac{10}{N}P$ units of the penalty, where, as we know, $N > 10$, which means $\frac{10}{N}P < P$, which means that he wins and it is beneficial for everyone to break the rules together. The Nash equilibrium condition is also satisfied in this case. Such an equilibrium is called in the theory of conflicts the equilibrium of collective action. Let's try to solve the control problem and create control rules that exclude the Nash equilibrium, which is undesirable for the management.

One of the simplest and most rational ways for the controlling entity to exclude the second type of equilibrium is to compile a *rating of employees of the enterprise*. Moreover, it is surprising, but the criteria by which the ranking of employees in this rating will be is not so important: it can be either the duration of the employee's work without violations, or the usual alphabetical order or sorting by date of birth. The main thing is that each employee of the enterprise knows exactly what number he is on this list. Then the controller announces to all participants in the game that in case of a massive violation of the regulations, he will identify and punish the violating employee with the lowest number in the rating.

By assumption, the controller always knows how many and which employees violated the regulations. This assumption is often not so far from the truth, because according to the reports and performance indicators of the enterprise and the dynamics of changes in these indicators, the manager can often very accurately identify how many violations and in which area of production they take place.

So, after the controller's announcement is made, each employee with number K will be sure that all the numbers in the rating before him ($K-1, K-2, \ldots, 1$) have decided to work without violations, because even if he and his higher number K doubts and is afraid of being detected and fined, then his colleagues with lower numbers $K-1, K-2, \ldots, 1$ have much more reasons to doubt and they will most likely decide not to violate. The same reasoning is logically attributed to the $K+1$ worker, and so on.

By recursion, we come to the disappearance of the second Nash equilibrium and the presence of the only possible way for workers—the first equilibrium based on universal work without violations.

The Problem of Checking with Probabilistic Strategy of Participants

Let's consider a more complicated situation, assuming that each employee who violates the rules chooses the probability with which he does it. In a practical sense, this can also be implemented using a probabilistic randomizer that performs, for example, the function of tossing a coin if the probability of violation is 0.5 or throwing a dice if the probability is 1/6 or any pair, four or five of its faces with probabilities 1/3 or 2/3 or 5/6, respectively. The randomizer is given a length segment equal to the selected value of the violation probability. If the generated random number in the interval [0,1] falls into this segment, the violation is considered recommended for execution. Otherwise, at this iteration of the production process, you will have to work without discrepancy with the regulations.

So, let $x_1, \ldots, x_n \in [0, 1]$ be the performance levels of n employees. All of them violate the regulations, this is known for certain, but each violates the rules of the enterprise with a different degree of "malice", that is, with a different probability x_i. Of course, the situation is an idealizing approximation, and the x_i estimate is not guaranteed to be accurate, however, it can often be estimated quite accurately using internal reporting analytics by the checking department, experience from previous iterations of control actions, interviewing enterprise employees, work experience, reputation and other factors that form "expert assessments" (Gudkov, 2013).

Suppose that in our model the sanction for violation occurs depending on the damage caused to the enterprise from the unfair work of the ith employee. As a rule, it seems possible to calculate the damage from this as well. You can evaluate the level of benchmark performance, quality, and other performance indicators of all employees for various periods of time. Assuming that we know that the ith employee violates the regulations with probability x_i, it is clear that we cannot know exactly at what point in time this violation occurs, at which operation and in which part of the production schedule. Otherwise, the violation would be easily recognized, and sanctions (fines) would be guaranteed to be imposed on the employee.

Without loss of generality, let's assume that there is only one controlling employee in the model, and he can do only one check in 1 work cycle (iteration of the controlling action). Let's also assume that we choose a set of probabilities for checking each employee depending on their "malice" $x_1, \ldots, x_n$. Then we select the set $p_1, \ldots, p_n$—the vector of verification probabilities with a known "maliciousness" of violators $x_1, \ldots, x_n$. The vector $p_1, \ldots, p_n$ enters into a one-to-one correspondence with the violator's "malice" vector, that is, each ith violator with a violation frequency x_i is checked by control with probability p_i, respectively. The choice of this vector is the main "rule of the game" of the administration when checking. It is announced to all employees and is open and accessible information. The most important condition for the efficiency of the principle of control should be the confidence of subordinates in the rigor of its implementation. So, it should at least be believable. Hence, at least the constraint condition: $\sum_{i=1}^{n} p_i \leq 1$ must be satisfied. If for some reason it can be guaranteed that there will be no check for them for some x_i, then the control has "free time" and $\sum_{i=1}^{n} p_i < 1$.

For example, if 6 people are subject to control in the organization and the management claims that the controlling employee will check each of them with a probability of 0.5, then the impracticability of this strategy will quickly be obvious to employees, because $\sum_{i=1}^{6} p_i = 3 > 1$. The probability of any event cannot exceed one.

If in this case the manager claims that it is planned to check each of the six employees with a probability of 10%, then $\sum_{i=1}^{6} p_i = 0{,}6$ that is, in 40% of cases, no one is simply checked, the likelihood condition $\sum_{i=1}^{n} p_i < 1$ is satisfied.

Let us remind you that, according to our assumption, employees, as participants in economic relations, are as rational as possible, aware of each other's actions and have a very high depth of planning. Consequently, they understand well their gain in each system of checks, i.e. for each choice of a set of control probabilities $p_1, \ldots, p_n$.

Let $b_i \in \{b_1, \ldots, b_n\}$ be how much the ith employee will gain (save time or earn, for example, when it comes to theft at work) if he violates the regulations and this violation is not detected by control.

Let T be the size of the sanctions of the controlling body in case of a violation. If this is a penalty, then it is advisable to fulfill the condition $\forall i : T \gg b_i$. The sanction can also be dismissal, criminal punishment, etc. In any case, every punishment can be evaluated, therefore, T is known and computable.

Even if the audit of the supervisory body still came to the ith employee, and this employee is a violator of the regulations, it is likely that the violation will not be detected, because it is at the time of the audit that the employee may be working honestly. This probability is calculated as (1 - x_i) because according to the condition, the employee will be caught in violation only with probability x_i.

In choosing the set $p_1, \ldots, p_n$, it is important to note that each p_i depends on the entire set $x_1, \ldots, x_n$, not just on one x_i. This is because the violation strategy of each employee cannot but depend on how honestly their colleagues work (recall the example with two Nash equilibria).

Consider what the total payoff of the *i*th worker consists of in each iteration of the verification procedure.

$$u_i(x_i, x_{-i}) = b_i x_i - T p_i(x_1, \ldots, x_n) x_i, \tag{1}$$

where x_{-i} is a convention used in game theory, it is the entire set of x except for x_i, i.e. $x_{-i} = \{x_1, \ldots, x_{i-1}, x_{i+1}, \ldots, x_n\}$.

Note that u_i depends not only on x_i, but also on the rest x_{-i} because the payoff of each *i*th worker depends, among other things, on the probabilities of violation of his colleagues. As in our first example: if everyone else does not violate at all (with probability of 0%), then he will be punished with a probability of 100% as the only violator.

The definition of all the conditions described above in game theory is called "designated game" and is denoted:

$$p : [0, 1]^n \rightarrow \Delta_n$$

Or, which is the same:

$$\forall p \in {\Delta_n}^{[0,\ 1]^n}$$

This means that for any set of player violations vector $[0, 1]^n$, in our case, this is a set of employees violating production regulations, we have uniquely determined the strategy Δ_n, that is, the set of probabilities with which we check each of them (Kremlev, 2016).

Results and Discussion

In the field of contact between game theory and management, three necessary conditions for successful management can be formulated:

Firstly, the control scheme proposed by the control system (rule, law, algorithm, etc.) according to which the control object (collective, people, department, organization, etc.) will function, must inspire confidence. This means that it, at least, should not contain obviously impossible operations (bluff, lies, contradictions). The object of management should not have any doubts that the manager will really be able to carry out what was planned (Kennedy, 2014). And, most importantly, will be able to impose sanctions (punish) for non-compliance with the rules.

Secondly, the implemented control scheme should be understandable and to a certain extent simple. Its implementation should not be divorced from reality. If the management scheme is too complex, then its implementation will not be clear and will discourage employees from sticking to it. For example, if the control system punishes the violator with a probability given by an overly complex formula with an abundance of non-obvious conditions and schemes, then the fulfillment of these conditions will be divorced from reality, it is not clear.

Third, the control scheme must be resistant to collusion. That is, it should not allow collective deviations by agreement. For example, in the case of mass violations discussed above, we had an undesirable Nash equilibrium that needed to be eliminated.

Solving the Problem of Control with Sanctions Against the Most Malicious Violator

Let's return to the problem of control. The first and most obvious assumption about the method for solving the problem of choosing an employee who should be subjected to control is to identify the most "malicious" violator and study his work for violations of the regulations. We choose the number of the most "malicious" violator $i : x_i = \max \{x_1, \ldots, x_n\} \Rightarrow p_i = 1$, and control is guaranteed to come to him. Then it is beneficial for all other employees to at least slightly reduce the probability of their violation so that it is less than the maximum. In this case, with this choice of the control scheme, their verification is excluded. Then, at the next step, when the previous "leader" also, for reasons of expediency, reduces the probability of violation, those who have reduced it will be the most "malicious" and all their colleagues will further reduce the probability of their violation already relative to them, etc. down to zero. We have, it would seem, the only and most favorable Nash equilibrium—as a result, everyone does not violate the gradual decrease.

This equilibrium is based on the principle of competitive behavior in the oligopolistic market model of Joseph Bertrand (1883).

But! This equilibrium, obviously, will not be strong, because if collusion occurs, and all employees decide to increase the level of their violations to the maximum so that $x_1 = x_2 = \ldots = x_n = \max \{x_1, \ldots, x_n\}$ then all will be "leaders" and control will be forced to be assigned equally likely: $\forall i : p_i = \frac{1}{n}$. Then it will be unprofitable for employees to violate the regulations only if the sanctions are very large $T > n \max_i b_i$. But this contradicts another important principle of governance—the principle of proportionality of punishment. It consists in the following: any sanction for violation of the rules of management must be commensurate with the damage that this violation caused to the system of management. If the maximum gain of an employee from an unnoticed violation of the regulations is estimated as b_i, the damage to the enterprise is commensurate with this value, and the punishment for the detected violation exceeds it by a number of times equal to the size of the team (n), then it turns out that one employee is punished for violations of the entire team, which is unacceptable and leads to the collapse of management.

Control with "Tolerance Threshold"

Let's assume that the gain of one worker from violation of production regulations is so large that it exceeds the amount of sanctions: $\exists j : b_j > T$. In other words, $b_j = T + \Delta$, where $\Delta > 0$ is the balance of the employee's gain after paying the fine. This situation may arise, for example, under severe pressure from the trade union, which does not allow the introduction of penalties and sanctions above a certain barrier value. Or this can happen due to the disproportion of damage for different employees of the enterprise in case of violation of the regulations. In this case, how much an employee will gain from the violation may be in his case proportional to the great damage caused to the entire enterprise, that is, it can also be very high. Such employees "benefit" from violation of the regulations significantly more than their colleagues.

It would seem that for these situations, any choice of control strategy and sanctions would be meaningless, because even having paid a fine T, the employee will still remain in profit by the value $\Delta > 0$ and it is beneficial for him to violate the regulations in 100% of cases. But even under these conditions, it is possible to reduce the percentage of violations of the regulations by such employees. The task is reduced to minimizing their profit, which, although it always exists, can have different values according to different choice of control strategies.

To do this, we introduce a forced measure—a certain value called the "violation tolerance threshold". That is, such a value of x_i below which the control will under no circumstances check the activity of this employee. If an employee violates the regulations less often than this value, then his work is not controlled and sanctions will not be applied to him, no matter how much the manager would like. A consequence of the first principle of management is that the management scheme should inspire confidence not only in terms of the inevitability and size of punishment, but also in the field of encouragement and fulfillment of promises given to subordinates.

This control should be as delicate as possible, psychologically verified and applied in isolated cases. It is often especially costly, as it uses a lot of the manager's resources, especially his time.

So, if an employee commits violations of the regulations with a probability below the threshold of tolerance, sanctions are not applied to him. In this case, management saves on control, which only needs to check that $x_i < \Pi$, where Π is the value of the very "violation tolerance threshold" that was agreed upon. If the employee's violation ("malice") probability exceeds this threshold, then, despite the fact that for each iteration of the production process, he still receives a profit no less than Δ, he is punished by the value T with probability p_i increasing proportional to x_i and as a result this profit in the long run is significantly less than the case when the frequency of violation is kept within the threshold value (0, П). Even if his gain with a percentage of violations above the "threshold" grows with x_i and reaches a maximum at the point $x_i = 1$, that is, with 100% violations of the regulations, the control task is to choose p_i so that this maximum $u_i(x_i = 1, x_{-i}) = b_i - Tp_i(x_1, \ldots, x_n)$ was still less than the gain $b_i(\Pi)$ at the threshold frequency of violations. Graph 1 shows that the value of $b_i(\Pi)$ is greater than $u_i(x_i, x_{-i})$ at $x_i = 1$, that is, with a 100% frequency of violations of the regulations. It means that, $u_i(x_i = 1, x_{-i}) < b_i(\Pi)$. This means that the payoff for constant ($p_i = 1$) violations is higher (it is equal to $b_i(1)$), but if we subtract the penalty for a probable check, then it becomes clear to a rationally acting employee that it is more profitable not to violate the regulations more often than the threshold value and without checking (the net maximum payoff is then $b_i(\Pi)$).

In this case, it is obvious that it is more profitable for employees to violate the regulations no more often than the allowed compromise value, that is, the task of significantly reducing violations has been solved. In addition, the honest work of any employee has a lot of advantages both in the psychological and social sense. One of the consequences of the main postulate of the theory of managerism (Mayo, 2014) is that it is always more pleasant for a person to work honestly, this is called "self-satisfaction and internal encouragement" of an employee (Fig. 1).

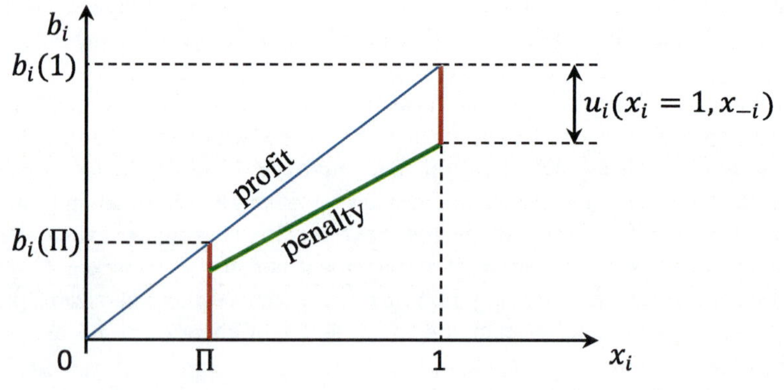

Fig. 1 Solution of the control problem with the "violation tolerance threshold". The y axis is – b_i, the "win" of the worker. On the x-axis – x_i, the probability of violating the regulations, that is, the level of quality of work or "malice" of the employee

Control with the Possibility of Collusion of Employees

Suppose in the situation under consideration, the regulations are violated by n employees with completely different levels of "malice", someone rarely violates, someone almost always. The management has chosen a strategy of equally probable control of all employees violating the regulations and any size of the sanction $T \geq 1$. Then $\forall i : x_i > 0 \Rightarrow p_i = \frac{1}{n}$. It is known that collusion can be established between employees. Suppose, in our case, employees violate the regulations in such a way that the condition imposed on the size of the "gain" of employees from violations is met: $\forall i \in \{1, \ldots, n\} \exists j \in \{1, \ldots, n\} : b_j < \frac{1}{i}$. Generally speaking, this means that there is a fairly significant spread of gains from violations of the regulations for different employees. Without loss of generality, for clarity, let's renumber employees in descending order of gain from violations: let $b_n < b_{n-1} < \ldots < b_1$. The condition imposed on the payoff now looks like this: $\forall j \in \{1, \ldots, n\} : b_j < \frac{1}{j}$. Then the situation will develop as follows: consider the strategy of the employee who is the least benefiting from his violations. He understands that he, like everyone else, will be checked with probability $p_n = \frac{1}{n}$. Then his profit from violations of the regulations will be: $u_n(x_n, x_{-n}) = (b_n - T/_n)x_n$. But this value is always negative based on the condition $b_n < \frac{1}{n}$ and $T \geq 1$. Consequently, with each new iteration, with each event of violation of the regulations, this employee will lose a little, his benefit will be negative, and he, acting rationally, will decide not to violate at all. This means that it will no longer be checked and for the remaining $n - 1$ the situation is repeated. Employee $n - 1$ is now the least profitable and the probability of checking for him (and for everyone else too) is $p_{n-1} = \frac{1}{n-1}$. Then for it $u_{n-1}\left(x_{n-1}, x_{-(n-1)}\right) = \left(b_{n-1} - \frac{T}{n-1}\right)x_{n-1}$. But this value, after the employee with number n refused to violate at the previous step, is also always negative based on the condition $b_{n-1} < \frac{1}{n-1}$ and $T \geq 1$. Of course, this worker with the number $n - 1$ also does not want to lose the resource with each new iteration and will decide not to violate.

Recursively, by similar reasoning, we prove that along the chain from n to 1, all n workers will stop violating. At the end of the recursion, the last worker understands that now $n = 1$, which means that the check will come only to him with probability 1 and it will be profitable for him to violate only in the case of the previous example, where $b_1 > T$, which is not true from the problem statement. The result—it also reduces violations to zero.

As a result, in this example, the possible collusion of employees played a negative role, because if they acted separately, then, most likely, it would be impossible to come up with such a set of probabilities $p_1, \ldots, p_n$, so that all violators would strictly comply with the regulations, reducing the "malice" to zero each. In the case of collusion, each subsequent employee took into account the fact that the previous one, who was less profitable from violations, had stopped violating the rules, the probability of checking him increased, and he also understood that now he also had no reason to do this.

Multi-stage Control Strategy

Consider the solution to the original problem proposed by Alexey Savvateev (2003) in his dissertation.

Let us introduce the so-called "class of sanctions strategies":

$$0 \leq z_1 < z_2 < \ldots < z_k < 1, z_{k+1} = 1,$$

where z_1 is the "tolerance threshold for violations", that is, what percentage of violations of the production regulations are we ready to forgive, acting similarly to the arguments of the Coase theorem on the external effect (Coase, 2007) that is, "shift this load" of the inevitable increase in costs to finished products. The threshold value for the majority of economic entities is, of course, computable according to the same logic by which we considered the values of the set $x_1, \ldots, x_n \in [0, 1]$ to be computable. Next, z_2 is introduced—the second "threshold". We, as managers of the enterprise, are no longer ready to endure it, and therefore we introduce some small probability of checking an employee who has violated this threshold, that is, violates the regulations more often and $z_1 < x_j \leq z_2$. Similarly, more stringent sanctions are imposed on violators who have crossed the threshold $z_2, \ldots, z_k$.

The principle of punishments is proposed to be chosen in the following way: the set of $\lambda_1, \ldots, \lambda_k$ is chosen like this: $\sum_{i=1}^{k} \lambda_i = 1$. In the payoff formula (1), we choose a set of probabilities $p_1, \ldots, p_n$ so that

$$p_i(x_i, x_{-i}) = {}^{\lambda_1}/_{m_1} + {}^{\lambda_2}/_{m_2} + \ldots + {}^{\lambda_k}/_{m_k},$$

where m_j is the number of violators, the probability of violation ("maliciousness") whose is higher then z_j. That is, the number of employees who "crossed the *j*th threshold of tolerance for violations."

We start with those employees x_j who have $x_j > z_1$. We found out that those employees who have $x_j < z_1$ in the sum $p_i(x_i, x_{-i})$ do not present, get our "forgiveness", because of the agreement of a "threshold of tolerance". The rest are controlled with some small probability ${}^{\lambda_1}/_{m_1}$. Now consider those who have crossed the next threshold: $x_j > z_2$. One more term is added to their probability of being checked: ${}^{\lambda_2}/_{m_2}$, that is, the probability of control increases with the increase in the "malignity" of their violations and becomes equal to

$$p_i(x_i, x_{-i}) = {}^{\lambda_1}/_{m_1} + {}^{\lambda_2}/_{m_2}.$$

Then similarly for $x_j > z_3$ and so on. That is, in the general case, if an employee has exceeded the threshold number l with his "malice" of violations, then the probability of checking his control employee will look like this:

$$p_l(x_i, x_{-i}) = \sum_{i=1}^{l} \frac{\lambda_i}{m_i}, \forall m_i \neq 0$$

It is worth noting that if at the next step $m_i = 0$ (no one exceeded the ith "violation tolerance threshold"), then $\frac{\lambda_i}{m_i} = 0$ is assumed. As Alexey Savvateev rigorously proved, with such a choice of control strategies, the situation leads to a strong Nash equilibrium, that is, the strategy is resistant to collusion of workers. It has also been proven that the solution works for any number of employees.

Conclusions

It is obvious that this solution to the problem of the frequency of control is rather complicated and unusual for the manager. However, it seems no less obvious that if it is applied, it will be possible to save significantly, moreover, both the manager's time and money and production resources. This can be especially clearly understood, considering, for example, a violation of the regulations as theft of materials or ingredients for the alloy during the production of means of transporting goods. If the alloy formula is violated, the properties of the material may change, and at the same time, the service life of the product will change and its competitive qualities will deteriorate, there will be shortcomings that were not taken into account in the design and production technology, etc. As a result, the reliability of vehicles using such parts is reduced, increasing the costs of the transport business and reducing its efficiency.

Strict certainty and algorithmization of checking, with possible difficulties in its implementation, in the process of work, eliminates all these disadvantages and saves resources for both the manager and the entire enterprise as a whole, regardless of the industry and scope of its activities.

The application of game theory methods in the checking of the transport business is relevant and requires in-depth research. Being multi-stage and complex, the transport business can obviously be divided into many processes, carried out by workers who need to be controlled. Services associated with transport tasks (such as loading, unloading, forwarding, temporary storage, documentary and legal support, etc.) also require rational management and, as a result, control. Moreover, the specificity of these related services often makes them the most "tempting" for violations of regulations, rules, up to outright theft and corruption.

Consider, for example, a trucking business. Some of the main target indicators of the production schedule are the delivery time between key points, fuel consumption and depreciation of equipment. All these indicators have a certain tolerance for violations, both objective (fuel system breakdown, traffic jams, weather conditions, bad road conditions, etc.), and subjective, that is, depending on the quality of work and the conscientiousness of employees (theft of fuel and spare parts, fictitious breakdowns, excessive rest, alcohol consumption, etc.) (Dorofeev, 2018).

Particularly subjective factors are subject to control in transport management. It is possible to allow a certain "threshold of violations", that is, to include some additional costs in the calculation of profitability. Further, with this "threshold", the controlling manager turns on the Game theory algorithms and achieves the calculated result, as in the problem with the result in the form of a strong Nash equilibrium. This can significantly save on the costs of the entire business or even save it from bankruptcy.

That is why there is an urgent need to optimize and improve the efficiency of checking operations in the transport business and introduce modern innovative technologies and mathematical methods into management.

Based on this, the problem of applying game theory in control in the transport business opens up a field for research and poses new tasks for optimizing algorithms and checking principles.

Thus, the proposed methods for applying game theory to checking problems in business for the transport business are of particular importance. This is due to the fact that the optimization of control leads to a reduction in the costs of the transport industry, which will inevitably have an impact on the stability of the economy as a whole and is of macroeconomic importance.

References

Becker, G. S. (1976). *The economic approach to human behavior* (Vol. 803). University of Chicago Press.

Bertrand, J. (1883). Book review of theorie mathematique de la richesse social and of recherches sur les principes mathematiques de la theorie des richesses. *Journal des savants, 67*(1883), 499–508.

Bolotova, L. S., Novikov, A. P., Surkhaev, M. A., & Nikishina, A. A. (2015). "Machines that imitate life" Features of the architecture of such software systems. *Journal of Applied Informatics, 10*(55), 114–140.

Coase, R. (2007). *The Firme, the market and the law*. New Publishing House.

Dorofeev, A. N. (2018). *Efficient road transport management (Fleet management)*. Dashkov and Co.

Fayol, A., Emerson, G., Taylor, F., & Ford, H. (1992). *Management is a science and art*. Republic.

Golovan, S. V., Guriev, S. M., & Makrushin, A. V. (2005). *Contract theory: A collection of problems with solutions*. Moscow.

Gudkov, A. V. (2013). The use of a rating assessment of the financial condition of organizations in economic analysis. *Scientific Opinion, 8*, 306–317.

Gudkov, A. V. (2022, January). Actual factors efficiency in the sphere of art management: Remote control specifications. In *Second conference on sustainable development: Industrial Future of Territories (IFT 2021)* (pp. 470–478). Atlantis Press. https://doi.org/10.2991/aebmr.k.211118.084

Kennedy, D. (2014). Hard management. In *Make people work*. Alpina Publisher.

Kremlev, A. G. (2016). *Basic concepts of game theory: Textbook*. Ural University Publishing House.

Mayo, E. (2014). *The social problems of an industrial civilisation*. Routledge.

Mazalov, V. V. (2010). *Mathematical game theory and applications*. Lan.

Perevedentsev, S. (2018). *Control. video lecture*. Retrieved from https://unimys.com/product/sergey-perevedentsev-kontrol-udalennyh-sotrudnikov-2020/

Savvateev, A. V. (2003). *Modeling of corruption and lobbying in transitional economies*. Candidate of Economic sciences Thesis (Central Economics and Mathematics Institute of the Russian Academy of Sciences).

Seligman, B. (1968). *Main currents in modern economic thought*. Progress.

Tsvetaeva, E. M. & Dzgoeva, A. A., (2019). Zugzwang in the strategy and tactics of top managers. *Economic Development Research Journal*. Retrieved from http://edrj.ru/article/14-03-2019

Regulatory Regulation of the Issue and Circulation of Digital Currencies in the Countries of the Modern World

Dmitry Andreevich Makarov, Nikolay Petrovich Mayurov, and Grigory Khoteevich Humenyuk

Introduction

The relevance of the studied issues is determined by a number of circumstances. Currently, crisis phenomena are increasing in the global financial system. They are caused by the growth of informatization of economic processes and an increase in the share of information products themselves in world production. The development of the Internet and the facilitation of communication between buyers and suppliers of goods, works and services necessitates fundamental changes in the existing system of their financing. To conclude a transaction for the purchase of goods, the performance of works and the provision of services by consumers and entrepreneurs, who are often separated by tens of thousands of kilometers, it is only necessary to press a few keys on a computer. These transactions can be valued from several dollars to billions of dollars. At the same time, the system of payments in the form of cash and non-cash settlements using credit institutions no longer corresponds to the realities of the modern world as inefficient and time-consuming. As a result, within the framework of the dynamics of the development of modern, not only global, but also domestic, trade, there is an objective need to simplify the settlement system. This circumstance has caused the emergence of digital payment systems that complement the existing cash and non-cash money circulation.

But at the same time, previously unknown problems of regulatory regulation of such digital calculations appear. We are talking about emerging gaps in the legal regulation of settlements using digital resources that do not belong to the category of cash or non-cash money. Among them, the following should be noted:

D. A. Makarov (✉) · N. P. Mayurov · G. K. Humenyuk
St. Petersburg University of Management Technologies and Economics, St. Petersburg, Russian Federation
e-mail: s.gumenuk@spbacu.ru

A. Rumyantseva et al. (eds.), *Challenges and Solutions in the Digital Economy and Finance*, Springer Proceedings in Business and Economics,
https://doi.org/10.1007/978-3-031-14410-3_23

1. the uncertainty of the legal status of cryptocurrencies created by various owners of information platforms and their use as a means of payment, circulation, and capitalization.
2. the ambiguity of the legal grounds for issuing digital money and their circulation as a means of payment within the framework of national and international settlements.
3. the commission of illegal acts for the purpose of legalizing illegally obtained funds and tax evasion.
4. undermining the value of national and regional currencies, as well as world money—the US dollar, by using payment systems that are not recognized as a currency under the legislation of a particular country, and many others.

The resolution of these problems of legal regulation is one of the urgent tasks of reforming financial legislation in the countries of the modern world. Legislation on the regulation of legal relations related to the use of digital money is currently insufficiently developed. In the leading countries of the modern world, including the USA, China, Japan, Germany, Russia, etc. there is either the development of draft laws in this area, or the adopted laws have not fully begun their effect. Only based on a deep and comprehensive scientific consideration of this issue by the scientific community is it possible to reform this branch of financial law dedicated to the regulation of legal relations related to money circulation.

Materials and Methods

Various scientific methods used in economics, law, statistics and other sciences were used in scientific research. Among them, it is necessary to note the methods of formal-legal, statistical, fact-gathering, etc. General scientific methods of analysis and synthesis, induction, and deduction, etc. are also used. The normative basis of the study is legislative acts and other sources of law that contain norms regulating legal relations related to the use of digital currencies in the countries of the modern world. Among them it should be noted: Resolution of the Cabinet of Ministers of Japan "On Virtual Currency Exchanges" (E-GOV, 2021), The US Law "On Uniform Regulation of Activities in the Field of Virtual Currencies" (The Blockchain Law Blog, 2022), California Law "Legal Money: Alternative Money", Wyoming Digital Assets Act, New Jersey Blockchain and Digital Assets Act, Notification of the People's Bank of China "On preventing the financial risk of Bitcoin", The law of Germany "Taxation of bitcoins and other so-called virtual currencies with sales tax (Bundesministerium der Finanzen, 2022), The Law of the Russian Federation "On Digital Financial Assets", Directive of the European Parliament 2009/110/EC (EUR-LEX, 2009) and others.

The theoretical basis of the study was the works of specialists in the field of financial and information support for the turnover of cryptocurrencies in the modern world. Among them, the following authors should be noted. Matthew Green is a specialist in computer security and cryptography, associate professor of computer science at Johns Hopkins University. Peter Van Valkenburg Director of the Research Department at Coin Center. Christopher Giancarlo chapter Digital Dollar Project (DDP), former head of the Federal Commodity Futures Trading Commission. The organization he represents is the world's largest scientific center for the study of digital currencies in the modern world. Daniel Gorfine Director CFTC. Ben Broadbent, one of the chairmen of the Bank of England, etc.

In general, the works of modern researchers are devoted to the following main problems of regulatory regulation of cryptocurrencies in the modern world, namely digital currencies as:

1. a de-dollarization tools.
2. securities.
3. an analogue of non-cash money that simplifies trade and money transfers.
4. a way to avoid taxation.
5. an alternative to using a global payment system Society for Worldwide Interbank Financial Telecommunications, etc.

Results

Digital currency is a form of currency existence along with cash and non-cash forms. It exists in a certain digital code and can only be used when making a payment via the Internet as part of operations on a certain virtual platform. Storage of digital currency is possible on special electronic wallets. The use of digital currencies is carried out on special portals related to the purchase of goods, works and services in social networks, gaming platforms, exchanges, etc. (Tsindeliani, 2019). Digital currencies can be regulated and virtual. Regulated digital currencies are issued by the central banks of states along with currency in cash and non-cash form (Shaen et al., 2017). This practice is currently declared, for example, in China and the Russian Federation (Bech & Garratt, 2017).

Virtual currencies are cryptocurrencies, and they are issued by private law entities, and not by the state or official issuing centers, for example, the Central Bank of the Russian Federation, the US Federal Reserve, etc. In Japan, the issue of cryptocurrencies for use in exchange trading requires state registration (Hagiwara & Nakamura, 2018). Cryptocurrencies are electronic products that arise because of the work of the program—a digital code. Blockchain computer technology is used, a special decentralized protocol is functioning (Nakamoto, 2018), because of which a special digital code appears that belongs to one or another owner. If the issuer is centralized, then we are talking, strictly speaking, not about cryptocurrency, but about electronic money. Cryptocurrencies are issued and stored in a decentralized manner (Song et al., 2019). The term "mining" is used. Cryptocurrencies, as a rule, are "mined" by certain persons as an initial method of acquiring ownership rights and only later they become the subject of turnover (Abdulhadi & Talabani, 2022). As a result, we are talking about a decentralized and often spontaneous emission, directly determined by the features of the digital protocol execution. The register of operations is also maintained in a decentralized manner (Financial Action Task Force, 2014). Cryptocurrencies are one of the computer programs and virtual objects that exist on the Internet as a kind of movable property. As a result, they cannot have a physical form of embodiment as paper or metal money and records in accounting documents like non-cash money, which have their own form of reflection in computer accounting programs. Swiss legislation explicitly states that cryptocurrency is not a legal tender (Der Bundesrat, 2014).

In turn, electronic money is both issued and stored centrally on a special server. The issuing center can both issue electronic money into circulation and withdraw it. Virtual currencies can be convertible i.e., freely exchanged for fiat money and non-convertible (Grinberg, 2011). An example of non-convertible currencies is banknotes used in computer games, social networks, etc.

The need to study these issues is initially based on the importance of cryptocurrencies as an exchange commodity. A slight spread of cryptocurrencies would mean a lack of interest on the part of the states of the modern world in solving issues of their regulatory regulation. Market capitalization, defined as the total market value of the circulating cryptocurrency supply, which corresponds to the current price, in rubles multiplied by the current supply. It is indicated in brackets of the description of the cryptocurrency. In particular, as of January 15, 2022, the following cryptocurrencies with a market capitalization of more than 1 billion rubles participate in exchange trading: Bitcoin—BTC (₽62,163,750,870,665); Ethereum—ETH (₽30,436,503,633,993); Binance Coin—BNB (₽6,362,571,229,803); Tether—USDT (₽5,979,279,466,941); Cardano—ADA (₽3,605,612,557,154); Solana—SOL (₽3,571,550,630,655); USD Coin—USDC (₽3,465,970,222,956); Ripple—XRP (₽2,829,476,052,297); Terra—LUNA (₽2,352,395,942,491); Polkadot—DOT (₽2,102,860,629,231); Dogecoin—DOGE (₽1,791,985,967,097); Avalanche—AVAX (₽1,735,622,628,671); Polygon—MATIC (₽1,301,697,293,881); Shiba Inu—SHIB (₽1,272,635,127,909); Binance USD—BUSD (₽1,101,069,142,292). The total number of cryptocurrencies used in trading exceeds a thousand, the number of organized trading platforms for more than one cryptocurrency as a commodity is estimated in the world by hundreds. The total number of cryptocurrencies issued for organized trading, including those for which trading is no longer conducted, is about 14,000.

The main global platforms for trading cryptocurrencies are the following exchanges with an indication in parentheses of the trading volume in billions of dollars within 24 h: Binance (22.5); OKEx (4.7); Huobi (3.5); FTX (1.5); GATE.IO (1.2); YoBit (0,429); Currency (0,117); EXMO (0,094); Bybit (0,083). As a result, the countries of the modern world cannot ignore the initially spontaneously emerging cryptocurrency market.

The instability of the current economic situation in the world requires the adoption of sufficiently prompt measures to strengthen the financial system. After the collapse of the Bretton Woods and Jamaican world monetary and financial systems, it became necessary to change the position of the world currency. The US dollar currently performs the function of world money as the national currency of this state. Cross-border settlements between both private sector economic entities and between citizens and states mainly use the dollar as a means of payment in their cross-border settlements (Stosic et al., 2018). The main volume of securities in circulation is denominated in dollars. The dollar is used in the calculations of the Visa Electron system and the global interbank platform Society for Worldwide Interbank Financial Telecommunications. Calculations in dollars are also carried out on world stock and commodity exchanges.

Currently, the US dollar is not a currency secured by the possibility of exchange for gold or the abstract wealth of the American economy. The US dollar is primarily world money, mediating global economic turnover in all sectors of world trade, as well as capital investment. It is economically based in its significance primarily on the energy market and exists as a kind of "petrodollar". An important factor is the direct military dictatorship of the United States in relation to many countries of the modern world as a measure of non-economic coercion to expand the dollar as world money. At the same time, the broad issue of the US dollar by the Fed, which is based, among other things, on the colossal US budget deficit, which annually exceeds the amount of $ 1 trillion, inevitably leads to a disproportionate increase in the money supply, which accelerates inflation not only in the US, but also in other countries of the modern world. To absorb the surpluses of the money supply that arise, in fact, numerous often purely financial instruments are used. There is a swelling of the stock market, which absorbs excess money supply. Cryptocurrencies themselves create problems associated with an increase in the money supply in the world. Because they are a kind of similarity of money and can be used as a means of payment. But at the same time, they are also instruments of capital investment. Cryptocurrencies act as a commodity that is bought and sold for dollars like securities circulating in the stock market. They are not such by their legal nature, but in terms of their use as a commodity, they can act as their equivalent.

The American state and the Federal Reserve System also have a desire to use digital money as a financial instrument in the cryptocurrency market. This desire is necessary primarily for the implementation of market mechanisms of currency regulation. The digital dollar being created, announced but not released on January 1, 2021, should create competition in the digital currency market and possibly displace some of them. At the same time, two concepts of the digital dollar are being implemented in the United States in parallel. American researchers Matthew Green and Peter Van Valkenburg, representing the Coin Center organization, based on the analysis of the bills proposed to the US Congress, identified their two main directions (Green & Valkenburgh, 2020). The legal status of the digital dollar can be defined either as centralized or as decentralized. In the case of a centralized dollar, it should be just one of the forms of the embodiment of the American currency, along with cash and non-cash dollars. The digital dollar will be a fiat currency in the same way as the cash and non-cash dollar since it will also not be secured by anything (Czapliński & Nazmutdinova, 2019). But it will be completely controlled by the Fed and American government agencies. In this regard, it will not be a cryptocurrency since it will be just a form of embodiment of an existing American currency. The advantage of using the digital dollar will be the ability to conduct financial transactions more quickly and provide financial assistance to American citizens using existing electronic applications. Transactions with such a digital currency will be completely confidential for third-party users but will be under full control from the Federal Reserve and US government agencies. Making transactions with such a digital dollar will be preferable for the state than with cash.

It is planned to use encryption technologies used in Facebook's Libra cryptocurrency services for a centralized digital dollar. Within the framework of these systems, it is planned to preserve the confidentiality of the sender of the payment, the recipient of the payment, and the destination of the payment while controlling the amount of digital money from the Fed, government agencies, the owner and the digital money accounting platform. This system has a number of advantages, in particular the possibility of making a payment, bypassing the instance in the form of a credit institution. But at the same time, the centralized American dollar in this system has a significant drawback—the lack of confidentiality from government and credit institutions. As a result, the scope of its application will be limited and mainly to various social payments carried out by the State directly. The Treasury should maintain and update the register of transactions of the digital dollar.

But the question arises that, in fact, there will be a simple duplication of non-cash money circulation. The issue of fiat currency is carried out on behalf of state bodies with subsequent state control without physical restrictions, both in terms of the number of issued digital currency and transactions. As a result, the possibility of a two-tier decentralized system of digital currencies in the United States seems quite possible, and it will not be replaced by a digital dollar.

In general, in the United States, the position of various states remains ambiguous and at the federal level there is only a recommendation act on the legality of the turnover of cryptocurrencies as an asset when exercising the right to receive funds in US dollars, provided that this right is recognized by the legislation of the state in which such transactions were made. The laws of the US states have different legal regimes for the recognition of cryptocurrencies. It can be considered as a security, a deferred payment, cash in circulation along with a non-cash dollar (Wawrosz & Lansky, 2021), a virtual commodity traded on the stock exchange, etc. As a rule, according to the legislation of most US states in which the use of cryptocurrencies is recognized as legal, they have the legal status of securities in circulation in an undocumented form, causing benefits in monetary terms in US dollars or other material benefits.

The UK's position is ambiguous. Laws and judicial precedents do not protect the rights of owners of cryptocurrencies, but at the same time, their use does not fall under the legislation on the legalization of proceeds from crime. The use of cryptocurrencies in Germany is recognized as legal and is subject to taxation. Legal tender is the use of cryptocurrencies in Canada. In Russia, China, the UAE, Switzerland, Bangladesh, Bolivia, Ecuador and many other countries of the modern world, the use of cryptocurrencies is prohibited. There is no legislation in Spain that clearly regulates the circulation of cryptocurrencies (Alonso, 2019).

In the states that have announced the release of electronic national cryptocurrencies, including the Russian Federation, China, the Bahamas, etc., a policy is being implemented to exclude cryptocurrencies from circulation at the same time. Electronic money should have a single issuing center, freely exchanged in an equal ratio for non-cash and cash national currencies. Anonymity of payments should be excluded (Osivand, 2021). The benefits of transactions are subject to taxation, etc.

Discussions

It is necessary to note the following features of the position of digital currencies in the modern world. They should be divided into two main categories:

1. digital currencies issued by government agencies or central banks of states as one of the forms of national currency along with cash and non-cash.
2. cryptocurrencies themselves, as a rule, do not have a single issuing center and originate from the use of computer programs and special platforms.

In the first case, in relation to the use of national (regional) digital currencies, the norms of legislation applicable to monetary circulation in each country (group of states) are applied by default. As of the beginning of 2022, the process of creating such currencies is just beginning, and it is not possible to draw unambiguous conclusions about the specifics of the turnover of such electronic money due to the lack of sufficient factual material. In the leading countries of the modern world, the USA, China, Russia, Japan, etc., the introduction of digital national currencies has just been announced.

In turn, cryptocurrencies have become widespread as a commodity and a payment instrument for goods, works and services in a number of countries of the modern world. Such countries should include the USA, Japan, Canada, etc. The daily turnover of the global cryptocurrency trade amounts to billions of dollars. But at the same time, many countries, including the Russian Federation and China, do not allow the use of cryptocurrencies on their territory, and especially their conversion. The reasons for this policy are quite obvious. The use of decentralized cryptocurrencies guarantees the anonymity of making payments, which contributes to tax evasion and the absence of state control and supervision over the legality of making payments.

It is important to note the already mentioned circumstance that cryptocurrencies are an exchange commodity that absorbs an excessive mass of dollars. In this regard, it becomes obvious the approach of legislators to understanding cryptocurrencies primarily as a security traded on the stock exchange. One can cite as an example the decision of the New York court regarding the Telegram cryptocurrency by P. Durov (United States Court of Appeals for the Fifth Circuit, 2021). In this regard, cryptocurrencies play an important role in stabilizing the position of world money—the US dollar.

Conclusions

As a result, it is necessary to draw conclusions based on the analysis on the prospects for the development of regulatory regulation of digital currencies in the modern world. The emergence and widespread use of national electronic currencies is a matter of the very near future. This process is objectively conditioned by the steady growth of the digitalization of the economy. The use of digital money is possible without using or even opening a bank account. For many millions of modern people, the use of electronic gadgets in everyday purchases is commonplace. Similarly, direct payments to citizens and organizations without the use of a cashless payment system significantly accelerate the process of turnover of goods, works and services. Cross-border payments with electronic money are much more profitable and operational than the cumbersome banking system of letters of credit. It is possible to predict a steady increase in the value of electronic national currencies along with a decrease in the value of non-cash and primarily cash payments. At the same time, the development of decentralized cryptocurrencies, which are surrogates of both money and securities, will have its limited significance. In many ways, it was the instability of the financial system and the need to absorb the excess dollar mass that caused their appearance. Along with the development of national digital currencies, their value will decrease. The state is not interested in the existence of a significant volume of anonymous payments, which become the basis for the legalization of proceeds from crime and tax evasion.

References

Abdulhadi, H., & Talabani, H. S. (2022). Bitcoin ransomware detection employing rule-based algorithms. *Science Journal of University of Zakho, 4*, 5–10.

Alonso, S. L. N. (2019). Activities and operations with cryptocurrencies and their taxation implications: The Spanish case. *Laws, 16*, 2–13. https://doi.org/10.3390/laws8030016

Bech M. & Garratt R. L. (2017). *Central bank cryptocurrencies* Retrieved from https://www.bis.org/publ/qtrpdf/r_qt1709f.html.

Bundesministerium der Finanzen. (2022). *Umsatzsteuerliche Behandlung von Bitcoin und anderen sog. virtuellen Währungen; EuGH-Urteil vom* 22. Oktober 2015, C-264/14, Hedqvist. Retrieved from https://www.bundesfinanzministerium.de/Content/DE/Downloads/BMF_Schreiben/Steuerarten/Umsatzsteuer/Umsatzsteuer-Anwendungserlass/2018-02-27-umsatzsteuerliche-behandlung-von-bitcoin-und-anderen-sog-virtuellen-waehrungen.pdf?__blob=publicationFile&v=1

Czapliński, T., & Nazmutdinova, E. (2019). Using fiat currencies to arbitrage on cryptocurrency exchanges. *Journal of International Studies, 12*(1), 184–192. https://doi.org/10.14254/2071-8330.2019/12-1/12

Der Bundesrat. (2014). *Bericht des Bundesrates zu virtuellen Währungen in Beantwortung der Postulate Schwaab (13.3687) und Weibel (13.4070)*. Retrieved from http://www.news.admin.ch/NSBSubscriber/message/attachments/35361.pdf.

E-GOV. (2021). 仮想通貨取引所について"日本の大臣の内閣の決議Retrieved from https://elaws.e-gov.go.jp/document?lawid=347M50004c00001

EUR-LEX. (2009). *Directive of the European Parliament 2009/110/EC.* Retrieved from https://eur-lex.europa.eu/legal-content/EN/TXT/?uri=CELEX:32009L0110

Financial Action Task Force. (2014). *Virtual Currencies - Key Definitions and Potential AML/CFT Risks.* Retrieved from https://www.fatf-gafi.org/media/fatf/documents/reports/Virtual-currency-key-definitions-and-potential-aml-cft-risks.pdf

Green M. P. & Van Valkenburgh. (2020). *Without privacy, do we really want a digital dollar? If a digital dollar is developed it must be just as "bearer" and private as physical cash.* Retrieved from https://www.coincenter.org/without-privacy-do-we-really-want-a-digital-dollar/

Grinberg, R. (2011). Bitcoin: An innovative alternative digital currency. *Hastings Science & Technology Law Journal, 5*, 159–160.

Hagiwara Y. & Nakamura Y. (2018). *Japan expands cryptocurrency crackdown after Coincheck hack.* Retrieved from https://www.bloombergquint.com/business/japan-expands-cryptocurrency-crackdown-after-coincheck-hack

Nakamoto S. (2018). *Bitcoin: A peer-to-peer electronic cash system.* Retrieved from https://genius.com/Satoshi-nakamoto-bitcoin-a-peer-to-peer-electronic-cash-system-annotated

Osivand, S. (2021). Investigation of metaverse in cryptocurrency. *GSC Advanced Research and Reviews, 09*, 125–128. https://doi.org/10.30574/gscarr.2021.9.3.0306

Shaen, C., Grace, M., & Andrew, M. (2017). The influence of central bank monetary policy announcements on cryptocurrency returns volatility. *Investment Management and Financial innovations, 14*(4), 60–72. https://doi.org/10.21511/imfi.14(4).2017.07

Song, J. Y., Song, J. W., & Chang, W. (2019). Cluster analysis on the structure of the cryptocurrency market via bitcoin–ethereum filtering. *Physica A: Statistical Mechanics and its Applications, 527*, 121–339. https://doi.org/10.1016/j.physa.2019.121339

Stosic, D., Ludermir, T. B., & Stosic, T. (2018). Nonextensive triplets in cryptocurrency exchanges. *Physica A: Statistical Mechanics and its Applications, 505*, 1069–1074. https://doi.org/10.1016/j.physa.2018.04.066

The Blockchain Law Blog. (2022). *Uniform regulation of virtual-currency businesses act offers states regulatory framework for the virtual currency industry.* Retrieved from https://www.burr.com/wp-content/uploads/2018/01/Blog_Uniform-Regulation-of-Virtual-Currency-Businesses-Act-Offers-States-Regulatory-Framework-for-the-Virtual-Currency-Industry.pdf

Tsindeliani, I. (2019). Financial regulation & digital money: How Russia dips its toes into the waters of cryptocurrency. *Global Jurist, 19*(2), 134–144. https://doi.org/10.1515/gj-2019-0014

United States Court of Appeals for the Fifth Circuit. (2021). *New York Court Decision Case 1:19-cv-09439-PKC Document 227 Filed 03/24/20* Retrieved from https://storage.courtlistener.com/recap/gov.uscourts.nysd.524448/gov.uscourts.nysd.524448.227.0.pdf

Wawrosz, P., & Lansky, J. (2021). Cryptocurrencies and corruption. *Ekonomický Casopis, 69*, 687–705. https://doi.org/10.31577/ekoncas.2021.07.02

Socio-Economic Factors Determining the Genuine Savings

Aleksandr L. Pastukhov, Sergey Yu Solodovnikov, and Yulia V. Meleshko

Introduction

The problems of linking the theory of economic growth and the environmental imperative are the subject of the work of many scientists. Problems of forming the equilibrium of the national economy in the context of changing the external economic environment with the help of financial mechanisms are considered in the works of Wolf (2015), Mundell et al. (2002), Adrian et al. (2018). Problems of attracting investments in environmental projects are resourced by Ljumović et al. (2021), Chan et al. (2021), Prędkiewicz and Kalinowska-Beszczyńska (2020), Zhang et al. (2021). Many scientists have achieved success in improving the methodological support for assessing sustainable development (El Alaoui, 2020; El Alaoui & Ben-Azza, 2017; Figge et al., 2018; Molina-Sánchez et al., 2018; Smol et al., 2017; Tantau et al., 2018; Zhao et al., 2017). At the same time, an exhaustive comprehensive assessment of the real costs (social, environmental, economic) incurred by society because of economic activity is still difficult today. The reasons lie not only in the fact that such a voluminous task is rarely set by researchers-economists, but also in the high impact on the sustainability of qualitative socio-economic factors that are difficult to quantify.

There are many rating systems of sustainable development, the most common of which are OECD Sustainable Development Indicators, UNECE Sustainable Development Systems, System of Environmental-Economic Accounting (SEEA). The UN recommended index of genuine savings (GSI) is one of the most well-known integral

A. L. Pastukhov
St. Petersburg University of Management Technologies and Economics, St. Petersburg, Russian Federation

S. Y. Solodovnikov · Y. V. Meleshko (✉)
Belarusian National Technical University, Minsk, Republic of Belarus

A. Rumyantseva et al. (eds.), *Challenges and Solutions in the Digital Economy and Finance*, Springer Proceedings in Business and Economics,
https://doi.org/10.1007/978-3-031-14410-3_24

indexes (indicators) of sustainable development. It allows you to give the most accurate assessment of sustainable development. The specificity of economic growth accounting based on the genuine savings lies in the potential ability to estimate how adaptive a particular system is without causing irreparable damage to itself and a person. However, the application of this index in practice is hampered by the lack of a unified approach to its calculation.

The authors of the article proceed from the hypothesis that since in order to accurately assess the results of economic activity over a relatively long period, it is necessary to use methodological support that allows taking into account all possible costs (economic, environmental, social). The existing system of indicators of economic efficiency needs to be supplement with the index of genuine savings. According to the authors, it is the index of genuine savings that have the necessary epistemological and practical potential to solve the problem of assessing the country's sustainable development.

The purpose of this article is to develop a methodology for calculating the index of genuine savings by identifying and meaningfully describing the socio-economic factors that determine the growth of genuine savings. This will make it possible to advance in the field of assessing the sustainable development, to move from the slogans of environmental protection to the use of methodological tools for assessing the actual result of economic activity in a given territory, taking into account all environmental resources.

Materials and Methods

The material of the study was the work of well-known scientists and economists concerning the problem of assessing economic growth, taking into account environmental and social costs. The methodological basis of the study is represented by the theoretical provisions of the concept of sustainable development. The starting point for the study was the index of genuine savings proposed by the United Nations. As a criterion for assessing socio-economic efficiency, the modified Kaldor-Hicks criterion is used.

To solve the task, the following were used: systematization and conceptualization of scientific ideas, subjective approach, as well as general scientific methods of analysis, synthesis, and qualitative assessment. The methods were applied in the system, but the choice of each of them was determined by the ability of the method to establish a causal relationship, to identify the internal structure, the driving forces for the growth of genuine savings, as well as its ability to act in conjunction with other methods.

Results

Many modern scientists pay attention to the role of socio-economic factors in solving environmental problems. Li et al. (2020, p. 556), point out that crowdfunding investors in agricultural products may not always be rational but can be somewhere between rational and irrational. Ljumović et al. (2021) «argue that backers (crowd-investors—authors' note) are often driven by normative or altruistic motives <...>, usually focus on the entrepreneurs' core values and ideas (such as sus tainability, social agenda and similar) instead of focusing on business plans <...>, and are initiated by intrinsic motives <...>». De Crescenzo et al. (2020) highlight the critical role played by social networks in the promotion of renewables and energy projects.

Socio-economic factors that determine the growth of genuine savings, have their impact not in isolation, but systemically. In other words, in the process of influencing the dynamics of genuine savings, these factors also influence each other, as a result of which not only synergistic effects arise and constantly change, but these factors themselves are transformed. The latter is expressed in the evolution of the ways, mechanisms and results of their impact both on the growth of genuine savings and on each other. This stems from the fact that all real (as opposed to virtual) social phenomena cannot be initially considered in statics, and then in dynamics, "since the origin, functioning, and change of any social objects is continuous, not ceasing for any instant process (called life), which can only be divided into some ideal periods or states for certain scientific purposes, with a number of reservations and for certain scientific purposes" (Solodovnikov, 2014, p. 17).

The action of socio-economic factors that determine the growth of genuine savings is predetermined by the entire system of socio-cultural, political, economic, ideological and other relations. The latter in the most general form can be described using the categories of culture, civilization, the economic system of society and the institutional matrix.

Socio-economic factors that determine the growth of genuine savings can lead to both gradual quantitative changes and qualitative changes in the latter, i.e. to their evolution. The basis for the evolution of genuine savings are the changes taking place in nature, civilization, culture, the economic system of society and the institutional matrix. Accordingly, the basic factors (prerequisites and conditions) that determine the growth of true savings are changes in: (1) nature, (2) civilization, (3) culture, (4) the economic system of society, (5) the institutional matrix.

Based on the objectives of our study, it is legitimate to consider civilization as an objective form of the structure of a society of divided labor, materialized in the form of a city as the next stage of social integration, the emergence of which correlates with the beginning of urban culture and with the Neolithic technological revolution. The literature emphasizes that "the socially integrative charge of the material civilization of the urban type had a radical impact on the spiritual sphere, which allows us to give a holistic analysis of various aspects of the life of a civilized society. With this understanding of the term "civilization", it will be possible to show the historically regular course of the emergence of a civilized society as the next stage of social integration" (Afanasiev, 1983, p. 5). This approach allows one to substantiate "a causal relationship between the demographic state of a society and the degree of complexity of the technology it practices, which makes it possible to explain the correlation of major demographic and technological revolutions in human history" (Afanasiev, 1983, p. 5).

The history of changing nature, the development of technology and the division of labor is of paramount importance for understanding the origin of civilization. Osokin, comprehending the reasons for the emergence of significant differences in European and Asian civilizations at the beginning of the first millennium of our era, gave such an example of their dependence on natural and climatic conditions: "If Europe consisted of such vast and smooth plains as Asia, then in the shortest possible time the barbarian tribes, who roamed chaotically for centuries between the Dnieper and the Rhine, would have managed to unite into one people. But in Central Europe, in a relatively small space, nature has given an amazing variety. The land occupied by the Germanic tribes, i.e. Central Europe, was at that time covered with impenetrable forests and swamps, cut through by mountains and hollows, which prevented the communication of neighboring tribes. Therefore, many Germanic peoples for centuries remained unknown to each other" (Osokin, 2005, p. 57).

Culture is a system of historically developing overbiological programs of human activity that are a condition for the reproduction and change of social life in all its main manifestations and are a social way of satisfying natural needs, usually repeatedly mediated. Such an understanding of culture makes it possible to connect it directly with the economic system of society. The economic system of a society is a cultural phenomenon that is a "single, stable, institutionalized, relatively independent, material and social complex, within which internally interconnected production, appropriation and socially significant consumption of material resources and goods are carried out to ensure the physical life of society, and also to create the material base necessary in all other spheres of public life" (Solodovnikov, 2020, pp. 13–14). The basis of the functioning of the economic system is labor relations based on the social division of labor. Simultaneously with the emergence of civilization, culture, the economic system of society and politics, institutional matrices arise and develop in close relationship with them (Solodovnikov, 2006).

In the most general form, the effect of all the above-mentioned basic factors (prerequisites and conditions) on the growth of true savings can be both unidirectional and oppositely directed. This is due to their relative independence. At the same time, it is genetically possible to build a sequence of their influence, namely: the peculiarities of natural and climatic conditions, the availability of natural resources, and geographical location have an impact on the emergence and development of civilization. At the same time, civilizational features, forming and reflected in culture, have a significant impact on the methods, ways and consequences of influencing nature. This impact is mediated by the economic system of society, which in turn acts both as a base and as an object of influence from the institutional matrix. As a result, each nation, each state forms a unique mechanism for the impact of changes occurring in nature, civilization, culture, economic system of society and institutional matrix on the dynamics of genuine savings.

The prerequisites and conditions for the growth of genuine savings have their impact on the latter not directly, but under the influence of other, more dynamic factors—the causes of genuine savings dynamics. As the latter, in our opinion, should be considered: firstly, the change in the place and role of the country in the international division and cooperation of labor; secondly, stable trends in global commodity and product markets; thirdly, the evolution of the social class structure; fourthly, a change in the accumulation and capitalization of social potential; fifthly, the emergence of new technological waves and a change in the ratio of dominant and expolar technological waves; sixth, the evolution of our management relations; seventh, the development of science about environment, human and society.

All socio-economic factors that determine the growth of genuine savings are closely related to each other and interdependent. In some cases, it is impossible to draw a clear line between them.

Discussion

Turning to the problem of true savings, it is impossible, simplifying this problem, to reduce it only to the dilemma of preserving nature for our descendants or wasting it thoughtlessly. In reality, society is faced with a much more difficult choice—the choice of a measure of acceptable damage to nature, if it is necessary to expand the reproduction of human and social potentials, as well as to fuel economic growth and a constant increase the standard of living. Even at first glance, these trajectories (vectors) of social, environmental and economic development are oppositely directed. Accordingly, in order to minimize possible risks and optimize the social, environmental and economic results of the growth of genuine savings, it is necessary to take into account the features of the functioning of the system as a whole.

To ensure sustainable human activity in natural environments, the government enacts various management policies and regulations. However, although these policies and regulations can achieve ecological protection, they also affect the development of the local economy and, in turn, the lives of residents. This one-way governance model usually faces certain challenges when implemented. For example, companies producing polluting products may choose to move to other cities with weaker environmental supervision rather than consider adopting an approach to co-existing with the local environment (Wang et al., 2021).

The above features of the interaction of societies with the environment, their ambiguity and multivariance, also led to the well-known logical internal inconsistency of the concept of sustainable development. The concept was based on the idea of minimizing the anthropogenic impact on nature, but over time it was expanded and supplemented by the ideas of social equity. As a result, there is terminological confusion in a number of works, and sustainable development begins to be understood as a kind of ideal system that allows both to preserve nature and observe social equity. As a result, the idea of the implementation of this concept becomes simplified and idealistic. So, for example, the description of sustainable global food supply chain in the article of Pietrzyck et al. (2021): "Sustainable FSCs in international trade activities have the potential to reduce environmental problems and the carbon footprint, mitigate greenhouse gas emissions, and promote responsible business and marketing practices, e.g., fair supply chains without the abuse of power".

Methodological uncertainty has led to an overly broad interpretation of sustainability. Some scientists consider it possible to apply this concept even at the micro level: Issues of climate change, geopolitics, labor conditions in emerging economies, and pressure from stakeholders and supply chain partners all play a role in shifting corporate focus toward the triple bottom line (TBL), the simultaneous achievement of environmental, social, and financial performance (Golicic & Smith, 2013; Waller et al., 2015). In our opinion, such an approach cannot be implemented in practice, since the coordination of such divergent interests as economic efficiency, social stability and environmental safety at the micro level is impossible in principle.

Economists know that when choosing one or another priority direction of development, one should proceed from the fact that economic resources, which are limited, will be directed to its implementation. If we fight poverty first, that is, improve equity by distributing wealth among different social classes through social policy or social revolution (the form in our case is not so important), then appropriate resources will be directed to this. Since, in our opinion, it is impossible to fight poverty only through the redistribution of already created wealth, but it is also necessary to carry out expanded reproduction (in other words, to ensure stable economic growth), and this will also require additional economic resources. If society sets itself the goal of "turning any type of economic activity into an environmentally friendly one", then additional resources should be redistributed for this. As rightly pointed out Zhang et al. (2021) "the single biggest problem for sustainable entrepreneurial firms is access to finance".

Modern science has not answered the question of how to achieve the harmonious development of society and environment. These directions of development have always been competitors, and the history of the development of mankind has not yet provided unambiguous evidence of what is more important at a particular historical stage, especially if we are talking not about humanity as a whole, but about individual social communities. As a result of such a semantic overload, the concept of sustainable development becomes an ideal utopian concept, losing its phenomenological identity and categorical specificity.

All countries today are faced with a difficult choice: what to fight for first of all—for economic growth and an increase in real incomes of the population, or for the preservation of the environment (growth of genuine investment). In the absence of stable economic growth rates, the main strategic goal of the country's leadership will be to enter the trajectory of economic growth and fight poverty, environmental problems will be solved on a residual basis. The development of the urban economy restricts the level of environmental governance (Amri, 2017). As the executors of environmental governance, local governments are also naturally restricted by their economic level (Li, 2017; Wang et al., 2018; Wu et al., 2020).

With 1–2% real GDP growth per year, the country will not be able to allocate significant funds for "green" development, and, accordingly, one cannot count on a significant increase in genuine savings. In essence, the real economic basis for the growth of genuine savings will be the growth of real GDP at a level of at least 5–6% per year, which in itself equally depends on external (fluctuations in the global market for finished products, dynamics of prices for energy carriers and raw materials, access to cheap financial resources, development of international economic integration) and internal (social stability, faster accumulation of social capital, the speed and consistency of technological modernization, the quality of training of skilled workers and engineers, etc.) factors. Of particular importance is the accumulation of social capital. In today's globalizing world, no state can ensure its sustainable social, ecological and economic development without the continuous growth of the social potential of society as a whole, groups and classes of its members, individual individuals. It is the increase in social potential at all levels that is the most important prerequisite for socio-economic, economic, political and environmental security (Solodovnikov, 2017). The main risks of the growth of genuine savings are the possible negative effect of these factors and the violation of the measure (ratio) between the growth rates of real GDP, household incomes and genuine savings.

Based on the contradictory nature of the interaction between society and its natural environment, the modified Kaldor-Hicks criterion was chosen as a national criterion of socio-economic efficiency for the purposes of our study. In accordance with it, all economic changes should be considered as increasing economic efficiency only if the beneficiaries can hypothetically compensate for the economic losses of the losers and still win, with minimal damage to the environment and health (physical and social) of a person, prevention of excessive property differentiation and effective state social policy.

Conclusion

The rethinking of traditional concepts of the possibilities for ensuring sustainable development made it possible to abandon the perception of nature as a free, gratuitous resource. The concept of sustainable development is distinguished by its historicity, dialectic, and internal social inconsistency. In modern conditions, the essence of this concept is the development of socio-economic mechanisms for managing socio-natural systems at the regional, national and global levels in order to ensure sustainable growth in the welfare of the population with minimal damage to the environment and human health.

For an accurate accounting of the results of economic activity over a relatively long period of time, it is necessary to use methodological support that allows you to consider all possible costs (economic, environmental, and social). To do this, the existing system of economic efficiency indicators should be supplemented with the index of genuine savings. In turn, for the possibility of its practical application, it is necessary to refine the methodology for its calculation in terms of socio-economic factors that determine the growth of genuine savings. Socio-economic factors that determine the growth of genuine savings can be classified as prerequisites, conditions (changes in nature, civilization, culture, society's economic system and institutional matrix) and causes (change in the place and role of the country in the international labor division; trends in global commodity and product markets; the evolution of the social class structure of society; change in the accumulation and capitalization of social potential; change in the ratio of dominant and expolar technological waves; evolution of labour management relations; development of science about environment, human and society). Each nation, each state forms a unique mechanism for the impact of changes occurring in nature, civilization, culture, the economic system and the institutional matrix on the dynamics of genuine savings.

To increase the reliability of economic growth forecasts, taking into account genuine savings, it is recommended to carry out a mandatory comprehensive one (with the involvement of specialists from various humanitarian and technical sciences), which will help to avoid errors and risks in making managerial decisions in the long and medium term. This specificity is as follows:

1. taking into account substationary-epistemological and real-ontological aspects;
2. the need to take into account the preservation and strengthening of such important characteristics of the economic system as: adaptability; resistance to external shocks; the ability to improve the standard of living; the ability to ensure the social potential growth; the ability to increase the competitiveness of national producers based on the application of the latest achievements of scientific and technological revolution;
3. the historically coming nature of scientists' thinking, their ideology and attitude, which directly affects the reliability of his (scientist's) judgments in economic growth accounting based on the indicator of genuine savings;
4. high probability of errors generated by subject-specific knowledge systems, which today include economic sciences, as well as the temptation to rely only on mathematical (economic and mathematical) tools for economic growth accounting based on the indicator of genuine savings, which is due to the complexity of the phenomenon of interaction of components within the system "human-environment-society" and a very high degree of abstraction of mathematics.

References

Adrian, T., Laxton, M. D., & Obstfeld, M. M. (2018). Advancing the frontiers of monetary policy. *International Monetary Fund*. https://doi.org/10.5089/9781484325940.071

Afanasiev, Y. N. (1983). *The concept of "civilization" in French historiography*. Civilization and historical process. (in Russian).

Amri, F. (2017). Carbon dioxide emissions, output, and energy consumption categories in Algeria. *Environmental Science and Pollution Research, 24*(17), 14567–14578. https://doi.org/10.1007/s11356-017-8984-7

Chan, H. F., Moy, N., Schaffner, M., & Torgler, B. (2021). The effects of money saliency and sustainability orientation on reward based crowdfunding success. *Journal of Business Research, 125*, 443–455. https://doi.org/10.1016/j.jbusres.2019.07.037

De Crescenzo, V., Baratta, R., & Simeoni, F. (2020). Citizens' engagement in funding renewable and energy efficiency projects: A fuzzy set analysis. *Journal of Cleaner Production, 277*, 124060. https://doi.org/10.1016/j.jclepro.2020.124060

El Alaoui, M. (2020). A fuzzy multiplicative performance indicator to measure circular economy efficiency. *International Journal of Mathematical, Engineering and Management Sciences, 6*(5), 1118–1127. https://doi.org/10.33889/IJMEMS.2020.5.6.085

El Alaoui, M., & Ben-Azza, H. (2017, April). Aggregation of performance indicators for supply chain and fuzzy logic extensions applied to green supply chain. In *International Colloquium on Logistics and Supply Chain Management (LOGISTIQUA)*, pp. 36–41. https://doi.org/10.1109/LOGISTIQUA.2017.7962870.

Figge, F., Thorpe, A. S., Givry, P., Canning, L., & Franklin-Johnson, E. (2018). Longevity and circularity as indicators of eco-efficient resource use in the circular economy. *Ecological Economics, 150*, 297–306. https://doi.org/10.1016/j.ecolecon.2018.04.030

Golicic, S. L., & Smith, C. D. (2013). A meta-analysis of environmentally sustainable supply chain management practices and firm performance. *Journal of Supply Chain Management, 49*(2), 78–95. https://doi.org/10.1111/jscm.12006

Li, Z. (2017). The impact of public participation on local government's environmental governance – An analysis of provincial data 2003-2013. *Chinese Public Administration, 8*, 102–108.

Li, Y., Du, J., & Fu, W. (2020). Thirty days are enough: What determines the crowd's cash time in Agri-food crowdfunding? *China Agricultural Economic Review, 12*(3), 553–575. https://doi.org/10.1108/CAER-10-2019-0176

Ljumović, I., Pavlović, V., & Knežević, G. (2021). Financing agri-food business in the Mediterranean area through crowdfunding: Do environmental issues matter? *New Medit, 20*(3), 101–117. https://doi.org/10.30682/nm2103g

Molina-Sánchez, E., Leyva-Díaz, J. C., Cortés-García, F. J., & Molina-Moreno, V. (2018). Proposal of sustainability indicators for the waste management from the paper industry within the circular economy model. *Water, 10*(8), 1014. https://doi.org/10.3390/w10081014

Mundell, R. A., & Zak, P. J. (Eds.). (2002). *Monetary stability and economic growth: A dialog between leading economists*. Edward Elgar Publishing.

Osokin, N. A. (2005). *History of the middle ages*. Harvest.

Pietrzyck, K., Jarzębowski, S., & Petersen, B. (2021). Exploring sustainable aspects regarding the food supply chain, agri-food quality standards, and global trade: An empirical study among experts from the European Union and the United States. *Energies, 14*(18), 5987. https://doi.org/10.3390/en14185987

Prędkiewicz, K., & Kalinowska-Beszczyńska, O. (2020). Financing eco-projects: Analysis of factors influencing the success of crowdfunding campaigns. *International Journal of Entrepreneurial Behavior & Research, 27*(2), 547–566. https://doi.org/10.1108/IJEBR-05-2020-0339

Smol, M., Kulczycka, J., & Avdiushchenko, A. (2017). Circular economy indicators in relation to eco-innovation in European regions. *Clean Technologies and Environmental Policy, 19*(3), 669–678. https://doi.org/10.1007/s10098-016-1323-8

Solodovnikov, S. Y. (2006). *Institutional matrices: Essence, personification and its genesis (political and economic essays)*. Institute of Economics of the National Academy of Sciences of Belarus. (in Russian).

Solodovnikov, S. Y. (2014). *Classes and class struggle in post-industrial society: Methodological foundations of political economy research*. BNTU. (in Russian).

Solodovnikov, S. Y. (2017). Theoretical and methodological foundations of the study of social capital as a political and economic phenomenon. *Jekonomicheskaja nauka segodnja (Economic Science Today), 5*, 6–56. https://doi.org/10.21122/2309-6667-2017-5-6-56. (in Russian).

Solodovnikov, S.Yu., (2020). Civilization, economic system of society, institutional matrices, foreign language and National Security: Hidden interrelationships. *Jekonomicheskaja nauka segodnja (Economic Science Today)*, 11, 12-17. doi: https://doi.org/10.21122/2309-6667-2020-11-12-17. (in Russian).

Tantau, A. D., Maassen, M. A., & Fratila, L. (2018). Models for analyzing the dependencies between indicators for a circular economy in the European Union. *Sustainability, 10*(7), 2141. https://doi.org/10.3390/su10072141

Waller, M. A., Fawcett, S. E., & Johnson, J. L. (2015). The luxury paradox: How systems thinking and supply chain collaboration can bring sustainability into mainstream practice. *Journal of Business Logistics, 36*(4), 303–305. https://doi.org/10.1111/jbl.12110

Wang, R., Wijen, F., & Heugens, P. P. (2018). Government's green grip: Multifaceted state influence on corporate environmental actions in China. *Strategic Management Journal, 39*(2), 403–428. https://doi.org/10.1002/smj.2714

Wang, J., Huang, J. C., Huang, S. L., Tzeng, G. H., & Zhu, T. (2021). Improvement path for resource-constrained cities identified using an environmental co-governance assessment framework based on BWM-mV model. *International Journal of Environmental Research and Public Health, 18*(9), 4969. https://doi.org/10.3390/ijerph18094969

Wolf, M. (2015). *The shifts and the shocks: What we've learned – And have still to learn – From the financial crisis*. Penguin Books.

Wu, L., Ma, T., Bian, Y., Li, S., & Yi, Z. (2020). Improvement of regional environmental quality: Government environmental governance and public participation. Science of the Total Environment, 717, 137265. https://doi.org/10.1016/j.scitotenv.2020.137265

Zhang, R., McCarthy, K. J., Wang, X., & Tian, Z. (2021). How does network structure impact follow-on financing through syndication? Evidence from the renewable energy industry. *Sustainability, 13*(7), 4050. https://doi.org/10.3390/su13074050

Zhao, H., Zhao, H., & Guo, S. (2017). Evaluating the comprehensive benefit of eco-industrial parks by employing multi-criteria decision making approach for circular economy. *Journal of Cleaner Production, 142*, 2262–2276. https://doi.org/10.1016/j.jclepro.2016.11.041

Translation in the Era of Digitalization: Slang Rendering in TV Series

Elena A. Sereda, Maria P. Efremova, and Anastasia S. Gerasimova

Introduction

As foreign film industry gains popularity in the Russian-speaking culture, translation is in high demand. However, not all language units lend themselves to a quality and equivalent translation. Slang as a specific language sphere can cause difficulties in it being translated since slang expressions are particularly expressive and distinctive.

Slang is one of the most controversial and not fully understood problems of linguistics. Society development is accompanied by an increasing number of slang units requiring incessant analysis. In addition, foreign TV series in Russia are gaining popularity, thus the need for quality translation increases. Now, slang is the core component of every modern TV series, in particular Spanish one. However, the issues of its translation are not developed sufficiently and require further consideration in finding possible slang translation options.

The study of the peculiarities of the translation of slang units within the audiovisual text tend to be of importance following the spread of audiovisual information in the modern digital society. Besides professional competencies laying translation requirements under ethical standards, professional translators should possess general professional competencies implying the ability to solve the translation issues and

E. A. Sereda (✉)
St. Petersburg University of Management Technologies and Economics, St. Petersburg, Russian Federation

M. P. Efremova
St. Petersburg State University of Economics (UNECON), St. Petersburg, Russian Federation

A. S. Gerasimova
Peter the Great St. Petersburg Polytechnic University (SPbPU), St. Petersburg, Russian Federation

A. Rumyantseva et al. (eds.), *Challenges and Solutions in the Digital Economy and Finance*, Springer Proceedings in Business and Economics,
https://doi.org/10.1007/978-3-031-14410-3_25

tasks in compliance with the linguistic and cultural norms of the source language and the target language (Gerasimova et al., 2022).

Slang has no single definition. Slang, being an integral part of language and, accordingly, speech, is one of the main issues of lexicology. Slang is examined in various language studies: philology, linguistics, lexicology, socio- and psycholinguistics, pragmatics, etc.

In Dictionary of the Spanish language of the Spanish Royal Academy, such terms as "jerga" (jargon) and "argot" (argot) are treated as synonyms, and the English word "slang" is absent in it: jerga—"Lenguaje especial y no formal"; argot—"1. Como jerga 2. Lenguaje especial entre personas" (Diccionario de la lengua española de Real Academia Española, n.d.).

Slang is a complex, difficult, and inevitable language phenomenon. Its emergence is always caused by historical, social and cultural tendencies of life of this or that language community (Izmaylova et al., 2017).

Slang is the utilization of exceedingly casual words, shortened forms, and articulations that are dismissed when asked to be taken as a significant aspect of the traditional language. The flood of online communications, for example, and microblogs administration made use of Internet slang practically omnipresent (Bindhu et al., 2021).

Now the definitions and understandings of the term "slang" are numerous, and, as it turned out, differ significantly. The contradictions lie in defining the scope of slang. Debates continue whether slang should comprise ironic and expressive words only, regarded as synonyms to standard language alternatives, or whether it should embrace non-standard vocabulary condemned by educated society. There is also a question regarding whether the concepts of "slang", "jargon", "argo", etc. can be interchangeable. But it is worth noting that at present, a kind of "common jargon" has formed—an understated style of speech that erodes already established linguistic norms and is not only used by a wide variety of segments of the population in everyday life, but also quite often sounds in television and print media (Kolmakova & Shalkov, 2021).

In this modern era, where technology rules the world, people's language develops the following technology. Trends in online social media can be reflected by the popularity of hashtags, topics or even neologisms such as Internet slang. The collective attention underlying popularity peaks indicate the participation of a massive number of individuals during the diffusion of the relevant information (Zhang et al., 2016). Thus, understanding of slang units is becoming extremely essential both while watching series or other audiovisual works and in online communication.

Slang is bound to a culture. It is not easy, therefore, to translate words categorized as slang. The more a translator is aware of complexities of differences between cultures, the better a translator s/he will be. Translators should be aware that their translation corresponds to certain culture, readership, and conveys the proper meaning (Sukmaningrum & Setyorini, 2017).

The present study is based on a definition developed through comparing existing approaches to slang. Slang is a historically non-standard, non-literary, colloquial (mostly in oral speech), emotionally colored, short-lived, mobile layer of vocabulary.

Material and Methods

The practical base of the research were *Money Heist*, youth Spanish heist crime drama television series, and *Elite*, Spanish thriller teen drama television series, their original and dubbed versions (dubbing made by *TVShows Studios*). 100 slang units were selected and reviewed.

The results obtained can be used in similar research for studying features and methods of translation of audiovisual texts. The results can also be applied in training of students of linguistics department for developing professional skills of future translators.

One of the main research methods is the cognitive-discursive method to analyze semantic diversity of utterances, and to review actions that contribute for the text pragmatics to reveal. Another method used is a communicative-pragmatic method for studying the communication processes for the national mindset to manifest. Among other methods was the method of scientific observation, an empirical, and the descriptive-analytical method to describe the research results obtained.

All these methods help to implement the purpose of the study, that is to review slang in linguistic and translation aspects, and to identify the main problems of translating slang unit from Spanish into Russian in *Money Heist* TV series (Spanish *La casa de papel*) and *Elite* (Spanish *Elite*).

Results

This article discusses the translation of slang revealed in the original Spanish TV series. Thus, this study examines the peculiarities of the translation of slang units in such type of translation as audiovisual translation. The peculiarities of this type of translation in comparison with other types of translation are mainly determined by the multidimensional semantic code of the object of this type of translation, i.e., audiovisual text. Both verbal and non-verbal information in an audiovisual text is transferred simultaneously through two channels—acoustic and visual ones, and the linguistic code ceases to play a decisive role. Viewers, including translators, perceive a series of codified signs connected in a certain way by the director (choice of angle and shooting) and the editor (editing).

With the spread of television and the Internet, audiovisual translation is becoming more and more in demand. We can surely assert that translation of feature films and documentaries, TV series, shows, programs of news and entertainment channel promotes mutual enrichment of cultures.

Depending on the order, audiovisual translation can combine the features of simultaneous, sequential, and written (literary) translation: voice-over translation, translation for subsequent dubbing, subtitling (preparation of subtitles). A distinctive feature of this type of interpretation is that audiovisual translation primarily relies on a video sequence and the performance of actors. When working with an audiovisual text, a translator needs not only to translate remarks of characters, but also to convey the meaning of the statements that was embedded in the original message, that is, to achieve the transfer of the author's idea as the extralinguistic, most often visual element, which has been added to the verbal media text in recent decades, still causes a lot of controversy (Lenkova, 2019). In this regard, the translator works not only with the original text and its translation, but also with such aspects as a dialogue context, sound effects, images and a shot as a whole.

This study examines the translation of slang in an audiovisual work for subsequent dubbing.

Discussion

La casa de papel, or *Money Heist*, known in Russia as *Бумажный дом*, is a popular Spanish youth TV series, which premiered on May 2, 2017. *Elite* is another Spanish-language series in the genre of thriller, drama, crime, released specially for the *Netflix* platform. The release took place on October 6, 2018. This study offers analysis of slang units from Spanish language series based on the dubbing made by studio *TV Shows studio*, which is more popular with the Russian-speaking audience.

86 slang units were selected through viewing the *Money Heist* (*Бумажный дом* in Russian) and *Elite* (*Элита* in Russian) TV series. The translator primarily used lexical transformations (70%), lexico-grammatical transformations were used much less often (20%), and only in 10% of cases slang units were translated through grammatical transformations.

The lexical transformations used comprise the selection of an analogue, specification, generalization, transcription/transliteration, loan translation (calque), omission, modulation, demetaphorization, and remetaphorization. Here are some examples:

Example. *Perdió el habla cuando supe que era un chivato*—Я офигела, когда узнала, что он *стукач* (selection of an analogue).

Example. *Ay, mira al figura, ¿que le pasa a este?* —Ой, глянь на этого *выродка*, что он о себе возомнил? (specification).

Example. No quiero ir a esa fiesta. No tengo ni un pavo.—Не хочу я идти на тусовку. Бабок нет (generalization).

According to the Multitran dictionary: *"pavo"*—sl. еврик. In this context, such translation will not be adequate due to the insufficiently widespread use of the slang unit еврик. Unstead, such slang units as *бабки, бабло, копейки*, etc. are more common in use with the Russian-speaking audience; so the translator's decision to hyperbolize the original meaning of the word *"pavo"* is highly successful.

Example. Pues, ¿qué eres?—Cocorocó.—Más alto.—Cocorocó.—Ну, кто это у нас?—Кокороко.—Громче.—Кокороко (transcription).

The translator's choice for this translation transformation is considered acceptable, since the meaning of the transcribed units is clear from the context, while the norms of adequacy and equivalence of translation are not violated.

Example. *Me meo en ese cabrón de mierda.*—Да плевать мне на этот *кусок дерьма* (loan translation (calque)).

Example. ¿Puedes dar vueltas por una vez y pensar en algo que valga la pena? ¡Qué estúpido idiota!—Ты можешь хоть раз пошевелить мозгами и придумать что-то стоящее? Тупица!(omission).

In this case, the translator omitted the unit *"estúpido"*, which is a good solution, since the words "stupid" and "idiot" are synonyms in Russian, therefore, there is no need to translate both units.

Example. ¿No la conoces? La sigo, esta chica está buenísima——Ты что, не знаешь ее? Я подписана на эту девушку в инстаграме, она крутецкая (modulation).

The Spanish verb *"seguir"* means "to follow, to continue" («следовать, продолжать»). In order to adapt the translation to a native Russian speaker, the translator had to convert the original lexical unit into a phrase logically derived from the meaning of the original unit. Thus, the recipient can understand the meaning of being a follower on a social network.

Example. *Quiero negociar con él, pero sabemos que es de puño cerrado.*—У меня к нему есть дело, но, как мы знаем, он *жлоб* (demetaphorization).

Example. Después de todo, si nos liaremos alguien la manta a la cabeza, encontremos una mina de oro.—В конце концов, если рискнем, сорвем куш (remetaphorization).

The grammatical transformations comprise replacement of a part of speech, replacement of a word form, replacement of a sentence member.

Example. Es hora de terminar la jarana. Vamos a hacer que todos vuelvan a casa.—Пора кончать отрываться. Let's get everyone home. The noun was replaced by a verb.

Example. *Vete a la mierda, ya lo tengo un morrón.*—Отвали, у меня свои *проблемы.* The singular noun has been replaced by a plural noun.

Example. *No me voy a perder la pelea sentada en un mustio club.*—Этот затхлый клубешник—не то место, в котором я намерена засиживаться. The circumstance of the place was changed into the subject.

Grammatical substitutions of different nature are quite common in translation from Spanish into Russian due to the significant differences in the grammatical structure of these languages. Therefore, the translator's decisions are quite reasonable and appropriate in all the examples given.

Lexico-grammatical transformations comprise compensation, antonymic translation, descriptive translation and holistic transformation were highlighted.

Example. *¿Cuántos impactos tengo?—Son tres, papá.—Menudo jaleo que tengo ahí abajo*Сколько раз в меня выстрелили?—Три, папа. Ну и месиво у меня там (compensation).

The Spanish word *"jaleo"* has the following meanings: general *confusion (сумятица), anxiety (тревога)*; jarg. *shucher (шухер)*; simple *high (хай).* The context prevents any of these meanings to be used in the target language, influencing the translator's decision. As a result, the translator chose *месиво* as the most appropriate lexical unit while it performing the same functions as the original slang unit *jaleo*, so the translation is considered adequate.

Example. Vamos a robar un banco e imprimir más de 200 millones de euros.—Suena superrico.—Мы собираемся ограбить банк и напечатать больше 200 млн евро.—Звучит неплохо (antonymic translation).

Example. El profesor es nuestro padrino. Hacemos lo que dice.—Профессор—глава нашей преступной группировки. Он говорит—мы делаем (descriptive translation).

Example. ¿Has oído algo de ella?—Sólo que su casa está a tomar por el culo.—Слышал че-нибудь о ней?—Только то, что живет хрен знает где (holistic transformation).

The examples in the current study show high importance for embracing different aspects of comprehensive multi-disciplinary approach towards slang units' translation in TV series. The rendering of the detected slang units proves to be successful and adequate bearing all features of its cultural, stylistic and other background.

Conclusion

Referring to the results of the analysis of slang units, it can be concluded that slang units belong to diverse theme spheres of different aspects of social life, from certain features and objects of reality to leisure. In addition, slang units can have a neutral, positive, negative assessment, depending on the context.

There is no single model for translating slang. A translator is forced to apply different transformation in each case depending on many factors: context; type, mood and features of the character uttering the remark; stylistic coloring and semantic load of slang expression; norms and requirements of the translating language, and so on. At the same time, a translator should consider both linguistic and non-linguistic factors when translating slang units in performing audiovisual translation.

References

Bindhu, V., et al. (Eds.). (2021). *International Conference on Communication, Computing and Electronic Systems* (Lecture Notes in Electrical Engeneering 733). Springer. 10/1007/978-981-4909-4_33

Diccionario de la lengua española de Real Academia Española. Retrieved from http://dle.rae.es/?w=diccionario&origen=REDLE

Gerasimova, A.S., Sereda, E., & Rubtsova, S. (2022). Transcreation as a creative tool of translation. In D. Bylieva & A. Nordmann (Eds.), *Technology, innovation and creativity in digital society.*

(PCSF 2021 Lecture notes in networks and systems, 345). Springer. https://doi.org/10.1007/978-3-030-89708-6_28

Izmaylova, G. A., Zamaletdinova, G. R., & Zholshayeva, M. S. (2017). Linguistic and social features of slang. *International Journal of Scientific Study, 5*(6), 75–78. https://doi.org/10.17354/ijssSept/2017/016

Kolmakova, V., & Shalkov, D. (2021). Anglicisms in the structure of modern youth slang: Genesis and Lexico-semantic models. In E.V. Toropova, E.F. Zhukova, S.A. Malenko, T.L. Kaminskaya, N.V. Salonikov, V.I. Makarov, A.V. Batulina, M.V. Zvyaglova, O.A. Fikhtner, & A.M. Grinev (Eds.), *Man, society, communication* (European Proceedings of Social and Behavioural Sciences) (Vol. 108, 1st ed., pp. 170–176). European Publisher. https://doi.org/10.15405/epsbs.2021.05.02.21

Lenkova, T. (2019). Tehree steps to understanding the creolized text. In N. I. Almazova, A. V. Rubtsova, & D. S. Byliaeva (Eds.), *19th PCSF 2019 professional culture of the specialist of the future* (Vol. 73, pp. 518–523). https://doi.org/10.15405/epsbs.2019.12.55

Sukmaningrum, R., & Setyorini, A. (2017). Cultural perspective in the translation of slang expressions in subtitle die hard 3. *ETERNAL (English Teaching Journal), 8*(2). https://doi.org/10.26877/eternal.v8i2.3047

Zhang, L., Zhao, J., & Ke, X. (2016). Who creates trends in online social media: The crowd or opinion leaders? *Journal of Computer-Mediated Communication, 21*(1), 1–16. https://doi.org/10.1111/jcc4.12145

Some Aspects of the Introduction of the Digital Ruble into the Russian Economic and Legal Space

Vladislava V. Kosareva, Anastasiya A. Bordyugovskaya, and Anna M. Belyatskaya

Introduction

The development of digital technologies, as well as the digitalization of the economy, are inevitable and natural processes. The use of digital technologies in the field of financial services reduces the need of society for non-cash payments and increases the need for non-cash payments. The result is a decrease in the cash issue. Sweden (since 2006), Great Britain and China (since 2013), France (since 2016), India (since 2016), South Korea (since 2017) are gradually abandoning cash (Soldatkin & Sigov, 2021). As a result, states are moving to alternative monetary forms, one of which has become a digital currency.

The special attention of banks to the issue of digital currency is moving from theoretical research to practical level. Many developed countries of the world have approached the issue of their national currency in text format. Just a few years ago, the idea of using a digital currency was unthinkable, but by mid-2020, the European Central Bank, China, South Korea and Sweden began using digital currency as an experiment. The banks of England and the USA are seriously discussing the use of a digital version of cash. The Bank of China will launch the Digital Currency Electronic Payment project in full format in 2022 (Gabov, 2021). In China, the development of a project on the use of the digital currency of the People's Bank began in 2014, and in 2020 its testing was launched. In mid-2021, the act "Progress of research and development of e-CNY in China" on the introduction of the digital

V. V. Kosareva (✉)
St. Petersburg University of Management Technologies and Economics, St. Petersburg, Russian Federation

A. A. Bordyugovskaya · A. M. Belyatskaya
State Institute of Economics, Finance, Law and Technology, Gatchina, Russian Federation

A. Rumyantseva et al. (eds.), *Challenges and Solutions in the Digital Economy and Finance*, Springer Proceedings in Business and Economics,
https://doi.org/10.1007/978-3-031-14410-3_26

yuan was adopted, according to which the process of its implementation has started quite successfully and continues (Konopiy & Borisov, 2021; Mikhailishin, 2021).

CBDC is the third form of state money, an electronic obligation of its Central Bank, denominated in the national monetary unit and serving as a means of payment, account measures, cost savings (Andryushin, 2021). Most of the designed CBDC are intended for general use, although some of them serve exclusively for wholesale payments and settlements between Central Banks. In relation to CBDC, three access models are used for their implementation: access only to financial institutions (Model Fl); access for all subjects of economic relations (Model EW); access for financial institutions, plus limited access to banks supported by CBDC (ModelFl+) (Hisamova, 2020).

The Central Bank of Russia did not stay away from the possibilities of using the digital ruble, which issued a corresponding report in which it offered to discuss with the financial sector, the expert community, as well as with all interested parties the key aspects of this important topic for the economy and society. The main purpose of introducing a national digital currency in the world is the emergence of non-state digital currencies (cryptocurrency). A non-state digital currency undermines the monopoly of national central banks and international national systems. Hence the catch phrase: "You can't win—lead" is more relevant than ever! Nevertheless, non-state digital currencies are more business projects of IT corporations, they are certainly a means of circulation, but not a measure of value. States consider cryptocurrency as one of the transitional links in the evolution of money from cash, non-cash and electronic money, revolving in such monetary buffers as PayPal, WebMoney, Qiwi, "Yandex.Money" (Fiapshev, 2020).

It should be noted that in most countries of the world a two-tier banking system is used. The first level assumes the sole use by the Central Bank of the issue of the national currency. Here, the Central Bank is the agent of the state for the placement of government debt obligations, as well as a lender for national banks. The second level is the national banks themselves, which conduct payment and settlement and credit and deposit relations with various organizations and with each other. This system allows you to set in motion two processes:

- Expenses and liabilities of one national bank may be assets of another bank (this is how transactions, including international ones, can be tracked)
- Non-cash circulation makes the issuer of funds not only the Central Bank, but also the second level of the system—national banks

The emergence of a digital currency will exclude national banks from issuers, while maintaining the Central Bank's monopoly on its issuance. With the advent of digital currency, the Central Bank becomes the creator, conductor, and custodian of digital money (Terentyev & Petrov, 2021).

Thus, the whole system becomes one-tier, and national banks will receive new functionality (control functions for transactions and the provision of a payment and settlement base for various transactions). This can also include cash, non-cash, digital forms of transfers, and the old business model will cease to exist.

Materials and Methods

A wide range of methods of scientific cognition was used in the preparation of the article, including such fundamental general scientific methods as analysis and synthesis, induction and deduction, formal logic techniques.

Results

The transition to a digital currency will allow countries to become more modern, and the products they offer in a market economy will be more in demand. As for the Bank of Russia (hereinafter referred to as the Central Bank of the Russian Federation), one of its priority goals is the development of the national payment system (hereinafter referred to as the NPS) and in this regard, the introduction of a digital currency into circulation—the digital ruble is a well-founded decision. The legal basis for the introduction of the digital ruble was Federal Law No. 259-FZ of 31.07.2020 "On Digital Financial Assets, digital Currency and Amendments to Certain Legislative Acts of the Russian Federation", which set the vector for the development of the third form of money in the state, as well as the Concept of the digital ruble, published on the official website of the Bank of Russia. The Concept defines a prototype of the digital ruble platform, which is scheduled to be launched in December 2021, and full testing of the system for 2022, after which a final decision will be made on the future of the digital currency in Russia. Since it is already planned to introduce appropriate amendments to the industry legislation (2022), it can be concluded that the issue of introducing the digital ruble has already been resolved: the digital ruble should be! In the Concept of the digital ruble presented by the Bank of Russia, it is assumed that the digital ruble as an additional form of the Russian national currency will be issued by the Bank of Russia in digital form and, as a measure of value, will be equivalent to 1 cash /non-cash ruble. The main advantage of the digital ruble over non-cash funds is to ensure the financial stability of the state since the redistribution of funds into foreign digital currencies will be limited.

It is also worth noting here that some analysts say that the use of the digital ruble will be able to reduce the negative consequences of the introduction of new economic sanctions by the United States and other Western European countries, since the use of the digital ruble will allow Russia to be from the dollar in international settlements and reduce dependence on the SWIFT (international payment system).

However, there is some skepticism about this assumption. If settlements between Russian organizations with a digital ruble seem quite feasible, then the ability to settle with external partners may be questionable, since the digital ruble does not yet have the right status, respectively, it is probably impossible to prove the possibility of circumventing sanctions on real transactions.

In addition, the question remains, what exactly will countries be tied to when making international payments with digital currency? This problem has already arisen with the introduction of digital currency in China. As you know, settlements between Russia and China are mainly carried out in euros and yuan, so the creation and use of digital platforms can improve and accelerate financial transactions on trade turnover. However, one way or another, but in international settlements, everything is tied to the dollar or euro. If countries want to get away from these currencies, there is still a need to search for some kind of single currency, whether it is the ruble or the yuan (Shchegoleva & Malsagova, 2021). In the meantime, one thing can be argued that the most likely or more predictable consequences of the introduction of the digital ruble is the strengthening of the role of the Central Bank of Russia. This is since all financial transactions will be carried out directly through the Central Bank of Russia, thereby the burden on commercial banks will be significantly less. On the other hand, the legal regulation can provide for another option when commercial banks will act as intermediaries in financial transactions. In any case, this issue requires legal regulation by the Russian Federation.

A reasonable question arises: what is the fundamental difference between the digital ruble and non-cash funds? Only by the method of use, namely the ability to pay with digital currency in offline mode, and by the method of storage—digital money is stored on "electronic wallets" in the Bank of Russia, and non-cash funds are stored on the corresponding accounts in commercial banks (Goltsov & Golovanov, 2018). Otherwise, the replenishment of the "electronic wallet", the transfer of one form of money to another—the digital ruble and non-cash funds are similar. That is, we can deposit money to a payment card and to an "electronic wallet" with digital rubles in three ways: from a bank account, from a bank card, by depositing cash. As for the transfer of funds from one form to another, everything is quite simple here: any form of money, whether cash, non-cash or digital, can be transferred to another, one of the above-mentioned forms without much effort.

There may also be a question about how the digital ruble differs from the cryptocurrency? Firstly, the cryptocurrency does not have a single issuer; secondly, the cryptocurrency does not imply guaranteed consumer protection, unlike the digital ruble, the only issuer of which will be the Central Bank of the Russian Federation; thirdly, the value of the cryptocurrency is very unstable and the range of cost fluctuations is quite wide (Gaponov & Balashev, 2021). Cryptocurrencies appeared as private money and the prospect of their widespread use in monetary circulation carries with it the threat of the emergence of the "digital Middle Ages" with its uncontrolled emission of various private virtual currencies (Sitnik, 2020). At the moment, there is a search for tools regulating decentralized currencies issued by private individuals: a complete ban (Vietnam, Ecuador); free circulation of cryptocurrencies, on a par with the national (Japan, Switzerland); the issue by the national Central Bank of its own virtual currency and the ban on the circulation of digital non-state digital currencies (China) (Dostov et al., 2018; Xie, 2019). Time will tell how these digital currencies will coexist, develop and interact in the future.

If we talk about the convenience of using the digital ruble, then this point is quite controversial—it's all about its infrastructure. The Central Bank of the Russian

Federation assumes that it will be possible to pay with a digital ruble, both in online and offline modes. And, if everything is clear with the online mode—it is necessary to have access to communication networks and use by analogy with electronic means of payment / mobile applications, then the offline mode assumes the advance reservation of a certain amount of money from your electronic wallet, which is necessary for certain calculations, as if a person took with him a certain amount of cash to another "wallet" that will be created on the client's mobile device, which is not a great convenience—using a plastic bank card in this case is much more convenient.

Discussion

Touching upon the issue of the risks associated with the use of the digital ruble, it should be noted that making payments using a bank card, today, seems more convenient than making payments through the use of digital currency. This is explained by the fact that from the point of view of usability, transferring money to pay for something from one wallet to another (off-line) serves as a kind of additional action on the way to receiving the final service, which can lead to the loss of potential users of the digital ruble. The digital ruble will be used by those citizens and private organizations who, for whatever reason, have lost confidence in commercial banks, and therefore have a positive attitude to storing money directly in the Central Bank.

The danger of digital currency for the state lies in the fact that by becoming the flagship of this digital instrument, the state gets the opportunity to provide its product to foreign organizations and individuals. Transactions will take place remotely on the issuing bank's servers. In addition, digital currencies of foreign banks may be more in demand than the currency of national banks, which may negatively affect the entire monetary circulation system of the country. The products of foreign banks may be more attractive (for example, loans with reduced rates) than the products of the Central Bank. Only competitive national products can overcome these dangers.

Among the risks of using the digital ruble, it is necessary to note the problem of ensuring cybersecurity. So, it will not be difficult for an attacker to hack a personal account to transfer digital currency from one wallet to another (Chekanov, 2021). However, the introduction of a blockchain system can help minimize these risks. A possible application of blockchain technologies is a distributed data storage system, where storage devices are not connected to a common server; this database stores the entire list of actions and records, they are called blocks. Each block has a timestamp and a link to the previous block. Such a record format (encryption) guarantees the safety of the record, since the user can only change the part of the database that he owns and to which he has access (keys), without access, recording is impossible. When editing data, information about this appears to all users (Guz, 2020).

Conclusion

Touching upon the topic of relevance and demand for the digital ruble, we must proceed from the habits and preferences of potential customers. Thus, the digital ruble can become one of the most popular payment instruments along with bank cards, payments using mobile devices, a system of fast payments, and, finally, cash. However, we should not lose sight of the fact that customer preferences can be quite different at the level of the subjects of the Russian Federation, as well as social groups of the population.

References

Andryushin, S. A. (2021). Digital currency of the central bank as the third form of state money. *Actual Problems of Economics and Law, 1*, 54–74.

Chekanov, P. E. (2021). Prospects and risks of issuing a digital ruble by the Bank of Russia. *Azimuth of Scientific Research: Economics and Management, 2*(35), 383–388.

Dostov, V. L., Shust, P. M., & Khorkova, A. A. (2018). Prospects of decentralized interbank settlements using blockchain. *Strategic Decisions and Risk Management, 2*(107), 22–25.

Fiapshev, A. V. (2020). The phenomenon of digital currencies as an alternative to fiat money. *Economy. Taxes. Law, 5*, 80–87.

Gabov, A. V. (2021). Digital ruble Central Bank as an object of civil rights. *Actual Problems of Russian Law, 4*(125), 55–65.

Gaponov, S. V., & Balashev, N. B. (2021). The development of digital money: cryptocurrencies and electronic money. *Scientific and Methodological Electronic Journal Concept, 2*, 112–118.

Goltsov, V. B., & Golovanov, N. M. (2018). Legal regulation of the use of electronic funds under the legislation of the Russian Federation. *Danish Scientific Journal (DSJ), 13*, 17–21.

Guz, N. A. (2020). Trends of digitalization of higher education. *The World of Science, Culture, Education, 2*(81), 236–237.

Hisamova, Z. I. (2020). The concept of the digital currencies of Central banks: the main risks in terms of compliance with AML requirements (the "AML") and KUS ("Know your customer"). *Actual Problems of Economics and Law, 3*, 508–514.

Konopiy, A. S., & Borisov, B. A. (2021). Legal policy of the Russian Federation and the People's Republic of China in the field of digitalization of the national currency. *Law and Politics, 7*, 16–26.

Mikhailishin, A. Y. (2021). Prerequisites for the emergence and world experience in the introduction of digital currencies of central banks. *Russian Journal of Economics and Law, 2*, 294–307.

Shchegoleva, N. G., & Malsagova, R. G. (2021). Digitalization VS internalization of the yuan in the global system of international settlements. *Problems of Theory and Practice of Management, 1*, 6–20.

Sitnik, A. A. (2020). Digital currencies of central banks. *Vector of Legal Science, 9*, 180–186.

Soldatkin, S. N., & Sigov, V. O. (2021). Digital ruble: features of the Russian model of creation and functioning of the national digital currency. *Bulletin of the Khabarovsk State University of Economics and Law, 2*(106), 62–69.

Terentyev, V. N., & Petrov, K. G. (2021). Digital ruble issue. *Bulletin of Samara State University of Economics, 7*(201), 80–87.

Xie, R. (2019). Why China had to "ban" cryptocurrency but the U.S. did not: A comparative analysis of regulations on crypto-markets between the U.S. and China. *Washington University Global Studies Law Review, 18*(2), 456–492.

Three-Dimensional Copy of a Three-Dimensional Copy: Database of Estampages of Proto-Bulgarian Inscriptions

Larisa Bondar

Introduction

The history of collecting information about the surrounding world, its systematization and preservation, as well as the preservation of historical memory has its roots in the Enlightenment era. According to the trends in science at the beginning of the eighteenth century. museum collections, cabinets of curiosities (*mirabilia*) (Zawartko-Laskowska, 2008: 80) were complex, universal and contained at the same time art objects, exhibits from the field of natural sciences, various curiosities and antiquities. A kind of program for the creation of such collections was outlined in the work of A.-J. Desalliers d'Argenville, which has become, in fact, a methodical guide for collectors (d'Argenville, 1727). It was an era that sought, according to T. F. Rosset, to name and inventory the surrounding world (de Rosset, 2014).

The largest of such collections demanded their cataloging early. Thus, the first museum in Russia, founded by Peter I in 1714, the Kunstkamera, after three decades of its existence, not only compiled, but also published in St. Petersburg in 1741–1745 a catalog of its collections: "Musei Imperialis Petropolitani". Cataloging, and then state accounting of museum and archival fonds have become an integral part of museum and archival legislation. Today, however, it is obvious to everyone that the time of catalogs and paper card files is irrevocably going into the past.

Computer databases help to provide prompt and targeted access to documentary information at a qualitatively new level, allow significantly expanding the scope of information use and the circle of users of the software product. The gradual creation of digital databases of museum objects and archival documents, formed according to

L. Bondar (✉)
St. Petersburg University of Management Technologies and Economics, St. Petersburg, Russian Federation

St. Petersburg Branch of the Archive of RAS, SPbB ARAS, St. Petersburg, Russian Federation

A. Rumyantseva et al. (eds.), *Challenges and Solutions in the Digital Economy and Finance*, Springer Proceedings in Business and Economics,
https://doi.org/10.1007/978-3-031-14410-3_27

various principles, makes it possible to obtain retrospective information without directly referring to a historical subject or document, contributes to improving the effectiveness of research by both users and museum and archival staff.

Digital databases of archival documents perform a double function: they create a custom copy that allows not to refer to the original document when working, ensuring its safety, and form an insurance fond in case of loss of the original.

Museums have a unified State catalog of the Museum Fond of the Russian Federation (https://goskatalog.ru). Archives have not yet created such a thing; in this regard, academic archives have advanced, creating a single platform with a fond catalog of archives under the methodological guidance of the Russian Academy of Sciences—the Information System "Archives of the Russian Academy of Sciences" (https://isaran.ru/?q=welcome). However, the existence of a unified database does not negate the need to create thematic electronic collections. The St. Petersburg Branch of the Archive of the Russian Academy of Sciences (SPbB ARAS) is actively working in this direction and in recent years has created several thematic databases. One of these databases concerns materials from the field of the history of writing.

Materials and Methods

The history of writing goes back centuries and millennia, D. Diringer suggests considering the middle of the IV millennium BC as the time of the issuance of its "birth certificate" (Diringer, 1968). The oldest written systems—Mesopotamian and Egyptian ones, and after them all the others—used different carriers: clay plates, papyri, parchment, paper, birch bark, etc. But among the earliest materials for fixing a written text, we must invariably include a stone.

The monuments of the designated in the title chronotope were created on stone. Proto-Bulgarian inscriptions of the ninth century were discovered and described during the expedition of the Russian Archaeological Institute in Constantinople (RAIC)—the only domestic scientific institution operating outside the Russian Empire in 1894–1914. The institute made archaeological excursions to Lebanon, Jordan, Syria, Macedonia, and in 1899–1900 the work was carried out on the territory of Bulgaria for the purpose of successfully realized excavations of the first Bulgarian capital—Aboba-Pliska. Describing the materials from Aboba-Pliska, director of the Institute Fedor Ivanovich Uspensky stated that among the artifacts found, in quantitative and qualitative terms, inscriptions are in the first place.

What was the fate of these stone documents in Bulgaria? By the time the RAIC arrived there and started its archaeological work, some of the inscriptions remained in their places, some were transported to the museums of Sofia and Tarnovo, and another part was moved from there, since the stones on which the inscriptions were made were of interest to local residents as building material. Future archaeological complexes were severely damaged during the construction of the railway in this area. In addition, merchants sold hewn stones from Aboba, sending them along the newly built railway. And only at the insistence of the RAIC, the trade in these stones was prohibited since 1895.

Stones with inscriptions, confidently identified as originating from Aboba-Pliska, were found in cemeteries (used as tombstones), adapted as structural parts of fountains and drains (for example, the middle of a column with the inscription was hollowed out and the column was adapted as a drain chute), used for the construction of private houses (Chernova et al., 2018: 817). However, the builders of the first Bulgarian capital themselves were equally utilitarian about the legacy of their predecessors: during the construction of the residence of the Bulgarian rulers, building material with traces of cultural life from other regions was transferred here. A considerable number of ancient inscriptions were found in this place, which came here with stones used as building material: a slab with the remains of a tombstone inscription (obviously from some Roman cemetery), a slab with the name of Marcus Aurelius (which served as the base of the column and lay on the ground on the side where the inscription was made), etc.

Even then, the problem of saving these monuments was obvious. When there was such an opportunity, they were moved to a safe place, under the supervision of the authorities, and then transferred to the National Archaeological Institute and Museum. But this was not always possible: this was excluded in cases where the stone with the inscription was an integral structural part of the building. In this case, not only it was impossible to guarantee its safety, but it was difficult to study it.

There were three ways to study the inscription: drawing the inscription, photo and estampage. An estampage is a paper print from an inscription or image on a solid material, which is an exact negative copy of the image. During the work of the RAIC, a significant number of estampages were created, which, after the liquidation of the institute in 1914, were exported to Russia and in the 1930s, by the decision of the commission established under the Academy of Sciences, were distributed among archival institutions. About 70 prints were received by the Archive of the Academy of Sciences of the Soviet Union (today SPbB ARAS), where in 2016 they were attributed and described by the archive researcher N. A. Pavlichenko and formed a separate inventory list as part of the fond No. 127 “Russian Archaeological Institute in Constantinople”.

Estampages, due to their accessibility, began to be used quite early as a means of copying a dimensional monument. The method of its making was so simple that, according to Emil Huebner, the author of the manual for the making of estampages, they are available for making for both men and women, and even for a not too clumsy apprentice (Hübner, 1881).

Estampage-making was not only a more accessible way of fixing epigraphic material compared to photography, but often the only possible way when the inscription was in a place where it was impossible to deliver photographic equipment or place it (Chernova et al., 2018: 815–816). So, in the early 1880s, the American epigraphist J. Sterrett, in order to make an estampage from the inscription on the stone used in the construction of the house, had to dig under this house (Pavlichenko, 2018: 467).

In addition, an estampage is the most accurate and most valuable copy compared to a photograph or drawing. In fact, an estampage performed the functions of a modern digital copy. And its primary purpose was to create an "insurance copy" of the epigraphic document, the relevance of which is confirmed by modern history. Thus, one of the Greek inscriptions of 197 AD, found in Maaloula (Syria), was most likely destroyed as a result of military actions in these territories in 2013–2014; it was not included in the compendia of epigraphic monuments, and the estampage of it stored in the SPbB ARAS is today, perhaps, the only facsimile copy of the currently lost monument (http://ranar.spb.ru/rus/vystavki/id/928/). At the same time, estampages are a well-preserved material (Pavlichenko, 2018: 466).

The second function of the printmaking, which is related to modern digitization is the creation of a "user copy": a paper estampage could be taken away and examined in office conditions. This is also important, since the work of reading the inscription can take decades.

To carry out the study of the inscription, it is important to have a 3d-copy of it. The limited number of estampages available (often made in one copy) also limit the circle of researchers of the inscription. Modern digital technologies allow to expand it, which make it possible to detail and speed up the study of documents. These include 3d-scanning.

SPbB ARAS organized the work by creating on the Internet an electronic database "Collection of estampages of the Russian Archaeological Institute in Constantinople". Having started working with the estampages, the archive staff found out that salvation is required not only for inscriptions, but also for the estampages from them. Over the past hundred years, the collection of RAIC estampages has been repeatedly moved from city to city and often stored without observing the temperature and humidity regime, so about 80% of the estampages required conservation and restoration, 20% were in a ruined state (Tunkina et al., 2020). In this connection, first of all, the scientific restoration of the collection was carried out in SPbB ARAS, for the first time a simple and convenient method of restoration of multilayer estampages was developed, which was tested by comparative studies.

In order to obtain detailed geometric characteristics, the prints were scanned with a high-precision manual 3d-scanner "Artec Spider". This method is successfully used in the areas of prototyping and reverse engineering. Due to the high accuracy (up to 0.05 mm), it is possible to obtain three-dimensional digital models of very high resolution—up to 0.1 mm, which fully meets the requirements when shooting estampages.

As a result, the database will provide information about 12 inscriptions discovered during archaeological excavations on the territory of Aboba-Pliska. Nine of them are Proto-Bulgarian monuments of the ninth century AD, the remaining three ones are Roman inscriptions in Greek and Latin.

Proto-Bulgarian inscriptions make up three groups. The first of them is the group characterized by F. I. Uspensky as "Columns with inscriptions in honor of statesmen and heroes". F. I. Uspensky lists nine such inscriptions and assumes that all of them were originally placed in Aboba, and then were moved to other places; there are five such inscriptions in the fond of the RAIC (according to SPbB ARAS, f. 127, l. 3, fl. 8–12). Such inscriptions, with minor exceptions, begin with the letter complex ΚΑΝΑΣΥVIGI (-ΒΙΓΗ, -ΒΗΓΗ, -ΒΥΓΗ), followed by the name of the Bulgarian ruler Omurtag (ruled in 814–831) or his younger son Malamir (ruled in 1831–1836). Today this complex is divided as κανα συβιγι and is interpreted as the proto-Bulgarian title of Omurtag and Malamir. The mention of the ruler's name gives the tombstone inscriptions the character of official acts. In all these inscriptions about the deceased it is said that he was θρεπτὸς ἄνθροπος of the ruler. The meaning of this concept remains unclear to the end. Obviously, if we are not talking about a title, then in any case this name is an indication of the high status of this person. The meaning of such inscriptions is determined by the fact that they contain an enumeration of names and titles, which provides information for the Proto-Bulgarian namesake, as well as characteristics of the service class in pre-Christian times. These inscriptions reflect the titulature of the ruler, from whose person the inscription was made; the titles and positions of the persons to whom the inscriptions are dedicated; the military merits of these persons and their origin.

The second group is characterized by F. I. Uspensky as "Fragments of treaties between Bulgarians and Greeks". Such inscriptions were found on the fragments of large columns—six fragments belonging to the beginning of the ninth century. They are of particular importance, since they contain information missing from the Byzantine chronicles: after the end of the war, the neighbors ceased to be an object of chronological interest for Byzantium. There are two such inscriptions in the SPbB ARAS (f. 127, l. 3, fl. 13–14): these are treaties between Bulgaria and Byzantium from 814 and 823, related to the 30-year peace (Canard, 1956: 61) that was concluded presumably in 814. F. I. Uspensky names just one another such inscription.

One inscription is about the construction by Omortag of a fortress and a bridge over the Ticha River (according to SPbB ARAS, f. 127, l. 3, fl. 15). And only one estampage represents a Bulgarian inscription of a later time—this is the tombstone inscription found in the Rila Monastery and published for the first time back in 1855 by the Czech scientist Jan Shafarik in 1855 ("Serbian and Bulgarian written monuments").

Results

The database of RAIC estampages created in SPbF ARAN has become a comprehensive resource. An extensive lemma is proposed for each estampage, including an archeographic description of the document (archive cipher; document type; time of creation of the estampage; material (paper type); the size of the estampage sheet; the presence of drawings of inscriptions, labels for estampages, old covers; digital copies of drawings, labels, covers; the presence of droppings, lines, drawings of letters on estampages, etc.); the context of the production of the estampage (the toponym of the time of making of the estampage and its modern name; the nature of the expedition / archaeological excursion, the region of the expedition / archaeological excursion, the date and composition of the expedition; the main goals and objectives of the trip, its results; published and unpublished materials of the expedition), an epigraphic description of the inscription (the toponym of the time of making of the inscription and its modern name, the place, date and context of the inscription, the place of storage of the original inscription, the size of the inscription, the size of the fields of the inscription, the size of the letters, the distance between the lines; the nature of the document; the dating of the inscription and its justification; the language of the inscription; the text of the inscription and its translation; the publication of the epigraphic document and the history of its publication; critical apparatus: various options for restoring the lost text, argumentation and commentary by the publishers of the inscription; historical commentary on the inscription) (Tunkina et al., 2020).

The descriptive and attributive information is accompanied by 2d-copies (photo fixation of the document before and after restoration) and 3d-copies (also in the state before and after restoration) in OBJ-format. This format, developed at *Wavefront Technologies* for their *Advanced Visualizer* animation package, is widely used for standardized files containing 3d objects—for storing 3d-models created by combining 3d-objects, grids, shapes, metadata, textures and much more. This format is convenient because OBJ-files can be exported and imported and opened by other 3d-programs available on a personal computer. The advantage of the format is the ability to easily edit files on any personal computer using available software: *Paint 3D, 3D Builder, Autodesk Meshmixer, Blender, MeshLab, Clara.io , FreeCAD*. The edited version of the OBJ-model can be saved in its own format or can be converted to some other 3D-file formats.

A WebGL-based viewer is defined for viewing. No other plugins or libraries are required to work with WebGL technology. Any platforms can be used: *Windows, Linux, Mac,* smartphones and ipads, game consoles. The main limitation is only the browser's support for WebGL technology. The history of the technology goes back to the experiments of the Serbian-born American developer Vladimir Vukićević—Director of Engineering for Mozilla, which led to the creation of a working group with the participation of developers of various browsers; as a result, the specification was presented in 2011 (https://web.archive.org/web/20110306042639/http://www.khronos.org/news/press/releases/khronos-releases-final-webgl-1.0-specification). Teamwork has led to the wide availability of this format, since it is supported by major browsers: *Mozilla Firefox* (from version 4), *Google Chrome* (from version 9), *Safari* (from version 6, WebGL support is disabled by default), *Opera* (from version 12, WebGL support is disabled by default), *IE*. Mobile browsers and platforms: *Android* browser (only on some devices, for example, on *Sony Ericsson Xperia* smartphones and some *Samsung* smartphones), *Opera Mobile* (starting with the 12th version), *IOS*, *Firefox* for mobile, *Google Chrome* for *Android*.

In the viewer, the end user will be provided with navigation, lighting, and texturing tools for prints. This approach allows you to document and preserve the available material as accurately as possible, make it more accessible for study and analysis.

Discussion

In the last decade, academic centers around the world have launched a program to digitize their archival collections. An international project is being successfully implemented in Russia—the electronic publication of the 3rd edition of the Corpus of Ancient Inscriptions of the Northern Black Sea region, *Inscriptiones antiquae Orae Septentrionalis Ponti Euxini graecae et latinae* (IOSPE[3]) (http://iospe.kcl.ac.uk/index-ru.html). Currently, the collection of estampages of the French Archaeological School in Athens is being digitized as part of the E-STAMPAGES program (École française d'Athènes, https://www.efa.gr/index.php/en/ressources-documentaires/les-archives/archives-estampages/le-programme-e-stampages). Similar projects were also carried out at the Center for the Study of Ancient Documents of the University of Oxford (http://www.csad.ox.ac.uk/CSAD/Images.html), at the Center for Epigraphical and Paleographical Studies of the Ohio State University (https://epigraphy.osu.edu/collections), at the University of Florida (University of Florida, http://www.digitalepigraphy.org /).

3d-technologies have become actively used in archaeology (Levy et al., 2010; Sideris et al., 2017). Specialized journals (*Digital Applications in Archaeology and Cultural Heritage; The Virtual Archaeology Review*) are devoted to the problems of preserving cultural heritage using additive technologies. Specialized centers have been established at research institutes and universities in Europe and the USA: ArcheoVision, University of Bordeaux, National Higher School of Architecture in Nantes, (France); The Center for Digital Archaeology, Dig@Lab (USA); Digital Archaeology, MayaArch3D Project (Germany); 3D DOM (Italy); Laboratory of Photogrammetry at the Institute of Archaeology Zinman (University of Haifa, Israel), etc. A popular direction is the 3d-copying of unique museum exhibits. The Smithsonian Institution has digitized and made publicly available about 100 of its museum exhibits and announced its intention to digitize and publish up to 1,000,000 items of storage (http://3d.si.edu). Google cooperates with several museums around the world on three-dimensional scanning of mobile art objects (http://artsandculture.google.com), etc.

The considered database of the SPbB ARAS provides specialists and a wide range of persons interested in Slavic history with full and multifaceted information about those Proto-Bulgarian inscriptions, estampages from which are among the documents of the SPbB ARAS. The important historical significance of these documents was noted already by F. I. Uspensky: Before the invention of the Slavic alphabet, all such acts of public and state significance were stamped on stone in the ninth century. Later, with the spread of writing on cheaper and more convenient material, these acts were intended for storage in archives. Inscriptions on stone were applied in pre-Christian Bulgaria to all needs, in other words, not only what needed to be preserved from oblivion for a long time, but also in general what needed to be made known. For the future history of state institutions in Bulgaria, this kind of monuments, the number of which is currently very small, are of paramount importance.

There are few Proto-Bulgarian inscriptions found on the territory of the first Bulgarian capital, and they are mostly published by Veselin Beshevliev (last edition of 1992 "Proto-Bulgarian inscriptions"). However, the study of their 3d-originals is limited by museum storage, while the preserved estampages allow expanding the circle of researchers. This circle is expanding even more (almost limitlessly) due to the availability of digital 3d-copies from these estampages. It was this task that SPbB ARAS faced when creating a digital database with retrospective information.

Another task that the database creation project allows to solve is to improve the technique of restoration of estampages by comparing 3d-models, namely, determining the presence of deviations between the surfaces of the two models that could arise during the restoration process. Being complex models with curved surfaces, they cannot be compared manually, but this becomes possible thanks to the use of modern CAD/CAM/CAE systems.

Conclusions

Narrative information about the early Bulgarian history has been preserved largely in the inscriptions left on the stone. If the inscriptions themselves can be preserved in museums, then their wide study becomes possible thanks to the production of 3d-copies from them in the form of estampages, from which, in turn, digital 3d-copies are made. These 3d-copies of 3d-copies become especially relevant when the original inscription is lost, examples of which are known. For this purpose, the database of SPbB ARAS estampages was created, which is an effective digital technology for documenting estampages. The advantages of 3d-technology in historical, archaeological and epigraphic research are expressed in obtaining a high-precision three-dimensional digital copy of the object, in the possibility of contactless work of researchers with monuments having different degrees of damage, in providing users with the opportunity to inspect virtual copies in any projections with varying degrees of accuracy, in the use of computer modeling methods, in the restoration of crumbling or disappeared objects, in preserving the model in volume with the possibility of subsequent replication, in creating digital repositories of 3d-objects, with the possibility of their systematization and comparative analysis. In a broader sense, these approaches fit well into the concept of the digital economy, which is being implemented in the Russian Federation today.

Acknowledgements The work was prepared with the financial support of a grant from the Ministry of Science and Higher Education of the Russian Federation in the form of subsidies, project No. 075-15-2020-786 "The History of writing of European civilization".

References

Canard, M. (1956). Les aventures d'un prisonnier arabe et d'un patrice byzantin a l'epoque des guerres bulgaro-byzantines. *Dumbarton Oaks Papers, 9*, 49–72.

Chernova, N. V., Pavlichenko N. A., & Bakhvalova E. V. (2018). Research and restoration of the print of a Greek inscription from the Jordanian city of Jarash (ancient Gerasa). In *Mueller Readings—2018* (pp. 815–822). https://doi.org/10.25990/archiveras.mc-2018.bqbv-qm07.

d'Argenville, A. J. D. (1727). Lettre sur le choix et l'arrangement d'un cabinet curieux. *Mercure de France, 2*, 1294–1330.

De Rosset, T. F. (2014). Kolekcja artystyczna—geneza, rozkwit, kryzys. *Acta Universitatis Nicolai Copernici. Zabytkoznawstwo i konserwatorstwo, 45*, 253–285. https://doi.org/10.12775/AUNC_ZiK.2014.010

Diringer, D. (1968). *The alphabet: a key to the history of mankind*. Hutchinson.

Hübner, E. W. E. (1881). *Über mechanische Copien von Inschriften*. Weidmannsche Buchhandlung.

Levy, T. E., Petrovic, V., Wypych, T., Gidding, A., Knabb, K., Hernandez, D., Smith, N. G., Schlulz, J. P., Savage, S. H., Kuester, F., Ben-Yosef, E., Buitenhuys, C., Jane, C. J. C., Najjar, M., & DeFanti, T. (2010). *On-site digital archaeology 3.0 and cyber-archaeology: Into the future of the past—New developments, delivery and the creation of a data avalanche. Introduction to cyber-archaeology* (pp. 135–153). Archaeopress.

Pavlichenko, N. A. (2018). Collection of estampages of Greek inscriptions in the fond of the Russian Archaeological Institute in Constantinople. In *Mueller readings—2018* (pp. 465–473). Renome. https://doi.org/10.25990/archiveras.mc-2018.e10b-jv65.

Sideris, A., Liritzis, I., Liss, B., Howland, M. D., & Levy, T. E. (2017). At-risk cultural heritage: new excavations and finds from the Mycenaean site of Kastrouli, Phokis, Greece. *Mediterranean Archaeology and Archaeometry, 17*(1), 271–285. https://doi.org/10.5281/zenodo.163772

Tunkina, I. V., Bondar, L. D., Ponikarovskaya, M. V., Pavlichenko, N. A., & Chernova, N. V. (2020). Prints of the Russian Archaeological Institute in Constantinople (1895–1914) in the collection of the St. Petersburg Branch of the Archive of the Russian Academy of Sciences: Scientific restoration, comprehensive study, creation of high-precision digital 3d-models. *St. Petersburg Historical Magazine. Research on Russian and Universal History, 4*(28), 194–203. https://doi.org/10/51255/2311-603X_2020_4_194

Zawartko-Laskowska, M. (2008). Mecenat naukowy Stanisława Augusta i jego przyrodnicze pasje. *Kronika Zamkowa, 1*, 65–112.

Development of Industry 4.0: A Practical Case Study from the Netherlands

Jacob Cornelis Bazen, Olga Dmitrovna Ugolnikova, and Irina Sergeevna Bazen

Introduction

The Fourth Industrial Revolution (Industry 4.0) is a concept introduced in Germany around 2011, as result of a project to describe and understand the trends and developments in manufacturing industry in Germany (Rojko, 2017). The financial crisis of 2008 as well as the ever-ongoing process of outsourcing and offshoring of industrial activity brought forward the need of thinking about in which way the German industry could remain competitive worldwide, as being on the forefront of innovation. As Vannevar Bush (1945) argued, scientific and technological leadership is crucial in maintaining an important and influential position on the world-stage. After its inception, the concept was well received and gained worldwide attention, not just in the research community, but also in the industrial society (Oztemel & Gursev, 2018). Industry 4.0 is about smart factories, able to produce autonomously, as the manufacturing system is IT driven, without the need for (much) human intervention. The reason for the development of the industry in this direction is the ever-increasing demand of customers towards individual tailor-made products. At this moment, in the most advanced industrial economies in Europe, so-called "high mix low volume" production systems are rapidly becoming standard

J. C. Bazen (✉)
School of Business, Building & Technology, Saxion University of Applied Sciences, Enschede, The Netherlands
e-mail: j.c.bazen@saxion.nl

O. D. Ugolnikova
St. Petersburg University of Management Technologies and Economics, St. Petersburg, Russian Federation

I. S. Bazen
Volksuniversiteit Enschede, Enschede, The Netherlands

A. Rumyantseva et al. (eds.), *Challenges and Solutions in the Digital Economy and Finance*, Springer Proceedings in Business and Economics,
https://doi.org/10.1007/978-3-031-14410-3_28

(Godinho Filho & Saes, 2013; Suri, 2020; Upton, 1995; Veza et al., 2015). This means that with this form of production rapid changes in products can be made, with probably less economies of scale, but relatively short delivery times. Such production systems allow for higher added values (Suri, 2020; Veza et al., 2015) giving industry a "raison d'être" in highly developed economies with high labour costs.

Often such "high mix low volume" products are assembled from different standardized components. At this moment such standardized industrial components are mostly outsourced to countries with lower wages and/or higher production capacities and because of the implementation of Enterprise Resource Management systems often delivered on an (almost) Just-In-Time basis (McLachlin, 1997). Industry 4.0 offers a promise for industry in more developed economies to (at least partly) take away the current cost price disadvantage (expensive labour) for production in the most developed economies as well as offering more balanced supply streams leading to less waste (Brozzi et al., 2020; Kovacs, 2018).

However, to this day Industry 4.0 remains largely a scientific concept without yet much adoption in practice (Oztemel & Gursev, 2018). Most of the available literature is also conceptual in nature and offers different types of frameworks, but rarely any examples of implementation of (near) Industry 4.0 projects (Zheng et al., 2021). This short chapter would like to do just that: to analyse the development of a proposed Industry 4.0 smart factory in the Netherlands, with the opportunities and challenges from a business perspective that this project has.

The aim of the study—to analyse the development of an Industry 4.0 smart factory in the Netherlands, with the opportunities, risks, challenges that this project has. The tasks of the work were defined: to provide a description of a highly innovative energy saving project in the housing sector; determine the prospects for the implementation of a project that is close to the requirements of Industry 4.0; specify the elements of the project that require improvement and full compliance with the requirements of Industry 4.0.

The object of research is a highly innovative INDU-ZERO project in the housing industry. This is a specific project that may lead to a smart factory close to Industry 4.0. The subject of the study is testing the project for compliance with the criteria of Industry 4.0. Conclusions are based on compliance with the list of Industry 4.0 criteria

The novelty of the study is as follows:

- A model of a smart factory close to Industry 4.0 has been developed.
- An interpretation of the energy neutrality model in the housing sector in the Netherlands, close to Industry 4.0, is presented.
- An analysis of the project's compliance with the criteria of Industry 4.0 is per-formed.
- The required investments are calculated, a business justification of the project's effectiveness is given.
- The problems, risks, and prospects of implementing projects close to Industry 4.0 are indicated.

Background

The materials for the study were scientific works that developed the concept of industrial revolutions. We will consider the background and context for the concept of Industry 4.0. The term "fourth industrial revolution" implies that there have been three previous industrial revolutions. The first industrial revolution is the mechanization of production from roughly the second half of the eighteenth century to the end of the nineteenth century. Toynbee (1884) introduced the concept of "Industrial Revolution" in the English language and placed the starting point for this industrial revolution around 1760, the time of the introduction of the "flying shuttle" in the weaving process and the subsequent introduction of the "spinning jenny", the first mechanized spinning machine. The textile industry is generally regarded as the first industry that industrialized, but many other sectors soon followed. The characteristic of this first industrial revolution is the use of coal to generate steam to power the machines. The start of the twentieth century gave rise to the second industrial revolution, the automated assembly line, and the further division of labour within the company. This second industrial revolution is usually linked to Frederick Winslow Tailor and Henry Ford, for respectively developing the first scientific theory on management and the application of the assembly line in automobile production leading to a large increase in production (Zheng et al., 2021). Characteristic of the second industrial revolution is the standardization of products as well as components to produce these products, with little attention for flexibility of the production process (Wang, 2018). Henry Ford's famous quote: "*You can choose any colour as long as it is black*" when talking about his T-ford, the first assembly line mass-produced automobile, quite accurately catches the concept of standardization belonging to Industry 2.0. The third industrial revolution came with the application of computers since the 1970s and digital computer aided design to produce more efficient and to use the digital techniques to allow for worldwide spanning supply chain networks (Oztemel & Gursev, 2018; Wang, 2018). Production in Industry 3.0 is more automated and sophisticated, offering much more flexibility in the production processes. The third industrial revolution saw the appearance of robots doing repetitive and/or risky tasks as replacement for human labour. As industry 3.0 comes with a lot of networking between firms, there is also increasing attention for cybersecurity, as such firms increasingly have to deal with risks of cyber-attacks (Wang, 2018). Still, even with automated production, necessary machine change-over times would significantly hinder further increases in flexibility of production. Human supervision and programming of machinery is still required, as well as in engineering customer orders to "producible" factory orders.

The fourth industrial revolution goes even further in terms of digitalization, it encompasses autonomous cyber-physical production systems and internet of things (Perales et al., 2018), cloud computing based on the usage of big data and the capability of machine learning where there is just minimal need for human intervention (Osterrieder et al., 2020; Wichmann et al., 2019). Industry 4.0 means that production systems in factories can operate on their own and make decisions for themselves about production batch sizes, specific types of products to produce as well as use machine learning to avoid previous mistakes and learn to adapt production towards better results in terms of customer needs. Such factories are also known as "smart factories", "dark factories" or "lights-off factories" (Oztemel & Gursev, 2018). The ability of systems to learn from dealing with previous situations and from human interventions will determine the competitive advantages of individual businesses (Wang, 2018), as well as helping with waste reduction (Kamble et al., 2018). In practice, Industry 4.0 will reduce machine changeover times to almost zero, allowing "economies of scale" even with a batch size of one (Oztemel & Gursev, 2018). Such a new production system also comes with major changes to the factory management (Piccarozzi et al., 2018).

The methodology is presented by the theoretical provisions that became the basis for the classification of Sony and Naik (2019): full industry 4.0 factories of meet the following criteria:

- Integration of cyber-physical systems in the production process of the factory.
- Automated data-management within the factory.
- Optimization of resource utilization (less waste).
- Production error reduction by means of machine learning.
- Automated supply-chain, including automated vehicles for transport.
- Large attention for cyber-security of the IT systems.
- Organizational mission and vison related to the adaptation of Industry 4.0.
- General scientific methods were used: modeling, classification, generalization, logical method, concretization, as well as the method of economic analysis; among the empirical ones—description, survey, measurement, practical modeling, the method of expert assessments.

Practical Implementation of Industry 4.0

We will indicate the problems and risks of implementing projects close to Industry 4.0.

It is important to note that Industry 4.0 is still a largely scientific concept and therefore, in practice there are so far only laboratory scale "autonomous factories" available (Sony & Naik, 2019). Autonomous machine learning is still in its infancy and it is expected that it will take several more decades to become fully mainstream (Oztemel & Gursev, 2018; Sony & Naik, 2019). One of the reasons for the limited current application of the industry 4.0 standard is the immense data infrastructure that is required to harness the power of machine learning by having the ability to handle big data. Furthermore, the technology that is available and would be possible to use to create a near-Industry 4.0 production environment is (still) very costly and would require large investments with, due to the newness of the technology, still uncertain results, making such investments highly risky. Also, there may be significant barriers related to staff of manufacturing companies, as very different work competences are needed. Staff members may very well be reluctant to work on Industry 4.0 implementation for the fear of the loss of their jobs (Horváth & Szabó, 2019; Stentoft et al., 2019). Kovacs (2018) mentions that in the USA and Europe around 50% of all jobs may be lost by computerization. Horváth and Szabó (2019) identify another barrier, namely the influence of legislation: among others the uncertainty coming from tender procedures, of which many (larger) industrial firms are dependent on.

Nonetheless, there are clear drivers of Industry 4.0 adoption which can be found in many studies of the subject (see for an overview Oztemel & Gursev, 2018). Industry 4.0 drivers usually originate from higher management, with the promise of more production control, a reduction of waste, an increase of time to market, a reduction of operating costs and a general shortage of a qualified labour force (Horváth & Szabó, 2019; Stentoft et al., 2019). In general however, Stentoft et al. (2019) find in their study among Danish manufacturing companies low Industry 4.0 readiness and practice levels, even though Denmark is at the forefront of technological development. Similar findings have been reported for Sweden (Truvé et al., 2019) and Germany (Bittighofer et al., 2018). The manufacturing sector in Finland and the Netherlands appear to be a bit more Industry 4.0 ready than those in most other European countries (Castelo-Branco et al., 2019). The next part of this chapter is therefore a case study of one Industry project in The Netherlands.

Industry 4.0 Case Study in the Netherlands

The experience of developing a specific model of energy neutrality in the housing sector, close to Industry 4.0, in the Netherlands will be presented.

Even though factories fully compliant to the industry 4.0 standard are perhaps still non-existent, this part is a case study on a specific project that may very well lead to a near-Industry 4.0 factory. Firstly, some background on the project will be provided, followed by the design choices and business model selection of this proposed smart factory. Secondly, the proposed factory with its design will be tested against the before mentioned criteria of Industry 4.0.

The project leading to the proposed smart factory is a project based on the Paris Agreement against climate change. If the goal of a significant reduction in carbon emissions is to be reached, the annual emissions have to be reduced tremendously during the next decades, given the situation that in 2021 an almost all time high amount of CO_2 equivalents was emitted (IEA, 2021). One of the key aspects of the policy plans of European countries in the North-Sea Area to reach the necessary reduction, is to retrofit the entire housing stock of these countries towards energy neutrality. This means that the net energy use of these houses should be zero, meaning that the houses themselves would provide all necessary energy for domestic appliances, heating, hot tap water etc. It is very well possible to build new houses and renovate existing houses towards this situation of energy neutrality: It is a combination of decentralised energy generation, usually by means of solar PV panels on the one hand, and energy saving, by adding extra insulation to the house on the other hand. Newly built houses and apartment blocks have to meet strict rules for energy usage in North Sea Region countries, and since these new houses are designed and developed basically from scratch, it is relatively easy to implement such energy requirements in the design itself (Smit, 2017).

However, the situation is very different for existing houses and apartment buildings. There is a huge diversity in size, and layout of different houses and apartments. To energetically retrofit the existing housing stock, houses must be measured individually to provide them later with an outer layer of insulation material, as well as the correct dimensions of solar PV panels on the roof. It is therefore evident that the energetic retrofitting of existing residential buildings is currently a very labour intensive and costly affair. Not only costs are problematic, given the current shortage of labour (which is not likely to end soon, given among others the demographic situation), but the speed of these necessary renovations is also too low as well. In the North Sea region alone, around 22 million residential buildings are in need of energetic retrofitting before 2050 (INDU-ZERO, 2021). With the current costs and speed of operation, this goal is unattainable.

Therefore, a consortium of university partners, governments and business has come together in a triple helix setting to tackle this issue, by focusing on finding a solution which both cuts the costs of renovations per dwelling by half and allows for a very significant increase of the speed of production of energetic retrofitting packages. The consortium named the project INDU-ZERO. The solution decided upon by the consortium was to develop a blueprint for a near Industry 4.0 "smart factory". This triple helix-based consortium was a very effective way to tackle the issue at hand, as seen in the previous section of the chapter, barriers of Industry 4.0 implementation can be found on legislative issues (government), technology and knowledge issues (university) as well as on business model/investment readiness issues (business).

Within the INDU-ZERO project, a fully automated digital driven production system has been designed, with three production lines, together able to produce renovation packages for 15,000 residential units in total yearly. Since every house and/or apartment is different, and labour saving is one of the main goals, each house must receive tailor made renovation packages to be placed as a shell around the house or apartment, by a team of four workers in 3 days. This means that work activities on site should be kept at a bare minimum: The factory should be able to produce unique and exactly fitting panels for every single dwelling which can be mounted to the dwelling within minutes (see for visual materials INDU-ZERO, 2021). Solar PV panels come integrated with the roof panels, taking away the need to mount them. Even original roof tiles don't have to be removed, the roof panels can be mounted right over them. Only existing chimneys must be taken away. Since there will be no wood or gas usage in the new situation, any existing chimney will be obsolete and can be taken off the roof in minutes with a special crane.

The factory itself, must be extremely flexible in terms of production: the market requires batch sizes of 1 (unique renovation packages for each single home or apartment). Therefore, the factory will be developed as an Industry 4.0 fully automated smart factory. Key to the development of such a factory is the development of the data stream, which would be the red thread throughout the production process. The starting point is the measurement of the dwelling, to find out the dimensions the renovation package needs to have, as well as the places where the windows and doors should appear. Traditionally this is done by manually measuring the distances and allowing for some slack space that can be made to measure during the final assembly of the insulation material. INDU-ZERO uses building scanners that produce a data point cloud, which is uploaded to the factory and translated to a digital twin of the building. With minimal engineering time (for example for adding extra visual attractiveness of the outside of the panels if so desired by the client), such digital twins are automatically used by the production system to produce the elements with the exact, millimetre accurate, dimensions. The overarching Enterprise Resource Planning (ERP) system automatically selects the necessary materials from the factory warehouse and orders additional materials whenever the inventory is below a certain safety level. Incoming trucks (an estimated 70 per day) are offloaded automatically by Automated Guided Vehicles (AGVs). Since every renovation package is unique, the ready-made product is picked up by a truck to be transported to the exact dwelling it was produced for. The on-site mounting process of the package, consisting of several panels for each side of the house, as well as the roof should be able to be mounted in the shortest possible time with the lowest possible amount of required labour. After the mounting process, the new installations must be connected and installed (the heat pump for heating and hot water and the solar PV panels, including inverters for providing energy). This takes up the bulk of the still necessary worktime for the on-site assembly of the renovation packages.

All in all, there are still 440 workplaces needed in the smart factory, for producing the 15,000 renovation packages yearly. This is a strong reduction from the current labour-intensive production situation (takt time of each produced panel is expected to be around 2 min), but still, it means that the proposed factory is not yet a fully Industry 4.0 "lights-off factory". The workforce is mainly occupied with quality supervision and solving machine breakdowns, as well as—as is expected—in some cases with rework of quality control rejected panels. A complete Industry 4.0 factory would also have machines automatically controlling the final quality of the produced panels. Even though there is certainly a good amount of quality control done automatically, still a final visual overall inspection has to be done by humans. In the blueprint, there is yet no machine learning tool foreseen, which can automate the final quality control. Consequently, also the rework section will require human workers to do any necessary rework activities. In a truly automated Industry 4.0 environment this would also be done by cyber physical systems with the capability of machine learning.

In terms of business model and required investment, the total amount of the investment is but is going to be more than 200 million Euro, which certainly is a considerable sum of money. Still, the business case allows for a relatively speedy return on investment, given the fact that the sales volume is forecasted at 15,000 units yearly (based on the demand from larger Dutch & German housing associations). In the city of Enschede, The Netherlands, three showcase houses have been

Fig. 1 Testcase houses of the INDU-ZERO project in Enschede, The Netherlands under redevelopment. Photo: Jacob Cornelis Bazen

completed with the help of the new on-site assembly process, designed in the project (Hellegers, 2021), see also Fig. 1.

Now of writing, the smart factory blueprint is in its final stages of preparation and upon delivery of the blueprint, with the development of the machines, the design of the factory and the design of the process of on-site assembly, the smart factory will be investor ready. When all goes according to plan and enough investors are willing to contribute, within 2–3 years the blueprint can be developed into a complete and working Industry 4.0 factory.

Conclusion

When testing the proposed blueprint against the Industry 4.0 criteria list (see section "Background" of this chapter), the factory scores as follows:

1. **Integration of cyber-physical systems in the production process of the factory**: The factory functions almost autonomously. The only human intervention in the normal production process is the final optical quality check of the panels, quality checks of different components are done automatically. When quality issues are detected, the panel is moved to the rework section of the factory, which

runs with physical labour. Therefore, one can argue that the blueprint is 90% Industry 4.0 ready.

2. **Automated data-management within the factory**: The factory is fully dependent on an automatic DataStream based on the point cloud scan of the dwelling. Some human interference can occur in terms of the design when customers would like to add patterns or other decorative elements to the panels. But, in principle, on this aspect the proposed factory fully meets the Industry 4.0 standard.
3. **Optimization of resource utilization (less waste)**: Feedback loops in the production are foreseen, to reuse cut-offs. It is currently untested, to which extent this will lead to optimization of resource use. Exact compliance with Industry 4.0 standards cannot be given at this point.
4. **Production error reduction by means of machine learning**: Currently, no machine learning tool is included in the design of the factory. Therefore, the proposed factory does not meet this part of the Industry 4.0 standards.
5. **Automated supply-chain, including automated vehicles for transport**: All in-house logistics are completely automated and will run without human intervention. This includes automatic unloading of supply trucks. The loading of the trucks with the readymade panels still must be done manually by the truck driver. Therefore, in terms of logistics, the proposed factory meets around 90% of the Industry 4.0 standards.
6. **Large attention for cyber-security of the IT systems**: There is a strong attention on cyber security, a relatively large part of the proposed workforce is foreseen as IT specialists, focusing on digital safety. For as much as can be seen now, on this aspect the factory will meet Industry 4.0 criteria.
7. **Organizational mission and vison related to the adaptation of industry 4.0**: Since the organisation is new, and the project team has focused on the design of a smart factory from the beginning, it could be safely assumed that in organizational sense, the factory will be Industry 4.0 ready.

Therefore, given the scores on seven important aspects of Industry 4.0, it can be concluded that even this highly innovative INDU-ZERO project, with its design of ground-breaking cyber-physical production systems does not fully meet the most important Industry 4.0 requirements. Even though this factory has many aspects of a "dark factory", it is still not fully running autonomously. Industry 4.0 is still a standard that needs significantly more time to become mainstream in production environments.

Acknowledgements This study on developing a practical Industry 4.0 factory blueprint was supported by the Interreg program of the European Union.

References

Bittighofer, D., Dust, M., Irslinger, A., Liebich, M., & Martin, L. (2018). *State of Industry 4.0 across German companies.* Paper presented at the 2018 IEEE International Conference on Engineering, Technology and Innovation (ICE/ITMC).

Brozzi, R., Forti, D., Rauch, E., & Matt, D. T. (2020). The advantages of industry 4.0 applications for sustainability: results from a sample of manufacturing companies. *Sustainability, 12*(9), 3647.

Bush, V. (1945). Science: The endless frontier. *Transactions of the Kansas Academy of Science, 45*(3), 231–264.

Castelo-Branco, I., Cruz-Jesus, F., & Oliveira, T. (2019). Assessing Industry 4.0 readiness in manufacturing: Evidence for the European Union. *Computers in Industry, 107*, 22–32.

Godinho Filho, M., & Saes, E. V. (2013). From time-based competition (TBC) to quick response manufacturing (QRM): the evolution of research aimed at lead time reduction. *The International Journal of Advanced Manufacturing Technology, 64*(5), 1177–1191.

Hellegers, B. (2021, November 16). Innovatief renovatiepakket als theemuts over woningen in Enschede: drie gewone rijtjeshuizen ineens internationale trekpleister. *Tubantia.* Retrieved from https://www.tubantia.nl/enschede/innovatief-renovatiepakket-als-theemuts-over-woningen-in-enschede-drie-gewone-rijtjeshuizen-ineens-internationale-trekpleister

Horváth, D., & Szabó, R. Z. (2019). Driving forces and barriers of Industry 4.0: Do multinational and small and medium-sized companies have equal opportunities? *Technological Forecasting and Social Change, 146*, 119–132. https://doi.org/10.1016/j.techfore.2019.05.021

IEA. (2021). *Global Energy Review 2021.* Retrieved from Paris: https://www.iea.org/reports/global-energy-review-2021

INDU-ZERO. (2021). Designing a factory for energy renovations. Retrieved from https://northsearegion.eu/indu-zero/

Kamble, S. S., Gunasekaran, A., & Gawankar, S. A. (2018). Sustainable Industry 4.0 framework: A systematic literature review identifying the current trends and future perspectives. *Process Safety and Environmental Protection, 117*, 408–425. https://doi.org/10.1016/j.psep.2018.05.009

Kovacs, O. (2018). The dark corners of industry 4.0—Grounding economic governance 2.0. *Technology in Society, 55*, 140–145.

McLachlin, R. (1997). Management initiatives and just-in-time manufacturing. *Journal of Operations Management, 15*(4), 271–292.

Osterrieder, P., Budde, L., & Friedli, T. (2020). The smart factory as a key construct of industry 4.0: A systematic literature review. *International Journal of Production Economics, 221*, 107476.

Oztemel, E., & Gursev, S. (2018). Literature review of Industry 4.0 and related technologies. *Journal of Intelligent Manufacturing, 31*(1), 127–182.

Perales, D. P., Valero, F. A., & García, A. B. (2018). Industry 4.0: a classification scheme. *Closing the gap between practice and research in industrial engineering* (pp. 343–350).

Piccarozzi, M., Aquilani, B., & Gatti, C. (2018). Industry 4.0 in management studies: A systematic literature review. *Sustainability, 10*(10), 3821.

Rojko, A. (2017). Industry 4.0 concept: Background and overview. *International Journal of Interactive Mobile Technologies, 11*(5).

Smit, S. (2017). *Op weg naar bijna energie neutrale gebouwen.* University of Twente.

Sony, M., & Naik, S. (2019). Key ingredients for evaluating Industry 4.0 readiness for organizations: a literature review. *Benchmarking: An International Journal, 27*(7).

Stentoft, J., Jensen, K. W., Philipsen, K., & Haug, A. (2019). *Drivers and barriers for Industry 4.0 readiness and practice: a SME perspective with empirical evidence.* Paper presented at the proceedings of the 52nd Hawaii international conference on system sciences.

Suri, R. (2020). *Quick response manufacturing: a companywide approach to reducing lead times.* CRC Press.

Toynbee, A. (1884). *Lectures on the industrial revolution in England: popular addresses, notes and other fragments*. Cambridge University Press.

Truvé, T., Wallin, M., & Ryfors, D. (2019). Swedish manufacturing SMEs readiness for industry 4.0: what factors influence an implementation of artificial intelligence and how ready are manufacturing SMEs in Sweden?

Upton, D. M. (1995). Flexibility as process mobility: the management of plant capabilities for quick response manufacturing. *Journal of Operations Management, 12*(3–4), 205–224.

Veza, I., Mladineo, M., & Gjeldum, N. (2015). Managing innovative production network of smart factories. *IFAC-PapersOnLine, 48*(3), 555–560.

Wang, B. (2018). The future of manufacturing: A new perspective. *Engineering, 4*(5), 722–728. https://doi.org/10.1016/j.eng.2018.07.020

Wichmann, R. L., Eisenbart, B., & Gericke, K. (2019). *The direction of industry: a literature review on Industry 4.0.* Paper presented at the proceedings of the design society: International conference on engineering design.

Zheng, T., Ardolino, M., Bacchetti, A., & Perona, M. (2021). The applications of Industry 4.0 technologies in manufacturing context: a systematic literature review. *International Journal of Production Research, 59*(6), 1922–1954. https://doi.org/10.1080/00207543.2020.1824085

The Transformation of Corporate Management Business Processes in the Context of the Digitalization of the National Economy

Vladimir V. Zemskov and Valeriy I. Prasolov

Introduction

The modern national economy has faced not only global economic challenges and threats but also internal ones that hinder its sustainable economic growth, such as the poor quality of measures taken to improve business processes to meet the requirements of the new model of development.

It is well-known that national economy consists of various frameworks. It is very important to have the elements of the digital economy infrastructure introduced into everyday economic and political life. The key elements of the digital economy infrastructure can be represented by:

- Digital corporate culture, equally perceived by all employees of the corporation
- Highly qualified personnel capable of implementing the skills required for using the digital economy infrastructure
- Identification of essential business processes required for digitalization
- Digital solutions that help shape the final product
- Business models that contribute to obtaining added value
- Bank of data used for digital solutions
- Digital infrastructure that provides uniform access to modern information technologies

Thus, the widespread introduction of modern information technologies ensures a sustainable transition to the service activity model through the mechanisms of interaction between various business entities.

V. V. Zemskov (✉) · V. I. Prasolov
Financial University, Moscow, Russian Federation
e-mail: VVZemskov@fa.ru; VIPrasolov@fa.ru

A. Rumyantseva et al. (eds.), *Challenges and Solutions in the Digital Economy and Finance*, Springer Proceedings in Business and Economics,
https://doi.org/10.1007/978-3-031-14410-3_29

At the same time, it should be considered that in the modern world, any corporate entity has various types of business processes, including financial, operational and credit cycles, that ensure efficient business activities, as well as product life cycle processes.

Materials and Methods

The literature on economy offers various interpretations for the concept of process/business process, as shown in Table 1.

As shown in Table 1, processes lie in the basis of the operational lifecycle related to producing products (activities, services), purchasing commodities and materials, selling finished goods, implementing marketing initiatives, creating cash flows, conducting investing and financial activities.

As the above definitions show, we can conclude that the following features define the concept of a business process:

- Activity type
- Availability of inputs and outputs
- Creating a product that is of value to the consumer
- Existing added value

Table 1 Classification of the existing concept of process

Author	Definition
Adam Smith (1776)	The division of labour, however, so far as it can be introduced, occasions, in every art, a proportionable increase of the productive powers of labour. The separation of different trades and employments from one another seems to have taken place in consequence of this advantage
GOST R ISO 9000-2001	Process is a set of interrelated or interacting activities, which transforms inputs into outputs
GOST R 57271.5-2016/EN 15221-5:2011	Processes include a set of activities in a certain order (work process) with defined starting and ending points, as well as clear inputs and required outputs (results)
Harrington (2006)	Process is a series of logically interconnected, related activities that takes an input, adds value to it, and produces an output to an internal or external customer
Zyryanova and Tarnovskaya (2012)	Business process is an execution of company goals through a certain complex of subsequent processes controlled at various management levels, with an input point, subsequent stream of actions, and output point
Patasotskaya (2010)	In a simplistic way, a business process can be regarded as a stream of alternating events and operations, limited by input and output

Source: compiled by the authors on the basis of sources (Harrington, 2006; Patasotskaya, 2010, Smith, 1776; Zyryanova & Tarnovskaya, 2012)

In our opinion, the main component of a process is the creation of added value. It should be noted that added value can only be obtained through introducing new advanced technologies, increasing labor productivity and, of course, transforming existing business processes to comply with the new development model.

Therefore, the implementation of such an ambitious objective requires decisive actions in rethinking the existing business processes to meet their requirements for a new development model in emerging markets for the products of the Industry 4.0, which is the Internet of things.

Results

It should be noted that while implementing the Industry 4.0 technology, it is necessary to consider the specifics of certain types of activities, including such categories as the availability of basic knowledge and competencies of personnel capable of successfully using modern technologies, the range of technical standards and GOSTs, the existing legal and regulatory framework governing technological processes, as well as the organizational maturity of the corporation in terms of introducing digital technologies into all business processes.

The specifics of the global economic environment determines the need for identifying innovative production methods and for searching new ways to improve management efficiency. And, as a result, first of all, at the level of the economic entity, it is necessary to ensure the identification and formalization, in the form of quantitative and qualitative indicators, of main and auxiliary business processes to improve the results of their functioning.

Currently, there are many standards and descriptions of modeling business processes in industrial enterprises that create the basis for customized software. They use various forms of business process visualization: from Pareto charts to flowcharts. Figure 1 shows a typical flowchart of the business process "Acceptance of raw materials, stock items, and primary commodities". The start of the business process is the receipt of raw materials and commodities from suppliers in accordance with the schedule; the exit is the closing of the shift task.

The effectiveness of management decisions largely depends on applied methods and tools, their specifics being determined by the nature of the ongoing production processes. Certain resources are always used for the execution of business processes: human, stock, technological, technical, informational, social, organizational, financial, etc. (Table 2).

The best global practice has developed and proposed many standards and descriptions of modeling business processes in industrial enterprises that create the basis for customized software. Many countries have chosen the transition from functional enterprise management to process management as a priority business process management strategy. The main goal of such a strategy is to stabilize financial and economic activities of the organization through optimizing business

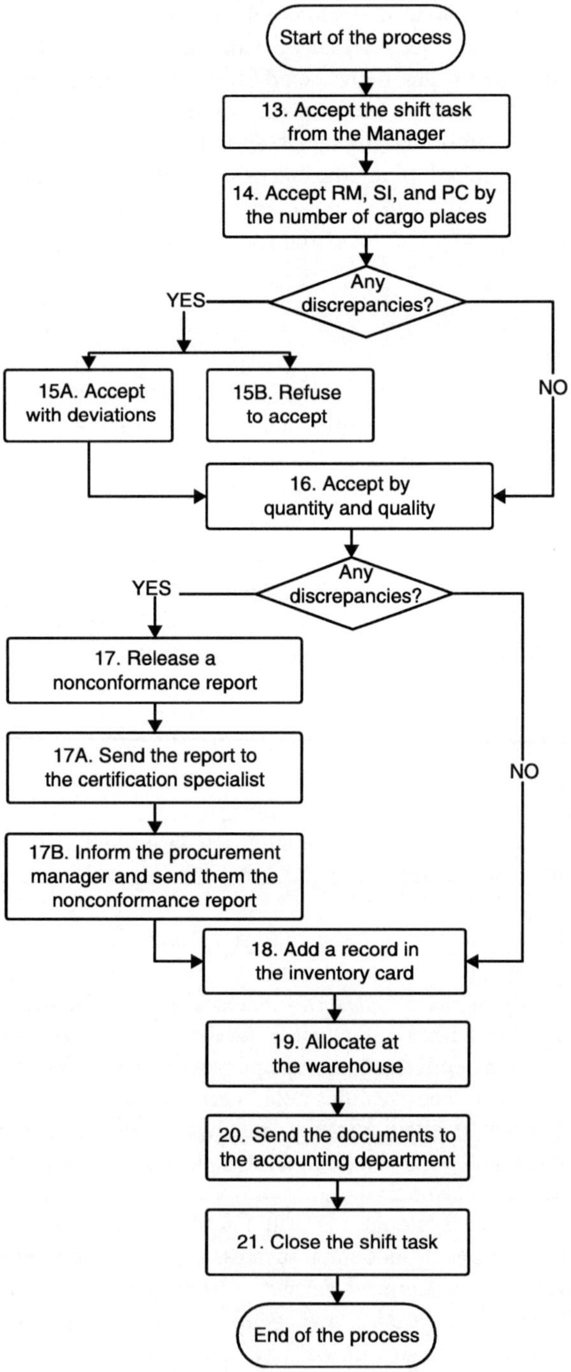

Fig. 1 Flowchart of the process "Acceptance of raw materials, stock items, and primary commodities"

Table 2 Business process "Acceptance of raw materials, stock items, and primary commodities"

Description of actions	Person responsible for actions	Document/record
Start of the process: Incoming delivery of raw materials, stock items, and primary commodities (RM, SI, and PC) from suppliers in accordance with the schedule		
Accept a shift task from the Head of Stock Department	Warehouse keeper	Shift task
Upon arrival of RM, SI, and PC, check for discrepancies in the number of cargo places	Warehouse keeper	
In case of discrepancies in the number of cargo places, refuse to accept or accept with deviations	Warehouse keeper	
In case of no discrepancies identified, accept by the quantity and the quality, using measuring equipment	Warehouse keeper	Consignment bill, invoice
If any discrepancies are identified, release a nonconformance report, send it to the certification specialist for review and approval and inform the procurement manager, handling them one copy of the nonconformance report	Warehouse keeper Certification specialist	Overage, Shortage and Discrepancy report (TORG-2)
In case of no discrepancies, based on receiving reports for RM, SI, and PC, add a record in the inventory card, create a receipt note	Warehouse keeper	Inventory card, receipt note in 1C
Allocate RM, SI, and PC at the warehouse in accordance with the storage scheme	Warehouse keeper	
Send the 2nd copy of accompanying documents, signed as accepted, to the Accounting Department	Warehouse keeper	Consignment note, invoice, OS&D report, additional documents
Close the acceptance shift task	Head of Stock Department	

process management and introducing the elements of digitalized infrastructure in the economies of developed countries.

The traditional model of business process management, shown in Fig. 2, describes the hierarchically functional operational scheme of an economic entity.

To improve management, it is necessary to classify business processes by various criteria, depending on the goals and objectives of management.

1. By the influence on the result of business activities, business processes are divided into main and auxiliary ones.
2. By functionality, it is possible to identify business processes of procurement (supply), production, distribution (sales) and settlements with buyers (customers or clients). The process of resource procurement is the "input"—a stage that ensures the implementation of the main business process of the enterprise. The settlement process is the "output", the final stage of the operating cycle at the enterprise (Smurov 2017).

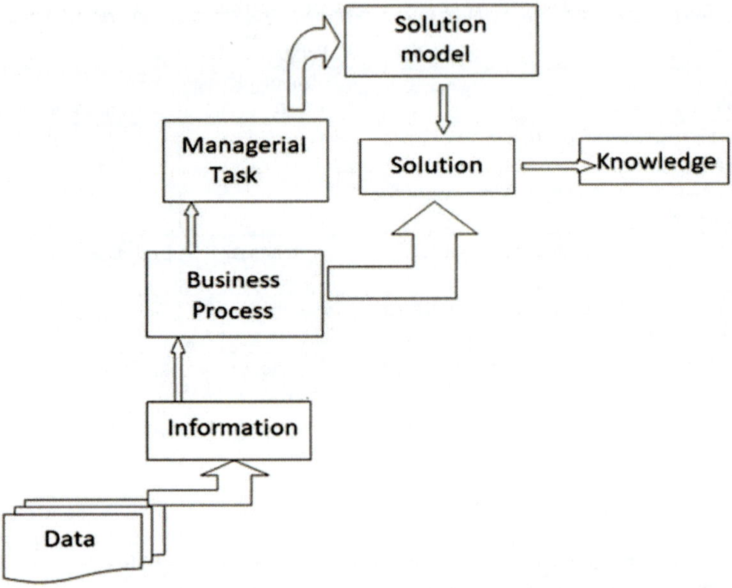

Fig. 2 Traditional model of business process management

3. By the nature of the activity, the business processes that transform the inputs received from procurement or supply processes into the outputs of implementation processes are considered as manufacturing processes.
4. By the area of functioning, there are external business processes and internal business processes that occur solely within the enterprise as an integral management system (Smurov 2017).

Discussion

The efficient management of business processes requires them to be identified and described objectively. In practice, the objective of formalizing business processes is achieved by respecting certain principles:

- Analysis and decomposition of each business process into its constituent elements
- Identification of key process parameters and their value for the industrial enterprise
- Formalization and documentation of operations that create the business process
- Clear understanding of the essence, specifics, and functions of all elements

The most important driver of growth in the application of process management models for business processes is the use of the "lean" method, which, in addition to changing the management concept, involves changing the type and the content of the analyzed information on business processes (Hawkins & Gravier 2019). The core of the "lean" method is to consume the absolute minimum of resources required to produce only such an amount of products that the market needs, and in this case everything that does not relate to the minimum amount of equipment, materials, spare parts, work and storage space and working time absolutely necessary to add value to the product, is regarded as losses. From this perspective, business entities exist to create value for customers. This leads to the alignment of activities around the process of creating value for their customers, despite occasional short-term financial losses. And this, in turn, means that each employee and each function are evaluated in the context of their value to consumers.

References

Harrington, J. (2006). *Process management excellence: The art of excelling in process management*. Paton Press.

Hawkins, T. G., & Gravier, M. J. (2019). Integrating COTS technology in defense systems: A knowledge-based framework for improved performance. *European Journal of Innovation Management, 22*(3), 493–523.

Patasotskaya, N. (2010). Business processes as a way to increase the efficiency of managerial decisions. *Vsyo dlya Buhgaltera (All for Accountant), 6*, 19–23.

Smith, A. (1776). *An inquiry into the nature and causes of the wealth of nations*. William Strahan.

Smurov, A. (2017). Problematic issues of executing state defense orders and potential solutions. *Gosudarstvennoye Regulirovaniye Ekonomiki (Government Regulation of Economy), 4*, 27–35.

Zyryanova, T., & Tarnovskaya, Y. (2012). Modelling of process approach for the purposes of management accounting. *Mezhdunarodniy Buhgalterskiy Uchet (International Accounting), 44*, 15–28.

Business Models in Clothing Market: Value Proposal Specific

Veronika V. Lizovskaya and Artem A. Moldovan

Introduction

The behavior of buyers and sellers has changed significantly over the past decades. Among the key trends in consumer behavior, Nielsens highlights sustainable consumption and the pursuit of a healthy lifestyle, new habits and behavior models due to a long period of self-isolation, the growing popularity of online services, focus on promotional offers and lower prices, expanding the shopping chart and willingness to buy new products and new brands. Buyers have become more demanding and pickier, they are ready for innovations in the offer that can refer to products and services as well as new approaches in customer care, they choose supplier with better service, better offers on the market and special offers (promo).

All the above determines the need to change approaches to doing business and use new strategies. Modern companies actively implement digital technologies already, trying to optimize costs and striving to increase business manageability, improve customer experience and offer innovative solutions and products. Omnichannel customer systems are becoming increasingly attractive as they respond to the need for online interaction with customers. The use of this channel has become one of the key elements for ensuring the competitiveness of companies during the pandemic and has determined the need to change the business model and the emergence of new ones.

High attention to the question of the business models started in the 1990s, when there were studies of approaches to business organization and the identification of

V. V. Lizovskaya (✉)
St. Petersburg University of Management Technologies and Economics, St. Petersburg State University of Industrial Technologies and Design, St. Petersburg, Russian Federation

A. A. Moldovan
St. Petersburg State University, St. Petersburg, Russian Federation

A. Rumyantseva et al. (eds.), *Challenges and Solutions in the Digital Economy and Finance*, Springer Proceedings in Business and Economics,
https://doi.org/10.1007/978-3-031-14410-3_30

key factors for the success of companies in the market. By the 2000s large number of cases, describing companies experience in management and market strategy setting was analyzed, that helped to form the concept of a business model and its connection with the company's strategy.

Some authors define the business model as part of the company's strategy (Magretta, 2002; Tikkanen, 2005). Within the framework of this approach, the strategy determines the direction of the company's development and links together all the components of the company's business model. Other authors consider strategy to be a part of the business model (Chesbrough, 2010; Tis, 2010). According to this approach, the strategy contains a vision of the position of the company within the external and internal environment. And the business model is the result of the strategy implementation. There is an approach that states that the business model and strategy are two independent constructs that can intersect (Casadesus-Masanell & Ricart, 2010).

Many researchers have attempted to formulate the concept of a business model. Shafer et al. (2005), Amit and Zott (2012), Hadjiheidari and Zarei (2012) define the business model as "architecture", "logic", "configuration", "set of elements".

Various authors consider a business model as a system of a set of certain elements. So, Gassmann et al. (2014) proposes to answer four questions to determine the company's business model: who may be interested in the proposal, what will increase the value for stakeholders, how, with what types of activities and resources the company can form this proposal and why it is worth doing, meaning definition of income flows.

A similar approach is proposed by Girotra and Netesin (2020), who also formulate four basic questions to form the business model: what to produce, who and when makes key decisions, and why this business can be interesting, meaning the motivation for the main parties of the business model.

Other authors also consider a four-element approach to characterize business model of the company, most often offering to formulate the value proposal for customers and other stakeholders, that helps to understand which resources and business processes can form the unique offer and the mechanism for generating financial results. Several models also propose to consider the main target segments and what benefits customers get from the company's value proposition.

Debelak (2006) proposes to consider the business model from six angles: attracting the right customers who can be of a high value for the company; offering customers significant personalized value; offering goods and services that provide sufficient profit; ensuring customer satisfaction; using the company's resources to improve its position in the market; sufficient funding.

Osterwalder identifies 9 blocks of business model elements: consumer segments, value proposal, distribution channels, customer relationships, income flows, key resources, key activities, key partners, cost structure.

The approach presented above, based on the definition of business model elements, is usually called static as an alternative to the dynamic approach, according to which the business model is defined through such terms as "process", "value creation", "activities" (Ason, 2019; Gavrilova et al., 2014; Kiel et al., 2017; Stephan von Delft & Yang Zhao, 2020; Tis, 2010).

In the early 2000s a new direction of research devoted to innovative development and business model change was widely spread (Amit & Zott, 2012; Casadesus-Masanell & Ricart, 2010; Chesbrough, 2010).

Gassmann et al. (2014) developed a system of innovative business models creation based on the recombination of 55 business model templates that they identified after analysis over the past 50 years. The authors note three main problems that hinder the creation of innovative business models:

- The need to think beyond the prevailing logic in the industry as focusing on already working models can inhibit the emergence of new ideas.
- The need to think thought categories of business models with abstract concepts rather than technologies and products, that can be easily seen and tested, it is more difficult to operate with abstract concepts that are perceived as business models.
- Lack of systemic tools that help creative thinking needed to find new solutions.

New solutions and new values can be largely associated with the development of Internet technologies.

All these changes in the market indicate the need to consider new approaches to doing business and corporate governance. Understanding and analyzing the business models of the largest companies in the market will give an opportunity to assess possible development directions and necessary changes in the business structures. Active development of e-commerce and online communication require businesses to introduce new business models and innovative solutions to existing ones.

Another approach combines the concept of elements and structure of business models and the concept of innovation. A study of Klimanov and Tretyak (2014) found the common concepts in the most frequently quoted research in business modeling that are "value" (76%) and "creation" (52%).

Thus, the issues of value proposal and value creation come to the fore when analyzing a business model and is the bases for analyzing the company structure and management as well as strategy decisions.

In a retail market that is highly influenced by current trends and changes in consumer behavior, the identification of key business models can provide a vision of promising areas for improving the work of companies and further necessary changes.

Materials and Methods

At the first stage of the study, the main market trends were analyzed. At the same time, the main emphasis was placed on the study of online commerce in Russia, comparing the main trends with other countries to determine the prospects and the need for Russian companies to prepare for a further change in the ratio of online and offline sales, which undoubtedly requires further changes in business models and may lead to the disappearance of some formats and the active development of others.

Further, the main groups of goods sold online were analyzed with a more detailed study of the group of goods “clothing and footwear”, which is developing most actively. In addition, the main companies in the clothing and footwear market were identified, which were assigned to separate groups, for which business models were further analyzed using the McKinsey approach (Berg et al. 2018; Gonzalo et al. 2020). After that, the traditional business model of a mono-brand store and the new marketplace business model were analyzed, using a four-element approach to characterizing the business model: key aspects in the value proposition, main resources and partners, sources of profit identified.

The source for the study was industry surveys, data from research agencies and consulting companies, and market statistics. The report “Rating of the TOP-100 largest online stores in Russia” by Data Insight (a research agency specializing in the e-commerce market) was the bases to analyze the elements of business models.

Results

The e-commerce market is growing rapidly all over the world. China is currently the largest e-commerce market with the turnover about $955 billion, followed by the US ($374 billion) and Japan ($187 billion) (Fig. 1).

According to the Association of Internet Commerce Companies (AKIT), in Russia, the share of e-commerce in 2020 was only 9.5%, which is almost 30% less than in the US and more than 60% less than in China and the UK.

26% of all online purchases in 2020 were held in Moscow and the Moscow Region, 16%—St. Petersburg. In addition, the regions with the maximum share of e-commerce include such as the Krasnodar Territory, Krasnoyarsk Territory and Samara Region (9% each), Rostov and Nizhny Novgorod Regions (8% each), Sverdlovsk Region and the Republic of Tatarstan (7% each).

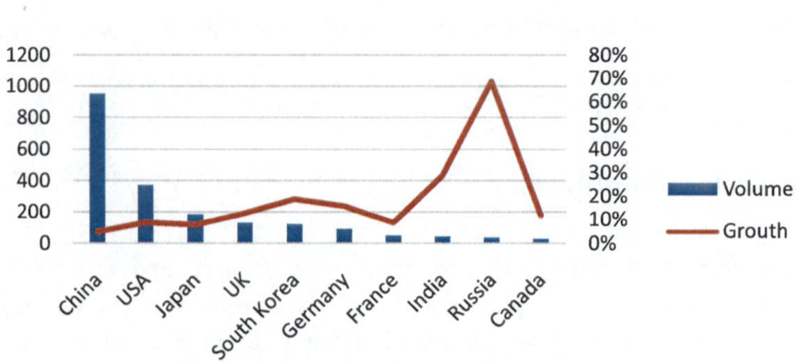

Fig. 1 Largest e-commerce markets in 2020

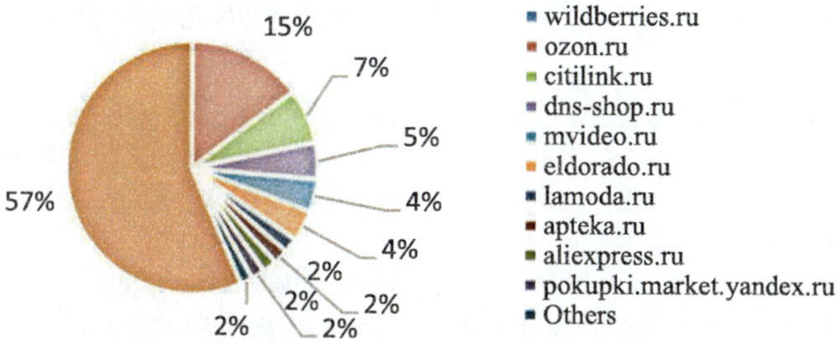

Fig. 2 Top 10 Russian e-commerce market leaders in 2020

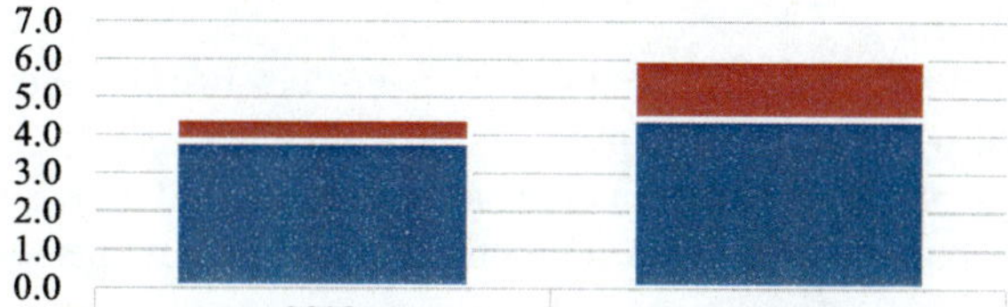

	2019	2020
Online sales	0.6	1.6
Offline sales	3.8	4.4

Fig. 3 Online and offline clothing market in 2019–2020

At the same time, the share of large businesses is growing extremely fast—from 32.9% in 2019, 45.3% in 2020, up to 57% in 2021.

According to the Data Insight research agency, the top ten Russian market leaders in terms of online sales in 2020 included Wildberries, Ozon, Citylink, DNS, MVideo, Eldorado, Lamoda, Apteka.ru, Aliexpress, Yandex.Market (Fig. 2).

The biggest share of online sales in the total volume of online commerce belongs to Wildberries ($5.4 billion), Ozon ($2.6 billion), Citylink ($1.7 billion).

The key drivers of digital transformation in retail include: the development of the Internet, the growing number of smartphone users, the growth in the quality of some components of e-commerce.

The qualitative structure of online stores in the domestic Russian market is dominated by electronics (30%), household goods (17%), clothing and footwear (13%).

At the same time, clothing and footwear (31.6%) and digital and household appliances (21.5%) occupy a significant share in the cross-border market of Russia, followed by furniture and household goods, food products, health and beauty products, auto parts, sporting goods, gifts and accessories.

The volume of online sales of clothing, footwear and accessories in 2020 increased by 37% (124 billion rubles or $1.7 billion), reached 457 billion rubles ($6 billion). The penetration of e-commerce into the fashion market almost doubled from 14.5 to 26.7% (Fig. 3).

According to Mckinsey, the following business models are typical for the clothing market:

- Vertically integrated direct-to-consumer (D2C) sales model associated with direct consumer sales of the brand's products. Such a clothing market player combines several steps in the value chain, may have its own physical retail store, and an online store
- Hybrid brand combines wholesale and retail. Such a company can have its own store, both on the Internet and offline, while it is also engaged in wholesales. Hybrid players such as Nike, Ralf Lauren, Tommy Hilfiger who work with retail customers directly online and offline as well as with wholesale clients
- Multi-brand online stores that sell their own and other company's products only via the Internet
- Online and offline multi-brand stores, that offer several brands in a physical store usually also sell their own products. Their goal is to create an integrated process that gives consumers an omnichannel experience and allows companies to maximize operational benefits in both physical and digital operations

As a result there have been identified two most successful business models in the clothing market: marketplaces and mono-brand omnichannel store.

Discussion

Many common solutions for all companies with different business models in the clothing market can be identified: pricing and discount offering, integrating the opportunities of responsible consumption, attention to ecological and social issues. In addition, a lot of attention in the market is paid to improving the customer service, the desire to make the product more accessible through the development of omnichannel interaction, ease of navigation in the store (both offline and online), the purchase process and payment. However, the differentiation of the mentioned elements depends on the bases of the value proposal in the clothing market.

The most popular business models in the clothing market are marketplaces and mono-brand stores with offline and online sales channels. These two approaches have significant differences, which can be seen first by analyzing the value proposition of companies. The main difference in customer value lies in the assortment (of one or many brands, respectively). This also determines that marketplaces rely on improving logistics and increasing the number of brands presented on the site, the breadth and updating of the assortment for any user, and the possibility of purchasing a set of goods.

Thus, the emphasis is on the brand value proposition of the marketplace itself, which becomes impossible to achieve without the use of big data to form a customized package offer for each client with their own consumption patterns and unique behavior aspects.

At the same time, mono-brand stores strive to create an exclusive offer: the uniqueness of the brand, visual image, and models. They are working to improve the shopping experience through the development of an omnichannel system, however, due to the limited range of products offered, they cannot develop a sales network as widely as marketplaces do.

As a summery, marketplaces are focused on mass character, accessibility, and maximum proximity to the client. The formation of a personalized offer and an increase in the number of goods purchased by the client is possible only based on big data analysis. At the same time, the basis for the development of marketplaces is close interaction with manufacturers and the development of their own logistics network.

Mono-brand stores, on the other hand, focus on exclusivity, excitement and experience associated with prestige and the ability to shape the image of the client through the goods and services supplied. They do not have the same opportunities as marketplaces to form their own logistics network and use the services of transport and logistics companies. Well-established interaction with the production structure (own or partner), the quality of materials used for production is highly important for mono-brand stores.

Strategic decisions of marketplaces and mono-brand stores should also be based on selected value proposals, and communication strategies should convey the selected values to customers. At the same time, the source of financial result for marketplaces is both the direct sale of goods and interaction with suppliers of goods, which generally forms an understanding of the profitable and costly part of the work on this business model. Mono-brand stores have income from sales, and the cost part of this business model is based on the creation of high-quality goods and services.

Conclusion

Analysis of the business model allows you to form a clearer vision of the business, determine the direction of development and the main factors hindering the development of the company, which must be corrected to improve financial results. Undoubtedly, such an analysis can make it possible to formulate a strategy for further business development. At the same time, considering the dynamics of changes in the market, a static approach to business modelling is no longer enough: the active introduction of digital technologies, changing models of consumer behavior and business interaction between stakeholders dictate the need for constant work to identify reserves for changes and directions of innovative development.

The key point for the formation of a business model, determining the sources of profit for the business and its financial model is the value proposal for customers and other stakeholders. And modern approaches to doing business are increasingly generating a profitable part of the business not only by customers, but also partners, shaping an ecosystem based on mutually beneficial partnerships.

Clothing market is developing rapidly leaning on trends: sustainable development and consumption, digital transformation and increased online interactions, which require a focus on an omnichannel approach with a changing role of traditional clothing stores.

Studying the main business models of companies in the clothing market and tracking their changes can make it possible to come to an understanding of the new model of this market, the formation of guidelines for its participants about strategical decisions and the development of specific solutions for companies to increase competitiveness and ensuring survival in a rapidly changing environment.

References

AKIT. (2020). E-commerce market in Russia Results of 2020 (AKIT). Retrieved from https://akit.ru/wp-content/uploads/2021/02/Аналитика-АКИТ-2020.pdf

Amit, R., & Zott, C. (2012). Creating value through business model innovation. *MIT Sloan Management Review, 53*(3), 41–49.

Ason, T. A. (2019). Innovative business modeling as a competitive advantage of a company on the international market. *Bulletin of Eurasian Science, 5*(11), 1–13.

Berg, A., Lobis, M., Hunter, E., Rolkens, F., Simon, P., & Yankelevich, H. (2018). Measuring the fashion world. McKinsey & Company Report. Retrieved from https://www.mckinsey.com/industries/retail/our-insights/measuring-the-fashion-world

Casadesus-Masanell, R., & Ricart, J. E. (2010). From strategy to business model and onto tactics. *Long Range Planning, 43*, 195–215.

Chesbrough, H. W. (2010). Business model innovation: Opportunities and barriers. *Long Range Planning, 43*(2–3), 354–363.

Debelak, D. (2006). *Business models made easy*. Entrepreneur Press, 240 p.

Gassmann, O., Frankenberger, K., & Csilk, M. (2014). *The business model navigator. 55 models that will revolutionise your business*. Pearson.

Gavrilova, T., Alsufyev, A., & Yanson, A.-S. (2014). Modern notation of business models: visual trend. *Foresight-Russia, 2*(8), 56–70.

Girotra, K., & Netesin, S. (2020). *Optimal business model. Four instruments of risk management*. Alpina Publisher, 216 p.

Gonzalo A., Harreis H., Altable S. S., & Villepele S. (2020). Digital transformation of fashion: now or never. McKinsey & Company Report. Retrieved from https://www.mckinsey.com/industries/retail/our-insights/fashions-digital-transformation-now-or-never

Hadjiheidari, N., & Zarei, B. (2012). Development and management of business models using the system dynamics approach. *Journal of Modeling in Management, 8*(2), 155–170.

Kiel, D., Arnold, C., & Voigt, K.-I. (2017). The influence of the industrial internet of things on business models of established manufacturing companies—a business level perspective. *Technovation, 68*, 4–19.

Klimanov, D. E., & Tretyak, O. A. (2014). Business models: the main directions of research and the search for a meaningful foundation of the concept. *Russian Journal of Management, 3*(12), 107–130.

Magretta, J. (2002). Why business models metter. *Harvard Business Review, 80*(5), 86–92.

Rating of the TOP 100 largest online stores in Russia in 2020. Data analysis. Retrieved from https://www.top100.datainsight.ru

Shafer, S., Smith, H., & Linder, J. (2005). The power of business models. *Business Horizons, 48*(3), 199–207.

Tikkanen, H. (2005). Managerial cognition, action and the business model of the firm. *Management Decision, 43*(6), 789–809.

Tis, D. J. (2010). Business models, business strategy and innovation. *Long-Term Planning, 43*(2), 172–194.

von Delft, S., & Zhao, Y. (2020). Business models in process industries: Emerging trends and future research. *Technovation, 105*(3).

World after COVID-19: five consumption trends: Nielsen analyses, 2020. Retrieved from https://nielseniq.com/global/ru/insights/analysis/2020/mir-posle-covid-19-5-trendov-potrebleniya/

Assessment of the Impact of the Economy Digitalization on GRP Dynamics: The Regional Aspect

Tatyana I. Bezdenezhnykh and Elena E. Sharafanova

Introduction

The digitalization of economic sectors is becoming an inevitable fact. There is an intensive introduction of digital resources in various sectors of the economy—industry, finance, healthcare, etc. The study of the problems of digitalization in Russia has become extremely relevant in recent years. It should be noted that the development of digitalization processes is one of the little-studied problems due to the short period of existence of this phenomenon. The term digital economy was first substantiated in the scientific works of Tapscott (1997) and Negroponte (1995). In the works of these scientists, the fundamental nature of information technologies was substantiated and the prospects for their influence on economic processes were considered.

Determining the role of various factors in changing the dynamics of economic development is given a significant place in scientific research. The scientific works of many foreign scientists indicate the conditionality of the development of the economy and institutions. These include the works of such scientists as De Soto (1989), Besley (1995), Jones and Romer (2010), Nonneman and Vanhoudt (1996) and others.

Scientists of all directions take an active part in the discussion of the problems of the influence of various factors on economic growth, among them: Lucas (1988), Rodrik (2008), Kaldor and Robinson (1965) and others. In the studies of representatives of Russian science such as Glazyev (2009) and others also reflect the relationship between economic growth and various socio-economic factors.

There are gaps in the scientific literature that our study aims to fill in relation to the analysis of the relationship between digitalization and economic growth at the

T. I. Bezdenezhnykh (✉) · E. E. Sharafanova
St. Petersburg State University of Economics, St. Petersburg, Russian Federation

A. Rumyantseva et al. (eds.), *Challenges and Solutions in the Digital Economy and Finance*, Springer Proceedings in Business and Economics,
https://doi.org/10.1007/978-3-031-14410-3_31

regional level. In addition, there is a predominance of qualitative studies of digitalization in the modern scientific literature which partly reduces the level of reliability in the analysis of the causal effects of digitalization from the standpoint of GRP growth. Despite the established connection between digital infrastructure and GRP growth in several foreign studies (Baquero Forero, 2013; Castellacci, 2011; Carbonara 2005; Evangelista et al. 2014).

The intensive spread of digital technologies has exacerbated the threat of territorial differentiation of regions in Russia. This problem exists to a greater or lesser extent in all countries of the world, however, in Russia, the indicators of territorial differentiation are extremely high (Bezdenezhnykh & Makenov, 2019; Nikolaev & Tochilkina, 2011). These processes have actualized the need to study the impact of digitalization on economic growth.

Despite investment efforts, Russia has not been able to achieve the expected economic prosperity associated with digitalization due to the persistent digital divide, including the lack of digital skills, the lack of ICT infrastructure in certain underdeveloped economic regions (Morozkina, 2020). Consequently, Russia must bring its economic potential, including its labor force, in line with the requirements of the new digital landscape.

The purpose of the study is to identify the key conditions for the digitalization of the economy and assess their impact on the volume and dynamics of GRP in the constituent entities of Russia.

The rapid penetration of digitalization processes into the Russian economy has determined the increase in the volume of scientific literature, which, using econometric modeling technologies, examines various aspects of the impact of digital resources on the economic development of the country and its individual regions and offers ambiguous results in explaining this relationship. Much of the writing suggests that the use of digital technologies stimulates the economy by facilitating communication, empowering individuals, creating jobs and spurring innovation.

An analysis of the intensity of the impact of digitalization on changes in the volume of GRP per capita has also become widespread in scientific research. Most of the work is based on the seemingly plausible assumption that the higher the level of digitalization, the higher the level of economic development of the region. Conversely, highly developed countries can afford the use of high-quality digital resources. Thus, the study "Digitalization as a global trend and growth factor of modern economy" (2019) shows that states with a high level of economic development have better access to the Internet, wide opportunities for obtaining information, and, accordingly, a higher level of digitalization. Digitalization acts as an essential factor in the growth of the modern economy.

The most common opinion is that the digitalization of the economy has a positive impact on productivity and economic growth. Let's take a look at these jobs. For example, studies (Czernich et al., 2011; Qiang & Rossotto, 2009) show that the use and expansion of broadband drives productivity and economic growth. Studies have shown that labor productivity increases with investment in computerization of business processes and the development of the Internet. The study (Katz et al. 2014; Katz & Koutrompis, 2013) shows that the prosperity of countries grows along with the digitalization index. The work graphically illustrates the dependence of the digitalization index on GDP per capita and vice versa. Research by Niebel (2014), Romer (1990) also showed a positive relationship between ICT and economic growth.

In the work of Cardona et al. (2013), based on an in-depth analysis of the empirical literature, the relationship is substantiated and the nature of the interaction between information and communication technologies and labor productivity is established (2013).

The seminal work of Evangelista et al. (2014) also found that the use of ICTs in business improves productivity.

Bezdenezhnykh (2019) conducted a detailed analysis of the current state of digital resources and identified the elements of digitalization that affect GDP. The study identified key variables that affect the dynamics of the Russian economy. The author made the following conclusion: there is a large correlation between all variables. The results of the study show that digitalization has a great impact on GDP.

The work of Khalin and Chernova (2018) defines the relationship between digitalization and the digital economy and identifies both possible positive consequences and risks and threats of digitalization for Russia. In their article, they conclude that digitalization leads to an increase in the efficiency of the economy.

The impact of digital technologies on economic growth in economically different countries is not the same. This problem is aggravated by the insufficient level of development of digital infrastructure in regions remote from the center, especially in rural areas. Some researchers found that there is a bipolarization in GDP growth rates between "ICT emerging economies" and digitally advanced countries. They found that the development of ICT had a positive impact on innovation and economic modernization, primarily in ICT-based emerging economies.

Let us also pay attention to very important aspects of research, the results of which can be applied to manage the formation of digital resources in Russian regions with an insufficient level of economic development. Niebel et al. (2014) prove that ICTs have a beneficial impact on the socio-economic development of developing countries—with insufficient digital infrastructure. This study shows that, for this type of country, ICT reduces the cost of doing business by facilitating access to both information and consumers via the Internet. Digital communications enable firms to maintain long-term relationships with customers. In addition, ICT empowers citizens by providing access to free online educational materials and news, enabling the growth of digital literacy.

There is a growing body of literature on the relationship between social media use and economic growth. In developing countries, social media has become an indirect driver of growth as new businesses are created through social media platforms. The researchers examined the impact of online and offline social media on branding and innovation as facilitators for the operation of firms in the hospitality sector using structural equation modeling. They found that the relationship between social media use and firm performance was very significant.

In the work of Kramin and Klimanova (2019) emphasizes the great role and proves the importance of digital infrastructure for the growth of the economy of the constituent entities of the Russian Federation. The authors put forward a hypothesis, which was further confirmed with the help of specific calculations, about the presence of elasticity of GRP and GRP per capita indicators for the defining indicators of digital infrastructure.

Research by Kadochnikova (2020) also shows a positive relationship for GRP per capita and digital inclusion. The positive impact of the increase in the number of subscribers of mobile broadband access to the Internet per 100 people on the increase in the growth rate of GRP was clearly manifested.

According to the research of Sukharev (2021) the interaction of digital and traditional (classical) economies takes place differently in regions with different functional orientations and different population densities. The author found that residents of some remote regions with relatively low-income levels (Jewish Autonomous Region, Altai, Chukotka Autonomous Region, etc.) nevertheless actively use digital resources (satellite access, etc.)

In the works of Russian authors, an assessment was made of the economic differentiation of regions, taking into account new factors of influence, and the spatial relationships of digitalization and industrial development were identified (Grigoriev et al. 2015; Naumov et al., 2020). In the work of Abroskin and Abroskina (2020) address specific topical issues about measuring GDP dynamics in the context of digitalization of the economy.

The scientific literature also reveals the critical role of the state in providing an enabling environment for access to and use of technology. It is especially valuable to get acquainted with the best practices of foreign experience in assessing the impact of the spread of digital resources on the development of the economy (Kazakova, 2013). The state is responsible for creating conditions for the acquisition or production of advanced technologies, as well as for creating and maintaining a legal framework that facilitates the use of ICTs.

Specific forms of digital technologies are considered and their role in global production networks is determined (Foster & Graham, 2016). The directions of penetration of broadband communications were determined and an assessment of its impact on the development of the economy was carried out (Ghosh, 2017). The issues of the relationship between the activities of e-government and the development of digital resources are not ignored (Twizeyimana and Andersson 2019; Zhao et al., 2015).

Researchers raise questions about the conceptualization and measurement of the digital economy (Bukht & Hicks, 2017). In an attempt to move away from the narrow concept of ICT from the infrastructure side, Evangelista et al. (2014) empirically examined the economic impact of digitalization on a group of 27 countries in the European Union from 2004 to 2008 using Arellano-Bond's GMM estimates. They concluded that of the three dimensions of the digitalization process, the use of ICTs and digital empowerment, rather than access to ICTs, are important for productivity and employment growth.

However, some researchers have noted that investment in ICT has a negative impact on labor productivity in SMEs, primarily due to the lagging effect between ICT use and productivity gains (cited by Evangelista et al., 2014). Another explanation for this negative relationship was that the use requires a skilled workforce and a learning curve to integrate the new technology, resulting in a potential technology and skills mismatch.

Other studies (Ishida (2015) draw attention to the reassessment of the relationship between ICT and economic growth. The author emphasizes that ICT in many cases did not have a significant impact on changes in real GDP in both the short and long term.

Thus, we can say that due to the ambiguity of the assessments of the results of the correlation between GRP and indicators characterizing the development of digital resources of the economy, our study is relevant. In this article, we test the hypotheses noted above on a much broader empirical basis.

Materials and Methods

The work was carried out on the basis of a critical understanding of the theoretical provisions, the collection, processing and interpretation of statistical data, as well as their graphical visualization, allowing to formulate conclusions about the subject of the study. To test the hypothesis that digital technologies are statistically significant growth factors of GRP (within a certain interval), we choose GRP per capita for the period from 2015 to 2019 as our independent variable and the digital technology scorecard from the database as our dependent variables. Digital technologies are understood as technologies for collecting, storing, processing, searching, transmitting, and presenting data in electronic form. Statistical information on the scale of distribution and directions of use of digital technologies by organizations is given for legal entities (excluding small businesses). Source of information—Statistical compendium "Regions of Russia"—Section - 18.

The level of use of digital resources in organizations located on the territory of a subject of Russia was assessed according to the following parameters (as a percentage of the total number of surveyed organizations): use of personal computers in organizations; use of "cloud" services in organizations; use of broadband access to the Internet; use of servers in organizations; use of local area networks in organizations.

In our article, we will consider in detail the data describing the dynamics of GRP. Indicators for the gross regional product are presented in section 9 of the statistical collection "Regions of Russia. Socio-economic indicators. 2021". The electronic version of this collection is posted on the official website of Rosstat in the heading "Publications". The section contains information on GRP and actual consumption of households in the territory of the constituent entities of the Russian Federation.

The data in the tables containing indicators by sectors of the economy are for 2017–2019. in accordance with the All-Russian classifier of types of economic activity.

More detailed information about GRP is available in the statistical collections and publications of Rosstat: "Russian Statistical Yearbook", "Russian National Accounts in 2013–2020" (https://rosstat.gov.ru/folder/210).

To ensure comparability and stability of the data, calculations were made for the period starting from 2015 (after the formation of the subjects of the Russian Federation—the Republic of Crimea and the city of Sevastopol) and up to 2019 (to exclude the impact of the COVID-19 pandemic).

Analysis of studies to identify correlations between GDP and digital technologies was carried out using a graphical method. The coefficient of variation was used as an indicator indicating the differentiation of the subjects of the Russian Federation. In evaluating the impact in this study, linear correlation coefficients were used to determine the relationship.

Results

As a result of the study, it was determined that the majority of Russian organizations use Internet resources in their work (89.6% of organizations in the business sector). However, in different regions, the use of the Internet by the population is different. In 2020, more than 95% of the population used the Internet in Moscow and the Moscow Region, the Khanty-Mansiysk Autonomous Okrug - Yugra, and the Yamalo-Nenets Autonomous Okrug, and 93.5% in Chechnya. While in the Oryol region this figure was 77.3%, in Tver region—77.9%, in Novgorod region—75.4%, in the Republic of Mordovia—76.2%.

The number of active subscribers of broadband access to the Internet (per 100 people) varies significantly by region—varies from 39.2 in the Novosibirsk region, 36.8 people in Moscow to 1.9 people in Ingushetia and 4.2 people in Dagestan.

If in Russia as a whole the indicator of the use of electronic document management in organizations (as a percentage of the number of surveyed organizations) was 70% in 2020, then in Chechnya it was only 37.8%, the Republic of Dagestan—41.7%, Chukotka Autonomous Okrug—62%.

Please note that local networks and own servers are not yet used in all organizations. "Cloud" services are used only in 25.7% of organizations. At the same time, in the Republic of Dagestan, this figure was only 11.1% in 2020, in the Chukotka Autonomous Region 14.4%, and in the Jewish Autonomous Region—15.3%.

The assessment of the impact of the introduction of digital technologies is carried out in our study in terms of correlation with the results of GRP growth. The criterion for the effectiveness of the digitalization process in our case is the acceleration of GRP growth in the regions and the country as a whole.

Before conducting a statistical analysis of the impact of digitalization on the differentiation of the gross regional product and its growth rates, two hypotheses were put forward.

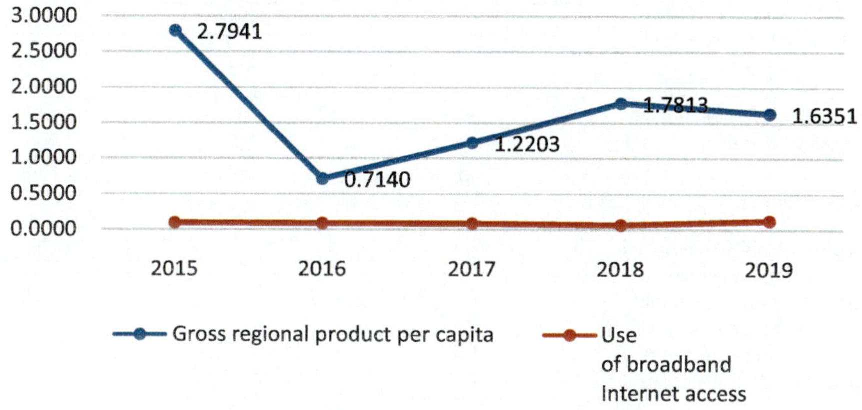

Fig. 1 Variation of the indicators "GRP per capita" and "Use of broadband Internet access" by constituent entities of the Russian Federation for 2015–2019

Research hypothesis 1: digitalization of the economy and management, contributing to the socio-economic development of the constituent entities of the Russian Federation, reduces the differentiation of regions in terms of growth rates of GRP per capita (dependent variable y1) and its level (dependent variable y2).

Competing hypothesis 2: the digitalization of the economy and management does not have a noticeable effect on reducing the differentiation of regions in terms of GRP growth rates and its level.

The results of the calculations presented in Fig. 1 indicate the heterogeneity of the sample in terms of the GRP per capita growth rate, which, in order to correctly assess the degree of dispersion in the numerical sequence, requires the elimination of minimum and maximum values.

However, in this study, only the trend is assessed, so sampling was not adjusted. Figure 1 does not demonstrate the coincidence of the dynamics of indicators between the two analyzed indicators. The correlation coefficient between them characterizes a weak relationship ($r = 0.06$).

For a more detailed analysis, we studied the relationship between GRP per capita of a subject of the Russian Federation and the prevalence of various digital technologies used in organizations located on the territory of this subject. The correlation matrix is presented in Table 1

As follows from the data in Table 1, there are no noticeable correlations between the growth rate of GRP per capita and the prevalence of the use of digital technologies in organizations. A similar result is obtained by checking the correlations between indicators of the use of digital technologies and the level of GRP per capita (Table 2).

The conducted study allows us to reject hypothesis 1 and support hypothesis 2, concluding that the spread of digital technologies in organizations of the constituent entities of the Russian Federation in modern conditions (2017–2019) does not

Table 1 Correlation matrix of indicators of the use of digital technologies in the constituent entities of the Russian Federation and the growth rate of GRP per capita

	y_1	x_1	x_2	x_3	x_4
y_1	1				
x_1	–0.10	1			
x_2	–0.10	0.62	1		
x_3	–0.22	0.61	0.72	1	
x_4	–0.08	0.25	0.32	0.37	1

Table 2 Correlation matrix of indicators of the use of digital technologies in the constituent entities of the Russian Federation and the volume of GRP per capita

	y_2	x_1	x_2	x_3	x_4
y_2	1				
x_1	0.10	1			
x_2	0.28	0.62	1		
x_3	0.14	0.62	0.72	1	
x_4	0.34	0.25	0.32	0.37	1

have a significant impact on the growth rate of GRP per capita, as well as on differentiation subjects of the Russian Federation by the volume of this indicator.

Discussion

The results obtained are not entirely consistent with the prevailing opinion about the positive impact of digital technologies on economic growth. However, considering the fact that statistical data were used for a very short period of time (5 years), it is reasonable to speak of the absence of such an influence with the following reservations:

- We are talking about the absence of statistically significant dependencies only in a short-term retrospective.
- Four characteristics observed by the state system of statistics and assessing the presence (but not the use and, moreover, the intensity of use) of digital technologies were used as explanatory variables.

It should be emphasized that the authors of some of the published works a priori assert the direct and positive impact of digitalization on economic growth (including the economy of the constituent entities of the Russian Federation), based on the provisions and theses of strategic documents in the field of digitalization and strategic planning. But you don't have to take wishful thinking. Regulatory documents set the vector of development, and the task of researchers is to identify, evaluate and interpret facts that indicate progress in a given direction. Given the limited availability of reliable data on the use of ICT, given the limited time for their intensive introduction into management practice, it is hardly possible to speak with confidence about the unconditional, direct and positive impact of digital technologies on the economic growth of the subjects of the Russian Federation and the reduction of their differentiation in terms of socio-economic development. The discussion of the results obtained by the authors made it possible to determine the direction of further statistical analysis, which hypothetically can reveal indicators of the effectiveness of economic development, the dynamics of which is explained by the spread of digital technologies: labor productivity, as well as characteristics of the innovative development of the constituent entities of the Russian Federation.

Conclusion

It should also be noted that the study did not take into account the sectoral affiliation of enterprises and organizations that used digital technologies in the analyzed period, as well as the sectoral structure of the region's economies, which, as can be assumed with a high degree of probability, affect both the volume and growth rate of GRP. per capita. Continued research in these areas will make it possible to refine and expand the results obtained.

References

Abroskin, A. S., & Abroskina, N. A. (2020). Dynamics of gross domestic product: actual measurement problems in the conditions of digitalization of the economy. *Bulletin of the University, 2*, 97–103. https://doi.org/10.26425/1816-4277-2020-2-97-103

Baquero Forero, M. D. P. (2013). Mobile communication networks and Internet technologies as factors of increasing technical efficiency. *Information Economics and Policy, 25*(3), 126–141. https://doi.org/10.1016/j.infoecopol.2012.11.004

Besley, T. (1995). Nonmarket institutions for credit and risk sharing in low-income countries. *The Journal of Economic Perspectives, 9*(3), 115–127.

Bezdenezhnykh, T. I., & Makenov, M. M. (2019). Investigation of the convergence process of the regions of the Russian Federation. *Economics and Management, 4*(162), 11–21.

Bukht R., & Hicks, R. (2017). Definition, conceptualization and measurement of the digital economy. Retrieved from http://www.informatics.manchester.ac.uk/news/latest-stories-updates/defining-conceptualising-and-measuring-the-digital-economy/

Carbonara, N. (2005). Information and communication technologies and geographical clusters: opportunities and dissemination. *Technovation, 25*(3), 213–222. https://doi.org/10.1016/S0166-4972(03)00095-6

Cardona, M., Kretschmer, T., & Strobel, T. (2013). ICT and productivity: conclusions from the empirical literature. *Information Economics and Policy, 25*(3), 109–125. https://doi.org/10.1016/j.infoecopol.2012.12.002

Castellacci, F. (2011). Narrowing the technology gap? *Review of Development Economics, 15*(1), 180–197. https://doi.org/10.1111/j.1467-9361.2010.00601.x. [Cross-reference] [Google Scholar].

Czernich, N., Falck, O., Kretschmer, T., & Woessman, L. (2011). Broadband infrastructure and economic growth. *Economic Journal, 121*(552), 505–553.

De Soto, H. (1989). *The other path: The informal revolution*. Harper & Row.

Evangelista, R., Guerrieri, P., & Meliciani, V. (2014). The economic impact of digital technologies in Europe. *Economics of Innovation and New Technology, 23*(8), 802–824. https://doi.org/10.1080/10438599.2014.918438

Foster, K., & Graham, M. (2016). Revisiting the role of digital technologies in global production networks. *Global Network, 17*(1), 68–88. https://doi.org/10.1111/globe.12142

Ghosh, S. (2017). Broadband penetration and economic growth: Does politics matter? *Telematics and Informatics, 34*(5), 676–693. https://doi.org/10.1016/j.tele.2016.12.007

Glazyev, S. (2009). The world economic crisis as a process of replacement of dominant technological structures. *Voprosy Ekonomiki, 3*, 26–38.

Grigoriev, R. A., Kramin, M. V., Kramin, T. V., & Timiryasova, A. V. (2015). Inequality of income distribution and economic growth in the regions of Russia in the post-crisis period. *The Economy of the Region, 3*, 102–113.

Ishida, H. (2015). The impact of ICT development on economic growth and energy consumption in Japan. *Telematics and Informatics, 32*(1), 79–88. https://doi.org/10.1016/j.tele.2014.04.003

Jones, K., & Romer, P. M. (2010). New facts of Kaldor. Ideas, Institutions, population and Human capital. *American Economic Journal: Macroeconomics, 2*(1), 224–245. https://doi.org/10.1257/mac.2.1.224

Kadochnikova, E. I. (2020). Convergence of economic growth and digitalization of households: spatial analysis of the relationship on regional panel data. *Actual Problems of Economics and Law., 14*(3), 487–507. https://doi.org/10.21202/1993-047X.14.2020.3.487-507

Kaldor, N., & Robinson, J. (1965). *Industrialization in developing countries*. Cambridge Overseas Study Committee.

Katz, R., & Koutroumpis, P. (2013). Measuring digitization: A growth and welfare multiplier. *Technovation, 33*(10–11), 314–319. https://doi.org/10.1016/j.technovation.2013.06.004

Katz, R., Koutrumpis, P., & Kallorda, F. M. (2014). Using the digitization index to measure the economic and social impact of digital programs. *Information, 16*(1), 32–44. https://doi.org/10.1108/information-10-2013-0051

Kazakova, M. (2013). The analysis of foreign experience in the field of decomposition economic growth on the basis of an assessment production functions.

Khalin V. G., & Chernova G. V. (2018). Digitalization and its impact on the Russian economy and society: advantages, challenges, threats and risks. *Administrative Consulting 10*, 46–63 (in Russian). https://doi.org/10.22394/1726-1139-2018-10-46-63

Kramin, T. V., & Klimanova, A. R. (2019). Development of digital infrastructure in the regions of Russia. *Terra Economicus, 17*(2), 60–76. https://doi.org/10.23683/20736606-2019-17-2-60-76

Lucas, R. E. (1988). On the mechanics of economic development. *Journal of Monetary Economics, 22*(1), 3–42.

Morozkina, A. K. (2020). Digital divide in BRICS countries: Problems of interregional inequality. *Bulletin of International Organizations, 15*(4), 70–90.

Naumov, I. V., Dubrovskaya, Y. V., & Kozonogova, E. V. (2020). Digitalization of industrial production in the regions of Russia. Spatial relationships. *The Economy of the Region, 16*(3), 896–910. https://doi.org/10.17059/ekon.reg.2020-3-17

Negroponte, N. (1995). *Being digital* (Vol. 243, pp. 1–12). Vintage Books. https://doi.org/10.25136/2409-8647.2021.1.34788

Niebel, T. (2014). ICT and economic growth comparing developing, emerging and developed countries. *SSRN Electronic Journal*. https://doi.org/10.2139/ssrn.2560771

Nikolaev, I. A., & Tochilkina, O. S. (2011). *Economic differentiation of regions: estimates, dynamics, comparisons. Analytical report*. FBK.

Nonneman, W., & Vanhoudt, P. (1996). A further augmentation of the Solow model and the empirics of economic growth for OECD countries. *The Quarterly Journal of Economics, 111*(3), 943–953.

Qiang, C., & Rossotto, C. (2009). *Economic impacts of broadband (information and communications for development: Extending reach and increasing impact)* (pp. 35–50). World Bank.

Rodrik, D. (2008). *One economics, many recipes: globalization, institutions, and economic growth.* Princeton University Press.
Romer, P. (1990). Endogenous technological changes. *Journal of Political Economy, 98*(5), S71–S102. https://doi.org/10.3386/w3210
Sukharev, M. V. (2021). Interrelation of digitalization levels and economic indicators in the regions of Russia. *Theoretical and Applied Economics, 1*, 1–12. https://doi.org/10.25136/2409-8647.2021.1.34788. https://nbpublish.com/library_read_article.php?id=34788
Tapscott, D. (1997). *The digital economy: Promise and peril in the age of networked intelligence.* McGraw-Hill. Retrieved from https://nbpublish.com/library_read_article.php?id=34788
Twizeyimana, J. D., & Andersson, A. (2019). The public value of E-government—A literature review. *Government Information Quarterly, 36*(2), 167–178. https://doi.org/10.1016/j.giq.2019.01.001
Zhao, F., Wallis, J., & Singh, M. (2015). The development of e-government and the digital economy: the relationship. *Internet Resource, 25*(5), 734–766. https://doi.org/10.1108/Input-02-2014-0055

Impact of the ESG Principles on the Corporate Financial Strategy

Anna Rumyantseva and Olga Tarutko

Introduction

As it is noted in Alda (2021) ESG, combining such categories as Environmental, Social, and (Corporate) Governance, reflects the basic principles of operations of socially responsible investors. ESG principles set the business to a qualitatively new level of intelligent financing, reorienting it from increasing current profits to significant social or environmental goals.

The process of financial strategy development is influenced by various factors; non-financial factors can be distinguished among them. These factors include a group of ESG factors; their growth of popularity has an impact on the growth of the significance of this group. In Yuan et al. (2022) the research on the influence of non-financial information disclosure to financial violations by companies proved the positive effect of ESG information, including the activities on preventing financial violations.

ESG financing is a relatively new direction, but it has already gained recognition and arisen interest in many countries. One of the fundamental differences between the ESG sphere and the generally accepted understanding of investing is the assumption of the possibility of donating part of the income in the name of achieving a good goal related to ecology, climate, or having a social orientation. In recent years, the situation has changed and there has been a strong influx of funds into ESG funds, against the background of which the return on shares of ESG funds exceeds the return on traditional ones. Thus, financing a business with a high ESG rating can provide a higher profit share (Chen et al., 2021). And the increase in the level of income and its capitalization, in turn, plays an important role in the development and

A. Rumyantseva (✉) · O. Tarutko
St. Petersburg University of Management Technologies and Economics, St. Petersburg, Russian Federation

A. Rumyantseva et al. (eds.), *Challenges and Solutions in the Digital Economy and Finance*, Springer Proceedings in Business and Economics,
https://doi.org/10.1007/978-3-031-14410-3_32

expansion of business. In turn, a high ESG rating creates additional competitive advantages for a company attracting financing. The ratings are formed under the influence of various indicators, including labour relations, corporate governance, level of transparency, ecological compatibility, and others (Clementino & Perkins, 2021). Approaches to ESG evaluation and its interrelation with financial results and companies' strategies are investigated in many publications, the most interesting of which are (Yoo & Managi, 2021) that proves the importance of non-financial information disclosure and (DasGupta, 2021) where evaluation of a lack of financial resources influence on dynamics and effectiveness of ESG principles is presented. Positive effect of ESG deals related to mergers and acquisitions, based on the sample test analysis, including 41 countries and 12 sectors of the economy, is presented in Barros et al. (2021).

Materials and Methods

The influence of ESG factors is steadily increasing and, becoming one of the analyzed components in determining the ratings of companies, they increase transparency, allow assessing and considering potential opportunities and risks. That is why the analysis of Green Bond Market trends and trends related to the increasing importance of responsible investment and sustainable development comes to the fore. Compliance of the company strategy with the goals of sustainable development makes the issuer more attractive to potential investors.

According to the Climate Bonds Initiative (2020), the Green Bond Market has shown an average annual growth of 95% over the 13 years of its existence. The first bond of this kind was issued by the European Investment Bank (EIB) and World Bank in 2007, and in November 2013 the first issue of corporate green bonds took place that can be interpreted as a turning point. The market had reached the value of US$100 billion at the end of 2015, then by December 2020 it has reached the value of US$1 trillion, thereby demonstrating exponential growth.

One of the market trends is a gradual increase in the share of large deals,[1] so at the end of 2020 the share of 33.59% accounted for deals worth more than US$1 billion, and another 29.64% are deals from US$0.5 to US$1 billion. For comparison, a year earlier these values were at the level of 27.57% and 31.77%, respectively. According to the Climate Bonds Initiative, in the first half of 2021 the share of transactions exceeding US$1 billion is at the level of 34.27%, and the share of transactions worth US$0.5–US$1 billion is 32.72% that is higher than the previous year.

The position of the Russian Federation in comparison with the activity of some other countries can be characterized as minor. Thus, the first issue of green bonds took place only in 2018 and the deal value is US$1.3 billion. For the sake of clarity,

[1] Prepared according to The Climate Bonds Interactive Data Platform. Available at: https://www.climatebonds.net/market/data/#country-map

Table 1 Dynamics of deals with green bonds for the period of 2018–2021

	2019	2020	First half of 2021			
Country	Deal value, billion US dollars	Amount of deals, pieces	Deal value, billion US dollars	Amount of deals, pieces	Deal value, billion US dollars	Amount of deals, pieces
Brazil	1.4	17	1.5	10	1.7	20
Germany	18.7	25	41.9	129	29.1	103
Spain	8.3	22	10.1	33	11.7	34
Canada	7.2	18	9.4	25	4.5	12
China	31.4	100	22.4	104	22	92
Russian Federation	0.6	2	0.5	3	0.3	1
USA	52.9	1134	52.1	747	37.6	495
France	30.8	61	37	57	23.7	21
Sweden	10.6	110	14.2	142	11.1	90

we present a comparison of transactions with green bonds of several countries for the period of 2018–2021, compiled according to [ibid.] in Table 1.

According to a survey compiled by Morningstar, in the third quarter of 2021 (Morningstar, 2021) the number of sustainable funds increased by more than 51%, reaching a value of 7486. The Morningstar methodology makes it possible to take into account ESG risks relative to comparable categories when compiling the rating that is ensured by a unique approach to evaluating and ranking the corporate and sovereign parts of the portfolio independently of each other. According to Bloomberg (2021), at the end of 2020 the value of assets that comply with the ESG principles reached a record US$37.8 trillion, and by 2025 will exceed US$53 trillion. Bloomberg analysts expect growth in the ESG area, as the assets of this area are diversified due to various subjects.

Results

This article identifies positive aspects for the corporate development, achieved through the introduction of ESG principles into the financial strategy. These aspects include:

– Growth of business attractiveness for foreign investors. The overwhelming volume of investments in Russian companies is accounted for by foreign capital, and many foreign investors take EGS factors into account, making investment decisions. There are funds with billion portfolios among those investing in the economy of the Russian Federation. In other words, ignoring the above principles, unwillingness to adjust the strategy in favor of sustainable development, lack of attempts to reduce the harm done to the environment can bring huge losses to the business if investments are not received due to the shift of investors' interests in favor of intelligent investment. And, on the contrary, operating and implementing the financial strategy from the position of social responsibility, under the steadily increasing influence of ESG factors and following the relevant principles, the corporation acquires a more attractive image for foreign investors. For example, the Enel Group has invested more than 200 billion rubles in the Russian economy, and the total investment for the construction of three wind power facilities in Russia, for which the company has received permission, will be about 495 million euros.[2]

- Growth of attractiveness of business for consumers. According to the EY Future Consumer Index, 43% of consumers are willing to pay more for goods and services, if they are produced by organizations that benefit society, and 64% are willing to change their consumer habits if it will benefit society (Rogers et al., 2021). The results of this study suggest that following the path of sustainable development will not have a significant negative impact on sales and, on the contrary, can attract more consumers to a company that follows ESG principles. In favor of this statement, it is also worth noting that young and ambitious investors born at the beginning of the twenty-first century look at the world in a completely different way; for most of them a healthy lifestyle and environmental care play a dominant role. Now there are very few such investors on the stock market, but their share will grow every year that will help popularize responsible investment.
- Increasing competitiveness. As shown in Kunin et al. (2018), the innovative potential of a corporation is an integral component of its competitiveness and reflects the growth potential, taking development prospects into account. In turn, innovation is an indispensable element that can ensure the development and technical re-equipment of companies in terms of environmental protection, reduction of harmful emissions into the atmosphere, improvement of waste processing, etc. Increasing competitiveness and building competitive advantages allow corporations to take higher positions in the struggle for partners, markets, and personnel.
- Opportunity to enter the international level and attract foreign capital. The desire and ambitions of the company owners to take it to a new level must be closely linked with global trends; it is necessary to choose the right way to enter international markets. ESG trends in the modern world cannot be ignored, they have penetrated into virtually all areas of business, and therefore a responsible attitude to the environment, climate and/or social orientation of doing business can bring tangible income to an entrepreneur. The inclusion of companies in ESG ratings, coverage of their activities in the media and various kinds of analytical reports provides an increase in image and recognition that, in turn, can attract

[2] Brochure Enel Russia. Available at: https://www.enelrussia.ru/content/dam/enel-russia/documents/ru/Enel-Country-Brochure.pdf

additional funding. In addition, increased transparency affects the status and trust of counterparties.

- Creation of the Sustainable Development Sector at the Moscow Stock Exchange. The initiative to create a Sustainable Development Sector at the Moscow Stock Exchange is, in fact, a motivating factor for companies to review strategies in favor of increasing social responsibility in order to increase the attractiveness of the financial instruments they have placed.
- Increasing the level of income and its capitalization. The possibility of obtaining a higher share of profit should be highlighted as one of the drivers in the implementation of ESG principles in the development strategy. The growing popularity of ESG funds stimulates demand and, as a result, an increase in prices for such assets. In addition, in recent years there has been an increase in technology sector companies that have a high share in the ESG indices that investors are targeting. According to Morningstar (Hale, 2020), at the end of 2019 ESG funds outperformed traditional funds.
- Reducing risks and ensuring a more stable cash flow. The article (Ashwin Kumar et al., 2016) shows the impact on ESG funds from such factors as low reputational, political and regulatory risks that leads to stabilization of cash flow and ensuring a higher level of profitability. It is also noted in Broadstock et al. (2021) that ESG funds have lower risk compared to other market offers.

Discussion

In recent years the popularity of tools that meet the principles of ESG has been steadily growing. Today we can say that there is a certain pool of investors, who are guided in decision-making not by financial indicators, but by the degree of compliance of the company strategy with environmental, managerial, or social goals (Mohammad & Wasiuzzaman, 2021). For example, according to Bloomberg (Taraldsen, 2021), Norway's sovereign wealth fund sold its portfolio of assets related to oil exploration and production, and in April 2021 Norges Bank Investment Management, which distributes the fund's portfolio, became the owner of a 50% stake in the Borssele wind farm (Norges Bank, 2021). As noted in Bhopal (2021), Norway's sovereign wealth fund calls for the development and implementation of concrete measures aimed at combating climate change.

Such examples can be given endlessly, and each of them will indicate a trend associated with the growing importance of sustainable development. Despite the rather dynamic development in this area, the Russian Federation is lagging behind the global trend. Today a large number of companies do not have a sustainable development strategy, a number of corporations are not aware of the prospects in terms of attracting financing from foreign investors. It should also be noted that in Russia there are no restrictions on financing the activities of companies, the ESG rating of which is low or the development strategy of which does not have the ESG focus. Russian investors show a rather restrained interest in ESG financing. At the same time, several companies are actively moving towards a sustainable development strategy. In a study of some Russian companies, the results of which are presented in Apenko and Fomina (2021), it is shown that companies that have developed and implemented a sustainable development strategy in their activities and form an institutional framework for it have a significantly higher level of maturity.

For example, Rosneft PJSC is striving to strengthen its position in international ESG ratings. Bloomberg notes that as of December 2021 Rosneft was among the best oil and gas companies in CDP's International Climate Rating (EQS Group, 2021). In December 2020 the Company presented a long-term carbon management plan until 2035. In the same year the Concept of Environmental Development for the same period was developed; both documents act as key elements of Rosneft's long-term environmental agenda.

Enel Russia PJSC that is a part of the international Enel Energy Group deserves attention among the energy companies. Enel Russia's priorities include ensuring sustainable development and a responsible attitude to the environment, as evidenced by the inclusion in the Dow Jones Sustainability World Index for the eighteenth year in a row (Enel, 2021).

According to the BCS GM survey, 70% of the blue chips that were surveyed intend to increase the resources they spend on ESG (Aris, 2019). But, despite the positive experience of a number of companies, there are barriers to dynamic ESG transformation, namely:

- The first and, perhaps, the main barrier to attracting ESG financing is the lack of transparency in terms of disclosure of information about ESG factors by issuers. The importance of such factors in the market is steadily growing and the disclosure of an exhaustive amount of information presented in an understandable and accessible form can please an investor and influence his decision in favor of acquiring an asset of such company. Since the disclosure of ESG information for Russian companies is voluntary, most of them place in the public domain only those indicators in respect of which the company has managed to achieve attractive results, leaving others undisclosed.
- Making decisions on financing the activities of a particular enterprise in terms of ESG is based on rating estimates that are of an informational nature (Avramov et al., 2021). These ratings are, in fact, subjective, and their values may be influenced by the first of these barriers, namely, insufficient transparency of non-financial indicators of corporations. In addition, at present, very few Russian companies appear in such ratings, which means that attracting funding from socially responsible foreign investors is practically inaccessible for other companies.
- Insufficient level of competence of Russian entrepreneurs in the ESG area. Thus, according to a study by AO Deloitte & Touche CIS (Deloitte, 2021), only 10% of banks have implemented some ESG practices in their activities. The process of transformation in the transition to sustainable development is quite long and complex and therefore requires certain competencies for the introduction and

implementation of such initiatives. The Russian business sector is largely characterized by insufficient development of corporate culture, lack of full understanding of the content and benefits of introducing ESG factors into their activities. The pursuit of improving non-financial indicators and the desire to follow the ideology of sustainable development goals, in the absence of an adequate level of competence, can lead to a decrease in the efficiency of companies, which in turn will affect their attractiveness in terms of financing.

Conclusion

In conclusion, it should be noted that despite the considerable number of barriers that arise in the way of the ESG ideology, it is possible to formulate a number of specific recommendations that can help expand the ESG influence, both in the world as a whole and in the Russian Federation in particular, including:

- Increasing the level of competence of entrepreneurs and creating an opportunity to attract additional funding provided through the assignment and further growth of the ESG rating can be achieved by popularizing the philosophy and ideology of sustainable development. The solution to this kind of problem can be provided through education. Moreover, we can talk about both educational programs of higher education, and about additional short- or medium-term programs. It seems more effective to achieve a synergistic effect through the introduction of the above ideology into all spheres of society, i.e. as a collaboration of the educational process, as well as the promotion of universal values. In Rumyantseva and Tarutko (2019) a number of promising changes and modern trends are formulated that are characteristic of the process of development of the educational sphere in the period of the economy digitalization. So, you can improve your competence on-the-job, for example, by taking courses on popular educational platforms.
- Ensuring transparency and comparability of information can be achieved through the development of standards for the provision of non-financial information on sustainable development and ESG indicators. Such initiatives exist in world practice, among them are the EU Directive on the disclosure of non-financial information, the Global Reporting Initiative, the Sustainability Accounting Standards Board and a number of others. There are cases where a business voluntarily prepares reports in accordance with several standards. In other words, the market has formed a request for the creation of a unified system of standards.
- Increasing the level of popularization of ESG principles can be achieved by promoting them not only at the federal, but also at the regional level, because the solution of global problems at the local level is a direct adherence to the ideology of sustainable development and the right way to improve the level of development of society and the quality of life.
- Involvement of small and medium-sized businesses in projects, the implementation of which is associated with the goals of sustainable development, can

enhance the effectiveness of implementing the ideology of sustainable development in the business sector, including through the possibility of developing intersectoral partnerships and accumulating financial resources.

- The involvement of employees in the process of achieving socially significant goals can become an important element on the way to improving the efficiency of the corporation. According to Acaroglu (2020), corporate employees expect the employer to be more involved in sustainable development goals. Thus, 83% of employees believe that their employers are not sufficiently involved in solving the problem of climate change, and 65% say that they are more likely to work in a company with a strong sustainable agenda. Social responsibility should be considered as one of the priority tasks of corporate governance since the rejection of it can be equated to the rejection of sustainable development initiatives (Belousov, 2020).
- Expansion of state support for financing corporations that implement the ESG principles could become a catalyst for this market that, in turn, would increase the amount of positive impact that was written about earlier.

In other words, business entities should change the way they work and their attitude to the importance of sustainable development. Orientation of the financial strategy of the corporation not only to current linear tasks, but also to a cyclical approach that offers the resources made from processed raw materials, and not only their consumption, seems to be extremely promising and capable of bringing such companies to qualitatively new positions at the world level.

References

Acaroglu, L. (2020). Employees want climate-positive action from companies. Here's how they can deliver. Available at: https://www.reutersevents.com/sustainability/employees-want-climate-positive-action-companies-heres-how-they-can-deliver

Alda, M. (2021). The environmental, social, and governance (ESG) dimension of firms in which social responsible investment (SRI) and conventional pension funds invest: The mainstream SRI and the ESG inclusion. *Journal of Cleaner Production, 298*, 126812.

Apenko, S., & Fomina, Y. (2021). Sustainability management of business projects in large Russian companies. In *SHS web of conferences* (Vol. 90, p. 01002). EDP Sciences. https://doi.org/10.1051/shsconf/20219001002

Aris, B. (2019). ESG is a must have in Russia, but adds no alpha. Available at: https://www.intellinews.com/esg-is-a-must-have-in-russia-but-adds-no-alpha-171358/

Ashwin Kumar, N. C., Smith, C., Badis, L., Wang, N., Ambrosy, P., & Tavares, R. (2016). ESG factors and risk-adjusted performance: a new quantitative model. *Journal of Sustainable Finance & Investment, 6*(4), 292–300. https://doi.org/10.1080/20430795.2016.1234909

Avramov, D., Cheng, S., Lioui, A., & Tarelli, A. (2021). Sustainable investing with ESG rating uncertainty. *Journal of Financial Economics*. https://doi.org/10.1016/j.jfineco.2021.09.009

Barros, V., Matos, P. V., Sarmento, J. M., & Vieira, P. R. (2021). M&A activity as a driver for better ESG performance. *Technological Forecasting and Social Change*. https://doi.org/10.1016/j.techfore.2021.121338

Belousov, K. (2020). Corporate social responsibility and sustainable development of modern Russian companies as a challenge of business globalization. In *SHS web of conferences* (Vol. 74, p. 06004). EDP Sciences. https://doi.org/10.1051/shsconf/20207406004

Bhopal, A. (2021). The Norwegian Oil Fund in a warming world: what are the interests of future generations? *Ethics, Policy & Environment*. https://doi.org/10.1080/21550085.2021.1940451

Bloomberg. (2021). ESG assets may hit $53 trillion by 2025, a third of global AUM. *Bloomberg Intelligence*. Available at: https://www.bloomberg.com/professional/blog/esg-assets-may-hit-53-trillion-by-2025-a-third-of-global-aum/

Broadstock, D. C., Chan, K., Cheng, L. T., & Wang, X. (2021). The role of ESG performance during times of financial crisis: Evidence from COVID-19 in China. *Finance Research Letters, 38*, 101716. https://doi.org/10.1016/j.frl.2020.101716

Chen, L., Zhang, L., Huang, J., Xiao, H., & Zhou, Z. (2021). Social responsibility portfolio optimization incorporating ESG criteria. *Journal of Management Science and Engineering, 6*(1), 75–85. https://doi.org/10.1016/j.jmse.2021.02.005

Clementino, E., & Perkins, R. (2021). How do companies respond to environmental, social and governance (ESG) ratings? Evidence from Italy. *Journal of Business Ethics, 171*(2), 379–397. https://doi.org/10.1007/s10551-020-04441-4

Climate Bonds Initiative. (2020). $1 Trillion mark reached in global cumulative green issuance: Climate bonds data intelligence reports: Latest figures. Available at: https://www.climatebonds.net/2020/12/1trillion-mark-reached-global-cumulative-green-issuance-climate-bonds-data-intelligence

DasGupta, R. (2021). Financial performance shortfall, ESG controversies, and ESG performance: Evidence from firms around the world. *Finance Research Letters*. https://doi.org/10.1016/j.frl.2021.102487

Deloitte. (2021). ESG banking in Russia. Research. Available at: https://asros.ru/upload/iblock/f6c/q0b1qh8xpem8vb3bsrh9oneb1241pnr1/ESG_banking-in-Russia.pdf

Enel. (2021). Enel confirms once again its global sustainability leadership with inclusion in Dow Jones Sustainability World Index. Available at: https://www.enelrussia.ru/en/all-news/news/2021/11/16112021

EQS Group. (2021). Rosneft Oil Company: Rosneft is among the best performing oil and gas companies in CDP's international climate rating. Available at: https://www.bloomberg.com/press-releases/2021-12-13/rosneft-oil-company-rosneft-is-among-the-best-performing-oil-and-gas-companies-in-cdp-s-international-climate-rating

Hale, J. (2020). US ESG funds outperformed conventional funds in 2019. *Morningstar*. Available at: https://www.morningstar.com/articles/973590/us-esg-funds-outperformed-conventional-funds-in-2019

Kunin, V., Rumyantseva, A., & Tarutko, O. (2018). Conceptual framework for managing the competitiveness of the entrepreneurial structures. In *MATEC web of conferences* (Vol. 193, p. 05071). EDP Sciences. https://doi.org/10.1051/matecconf/201819305071

Mohammad, W. M. W., & Wasiuzzaman, S. (2021). Environmental, social and governance (ESG) disclosure, competitive advantage and performance of firms in Malaysia. *Cleaner Environmental Systems, 2*, 100015. https://doi.org/10.1016/j.cesys.2021.100015

Morningstar. (2021). Global sustainable fund flows: Q3 2021 in review. Available at: https://www.morningstar.com/content/dam/marketing/shared/pdfs/Research/Global-ESG-Q3-2021-Flows.pdf?utm_source=eloqua&utm_medium=email&utm_campaign=none&utm_content=27223

Norges Bank. (2021). First investment in renewable energy infrastructure. Available at: https://www.nbim.no/en/the-fund/news-list/2021/first-investment-in-renewable-energy-infrastructure/

Rogers, K., Cosgrove, A., & Copestake, J. (2021). The CEO Imperative: Make sustainability accessible to the consumer. Available at: https://www.ey.com/en_gl/consumer-products-retail/make-sustainability-accessible-to-the-consumer

Rumyantseva, A., & Tarutko, O. (2019). Modern digital educational environment as a tool for the education development in Russia. *Advances in Social Science, Education and Humanities Research, 315*, 94–97. https://doi.org/10.2991/icpcs-19.2019.22

Taraldsen, L. E. (2021). Norway wealth fund dumps oil stocks amid $10 billion loss. Available at: https://www.bloomberg.com/news/articles/2021-01-29/norway-wealth-fund-dumps-oil-stocks-amid-10-billion-2020-loss

Yoo, S., & Managi, S. (2021). Disclosure or action: Evaluating ESG behavior towards financial performance. *Finance Research Letters*. https://doi.org/10.1016/j.frl.2021.102108

Yuan, X., Li, Z., Xu, J., & Shang, L. (2022). ESG disclosure and corporate financial irregularities—Evidence from Chinese listed firms. *Journal of Cleaner Production, 332*, 129992. https://doi.org/10.1016/j.jclepro.2021.129992

Students' Attitude to Digital Entrepreneurship

Olga O. Vasileva, Elena G. Pozdeeva, and Olga D. Shipunova

Introduction

One of the most important youth' characteristics is their readiness for active productive work aimed at positive changes. The institutions' system responsible for student's personal formation, intellectual development, adaptive abilities and entrepreneurial skills is aimed this competence formation.

Digital economy, bloggers' success and entrepreneurial activities results create prerequisites for students' interest in Internet business, which allows them to maximize their life goals.

Universities strive to expand the proposed activities' range within the framework of educational programs development, provide youth initiatives support, projects that set competence guidelines in accordance with economy digitalization dynamics (Razinkina et al., 2018).

Today student's personal readiness for entrepreneurship has been developed to a greater extent in socio-psychological and pedagogical approach, rather than within the sociological and economic approaches.

This problem necessitates socio-economic research for universities to adjust the educational path, allowing students to improve necessary competencies for online business.

Comprehensive study relevance of student's entrepreneurial intentions is also associated with the observed high dynamics of Russian students' interest in entrepreneurship, dictated by their desire for self-realization and promising professional trajectories development (Pokrovskaia et al., 2019).

O. O. Vasileva (✉)
Herzen State Pedagogical University of Russia, St. Petersburg, Russian Federation

E. G. Pozdeeva · O. D. Shipunova
Peter the Great St. Petersburg Polytechnical Univesity, St. Petersburg, Russian Federation

A. Rumyantseva et al. (eds.), *Challenges and Solutions in the Digital Economy and Finance*, Springer Proceedings in Business and Economics,
https://doi.org/10.1007/978-3-031-14410-3_33

The growing young people's interest in digital entrepreneurship during their studies is gaining power.

Students share among 18–24-year-olds wishing to start their own business is 52%, decreasing with age.

Youth polls results about hindering entrepreneurial activity problems showed that most of them are concerned by start-up capital lack (56.4%) and poor knowledge in business creation areas (22.3%) (Merkulov, 2017).

Actually, student's business readiness is analyzed in terms of entrepreneurship essence, entrepreneur's personality and characteristics, formed in learning process, factors influencing responsible behavior readiness (Hisrich & Peters, 1991).

Creativity, conscious life position, activity in relation to external environment objects, ability to work together are structural components of individual readiness (Magomedova & Khanova, 2011).

Complex structure of socio-psychological readiness is revealed as an integral formation that has motivational, emotional, personal and integrative components. (Punko, 2017).

The problem associated with person's propensity formation for entrepreneurial activity is stable personality characteristic achievement and consolidation of long-term readiness, revealed through a positive attitude towards entrepreneurial activity combined with motivation, abilities, knowledge and skills.

Students' entrepreneurial personality type is characterized by a willingness to take risks and responsibility for their activities results, ability to go beyond the established organization forms, to form and satisfy digital economic needs (Solntseva, 2019).

Entrepreneurial behavior is determined by the action of a public institutions complex of to a certain extent, therefore, important problems are the incomplete process of institutionalization of the business environment, insufficient development and consolidation of norms and codes of conduct, lifestyle and image aspects of an entrepreneur (Moskovtsev & Yurova, 2010).

Business readiness in behavioral concepts from the standpoint of planned behavior theory is justified through intentions (Krueger et al., 2000), formed under influence of three factors: personal attitude to business behavior, norms' understanding and perception of behavioral control.

This approach emphasizes subject's choice independence and construction of his own life path. Personal business readiness system has complex structure (Khashchenko, 2011), its elements are: psychological base formed by the subject of entrepreneurship choice as self-sufficient and responsible management of one's own life way; orientation subsystem, reflecting how the subject sees himself in it; subsystem integrating knowledge, skills and abilities necessary for business activities, business communication and provides flexibility and multi-variant behavior. The external environment, collective and individual experience influence entrepreneurial readiness formation. This sets the direction for the readiness study in a combination of three aspects: interaction with the external environment; awareness of specifics and readiness to engage in entrepreneurial activity; presence of special personal characteristics (Geranyushkina, 2014).

Psychological readiness can be defined as the mobilization of all individual resources for the successful implementation of entrepreneurial tasks (Khanova, 2014).

Readiness potential arises at action intersection of three personality's psychological space fields—achievement motivation, self-development attitude and behavioral readiness for online business. It is clear that digital entrepreneurship readiness implies a rational type of attitude towards people, as well as a communicative turn towards recognizing partnership value, self-esteem and self-improvement, acceptance of responsibility's meaningful role.

Materials and Methods

Research methodological basis is a factor analysis of students' attitudes towards digital entrepreneurial activity.

Four factor groups influence students' entrepreneurial activity: personal motives, educational environment, family and socio-economic context of digital entrepreneurship development in general. According to research, family business existence increases the positive relationship of entrepreneurial intentions with students' activity in business creating.

Therefore, it is important to link the personal component, the behavioral basis (models of entrepreneur's behavior) and the contextual aspect—external factors influence (family, university) in students' entrepreneurial intentions analysis. Psychological readiness trigger is risk readiness, and the social side of readiness is associated with environment influence.

University's educational environment has influence power for students, which creates the value regime of entrepreneurial initiative priority, supporting conditions, contributes to necessary knowledge and skills acquisition, develops behavioral attitudes, emphasizing successful entrepreneurial and innovative activities digital examples.

The conducted research was aimed at testing the hypothesis about university's educational environment influence on entrepreneurship readiness formation.

Students' intentions to acquire competencies in digital entrepreneurship field were analyzed and students' real activity was assessed.

Students of the Peter the Great St. Petersburg Polytechnic University became research object.

Survey results conducted among students of Peter the Great St. Petersburg Polytechnic University (September-October 2021, methods—online questionnaires and focus group) made it possible to establish students' readiness potential, associated with the growing students' interest in digital entrepreneurship, the need to obtain relevant business competencies.

Online questionnaires and focus group were applied as empirical methods. The online survey method allowed students to take part in the research by answering questionnaire points at a convenient time for them in a comfortable environment and while maintaining anonymity.

Results

Survey results conducted among students of Peter the Great St. Petersburg Polytechnic University (September-October 2021, methods—online survey and focus group) made it possible to establish students' readiness potential, associated with the growing students' interest in digital entrepreneurship, the need to obtain relevant business competencies.

The survey was conducted from September to October 2021. The sample was 380 people (1.7% of the total number of full-time higher education students at SPbPU; 95% confidence level). 60% boys and 40% girls participated in the survey. The distribution by courses and education levels is presented in Fig. 1.

More than half of respondents (66%) answered "positive" and "rather positive" to the question "How do your attitude about people who are engaged in digital entrepreneurship".

31% of respondents in their answers to the question about associations with entrepreneurship associate it with material possessions—"money, profit, profitable business, profitable deals, stable income, high income, wealth, financial sufficiency, "beautiful" life, goods and services".

23% of students associated digital entrepreneurship with the desire to succeed and social acceptance, as well as with leadership, prestige, useful connections, experience, self-fulfillment, rich life and development.

18% of students considered among the entrepreneurial competencies: responsibility, resolve, communication skills, great working capacity and hardiness, venture ability, stress resistance, creativity and mental flexibility, a special mindset type, inventiveness, self-confidence.

14% of students in the associative range named outgoings associated with entrepreneurial activity: difficulties, big risks, instability, sudden issues, hard work, and nerves, lack of rest and time, unpredictability, high taxes, graft, obligations to the government.

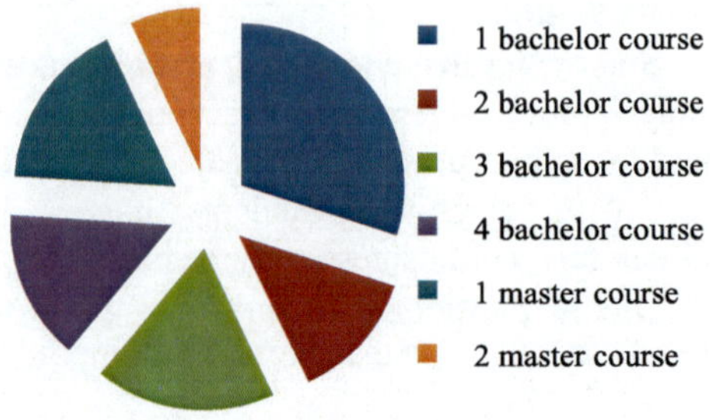

Fig. 1 Respondents' distribution in accordance with courses and educational level

Table 1 Two questions' comparison results

	Do you know people who have their own business?	Total		
Do you have experience in digital entrepreneurship?	Yes, this is my friend/ relative	Yes, I am an entrepreneur myself	I don't know entrepreneurs	
Yes	13.5%	2%	8.5%	24%
No	35.0%	–	41.0%	76%
Total	48.5%	2%	49.5%	100%

At the same time, 10% of the answers from associations' total number are related to entrepreneurship activity aspect, an active focus on transforming the external environment: invent, manage all resources, propose and implement ideas, improve the business, plan a budget, market monitoring, make deals, cooperation, environment care, offer users the best.

Small association's share (4%) describes autonomy when engaging in digital business activities: independence, freedom, self-employment, autonomy, lack of superiors.

In general, students perceive digital entrepreneurship with advantages rather than disadvantages.

51% of respondents answered that there were no businessmen among relatives and friends.

Answering the question "Do you know people who have their own business?" 26.5% of respondents answered that entrepreneurs are among their acquaintances; 20.5% of respondents said that their families have entrepreneurs.

Today 2% of students are already engaged in Internet business development and classify themselves as businessmen.

At the same time, 34% of respondents noted that familiar businessmen are an example for them.

24% of respondents noted business experience, which involves not only their own business, but participation in any business project, assistance in developing an idea, concept, team selection, etc. Great majority of students (76%) do not have entrepreneurial experience yet.

A statistically low correlation was found between business experience and businessmen existence among relatives and friends (Cramer's V-coefficient = 0.248) (Table 1).

40.5% of respondents noted plans to organize/continue their digital business, 28% are still undecided and 31.5% answered negatively to the question "Do you plan to organize/continue digital business in the future?".

The intention to engage in entrepreneurial activity in the future was mostly expressed by those who are acquainted with entrepreneurs.

Among those who plan to develop their business in the future, 11% do not yet know when they will start this, and 10.5% of students are already making efforts, 10% supposed to be engaged in entrepreneurship while studying at university.

Table 2 Response distribution to a question "What qualities do you already possess in order to do business?"

Qualities and skills	Respondents' share (%)
Goal orientation	45
Communication skills	40
Resourcefulness	31
Prudence	28.5
Risk readiness	28
Leadership skills	27
Foresight	27
Honesty	26.5
Stress resistance	26
Professionalism	23
Creativity	21.5
Ability to convince	19
Readiness to constantly develop	8
All qualities and skills	3

6% of students will start developing their business after university graduation and 3% want to complete specialized courses first and then start business creating and developing.

59.5% of respondents do not plan to engage in digital entrepreneurship.

Respondents who plan to do business were asked about goals definition of business activities.

The results were divided as follows:

I want to realize my ideas and develop—31%.
I don't want to work for hire—22%.
I want to be financially independent—21%.
I want to make my dream come true—17.5%.
I want to change people's life for the better—9.5%.
Entrepreneur is prestigious—2.5%.

About 45% of all respondents answered open question "Who is a successful entrepreneur example for you?", that they do not have an example of a successful entrepreneur. At the same time, 19% of students exemplified their relatives and familiars, explaining that these people have succeeded alone without anyone's help. 13.5% of students named Elon Musk, Steve Jobs and Pavel Durov as an example of a successful entrepreneur.

The question about necessary qualities for a successful entrepreneur revealed following lowest rank qualities: honesty, professionalism, prudence, ability to convince, creativity. Ability to be responsible for one's own actions, goal orientation and stress resistance were the most important qualities. In the middle ratings ranges were communication skills, risk readiness, resourcefulness, and foresight.

The survey allowed determining students' self-concept of development level and possession of digital entrepreneurship necessary qualities (Table 2).

Table 3 Response distribution to a question "What knowledge do you lack to get into digital entrepreneurship?"

Knowledge area	Respondents' share (%)
Law	42.5
Digital business basics	32.5
Marketing and Management	23
Accounting	19
Foreign languages	14
All knowledge areas	18.5

However, students are worried about knowledge gap for digital entrepreneurship in certain areas: law (42.5%), digital business basics (42.5%), marketing & management (23%), accounting (19%), foreign languages (14%).

All the above are required to 18.5% of students, and a quarter are not interested in mastering the necessary skills in business (Table 3).

Students separately noted that they would like to be more professional in the field of graphic and web design, internet ad settings, and they also lack self-presentation, conducting business negotiations and professional networking skills.

Key factor question concerning the psychological business readiness—the ability to take risks showed that respondents' half is ready for risk, 23%—not ready and 27% could not give a definite answer.

At the same time, 17% of respondents agreed with following statement "I'd rather prefer to quit new business if there's a failure", 37% flatly rejected. This result indicates sufficient psychological students' readiness. 90.5% of respondents sure that Internet business involves personal responsibility. Students associate professional responsibility with discipline, organization and the ability to meet deadlines (38.7%), estimate action and behavior aftermath (32.3%), high work quality, independence and ability to use the resources available (32%).

Students were asked to choose closest style of business approach to them. Students highlighted following competencies in entrepreneurial responsibility structure: to keep everything under control—51%; to distribute responsibilities among members of a team, group, network members—45.3%; to foresight the decisions' aftermath—37%; strong experiencing ups and downs, successes and failures—29%; training and improving skills to minimize possible failures—28.5%; helping others readiness—18%; trust—8% (Fig. 2).

Determinants that set personal and social entrepreneurial maturity degree are future to engage in entrepreneurial activities and business behavior types. In this regard, connection strength between two variables was tested (based on Cramer's V-coefficient = 0.398), which demonstrates a clear relationship between business plans with independence and responsibility principles.

Students who do not plan to develop their business indicated preference to work according to certain algorithm and follow other people's directions, or choose other, less businesslike strategies. Questions block about external factors influence was important for the research (Fig. 3). Governmental support (37.5%), tax system in Russia (27%), fair competition in a certain industry (25%) has the greatest influence on desire to involve in business.

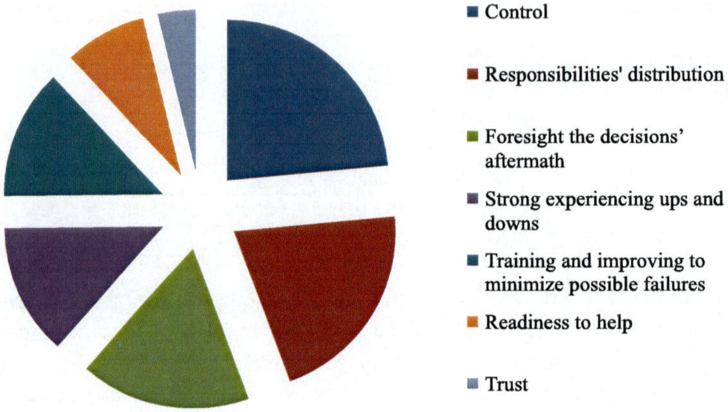

Fig. 2 Competence structure of entrepreneurial responsibility

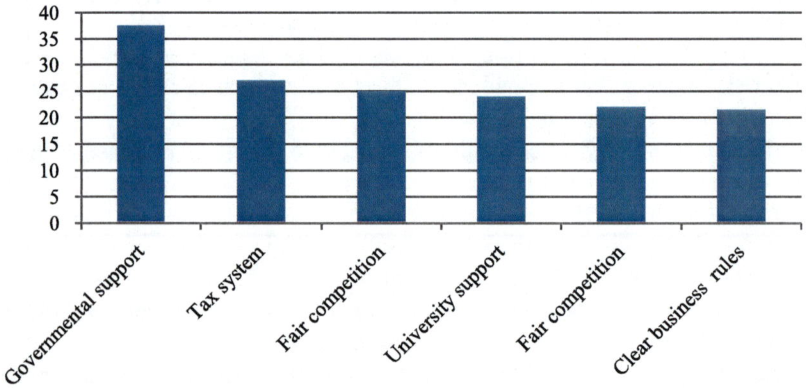

Fig. 3 External factors influenced on desire to involve in business

Following external factors influence is also significant: university support (24%), country economic development level (22%), clear rules existence for doing business (21.5%).

Respondents noted negative factors that affect students' readiness for entrepreneurial activities: lack of necessary knowledge, skills, and abilities (36%) experience (32%), financial resources' insufficiency (35.5%), time (29%) and ideas (26%) pressure, high taxation (25%) (Fig. 4).

According to respondents' answers to an open question, the negative impact of following factors can be identified: bureaucracy, high loan interest rates, lack of influential acquaintances to assist and support at the beginning.

Four focus groups with students of humanitarian and technical study fields confirmed survey results in addition. Students are positive to digital business, considering it active, creative, innovate and making a valuable contribution to national economy. According to students, entrepreneurship requires following

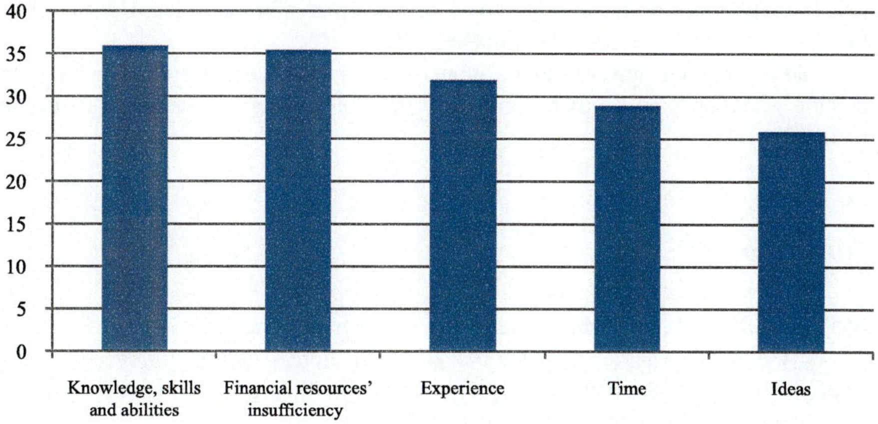

Fig. 4 Response distribution to a question "What negative factors affect your readiness for entrepreneurial activities?"

qualities: responsibility, resolve, sociability, great work capacity and hardiness, ability to take risks, stress resistance, creativity, enterprise, a special type of thinking, resourcefulness, self-confidence. Some factors that influence students' intention to engage in entrepreneurial activities were formulated in a group discussion: unique idea; starting capital; leisure time; personal characteristics.

Discussion

Digital space further exploration from innovative trends' standpoint in university activity is important in the context of educational offers for open educational courses, which will enable youth to acquire relevant knowledge and competencies and receive an adequate assessment of their intellectual and professional potential (Sorokin et al., 2021).

Professionally oriented communicative digital services' research is also a promising trend. Such platforms significantly contribute to young people's soft skills improvement, help to create communication networks being basis of entrepreneurial orientation's implementation.

Sociologists note that young people's abstract ideas about personal competitiveness are transformed into competitor focus. This focus contributes the distribution among youth a business vision and action readiness, turns their existing practical experience, educational and innovative potential into a key competitive advantage (Shafranov-Kutsev & Cherkashov, 2020).

Further research in to trending practice-oriented approach in higher education is also important, it correlates with student's entrepreneurship development tasks, allows to construct graduate competencies system based on digital culture development, focus training for practical skills on a market request and maintaining initiative in project activity (Starodubtsev, 2021).

Practice oriented approaches of modern universities create student's basis to develop confidence in its own competitiveness in the digital business (Pozdeeva et al., 2018).

Conclusion

Survey results conducted among students of Peter the Great St. Petersburg Polytechnic University (September-October 2021, methods—online survey and focus group) proved students' readiness high potential, growing interest in digital business and necessity to obtain relevant business competencies.

The poll showed that almost half of students plans to do business, a quarter of those surveyed already had experience in running own online business.

University actively promotes the skills development in working with information, stress resistance, diligence and purposefulness, the ability to be responsible for one's own actions in the opinion of students.

However, university contribution to stimulating such qualities as the ability to persuade, readiness to take risks and foresight seems to be minimal.

Students are still cautiously assessing their entrepreneurial attitudes; some choose other career paths.

Shortage of bright examples in local entrepreneurial practice and insufficient media attention to career technologies in digital entrepreneurship field influence a cautious attitude.

However, the interest and efforts to acquire digital economy demanded knowledge is supported by students' desire and their willingness to try themselves in business, relying to digital technologies advantages.

References

Geranyushkina, G. P. (2014). Theoretical approaches to the problem of readiness for entrepreneurship. *Psikhologiya v ekonomike i upravlenii, 1*, 50–53.

Hisrich, R., & Peters, M. (1991). *Entrepreneurship: Starting, developing, and managing a new enterprise*. Progress-Univers.

Khanova, Z. G. (2014). Psychological characteristics of personal and professional self-determination of businessmen. *Vestnik universiteta, 17*, 342–346.

Khashchenko, T. G. (2011). The technology of formation of psychological readiness for entrepreneurial activity in an agricultural university. Empirical substantiation, Vestnik TGU, 4. https://cyberleninka.ru/article/n/tehnologiya-formirovaniya-psihologicheskoy-gotovnosti-k-predprinimatelskoydeyatelnosti-v-agrarnom-vuze-empiricheskoe-obosnovanie. Accessed 29 Dec 2021.

Krueger, N. F., Reilly, M. D., & Carsrud, A. L. (2000). Competing models of entrepreneurial intentions. *Journal of Business Venturing, 5–6*, 411–432. https://doi.org/10.1016/S0883-9026(98)00033-0

Magomedova, K. N., & Khanova, Z. G. (2011). Factors affecting the formation of students' psychological readiness for entrepreneurial activity. *Nauchnyy potentsial, 4*(5), 74–80.

Merkulov, P. A. (2017). Youth entrepreneurship as a factor of sustainable socioeconomic growth. *Srednerusskiy vestnik obshchestvennykh nauk, 3*, 42–52.

Moskovtsev, A. F., & Yurova, O. V. (2010). Entrepreneurship: Key features and behavioral strategies. *Izvestiya vysshikh uchebnykh zavedeniy: Povolzhskiy region, 4*, 102–110.

Pokrovskaia, N. N., Ababkova, M. Y., & Fedorov, D. A. (2019). Educational services for intellectual capital growth or transmission of culture for transfer of knowledge—Consumer satisfaction at St. Petersburg universities. *Education Sciences, 9*(3), 183. https://doi.org/10.3390/EDUCSCI9030183

Pozdeeva, E. G., Trostinskaya, I. R., & Evseeva, L. I. (2018). Problems of forming the professional responsibility of the engineer in the conditions of modern innovative activity. *Nauchno-tekhnicheskiye vedomosti Sankt-Peterburgskogo Gosudarstvennogo politekhnicheskogo universiteta, 3*, 57–65. https://doi.org/10.18721/JHSS.9306

Punko, N. Y. (2017). Structure of social and psychological readiness of professional formation of students in higher education institution. *Mir nauki, 2*. http://mir-nauki.com/PDF/35PDMN217.pdf. Accessed 29 Dec 2021.

Razinkina, E., Pankova, L., Trostinskaya, I., Pozdeeva, E., Evseeva, L., & Tanova, A. (2018). Student satisfaction as an element of education quality monitoring in innovative higher education institution. In D. Safarik, Y. Tabunschikov, & V. Murgul (Eds.), *E3S web of conferences* (Vol. 33, No. 03043). https://doi.org/10.1051/e3sconf/20183303043

Shafranov-Kutsev, G. F., & Cherkashov, E. M. (2020). Youth orientation towards competition and entrepreneurship. *Sociological Studies, 4*, 117–123. https://doi.org/10.1016/S0883-9026(98)00033-0

Solntseva, G. N. (2019). Entrepreneurship: social functions, features of activity and psychological determinants of success. *Psikhologiya. Zhurnal VSHE, 4*. https://cyberleninka.ru/article/n/predprinimatelstvo-sotsialnye-funktsii-osobennosti-deyatelnosti-i-psihologicheskiedeterminanty-uspeshnosti. Accessed 29 Dec 2021.

Sorokin, P., Povalko, A., & Vyatskaya, Y. (2021). Informal entrepreneurship education: Overview of the Russian field. *Foresight and STI Governance, 15*(4), 22–31. https://doi.org/10.17323/2500-2597.2021.4.22.31

Starodubtsev, V. A. (2021). Practice-centered education in higher school. *Vyssheye obrazovaniye v Rossii, 30*(5), 75–87. https://doi.org/10.31992/0869-3617-2021-30-5-75-87

Organizational-Economic Mechanism for Monitoring University Activities Under the Conditions of Digitalization

Natalya L. Ketoeva, Meri T. Zargaryan, and Ekaterina A. Sysoeva

Introduction

Modern education has been influenced by economic, social, and political factors that have led to a number of major changes. Development of the educational services market, several economic conditions, the development of digital economy and digital technologies, the formation of the education sector as another sector of the economy have an impact on the development of the modern system of higher education. Especially, in the digital economy, the competitiveness of higher education institutions is ensured by the professional management and leadership qualities of its leaders, which imply certain managerial training, the development of certain competencies.

The development of the digital economy in the Russian Federation is aimed at creating the necessary conditions that lead to the emergence of new computer and digital technologies. Personnel training is one of the main directions of the program "Digital Economy of the Russian Federation" (2018). It helps to implement research, including digital, competencies among students and young scientists, to train highly qualified specialists in the field of the digital economy, and to develop scientific and technical capabilities of enterprises (Ketoeva et al. 2020b).

The aim of the study is to model an organizational and economic mechanism for monitoring the effectiveness of the complex activities of departments as business units of a higher educational institution.

To achieve this goal, the following tasks were set and solved:

- a review of modern scientific literature on research issues was carried out.
- analysis of the mechanism of management of departments in universities.

N. L. Ketoeva · M. T. Zargaryan · E. A. Sysoeva (✉)
National Research University "Moscow Power Engineering Institute", Moscow, Russian Federation
e-mail: ketoyevanl@mpei.ru; ZargarianMT@mpei.ru; sysoyevayeka@mpei.ru

A. Rumyantseva et al. (eds.), *Challenges and Solutions in the Digital Economy and Finance*, Springer Proceedings in Business and Economics,
https://doi.org/10.1007/978-3-031-14410-3_34

- the departments' structure of management of business processes was analyzed.
- we developed a mechanism for monitoring the complex activities of departments based on the formation of competitive business units.

The scientific novelty of the study is the development of an organizational and economic mechanism for monitoring the effectiveness of the integrated activities of departments as business units of the university in the context of digitalization.

Materials and Methods

The basis of the study was the following methods: dialectical scientific knowledge and private scientific (analysis, synthesis, comparison, logical and system-structural analysis, formalization, analysis of legal documents), modeling.

Results and Discussions

The effectiveness of the complex activities of the departments as business units of the university.

In the modern world, education is increasingly viewed from the perspective of the market, and higher education as a service of the university. The market of educational services as a product offers the author's developed educational programs for consumers. At the same time, as in the labor market, graduates (specialists, bachelors, masters) who have been trained in a specific educational program act as an offer (Ivanovskaya et al. 2019). And in the market of scientific and technical products and services, an educational institution presents the results of research and innovation activities in the form of patents, know-how, software, etc. (Emelina and Kiseleva 2019).

In this study, the activities of the department were considered as the activities of a business unit of a higher educational institution. The department at the university is a link between the market of educational services, the market of scientific and technical products/services and the labor market. By increasing the effectiveness of the department, the ranking of the entire university increases (Zaitseva 2020; Vetluzhskikh 2014). The study was conducted on the example of MPEI within the framework of the grant "Mechanism for monitoring the integrated activities of the departments of MPEI in order to ensure the competitiveness of a higher educational institution.

The effectiveness of the university's interaction with consumers of educational services and received intangible assets depends on its performance results, places in Russian and international rankings. High performance can be achieved through the organization of effective communication with students and graduates, as well as the possibility of realizing the potential of long-term relationships with them

(Sundukova et al. 2019). The main link of the relationship system built by the university is the departments that implement complex activities, including educational, educational, cultural, social, research and innovation.

Practical experience and the requirements of management science show that the success of a higher educational institution is achieved by the interested participation of each link, including each employee. A special place in the system of higher education belongs to the department (Andreeva and Mizova 2018). Since at the level of the department there is a main relationship with students and graduate students, young scientists and researchers and a direct educational, cultural, social impact on them. The efficiency of the university directly depends on the activities of the department and its head, this is especially important in today's rapidly changing world, aimed at the development of information technology (Ketoeva et al. 2020a).

The activities of the department aimed at training specialists in the field of digital economy unite teachers, candidates of sciences, doctors, and professors in the field of digital, corporate and sectoral economics, finance, accounting, law, practical psychology and business informatics (Tumbas et al. 2019). The goal of the department in the digital economy is to develop digital skills and entrepreneurial skills among students and graduates. A feature of the department of the twenty-first century is the combination of science and innovation in its activities.

Department management is an important priority in the activities of its head. The leader is a professional executive and initiative organizer, able to take responsibility in making decisions on various issues within their competence (Sudlow 2019).

The leadership of the department has several features. The main goal of the department is to provide educational, methodological, scientific, educational work. The activity of the department is aimed at training specialists who have deep theoretical knowledge, practical skills, and abilities, who combine high moral qualities and general culture.

Solving its tasks, each department should consider the strategic objectives of the national education system (Pathak et al. 2021). In the modern world, one must constantly adapt to the changes in the internal and external environment in the digital economy.

Thus, the department is the main link in the process of building relationships between market participants. Its effective activity affects the creation of competitive advantages of the university in the strategic perspective (Tarasov et al. 2021).

An important task of Russian universities is to strengthen their positions in international rankings. Modern trends, such as the digitalization of education, the globalization of educational processes, the development of the Internet, university websites, social networks, contribute to the competition of universities not at the regional level, but at the national and global level.

Architecture of business processes of a higher educational institution

Organizational and economic processes in the university include organizational, educational, managerial, technological, financial, scientific, social, and other management actions. The use of the organizational and economic mechanism in the process of monitoring the effectiveness of the integrated activities of the departments of the university will lead to a qualitative change in educational and scientific

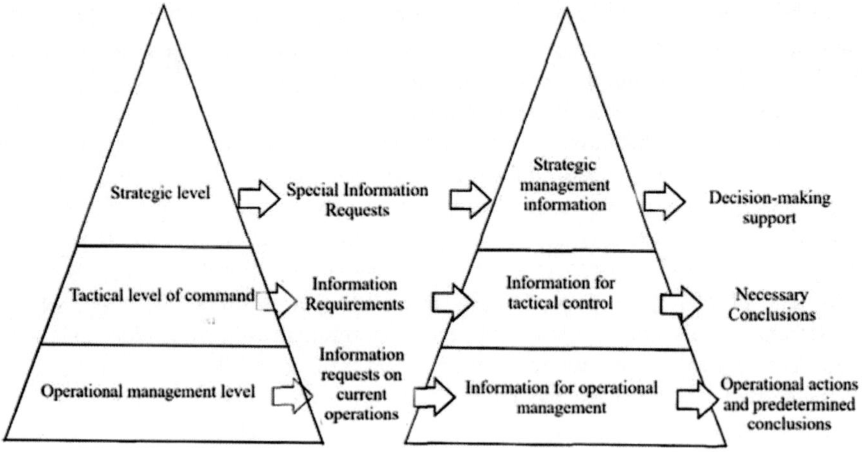

Fig. 1 Levels of management and methods of management influence

characteristics, thereby forming a stable position of the department as a business unit of the university for the provision of educational services, as well as the competitiveness of the university relative to competing universities. The mechanism chosen by the University for monitoring the complex activities of the departments determines the way in which it performs certain types of its activities, and the entire chain of product value formation. The optimal interconnection of all business processes within an organization forms the system level, or business system. In the university, one can single out educational, research activities, which include different functions, different sets of business processes aimed at achieving different goals. Thus, all business processes can be classified by management levels Fig. 1.

The strategic level of university management includes business processes of development and management (rector's office, vice-rector for educational activities, vice-rector for scientific activities, vice-rector for economics). It is here that a new quality of business processes arises, which is absent at other levels—a strategic plan, including the mission of the organization, the main strategic goal.

Strategic decisions made at the highest level set the target orientation, orientation of the entire set of business processes of the organization, forming priorities and basic requirements for the results they receive.

The tactical management level of the university can be attributed to supporting business processes (educational management, educational department, directors of institutes).

And the operational level is the main business processes, i.e., the direct activities of the departments (the head of the department, the teaching staff). Based on the foregoing, it is possible to define the department as part of the university system, a kind of strategic business unit in the educational services market.

The department is the main structural subdivision of the university, carrying out educational, methodological and research activities in one or more areas, while realizing its main task—training highly qualified personnel.

Each of the above activities ultimately creates a unified management strategy, the result of which is the formation of a general rating of the department and the university.

New scientific results of the study: organizational and economic mechanism for evaluating the effectiveness of the integrated activities of departments as business units of the university.

The main goal of the organizational and economic mechanism for monitoring the integrated effectiveness of the activities of the departments as business units of the university is to ensure the competitiveness of the business unit of the educational institution, the competitiveness of the implemented educational programs, as well as the products of scientific, research and innovation activities of the departments. The mechanism involves continuous monitoring of the effectiveness of the activities of the departments, and the prompt adoption of managerial decisions to increase the efficiency of the business units of the university and minimize the level of risk in the framework of the implementation of complex activities.

The mechanism involves monitoring considering the following stages:

Competitive analysis—analysis of external influence factors, as well as a detailed analysis of competitors and the competitive position of the university based on the use of modern methods of competitive analysis, analysis of the educational services market and related markets, the requirements of individuals for the education services of universities. As well as the further formation of the financial system of the university: the mechanisms and structures of financing the university.

Collection of initial information about the control object, in this case, the educational and research activities of the university act as the control object. Collection of the necessary information to assess the effectiveness of the integrated activities of the departments as business units of the university, the establishment of criteria for evaluating the educational and research activities of the departments, the formation/adjustment of the questionnaire for the heads of the departments and the direct filling of the questionnaire by the heads of the departments.

At this stage, a direct managerial impact is assumed with the help of a control mechanism. In this model, the control mechanism is the evaluation of the effectiveness of the departments, the formation of the rating of the departments and the determination of standard values for each indicator of activity. The management mechanism involves an assessment within the digital platform of the RUR RMP.

The final stage of monitoring the complex activities of the departments as business units of the university is the adoption of certain management decisions aimed at developing possible measures to improve the efficiency of the integrated activities of the departments, including by individual indicators, as well as determining the tasks and priorities for the development of the departments for the next 5 years.

The result of the adopted management decisions should be to increase the competitiveness of the department as a business unit or to maintain competitive leadership.

This mechanism (Fig. 2) was developed based on National Research University "Moscow Power Engineering Institute" (hereinafter referred to as MPEI) and it involves increasing the efficiency and competitiveness of not only individual business units of the university, but in a synergistic effect in the formation of a competitive position of the university in the educational services market.

MPEI plays an important role in the training of highly qualified specialists, using breakthrough digital technologies for working with talented youth, and in the innovative development of the energy industry. The leading positions of MPEI in the educational and scientific fields are marked by awards from foreign countries, domestic and international ratings, and the high reputation of MPEI graduates among large Russian and foreign energy companies.

Since 2014, MPEI has been included in the list of the best universities according to the world's leading rating agency Quacquarelli Symonds (QS), and in 2018 MPEI was included in the top 100 universities in the BRICS countries (Collier 2021). MPEI is included in the list of nine Russian universities that have a QS audit score of "4 stars" and above. In 2021, in the Times Higher Education (THE) ranking among universities in emerging economies, it was in the group of 351–400. Annually participates in the rating "English-language sites of Russian universities", actively improving its Internet portal and carrying out a global transformation of the content of the English-language site (Timofeev et al. 2021).

Conclusions

MPEI annually shows a high dynamic of growth in indicators that determine the position in the global university rankings, and also plays a significant role in international projects, allowing it to increase its international academic reputation (Kuzib et al. 2020).

Such a transition of the organization of the integrated activities of MPEI's departments to the process approach made it possible to ensure a consistently high level of qualitative and quantitative indicators of the activities of the departments. It will also lead to sustainable competitive success of the university.

Thus, by applying the results of this study, the problems of ensuring the competitiveness of departments, increasing the efficiency of educational, innovative and research activities of departments, as well as increasing the competitiveness of a higher educational institution in the digital economy are solved. This will ensure the market orientation of educational and research and innovation activities, satisfying the needs of consumers.

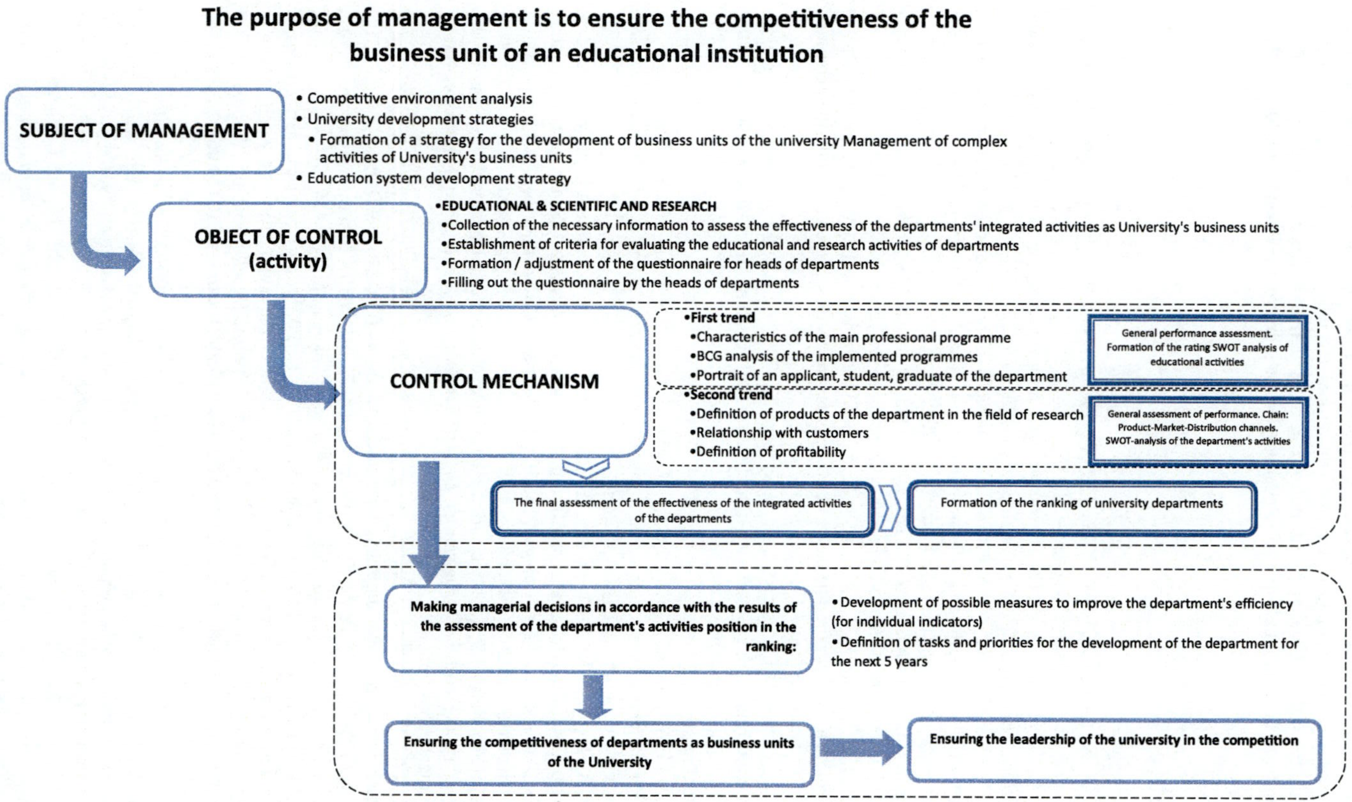

Fig. 2 Modeling the organizational and economic mechanism for monitoring the effectiveness of the integrated activities of departments as university business units in the context of digitalization

Acknowledgements The investigation was carried out within the framework of the project "The mechanism for monitoring the integrated activities of the department of MPEI to ensure the competitiveness of the business unit of the educational institution" with the support of a grant from MPEI for implementation of scientific research programs "Energy", "Electronics, Radio Engineering and IT", and "Industry 4.0, Technologies for Industry and Robotics in 2020–2022.

References

Andreeva, A. N., & Mizova, E. M. (2018). Digital economy: New business opportunities. *Economy and Business: Theory and Practice, 4*, 19–21.

Collier, S. (2021). *Top 10 universities in Russia 2018*. https://www.qschina.cn/university-rankings-articles/brics-rankings/top-10-universities-russia-2018

Emelina, O. S., & Kiseleva, M. A. (2019). Improvement of the model of interaction between the subjects of the scientific and educational process in technical universities. *Economics and Management of Management Systems, 33*(3), 29–35.

Ivanovskaya, I. I., Malyarchuk, N. N., & Chumicheva, R. M. (2019). *Problems and prospects for the development of science and education in modern conditions*. New Science.

Ketoeva, N. L., Kiseleva, M. A., Zargaryan, M. T., & Sysoeva, E. A. (2020a). Academic mobility of a student as a factor in the formation of key competencies of a graduate in a digital economy. *Journal of Economics and Entrepreneurship, 8*, 1297–1301.

Ketoeva, N., Kiseleva, M., Zargaryan, M., & Sysoeva, E. (2020b, April). Project management in the higher education institution activity in terms of digital economy. In *III International Scientific and Practical Conference "Digital Economy and Finances" (ISPC-DEF 2020)* (pp. 74–78). Atlantis Press.

Kuzib, E. V., Litvinyuk, A. A., Gulicheva, E. G., Ledneva, S. A., Osipova, M. S., & Sysoeva, E. A. (2020). *Problems of attracting talented youth to the sphere of science, high technologies, and higher*. First Economic Publishing House.

Pathak, D. K., Shankar, R., & Choudhary, A. (2021). Performance assessment framework based on competitive priorities for sustainable freight transportation systems. *Transportation Research Part D: Transport and Environment, 90*, 102663. Accessed November 14, 2021, from https://doi.org/10.1016/j.trd.2020.102663

Presidium of the Presidential Council for Strategic Development and National Projects. (2018). *Passport of the national program of Digital Economy of the Russian Federation*. http://government.ru/info/35568/

Sudlow, B. (2019). Review of Joseph E. Aoun (2017). Robot Proof: Higher Education in the Age of Artificial Intelligence. *Postdigital Science and Education, 1*(1), 236–239.

Sundukova, G. M., Bobyleva, N. V., & Derevyagina, L. N. (2019). Strategic management of a university in a digital economy. *Bulletin of Eurasian Science, 11*(3), 1–15.

Tarasov, A., Sysoeva, E., Gulicheva, E., & Hoa, M. N. T. (2021, July). Strategic partnerships of universities as a tool for intensifying the development of the regional economy and society in digital economy. In *3rd International Conference Spatial Development of Territories (SDT 2020)* (pp. 372–378). Atlantis Press. https://doi.org/10.2991/aebmr.k.210710.063

Timofeev, I. N., Bakumenko, O. A., Karpinskaya, E. O., & Bocharov, I. A. (2021). *Electronic internationalization: English-language Internet resources of Russian universities (2021)*. Report No. 72/2021. Russian International Affairs Council (RIAC). NP RIAC. https://russiancouncil.ru/activity/publications/elektronnaya-internatsionalizatsiya-angloyazychnye-internet-resursy-rossiyskikh-universitetov-2021-g/#short

Tumbas, P., Sakai, M., Pavlicevic, V., & Rakovic, L. (2019, March). Digital competencies in business informatics curriculum innovation. In *Proceedings of the 13th International Technology, Education and Development Conference (INTED), Valencia, Spain* (pp. 11–13).

Vetluzhskikh, E. (2014). *Remuneration system: How to develop goals and KPIs*. Alpina Publisher.

Zaitseva, T. V. (2020). *Motivation of labor activity: A textbook*. INFRA-M.

Transformation of Financial Relations Under the Digital Economy in the Forest Industry

Snezhana N. Bareyko, Tatyana S. Bazhaeva, Kseniya A. Kozhukhina, and Elena V. Ivanova

Introduction

The active introduction of digital technologies in recent years has significantly changed the face of key sectors of the economy. More and more entrepreneurs are transferring their business processes to the digital environment, thereby essentially reducing transaction costs, and significantly increasing the volume of economic activity. The digital economy sets the direction for the transformation of traditional sectors of the economy, the emergence of new markets and niches.

In the modern information environment, a "new digital" market has formed that has the following characteristics:

- competitiveness of enterprises.
- optimization of business processes.
- high customer focus.

Modern entrepreneurs must respond many times faster to market changes and introduce new products and services to the market through electronic sales channels. The creation of new product value is influenced by high-speed processing of big

S. N. Bareyko (✉)
Luga Les LLC Group of Companies, St. Petersburg, Russian Federation

T. S. Bazhaeva
Leningrad State University named after A.S. Pushkin, St. Petersburg, Russian Federation

K. A. Kozhukhina
University at Interparliamentary Assembly of EurAsEC, St. Petersburg, Russian Federation

E. V. Ivanova
St. Petersburg University of Management Technologies and Economics, St. Petersburg, Russian Federation
e-mail: pavlova@spbacu.ru

A. Rumyantseva et al. (eds.), *Challenges and Solutions in the Digital Economy and Finance*, Springer Proceedings in Business and Economics,
https://doi.org/10.1007/978-3-031-14410-3_35

data. Such analysis makes it possible to identify the new sources of value creation based on the interests of consumers and their behavior patterns. The possession of such information becomes a “valuable digital asset” of the company and allows it to be more competitive. This fact is confirmed by the fact that digital giants such as Facebook, Google, Microsoft, Apple, Amazon have reached record market capitalization.

Digital technologies make it possible to optimize the collection and storage of information, and modern information technologies make it possible to process information better, develop behavioral algorithms and predictive models.

For example, commercial enterprises actively use technologies related to the development of an algorithm that describes the customer’s consumption model, and then automatically delivers goods to him based on the predicted need, one of them Tesco.

Today’s business models are becoming more customer oriented: from value proposition that addresses the customer’s predicted need and just-in-time delivery to revenue streams based on how long a customer uses a product. This orientation allows business structures to become more competitive, i.e., enables a quick introduction of a new product to the market (time-to-market). Timely business response to changing customers’ needs is possible with continuous analysis of the results obtained, and modern digital platforms greatly facilitate this process, as Tesla launches new options and corrects claims in real time, remotely through software updates.

However, the digitalization of business processes in various sectors of the economy is not equally developed. For example, banking, services, and trade have moved to digital platforms for a long time, and the pandemic has only intensified this transition.

For example, in 2013 Uber emerged as one of the fastest growing companies using new forms of taxi services. Recently, similar platforms have emerged in the transport sector (Gett, Belka, YouDrive, Delimobil, etc.), professional services (YouDo, etc.), listing (Avito, etc.), education (OpenDo, Intuit), mutual lending (Fingooroo, Vdolg), rental of household items (Rentmania, Arendorium) and freelancing (Freelancer, etc.).

Yandex and Mail.Ru can easily compete with such major companies as Amazon, Apple, Facebook, etc., including by equity investments of high-tech start-ups.

Sberbank is improving its digital platform by developing e-commerce and providing customers with unified access to all platforms through the bank’s mobile app (Abdrakhmanova et al. 2021).

Figures 1, 2 and 3 show which industries have been using information technology most intensively in recent times (Malkov et al. 2018), where:

1. Mining
2. Manufacturing
3. Providing electric energy, gas and steam; air conditioning
4. Water supply; wastewater disposal, waste collection and disposal, pollution elimination
5. Construction

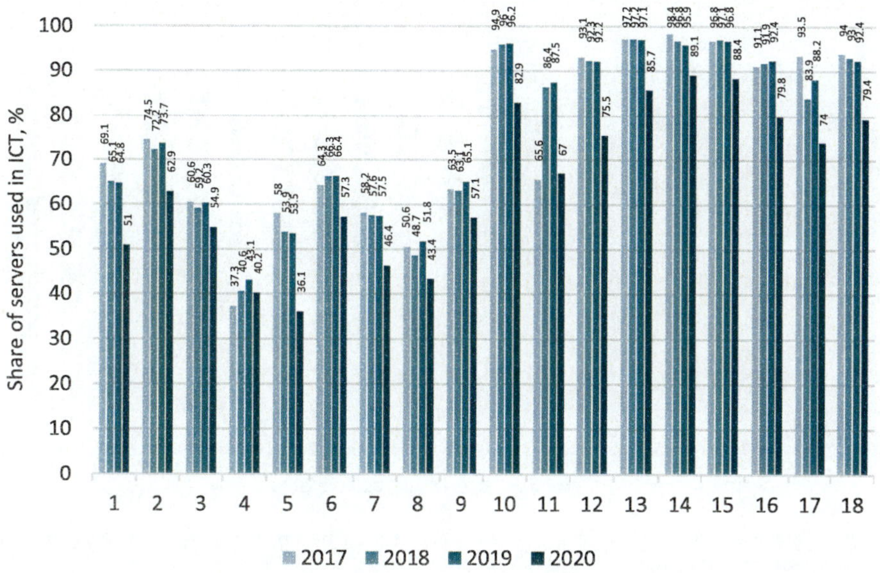

Fig. 1 Share of servers used by organizations in information and communication technologies by type of economic activity, %

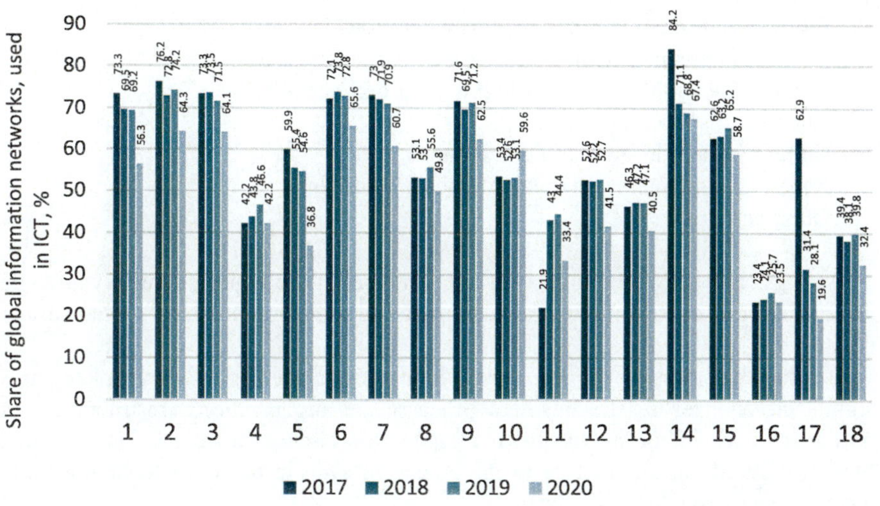

Fig. 2 Share of global information networks used by organizations in information and communication technologies by type of economic activity, %

6. Wholesale and retail; repair of cars, motorcycles
7. Transport and storage
8. Hotels and catering organizations
9. Information and communications

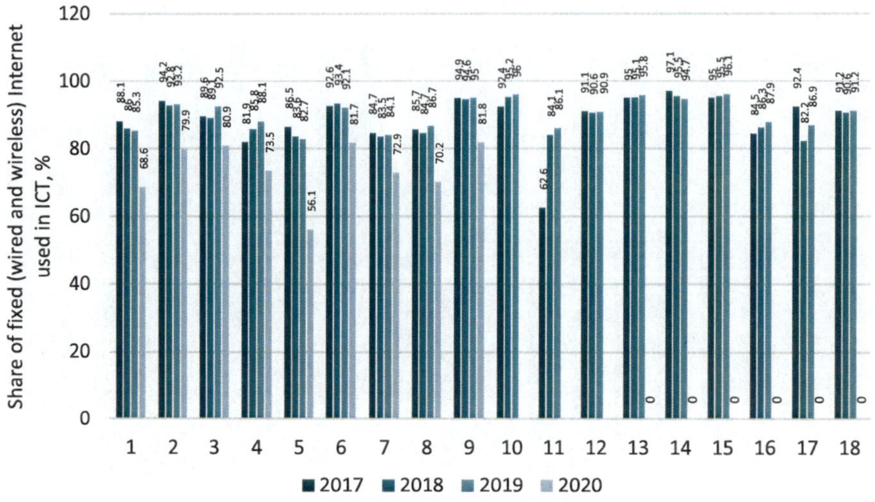

Fig. 3 Share of fixed (wired and wireless) internet used by organizations in information and communication technologies by type of economic activity, %

10. Finance and insurance
11. Real estate
12. Professional, scientific, and technical activities
13. Public administration and military security; social security
14. Higher education; training of highly qualified personnel
15. Health and social services
16. Culture, sports, leisure, and entertainment
17. Repair of computers, personal and household items
18. Other activities

As Figs. 1–3 show, the following industries reach the highest values: wholesale and retail, repair of cars, motorcycles, information and communications, finance, and insurance.

But the transition of "conservative" industries to the digital platform requires significant changes in the business processes of organizations, restructuring of traditional ways of doing business. Despite this circumstance, experts note that there is virtually no alternative to the transition even in the most technologically inert industries to digital platforms.

As Table 1 shows, in recent years the enterprises have most actively begun to use special programs for doing business.

Despite the success of many enterprises in the automation of production processes, the introduction of distributed control and monitoring systems, most companies still do not fully realize the potential of big data analytics and decision-making algorithms based on artificial intelligence. At the same time, information technology has the greatest transformational potential in the industry that is especially important

Table 1 Organizations using special software tools, %

Software tools	Total in 2015	Total in 2018	of them using Russian software	Total in 2019	of them using Russian software
Electronic document management systems	62.7	68.6	60.0	70.0	61.6
Software tools for financial settlements in electronic form	55.1	56.4	49.2	57.1	50.0
Software tools for solving organizational, managerial and economic tasks	52.3	54.9	45.3	54.8	45.1
Electronic legal reference systems	52.3	53.2	47.6	53.2	48.2
Software tools for managing the procurement of goods (works, services)	38.4	38.3	31.0	39.0	31.5
Software tools for providing access to databases via global information networks, including the Internet	31.5	31.9	23.3	32.0	23.1
Software tools for managing sales of goods (works, services)	21.9	25.9	19.5	26.0	19.3
Software tools for the management of automated production and/or individual technical means and technological processes	15.1	16.7	11.1	16.5	10.7
Tutorials	14.3	16.9	12.7	16.4	12.3
ERP systems	9.3	13.8	7.5	14.8	7.8
Design software	11.0	13.0	6.8	13.0	6.5
CRM systems	9.9	13.2	7.8	13.9	8.3
Editing and publishing systems	5.3	6.5	3.0	6.9	2.9
SCM systems	4.3	6.4	2.5	6.6	2.4
Scientific research software	3.9	4.5	2.0	4.6	2.0

Source: Information Society of the Russian Federation in 2020. Statistics Digest (Feoktistova 2020)

for companies with significant tangible assets. According to the survey, almost 50% of industrial companies rated information technology as a critical element on their path to success over the next 5 years. The development of an appropriate infrastructure requires the creation of innovation centers at the enterprises, the involvement of highly qualified specialists, and a significant increase in investment in cybersecurity.

Materials and Methods

The article uses information on the digitalization of business in the Russian Federation, the source of which is Form No. 3-inform in accordance with the Order of Rosstat dated July 30, 2021, No 463 that is provided to the territorial body of Rosstat by March 25 of the year following the reporting one. The Form includes 21 sections that reflect the technologies used, the costs of their use.

The methodological basis of the study is system, generalized, analytical and comparative approaches, as well as a set of methods of economic and statistical analysis.

Results

To ensure the economic security of Russia, national programs for the development of a new generation economy, including the implementation of innovative technologies, are tasks of strategic importance. Digitalization at the present stage of development is a global trend. The end-to-end penetration of information technologies into the Russian economy creates the basis for the formation of new markets and new conditions for the functioning of traditional production systems, including the timber industry (Bareyko and Kozhukhina 2019). The possibilities of achieving key indicators of the socio-economic development of Russia are associated with the development of digital technologies. The rapid development of information technologies and the digital economy requires new approaches to ensure the economic and information security of Russia.

Forestry has one of the lowest digitalization rates among most sectors of the Russian economy, although the implementation potential is huge. The most important document of the timber industry today is the Strategy for the Development of the Forest Complex. The new edition was finalized considering the national development goals of Russia, approved by the President last year. The main goals are to create a comfortable and safe environment; the new document considers the forestry problems and tasks, solves them without compromising the environmental component. It calls for making the forest more profitable, increasing its contribution to GDP, with special attention paid to the conservation function of the forest.

Effective and sustainable development of forestry is impossible without reliable, high-quality, and complete information about forests. The lack of such information is evidenced by numerous discrepancies in data in official sources, including those published on the websites of state forest management bodies and voiced by officials, as well as those contained in the state forest register (Akyulov 2018).

The forest industry has recently been under the spotlight of both government agencies and business community. As practice shows, logging and processing are most often disparaged: gray loggers, corruption, wood shortage, rapid price growth, export of round timber. At present, the state has become more active in relation to the forest industry and has taken an active course towards its control and transparency. In accordance with the Decree of the Government of the Russian Federation dated December 6, 2021 No 2214 "On approval of the form of an electronic accompanying document for the transportation of wood and products of its processing, the scope of the information included in the electronic accompanying document for the transportation of wood and products of its processing, as well as requirements for the format and procedure for filling out an electronic accompanying document for the transportation of wood and wood products" and in accordance with Part 8 of Article 50 of the Forest Code of the Russian Federation, the following parameters are approved:

- the form of an electronic accompanying document for the transportation of wood and products of its processing.
- the scope of the information included in the electronic accompanying document for the transportation of wood and products of its processing.
- requirements for the format and procedure for filling out an electronic accompanying document for the transportation of wood and products of its processing.

As a result of the innovations, 2022 is designed to bring the forest market to the electronic level of interaction the transportation of timber will be possible only if there is an electronic accompanying document. It will allow transferring the industry to a digital format, moving away from paper documents, establishing electronic interaction, digitizing the forest declaration, as well as stimulating timber processing.

From January 1, 2022, the export of unprocessed and roughly processed softwood and valuable hardwood timber from the territory of Russia has been stopped. By the Decree of the President of the Russian Federation, the forestry was given a task of preserving forests. The impossibility of exporting round timber should stimulate the development of timber processing.

In 2020 the Forest Industry Projects program was launched. Within the framework of this program, concessional loans are provided to SMEs at 1% or 3% per annum for the modernization of wood processing capacities by purchasing equipment from 20 to 100 million rubles. To implement larger-scale investment projects, enterprises engaged in the timber industry can use the IDF Development Projects program and receive a loan in the amount of 50–500 million rubles.

In this regard, planning the forestry transformation, it is required to draw up a list of tasks that will consider the industry needs, and at the same time conform with the target state in each area of the national economy digitalization and meet the deadlines.

In this regard, it should be noted that each industry and even, possibly, each enterprise should independently formulate its attitude towards digitalization and its implementation. Development programs for individual industries should correspond the digitalization program in the Russian Federation as a whole (Akyulov 2018).

Possible directions of digitalization and informatization of the Russian forestry are considered below and changes that can serve the development of the industry to ensure the national economic security are found out.

The development of digitalization and informatization in the forestry of Russia provides for the following activities (Fig. 4).

To optimize the development of informatization in the forestry of Russia, it is necessary to create a unified automated information system as a single platform for providing information and analytical support. The introduction of this system will allow receiving, processing, storing and using information about the state of forests, their quantitative and qualitative characteristics, about their use, protection and

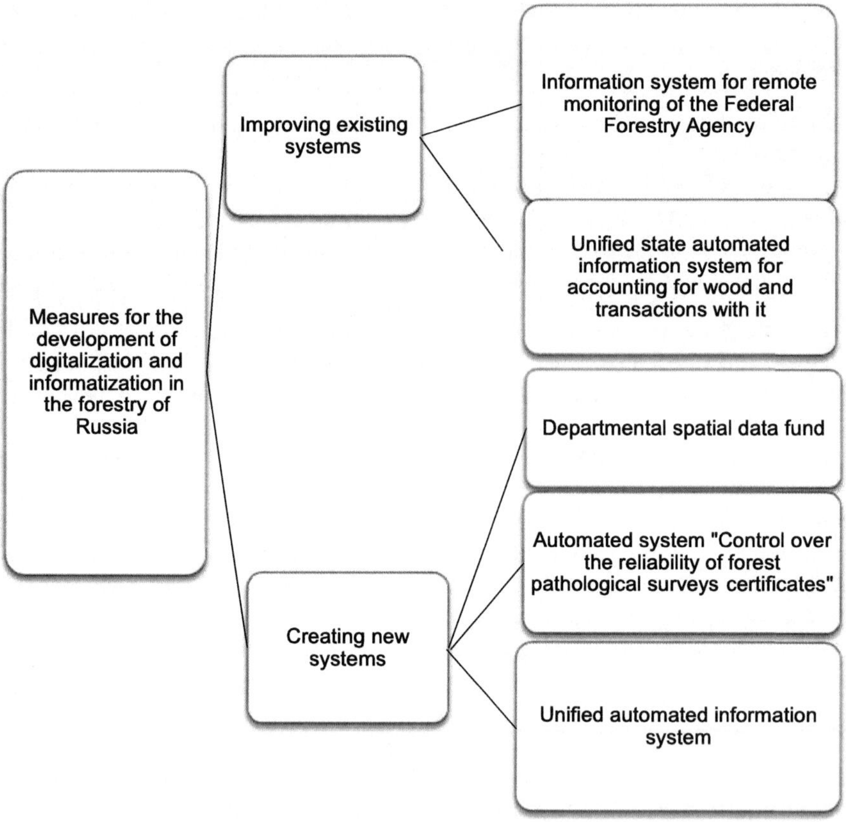

Fig. 4 Measures for the development of digitalization and informatization in the forestry of Russia

reproduction that in turn will increase the reliability and efficiency of providing information for making managerial decisions (Bareyko and Kravchenko 2021).

An example of the effective implementation and use of information technologies in the timber industry is Finland, where in recent years several successful public and private digitalization initiatives have been implemented at all levels of forestry, starting with the widespread use of laser scanning, state geoportals and federal-level forest databases (for example, Metsään.fi—the state geoportal of open field data and plans) and ending with local developments used in private farms.

Forest owners, using the Virtual Forests service, manage forests from anywhere in the world. Using this service, timber producers can take a virtual walk through their forests, plan forestry activities, and calculate future income from them. The service is designed on a digital model of forests, created based on publicly available forest inventory materials posted on the Internet, carried out at the expense of taxpayers.

The experience of Finland allows carrying out forest inventory using automatic tree-by-tree forest inventory with the help of drones. This technology makes it possible to create digital copies of forest ranges for more accurate forest management and has proven itself well in Finland, and in recent years some companies have been testing it in different regions of Russia.

Digital inventory has the following benefits:

- minimization of errors that occur with selective taxation methods.
- the ability to quickly access data from mobile phones and tablets.
- optimization of costs in obtaining data with the correct organization of all processes.
- increasing the speed of data acquisition (with the use of tree-by-tree valuation technology and appropriate software systems for data processing, up to 5000 ha tree-by-tree can be described per night).
- application of high-precision planning, which makes it possible to increase investment income.

The development and widespread use in forest management of such digital technologies as forest shooting from drones, laser scanning, remote sensing of forests can improve the accuracy and completeness of information about forests, as well as make it more accessible. In particular, the introduction of digitalization in forestry will ensure both internal and independent control of forest management processes and reduce the number of errors that lead to inefficiency of forest resources.

Rational forest management, based on modern approaches to accounting for environmental processes in forests for each specific area, will be able to optimize all managerial decisions in accordance with individual indicators of soil type, terrain slope, species diversity of trees. Automation of operations at the logging site contributes to solving the problem of a skilled labor shortage. Creation of tree growth models based on a combination of spatial data improves the efficiency of management in the industry.

Discussion

The problems of the forestry digitalization lie in the impossibility of long-term planning in the industry due to the volatility of the economy, the unfinished and frequently changing forest legislation, as well as in attracting additional investments for the modernization of technologies.

To solve the problems of the forest industry digitalization, first, it is necessary to determine the main priorities at the state level, move to forest planning at all levels (including its economic component), and choose to improve the quality and openness of information about forests as one of the priorities.

To implement such plans, it is necessary to develop a forestry information standard that will ensure the unification and comparability of different forest data sets. In Finland all market participants were involved in the development of such standard.

It is necessary to legalize data on forests obtained with the help of digital technologies in new digital formats, because the concepts of digital aerial photography and laser scanning have not yet been officially registered and fixed anywhere. A certain potential lies in improving the accuracy of forest inventory data in the event of abandoning such a taxation production standard as the minimum area of a plot and the number of allotments in a quarter. Finally, it is advisable to add metadata to the list of required parameters: an assessment of the quality of the information received, information about field samples. Now in the forest management materials, at best, it is indicated only who performed the work.

Numerous examples and foreign experience, as well as the accumulated experience of testing technologies in Russia, indicate that the development of modern digital technologies for obtaining and processing data and the widespread introduction of digitalization in planning, procurement, logistics, accounting, and monitoring contribute to the transition to high-precision, economically efficient and sustainable forestry. At the same time, it is important not only to obtain inventory data on forests; one of the priorities should be the receipt and processing of data to make successful managerial decisions on their basis that increase the efficiency of forestry. Decisions based on outdated data containing many errors in some cases lead to the opposite result.

Conclusion

Among the many problems of the social and economic development of Russia in the context of the formation of a global post-industrial society, a prominent place is occupied by the organization of sustainable functioning and safety of using information systems and information and communication networks that ensure economic activity. It is revealed that the Russian economy has the potential for digital transformations.

The prospects for financing digital transformation in Russia are obvious (Kuchumov et al. 2021; President of the Russian Federation 2020):

1. increasing the competitiveness of Russian enterprises at the international level.
2. increasing the share of Russian software in the activities of enterprises.
3. strengthening the presence of Russian digital developments in world markets.
4. the growth of the attractiveness of Russian education and scientific enterprises conducting research and development.

There is also a desire to achieve results.

However, the source of risks for the forest industry lies in lagging behind the world level of the main Russian technologies used in the timber business. As the information infrastructure of the business becomes more complex, the influence of this factor on the results of the activities of the timber industry organizations will increase. The adopted program documents for the development of the Russian forestry should be considered from a unified system position and considering the interests of not only the state, but also business (Akyulov 2018).

References

Abdrakhmanova, G. I., Bykhovsky, K. B., Veselitskaya, N. N., Vishnevsky, K. O., & Gokhberg, L. M. (2021). Digital transformation of industries: Starting conditions and priorities. In *XXII International Scientific Conference on Problems of Economic, and Social Development, Moscow*.

Akyulov, R. I. (2018). Modern technologies of the digital economy in the management of the Russian forestry. *Debate, 4*, 24–31.

Bareyko, S. N., & Kravchenko, S. K. (2021). *Risks and threats to Russia's economic security in the context of the economy digitalization. Socio-economic processes of modern society: Theory and practice* (pp. 16–18).

Bareyko, S. N., & Kozhukhina, K. A. (2019). Economic and information security of Russia in the digital economy. *Nauka Krasnoyarya, 8*(5), 7–18.

Feoktistova, I. V. (2020). *Selected federal statistical observation on the use of information technologies and information and telecommunication networks by the population*. Federal State Statistics Service. https://gks.ru/free_doc/new_site/business/it/ikt20/index.html

Kuchumov, A., Testina, Y., & Kozhukhina, K. (2021). Experience of creation and development of tourist and recreational zones in Russia. *Bulletin of the National Academy of Tourism, 60*(4), 17–19.

Malkov, P. V., Shapoval I. N., Smelov, P. A. ... Feoktistova, I. V. (2018). *Information society in the Russian Federation. 2020: Statistical collection* [Electronic resource]. Moscow, NRU HSE, 2020. https://rosstat.gov.ru/storage/mediabank/lqv3T0Rk/info-ob2020.pdf

President of the Russian Federation. (2020). *On the national development goals of the Russian Federation for the period up to 2030*. http://www.kremlin.ru/acts/news/63728, 16.

Artificial Intelligence Technologies in the Implementation of the e-Navigation Concept

A. A. Burykin, M. N. Grachev, and S. V. Kolesnichenko

Introduction

The constant technological development of the transport industry is associated with the continuous development of IT technologies. Trends in the development of IT technologies make it possible to determine the ways of implementing the concepts of e-Navigation in modern and promising automated control systems (ACS) of a ship, including integrated bridge systems (IBS), considering the safety of navigation in difficult metrological conditions (Bezborodov and Ismagilov 2017).

Its ideas e-N, formulated in 2005 at the 81st and in 2014 at the 94th sessions of the Maritime Safety Committee (MSC) of the International Maritime Organization (IMO) and set out in the form to be solved within the framework of the Strategic Implementation Plan of e-N (SIP), are taking possession of the minds of an increasing number of participants in maritime transport (Rivkin 2019; Patraiko 2007a, b; Mitropoulos 2007; IMO 2021). In this plan, problems are formulated, and specific

A. A. Burykin (✉)
Admiral Makarov State University of Maritime and Inland Shipping, St. Petersburg, Russian Federation

St. Petersburg University of Management Tecnologies and Economics, St. Petersburg, Russian Federation

Russian Academy of Transport, Moscow, Russian Federation

M. N. Grachev
PPF Meridian JSC, St. Petersburg, Russian Federation

S. V. Kolesnichenko
Admiral Makarov State University of Maritime and Inland Shipping, St. Petersburg, Russian Federation

St. Petersburg University of Management Tecnologies and Economics, St. Petersburg, Russian Federation

A. Rumyantseva et al. (eds.), *Challenges and Solutions in the Digital Economy and Finance*, Springer Proceedings in Business and Economics,
https://doi.org/10.1007/978-3-031-14410-3_36

tasks are listed that must be solved in the design and creation of equipment that meets the requirements of e-Navigation (Rivkin 2015). The documents posted on the IMO website directly address the problems of international merchant shipping.

Materials and Methods

If in 2014 it was assumed that the implementation of products created during the implementation of SIP would begin in 2020, then in fact this deadline is being pushed back to the 2025s–2030s. The drivers of the introduction of the concepts e-N into practice are the countries of the EU and the South Asian region.

In Russia, the subject of e-Navigation was developed within the framework of the Federal Target Programs "Development of civil marine technology for 2009–2016", "Maintenance, development and use of the GLONASS system for 2012–2020" and "Development of Shipbuilding for 2013–2030", as well as the strategies "Water Strategy of the Russian Federation until 2020" and "Transport Strategy of the Russian Federation for the period up to 2030", and within the framework of the implementation of the action plan (roadmap) "Marinet" (MariNet) of the National Technological Initiative, approved in 2017 by the Presidential Council for Economic Modernization and Innovative Development of Russia (Pinskiy 2016).

At the same time, the task is to integrate the Russian segment of the Maritime Connectivity Platform (MCP) being created with the foreign segments of the Baltic Sea and the Arctic zone of Europe, created within the framework of the EfficienSea2 project, and to extend the availability of the services provided to other marine areas, including the Russian Northern Sea Route, and to create a coastal station of an automatic data exchange system, which is an MCP telecommunications facility providing broadband communications via the «coast-satellite line».

Recently, the MSC and the Navigation, Communications, Search and Rescue, (NCSR) (Wingrove 2019a, b, c) have been discussing proposals aimed at unifying the development of IBS. NCSR which met in 2019, presented draft guidance documents on the harmonization of IBS equipment, electronic cartographic navigation and information systems, radars and other navigation equipment with an electronic interface. It is important that these documents contain proposals for the unification of the presentation of all navigation parameters displayed on screens and recorded in such equipment.

The components of the IBS providing navigation safety of navigation, the tasks of maneuvering, control of the passage of the route is constantly being improved (Kopanev and Muzychenko 2002). The state and trends in the development of information technologies, the experience in the development of elements of navigation equipment indicate the need to include in the IBS of a vessel (icebreaker) a software and hardware component that implements the intellectualization of the complex's functions. Moreover, these functions should be performed in automatic, semi-automatic and manual modes, depending on the emerging navigation and ice conditions at sea and on the decision of the ship's (icebreaker) captain.

The basic elements of such a ship's ACS subsystem are electronic positioning aids with a high level of integration—integrated navigation systems (INS), electronic navigation maps (ENM), analytical functions that minimize the influence of the human factor and allow reduce the burden on the ship's captain (Grachev et al. 2019).

In addition, within the framework of e-Navigation SIP, there is a need to implement a unified maritime data structure (EMCD, CMDS), using the universal model for presenting hydrographic data S-100 (further development of S-57), developed by the International Hydrographic Organization, IHO).

The next issue that needs to be resolved within the framework of e-Navigation SIP is the continuous provision of vessel masters (boat masters) with the necessary data. Therefore, the problems formulated in e-Navigation SIP are fair and even more relevant for ships sailing in the circumpolar regions, in the Arctic regions, including those following the Northern Sea Route (the NSR is a historically established national unified transport communication of Russia in the Arctic). For this purpose, it is planned to develop additional functions that must be implemented, taking into account the peculiarity of the Arctic navigation area (Burykin and Grachev 2021a):

- Ice reconnaissance by an unmanned aerial vehicle (helicopter) to clarify the previously obtained forecast along the route of movement of ships.
- Control of an unmanned aerial vehicle (UAV) in the process of ice reconnaissance, its takeoff and landing on the icebreaker.
- Receiving data on ice conditions from UAVs, processing them in real time, issuing recommendations to the captain of the icebreaker.
- Laying (adjusting) the route considering the real meteorological situation, including ice, to ensure the maximum level of navigation safety of ships.
- Construction of a wake column of a convoy of ships to ensure navigational safety during movement and joint maneuvering of ships in the convoy.

Results and Discussion

The computation model for automating the tasks of choosing and making decisions during joint maneuvering, laying, and controlling the route of a convoy of ships behind an icebreaker is based on production rules representing knowledge in the form of rules "IF condition, then action" (Burykin and Grachev 2021b). Knowledge-based systems or expert systems (ES) began to be called intelligent systems (Grachev and Zaitsev 2016), hence the name "intelligent ship (ship) autopilot". A distinctive feature of intelligent systems (including expert systems) based on production rules is the comparative simplicity of describing the dependencies and limitations of the subject area, as well as their sufficient understanding by the user (President of the Russian Federation 2019).

Ready-made expert systems are not universal in terms of use since they work only in one specific subject area. The experience of developing such systems shows that

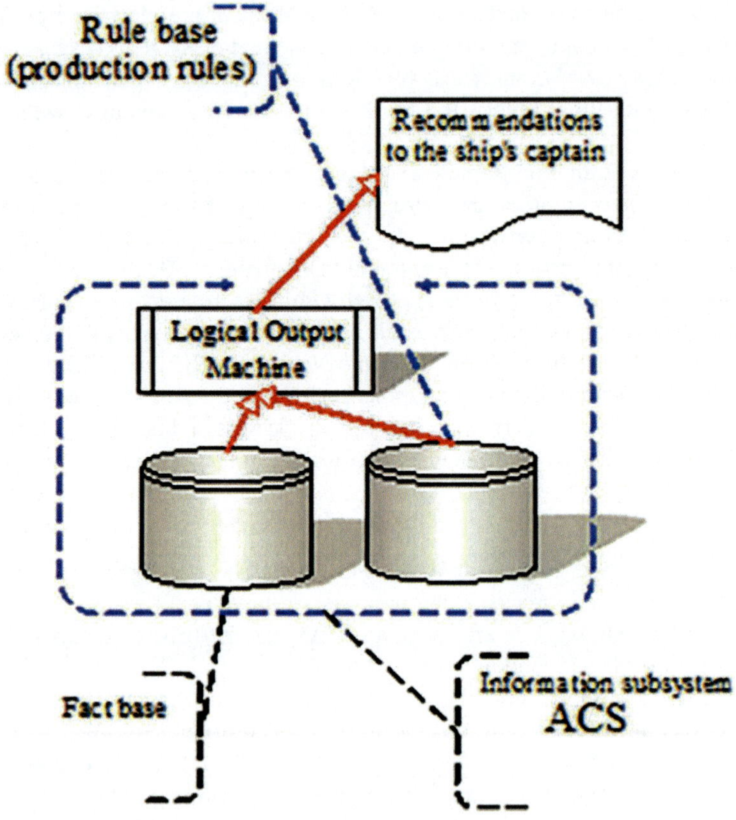

Fig. 1 Generalized structure of the hybrid expert system

in a machine of direct inference (interpretation of production rules), production rules can be implemented in the form of stored procedures database management system (DMS) of the system being developed. Below is a generalized structure of a hybrid expert system, components of an intelligent IBS that implements additional functions (Fig. 1) (Burykin and Grachev 2021a, b; Grachev and Zaitsev 2016; Ivanova and Avdoshin 2018).

Development experience shows that it is necessary to apply the modern concept of MVC (Model-View-Controller). This concept makes it possible to significantly simplify the ACS software. The concept of grid systems is used for the interaction of ships in a group.

In general, and briefly, the architecture of Grid systems is a progressive technology created for geographically (and globally) distributed computing resources to create a secure, coordinated computing space of organizational units for the joint use of computing and information resources of these organizational units in their common interests. The architecture of Grid systems is based on the concept of service-oriented architecture (SOA).

In our case, the organizational units are the vessels that are part of the caravan, and their common interests are expressed in its optimal construction and retention by each vessel of its place behind the icebreaker. Very briefly, the "intelligent" component of the intellectual IBS of the icebreaker calculates the point where the vessel should be and transmits the appropriate command to the IBS of this vessel, which automatically performs the corresponding maneuver using the computing resources of its IBS. The key element in the approach under consideration is the "Unified up-to-date information model of management facilities (caravan vessels and icebreakers)", which should reflect the current composition, condition, and position of vessels, as well as environmental factors (meteorological and ice) that have a significant impact on the navigation safety of the route.

It should be noted that the mentioned unified distributed automated control system by a vessel (a group of vessels, including an icebreaker) is built according to a client-server architecture. Here, from the point of view of IT technologies, the server is the intellectual IBS of the icebreaker, with the help of which ice exploration is carried out, the "clients" are intellectual IBS of the caravan vessels, which use the icebreaker's resources, namely information about the ice situation. The analysis of the obtained data on the ice situation for the issuance of recommendations on the route should be carried out based on "neural networks" technology.

Conclusion

In conclusion, as well as conclusions from the material considered, it should be noted that in the article, out of the three basic components of e-Navigation, namely: ship systems, onshore systems, and communication infrastructure, only the ship component was considered. Due to the limited volume, the article highlights only the most fundamental aspects of the implementation of E-NAVIGATION with AI elements during the escort of convoys of ships of the NSR. This approach was tested on a software model of the automated control system, which confirmed its effectiveness. Therefore, the proposed approach to the implementation of the principles and requirements of e-Navigation, as an integral part of the ship's ACS, is considered by the authors to be quite promising.

References

Bezborodov, G. I., & Ismagilov, M. I. (2017). From the strategy of e-Navigation to the concept of the i-Water area. *Morskoy Vestnik, 1*(61), 77–82.

Burykin, A. A., & Grachev, M. N. (2021a). Implementation of E-Navigation principles using artificial intelligence technologies. *International Conference "The Arctic. Sea cargo transportation: Problems and solutions – 2021"*

Burykin, A. A., & Grachev, M. N. (2021b). Implementation of the elements of artificial intelligence technology in promising automated control systems for a surface ship and automated control systems for the temporary formation of naval forces. *Military Thought, 4*, 50–57.

Grachev, M. N., Prirez, V. A., & Kasyanov, V. V. (2019). Implementation of the principles of E-Navigation in the construction of advanced automated control systems of the ship. *Marine Information and Control Systems, 2*(16), 26–31.

Grachev, M. N., & Zaitsev, A. N. (2016). Directions of development and perspective architecture of the integrated combat control system of a surface ship. *Marine Bulletin, 3*(59), 53–55.

IMO. (2021). *E-navigation*. https://www.imo.org/en/OurWork/Safety/Pages/eNavigation.aspx.

Ivanova, G. S., & Avdoshin, A. P. (2018). Production machine of automatic inference for the creation of expert systems. *Economy and quality of Communication Systems, 1*.

Kopanev, A. A., & Muzychenko, O. N. (2002). Prospects for the development of BIUS for ships of small and medium displacement on the basis of the "Requirement-M" system. *Marine Radioelectronics, 2*.

Mitropoulos, E. (2007). E-navigation: A global resource. *Seaways, March*, 7–9.

Patraiko, D. J. (2007a). e-Navigation concept. *European Journal of Navigation, 5*(5).

Patraiko, D. (2007b). Introducing the e-navigation revolution. *Seaways, March*, 5–9.

Pinskiy, A. S. (2016). E-Navigation and bezekipazhnoe navigation. *Transport of the Russian Federation, 4*, 50–51.

President of the Russian Federation. (2019). *National strategy for the development of artificial intelligence for the period up to 2030, approved by the Decree of the President of the Russian Federation October 10, 2019*. No 490.

Rivkin, B. S. (2015). E-navigation–ten years. *Gyroscopy and Navigation, 91*, 173–191.

Rivkin, B. S. (2019). E-navigation on the march. In *XIII All-Russian meeting on management problems* (pp. 1–7).

Wingrove, M. (2019a). Ship-port e-data exchange enters into force. *Marine Electronics & Communications*.

Wingrove, M. (2019b). New performance standards to improve navigation. *Marine Electronics & Communications*.

Wingrove, M. (2019c). IMO finalise guidance on polar navigation and communications systems. *Marine Electronics & Communications*.

Application of Digital Simulation Models of Railway Transport Polygons

E. V. Malovetskaya, R. S. Bolshakov, and A. B. Chernykh

Introduction

The reform of the Russian railway industry has been going on for about 20 years. All the ongoing transformations, both positive and negative, have a significant impact on the main performance indicators of the industry (Macheret and Ledney 2019; Sotnikov and Schoenfeld 2011). More often, industry analysts and experts focus on negative changes.

So once again the question of the admission of private locomotives to public infrastructure is raised. This is not the first time this proposal has been received from the Union of Railway Transport Operators (SOJT), lobbying for the liberalization of locomotive traction for freight rail transportation and motivating this proposal with "a shortage of locomotive traction and the presence of a barrier for timely delivery of cargo to the consumer" (Kozlov and Kolokolnikov 2018; Colin 2017; Malovetskaya 2019a).

The result of this work is to identify possible infrastructural and technological limitations in the implementation of the program for the admission of private locomotives to public infrastructure.

E. V. Malovetskaya (✉) · R. S. Bolshakov
Irkutsk State Transport University, Irkutsk, Russian Federation
e-mail: Malovetskaya_EV@irgups.ru

A. B. Chernykh
St. Petersburg University of Management Technologies and Economics, St. Petersburg, Russian Federation

A. Rumyantseva et al. (eds.), *Challenges and Solutions in the Digital Economy and Finance*, Springer Proceedings in Business and Economics,
https://doi.org/10.1007/978-3-031-14410-3_37

Materials and Methods

For a more adequate assessment of what is happening, it should be noted that in recent years, namely from 2012 to 2020, loading on the Russian Railways network increased by 6.9% to 1288 million tons (Fig. 1), but the dynamics was characterized by different trends.

Freight turnover on the Russian Railways network has been steadily growing from 2012 to 2020 (except for a decrease in 2013). In 2020, it amounted to 2598 billion t-km, which exceeds the level of 2010 by 29.7%.

It is worth noting that in 2018, a historical record of cargo turnover was set for the entire history of railway transport in Russia, while the increase in cargo turnover is due to an increase in the range of transportation.

The growth in the volume of traffic and, therefore, the increase in the unevenness of car traffic causes significant losses in railway transport, including those associated with a shortage of locomotive fleet for the implementation of timely delivery of goods to the consumer (Malovetskaya 2019b; Malovetskaya et al. 2020). The problem of unevenness at the junctions of railways and when loading wagons in the direction of the ports of the Far East is particularly relevant, the functioning of which is characterized by fluctuations in the volume of work within significant limits (Titov 2013). In connection with the above, it is necessary to evaluate the effectiveness of the admission of private locomotive traction to public infrastructure.

The analysis of the incoming car traffic and the possibility of introducing private locomotive traction to the public infrastructure is most appropriate to consider on the example of the Eastern polygon—the most advanced on the network of JSC "Russian Railways" and having direct access to the seaports of the Far East.

The presented work reflects the features of the description of the technology of operation of one of the junction points of the railways of the Eastern Polygon when implementing the introduction of private locomotive traction on public infrastructure in the AnyLogic simulation system. In particular, the technology of operation of the Petrovsky Plant station, which is a junction point of the Trans-Baikal Railway, in two modes of operation was considered:

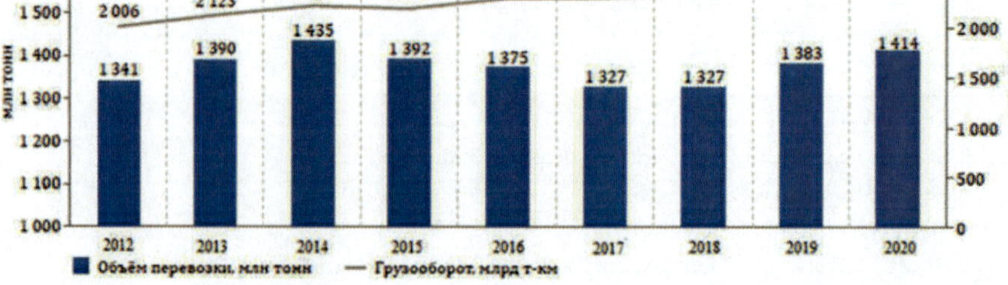

Fig. 1 The volume of cargo transportation and cargo turnover on the network of JSC "Russian Railways"

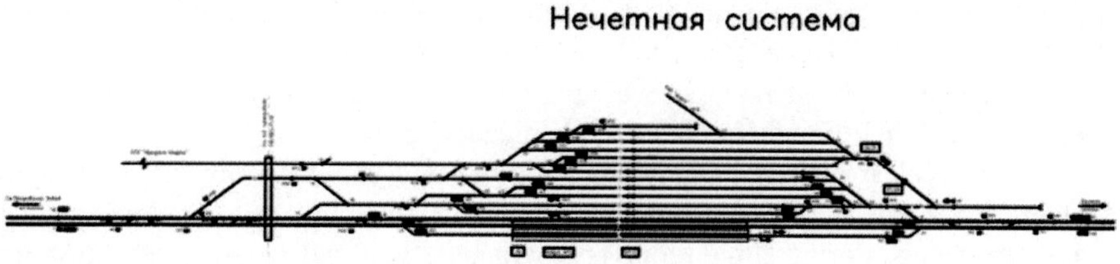

Fig. 2 Scheme of the odd system of the station P.z.

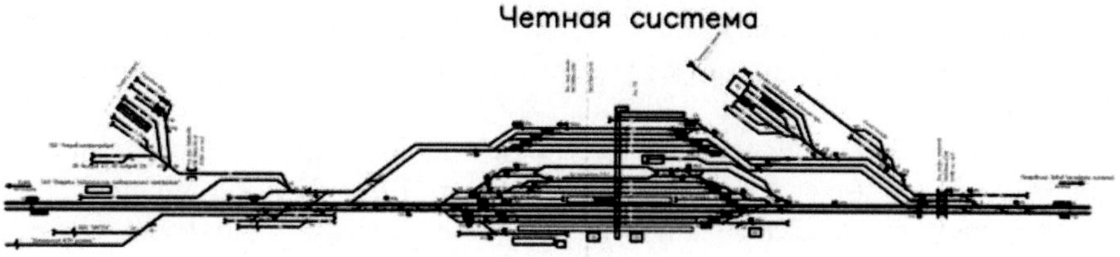

Fig. 3 Scheme of the even system of the station P.z.

- basic—when working with the locomotive fleet of the infrastructure owner.
- simulation—when implementing the technology of introducing private locomotive traction to public infrastructure (several private transporters).

Petrovsky Zavod station has a rather complicated track development, consisting of even and odd systems, which allows you to more fully implement the modeling process and get more adequate results.

The station carries out work on the processing of freight trains:

- on reception, departure and passing.
- arrived to disband.
- trains of their own formation.
- prefabricated trains.
- trains with changes in weight and length.

The diagrams of the even and odd station systems used in the simulation are shown in Figs. 2 and 3.

The AnyLogic simulation system, which has a railway library, was chosen as a research tool. The AnyLogic railway library allows you to efficiently model and visualize the operation of railway junctions and transport systems of any complexity and scale. Allows you to implement the processes of operational planning, management of the rolling stock fleet, maintenance, and repair of vehicles.

In railway models of the AnyLogic system, trains move in accordance with the logic specified by the process diagram, and at the same time, each railway car and train in the model is an agent with its own states and properties. This, along with the ability to interact with other libraries, allows for the accurate modeling of complex rail systems. The agent type can have parameters. Parameters are usually used to set the static characteristics of an agent. Here you can set different parameter values for different agents of the same type, which is required in cases where agents have the same behavior, but they differ in some characteristics.

There is a clear difference between variables and parameters in the AnyLogic system. A variable is typically used to set the changing state of the model, and its value can change during simulation. The parameter is usually used to set the static characteristics of the model. Parameter values can be changed while the model is running. If necessary, you can specify the actions that should be performed when the parameter value changes by writing an event handler for this event. Top-level agent parameters can act as input for models that are exported to AnyLogic Cloud. The «Run Configuration» editor allows you to select the input data to be exported with the model.

Like many other simulation tools, AnyLogic supports simple types of parameters: real (real), integer (integer), and boolean (boolean). But only AnyLogic provides a unique opportunity to create parameters of any Java classes. In the model, you can create a String class parameter to store string constants, or an ArrayList class parameter that will store a dynamic array of objects. You can create a parameter of type Object (Java base class) and assign this parameter an object of any Java class as its value. Since all elements of the AnyLogic model are defined as instances of Java classes, you can create parameters for the corresponding classes and use model elements as parameters.

Results

Based on the results of the analysis of the flow of wagons, according to the obtained statistical characteristics, as well as the appearance of the histograms, a hypothesis was put forward about the normal distribution law.

The probability density for the normal distribution law is expressed by the formula:

$$f(x) = \frac{1}{\sigma\sqrt{2\pi}} \cdot e^{\frac{-(x-M(X))^2}{2\sigma^2}} \quad (1)$$

where σ is the root mean square deviation of a random variable from its mathematical expectation; x is the current value of the random variable; M[X]—mathematical expectation.

Since the random variable is the power of individual appointments—N, then the distribution function can be expressed

$$F(N) = \frac{1}{\sigma_1 \cdot \sqrt{2\pi}} \int_{-\infty}^{N} e^{\frac{-(N-M_1)^2}{2\cdot\sigma_1^2}} dN \quad (2)$$

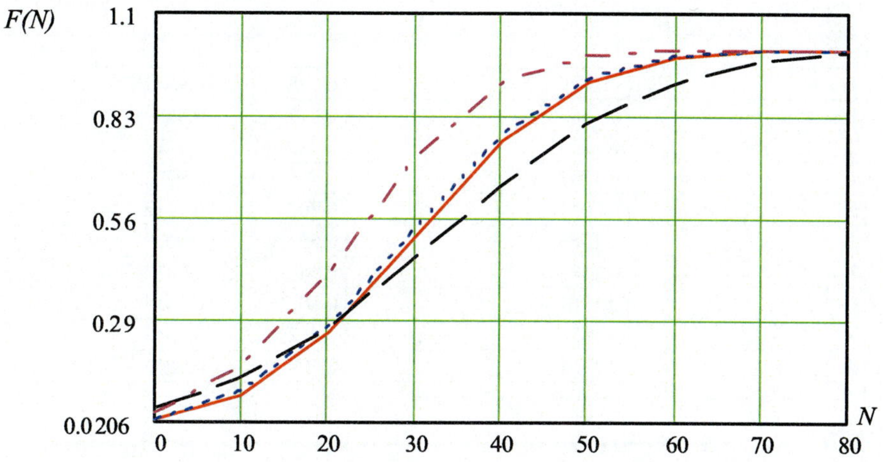

Fig. 4 Graphs of destination power distribution functions for stations 1, 2, 3, 4

To assess the degree of closeness of the theoretical distribution to the statistical distribution, the goodness-of-fit criteria were tested. The verification was carried out according to the criterion of Kolmogorov consent (Fig. 4). The probability that the largest deviation of Fn(N) from F(N) will exceed a given number can be calculated by the formula

$$P(\lambda) = P\left(max\,(F_n(x)) \geq \frac{\lambda}{\sqrt{n}} \right) = 1 - \sum_{k=-\infty}^{\infty} (-1)^k \cdot e^{-2k^2\lambda^2} \quad (3)$$

Using the scheme of applying the Kolmogorov criterion, comparing the known values of the accumulated particulars Fn(N) with the calculated values of the integral function of the theoretical distribution F(N), the largest values D of the module of the difference between them were found, from the values of which the corresponding values of the probability P(λ). The main elements of the calculations are summarized in Table 1.

According to the results of the analysis, it was found that the fluctuation in the magnitude of the power of appointments is subject to the normal distribution law (Figs. 5 and 6).

In addition, based on the structure of the incoming flow of wagons, it was revealed that the non-routed laden wagon flow, considering short-range traffic, is more than 50% of the total loading.

To exclude negative values of the time interval, which is possible when using the normal distribution law of a random variable, a transition was made from the normal distribution law of a random variable to a truncated normal distribution law.

The distribution density for the truncated normal distribution is given by

Table 1 Basic elements of calculations

Destination station	Difference modulus D	λ	Probability value P(λ)
1	0.1149	0.398	0.9972
2	0.1557	0.517	0.9639
3	0.2106	0.729	0.7112
4	0.1237	0.426	0.9972
5	0.2027	0.702	0.7112
6	0.1553	0.501	0.9639
7	0.1121	0.388	0.9972
8	0.1461	0.506	0.9639
9	0.1854	0.642	0.8643
10	0.1181	0.409	0.9972

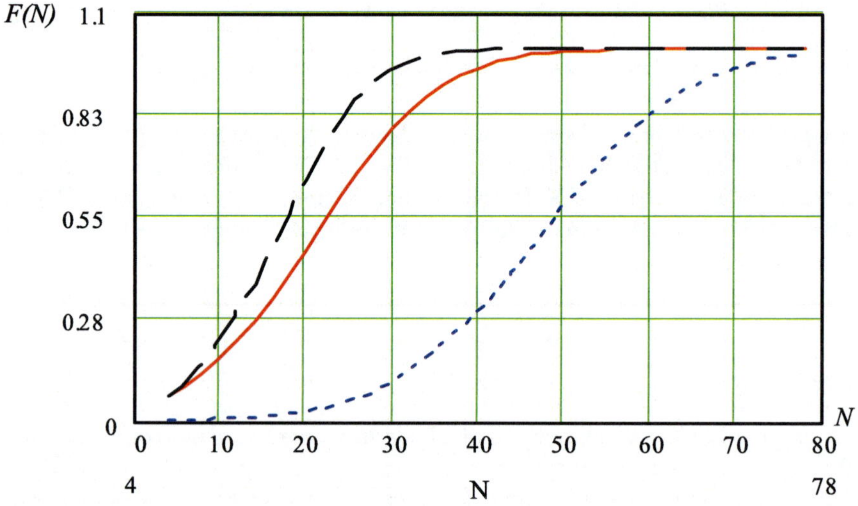

Fig. 5 Plots of destination power distribution functions for stations 5, 6, 7

$$f(x) = \frac{c}{\sigma\sqrt{2\pi}} \cdot e^{\frac{(t-M_t)^2}{2\sigma^2}} \tag{4}$$

where Mt is the mathematical expectation; σ is the root mean square deviation of a random variable from its mathematical expectation; c = 1/F0(Mt/σ) for Mt > 2σ the coefficient c is very close to unity.

Subsequently, the unevenness of the arrival of trains in the station operation model was set by a truncated normal law of the distribution of the time interval between the moments of arrival of trains.

To characterize the flows of cars arriving at the station, it is necessary to establish patterns of power fluctuations for various purposes and the time of their arrival. In order to establish these dependencies, the relevant statistical material was analyzed

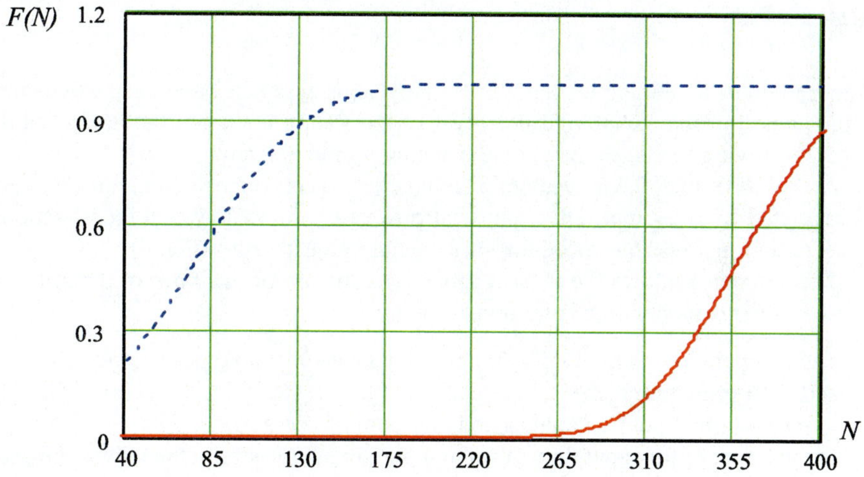

Fig. 6 Plots of destination power distribution functions for stations 9, 10

Table 2 Parameters of intraday fluctuations in the power of a non-routed train flow

Destination station	Mathematical expectation M(N)	Dispersion D(N)	The average quadratic deviation σ(N)	The coefficient of variation ν(N)
1	29.834	213.12	14.598	0.489
2	28.932	208.88	14.45	0.499
3	32.55	400.21	20.05	0.615
4	22.7	162.96	12.77	0.562
5	14.96	147.87	12.16	0.491
6	356.23	1509.39	38.85	0.109
7	77.33	1987.34	44.58	0.576
8	22.1	129.04	11.36	0.538
9	47.23	190.58	13.8	0.29
10	16.65	68.19	8.26	0.495

and processed. Fluctuations in the power of individual destinations were considered, as well as intervals for the arrival of trains at the station.

To identify the distribution laws for the magnitude of the power of assignments, an analysis was made of the decomposition of trains according to the assignments of the train formation plan (TFP). To establish the main parameters of the intraday fluctuation, the arrival of trains at several technical stations was analyzed. A data sample was taken from 40,000 full-scale sheets (Table 2).

Discussion

The method of simulation modeling in the AnyLogic system is based on the dynamic reproduction of the technological process of the station operation by manipulating objects (moving trains and cars) on the station mimic diagram.

At the same time, the movement of simulation objects on the mnemonic diagram is regulated by rules that reflect physical processes in a real system (acceleration, deceleration, permissible speed for infrastructure objects, etc.) (Fig. 7).

This approach allows the most reliable reproduction of the work of the railway station in various operating conditions, namely:

- station operation technologies (train maintenance technologies, traffic schedule and train formation plan).
- infrastructure at public facilities and non-public access roads.
- the structure of the input flow of wagons and trains (the size of the movement and the technology of working with these flows).

The main components of the basic data used in the AnyLogic system models include the infrastructure complex, the complex of technological processes, as well as data on resource provision.

The duration of all technological operations was set in the model in accordance with technological work schedules, random deviations in the duration of technological operations, as well as the intensity of train flows, were set by the distribution laws of random variables. The model consists of "flow diagrams" imitating the technological chains of train and locomotive processing. The main objects (cars and locomotives) with which operations are performed in the digital model of the station are formalized in the form of elementary groups of cars and elementary groups of locomotives.

An elementary group of cars is a group of cars with which a unique sequence of technological operations is performed at the station, and which must be allocated to a separate technological group within the considered section. In fact, an elementary group of wagons is a set of features by which the program determines whether a wagon belongs to a technological group that has a similar processing technology. That is, the program tries to identify all the arriving cars within the Polygon by several signs—what each car is like in the train and compare it with the appropriate technological chain—the sequence of operations that must be performed with it.

To determine the average idle time of trains waiting for a locomotive, the following interrelated model objects are used: "TimeMeasureStart"—remembers the moment when the train arrives at a certain point in the flow diagram; "TimeMeasureEnd"—calculates for each request received in it the difference between the current moment of time and the moment stored in the "TimeMeasureStart" object. Based on the results of collecting statistical data on the time spent by each mobile unit in the model, histograms are built, and statistical characteristics are calculated (Fig. 8).

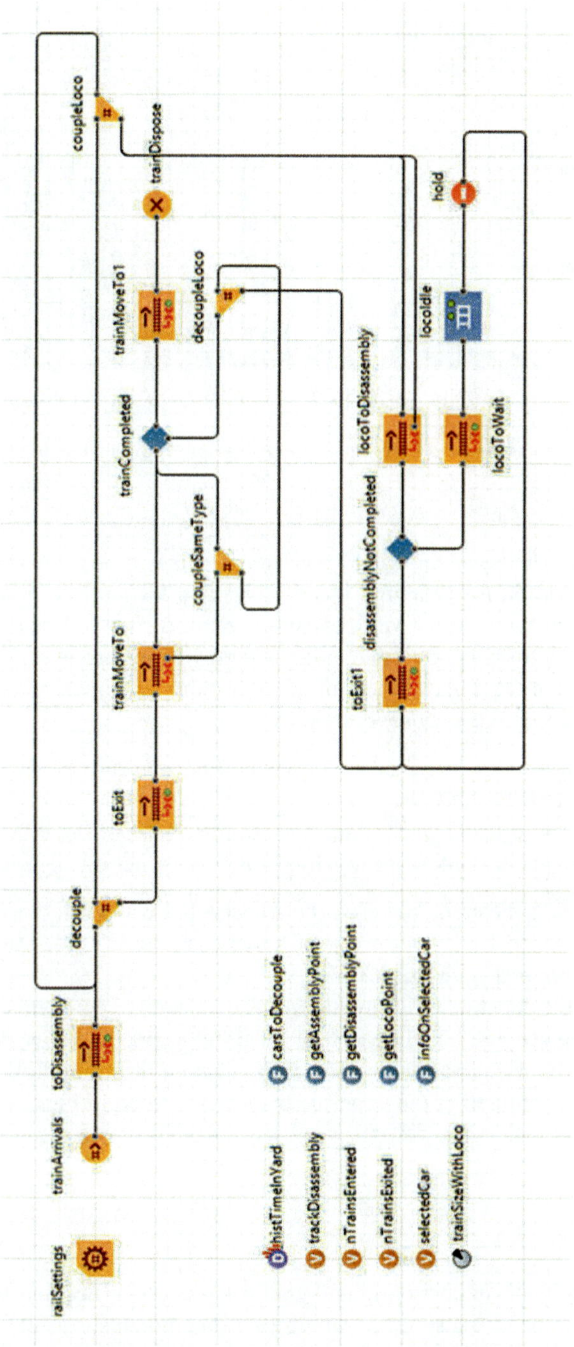

Fig. 7 Diagrams of the simulation model of the operation of the railway junction station in the AnyLogic system

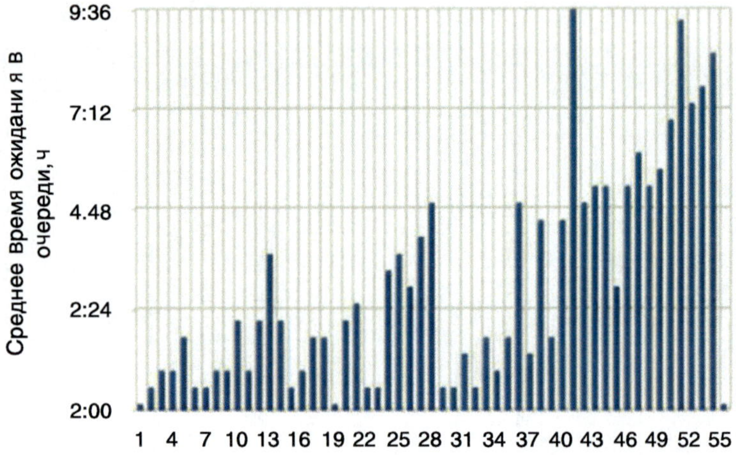

Fig. 8 The results of measuring the duration of the waiting time of the locomotive in the simulation model

The presented scheme of the simulation model makes it possible to evaluate the operation of the station for changing locomotives and locomotive crews in various operating modes of the locomotive fleet (by the main carrier and when introducing private locomotive traction) by conducting a series of experiments with different values of the intensity and unevenness of the train flow. The operation of the station in the model was considered in several options:

- when operating as a locomotive fleet of one carrier (owner of the infrastructure of Russian Railways OJSC).
- when two private carriers work together with the owner of the infrastructure.
- when three private carriers work together with the owner of the infrastructure.

At the same time, a different share of the transported car traffic was considered, transferred for service by locomotives of private carriers (25%, 50% and 75%, respectively, of the total train traffic) (Fig. 9).

According to the constructed simulation model, several experiments were carried out to estimate the average idle time of trains waiting for locomotives, depending on the intensity of arrival and various values of the uneven incoming flow of cars. The results of the data obtained in the experiments are presented in Figs. 8 and 9.

Conclusion

Based on the results of the modeling and analysis process, it becomes obvious that the optimal downtime of transit trains waiting for locomotives is achieved under the conditions of a monopoly of one carrier. With an increase in the average idle time of

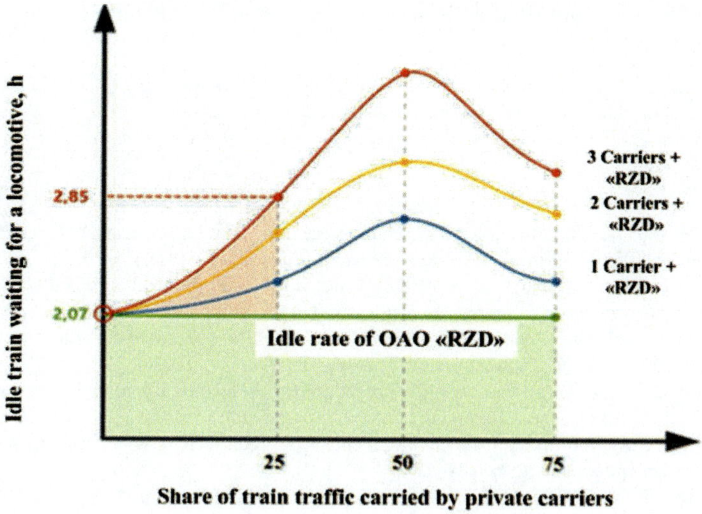

Fig. 9 Nomogram of changes in the average idle time of trains waiting for a locomotive in various operating modes

trains at the station, it is possible to ensure an average level of capacity through the construction of new infrastructure.

The introduction of the institution of private carriers has a high probability of the implementation of technological, social risks and consequences associated with ensuring traffic safety.

The involvement of private carriers in the operational work of the railway network will lead to operational losses and an increase in the need for the operating fleet of locomotives to develop the base volume of work.

The experience of the past stages of reform in the industry has shown that the withdrawal of the wagon fleet from the control of the network carrier has led to an increase in the number of freight cars and empty car traffic, which is not synchronized with the capabilities of the railway infrastructure.

World practice confirms that the competition of carriers on a single infrastructure is possible only in conditions of a surplus of capacity, while in Russia there is a significant shortage of them.

The admission of private traction rolling stock to the infrastructure of public railway transport in the Russian Federation can be considered only in two aspects:

- admission of private locomotives provided to the carrier by Russian Railways on lease or on other civil law grounds for the performance of cargo transportation services.
- formation of private carriers providing cargo transportation services to final cargo owners using the public infrastructure.

The first form of admission of private locomotives is already partially used on Russian railways: firstly, by leasing locomotives from private owners in some regions of Russia from Russian Railways; and secondly, when organizing their own train formations (SPF).

References

Colin, A. V. (2017). Liberalization of locomotive traction: "For" and "against", Transport of the Russian Federation. *Journal of Science, Practice, Economics, 5*(72).

Kozlov, P. A., & Kolokolnikov, V. S. (2018). Calculation and optimization of railway polygons. *Bulletin of Rostov State University of Railways, 3*, 113–120.

Macheret, D. A., & Ledney, A. Y. (2019). Traffic volumes the key factor in the efficiency of use and development of transport infrastructure. *Railway Economy, 4*, 28–38.

Malovetskaya, E. V. (2019a). Updating the procedure for calculating the plan for the formation of trains and clarifying the list of indicators of operational work on the basis of simulation modeling taking into account the development of polygon technologies. *Modern Knowledge-Intensive Technologies, 1*(53), 184–188.

Malovetskaya, E. V. (2019b). The relevance of the use of simulation modeling in calculating the train formation plan considering the development of polygon technologies. *Journal "T-Comm – Telecommunications and Transport", 4*, 48–53. https://doi.org/10.24411/2072-8735-2018-10261

Malovetskaya, E. V., Bolshakov, R. S., Dimov, A. V., & Byshlyago, A. A. (2020, February). Planning of qualitative indexes of railroad operational work in polygon technologies. In *IOP Conference Series: Materials Science and Engineering* (Vol. 760, No. 1, p. 012041). IOP Publishing. https://doi.org/10.1088/1757-899X/760/1/012041

Sotnikov, E. A., & Schoenfeld, K. P. (2011). Uneven freight transportation in modern conditions and its impact on the required throughput of sections. *Bulletin VNIIZHT, 5*, 3–9.

Titov, G. B. (2013). Methodology for assessing the probability of timely delivery of cargo by rail. *Izvestia of St. Petersburg University of Railways, 2*(35).

Development of a GIS Model "World Heritage" to Assess the Tourist and Recreational Potential of the Territory

Alexander S. Baranov, Sofiia A. Baranova, and Inga G. Filippova

Introduction

The twenty-first century opens up new technological horizons for the development of mankind. The era of informatization and digitalization allows processing huge amounts of data that were previously unthinkable and unavailable. One of the vectors of this process is the emergence of special software—geoinformation systems, which include components for collecting, transmitting, storing, processing and issuing information about the territory.

Tourist and recreational design are carried out on the basis of accounting for tourist and recreational resources. The latter have a decisive influence on the territorial organization of tourist and recreational activities. On the principle of their capacity and combinations, tourist and recreational areas and centers are formed. They also determine their specialization and economic efficiency (Amelkina 2020).

The use of tourist and recreational resources refers to expensive, but highly effective in socio-economic terms activities. Quality assurance of this activity requires careful inventory and assessment of the state of natural and natural-anthropogenic geosystems, bodies and natural phenomena, artifacts from the design (Grishin 2014, p. 24).

An important aspect of tourist and recreational design is taking into account the spatial organization of territorial systems (Yavorska et al. 2018). A correct estimation of the potential opens the way for optimizing activities through the use of

A. S. Baranov (✉) · S. A. Baranova
Herzen State Pedagogical University, St. Petersburg, Russian Federation

I. G. Filippova
St. Petersburg University of Management Technologies and Economics, St. Petersburg, Russian Federation

A. Rumyantseva et al. (eds.), *Challenges and Solutions in the Digital Economy and Finance*, Springer Proceedings in Business and Economics,
https://doi.org/10.1007/978-3-031-14410-3_38

territorial combinations of individual and group properties of objects of tourist and recreational space.

The high tourist and recreational potential in itself do not guarantee the region the appropriate level of the tourism industry. For its development, it is necessary to rationally and effectively organize tourist and recreational activities based on the use of promising technologies and available material and information capabilities (Baranov and Fedorova 2012, p. 27).

Such an opportunity is provided by geoinformation technologies. In the case of tourist and recreational design, they allow for spatial modeling and cartographic inventory of resources, optimize the information support of tourism entities and qualitatively organize specialized activities in the region.

The development of tourism is unthinkable without the use of certain types of resources. Traditionally, it is invariable to allocate natural components and their cultural analogues among tourist resources. Under natural resources, it is necessary to understand the active processes of the biosphere, and under cultural ones—the technosphere (Gumilev 2008, p. 340). However, we have to admit, by relying on the works of V.I. Vernadsky, L.N. Gumilev, etc., that biospheric processes are in more harmonious relations, unlike technospheric ones, because the latter exist only as long as a Person spends his own energy on their "maintenance"—muscular strength, finances, etc.

However, a closer study of tourist resources suggests that all of them, without exception, belong to the technosphere. After all, practically no natural (ecological) landscape unchanged by man has been preserved nowadays, but there are only anthropogenic ones.

Consequently, it can be seen that the technosphere includes objects of tourist exposition. Then, to maintain them, it is necessary to increase a certain amount of energy (labor, money and knowledge). Otherwise, the materials removed from the biosphere will eventually return to the bosom of nature. A typical example of this kind of phenomena can be considered the list of "7 wonders of the ancient world", of which only The Great Pyramid of Giza has survived to the presence of this day. From the rest, at its best—ruins (the Lighthouse of Alexandria, the temple of Artemis of Ephesus), or—memories.

Human handiwork from both bone and living matter (tools, works of art, pets, cultivated plants) ... can either be preserved, or, if not preserved, destroyed... The pyramids stand for a long time, the Eiffel Tower will not stand for so long. But neither one nor the other is eternal. This is the fundamental difference between the biosphere and the technosphere (Gumilev 2008, p. 48).

It is precisely with this feature—the extreme degree of vulnerability of objects of tourist attractiveness, that the problem of estimating the tourist and recreational potential of the territory is connected.

It is precisely with this feature—the extreme degree of vulnerability of objects of tourist attractiveness, that the problem of assessing the tourist and recreational potential of the territory is connected.

To improve the quality and efficiency of assessing the significance of the territory for tourist activity, the authors have developed a model of a geoinformation system

(Baranov 2005, p. 96). The phenomenon of the UNESCO World Heritage Site was chosen as an object of research (Baranov 2017, p. 23). On the one hand, these objects have the highest level of attractiveness, on the other they exist everywhere (Baranov et al. 2020, p. 12).

Materials and Methods

The GIS model "World Heritage" consists of the following blocks: cartographic, informational, statistical, search and effective, Fig. 1.

A cartographic block is used to study the features of the geographical location of World Heritage Sites. Here you can also find information about the country which they are located in.

To study a brief description of objects, an information block is needed. Here you can also find the criteria according to which they were included in the UNESCO World Heritage List.

To obtain generalized digital information about the World Heritage System, a statistic block can be used. It can be used to conduct statistical analysis, compare regions and countries, monitor the dynamics of changes in the World Heritage List, etc.

Much attention in the developed model is paid to search activity. The block of the same name allows you to quickly navigate to any object that satisfies the user's request.

However, the key block is the result block. Here you can develop scenarios for reporting. Since GIS allows you to specify many additional criteria and characteristics for each object. These criteria are systematized and presented in a specially created classifier tool.

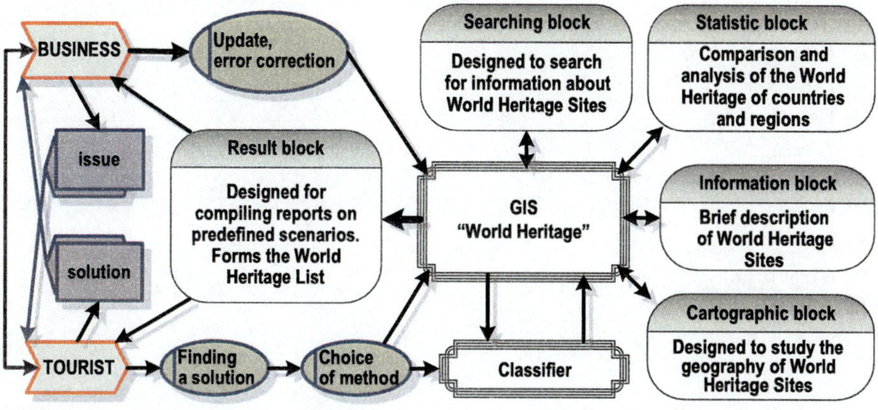

Fig. 1 The GIS model "World Heritage"

Fig. 2 Classifier of UNESCO World Heritage sites

Classifier is a structured list of characteristics that allows you to find each object its own place and a specific designation (Baranov 2005, p. 35). Typological features identified using the classifier are group features (common to several World Heritage Sites), and unique features (specific to a particular object, for example, underwater heritage (Asadov et al. 2021)) can be identified by comparing different objects, Fig. 2.

In the methodology proposed by the authors for assessing the tourist and recreational potential of the territory, an integrated approach is proposed based on a systematic analysis of the imaginality (Asadov et al. 2017, p. 62) of the main geographical spheres of the Earth (atmosphere, hydrosphere, lithosphere and biosphere) on the one hand, and spheres of society (theosphere, politosphere, economosphere and sociosphere) on the other (Baranov and Baranova 2021, p. 155).

*the evaluation result is determined in the range from 0 to 10 according to the following criteria:

1. Missing
2. Extremely low
3. Low
4. Below average
5. Average
6. Above average
7. Good
8. High
9. Very high
10. Absolute

Results and Discussion

Visualization of the results of valuating the tourist and recreational potential of the national park "Curonian Spit" (Baranov et al. 2020; Asadov et al. 2019) is shown in Fig. 3.

The rating of 9 points for the atmosphere is explained by the fact that there are no large industrial enterprises in the area of the national park, and therefore the air is quite clean and saturated with oxygen and iodine. The smell of the sea and coniferous forest is felt at every step. Minimized amount of exhaust fumes. The disadvantages include the seasonal factor—strong winter winds do not contribute to the formation of favorable conditions for recreation.

The score of 7 points for the hydrosphere is explained by the fact that the water in the sea is clean, well iodized. However, garbage is often thrown out on the seashore, and the bay "blooms" in the summer. The large length of sandy beaches creates many cozy places to relax. But the extremely poor variety of inland waters does not allow us to put a higher estimate.

The rating of 9 points for the lithosphere is explained by the fact that there are unique landscapes on the territory of the national park. The modern relief was formed in the Holocene epoch, after the melting of the last glacier (Rosentau et al. 2021). The visiting card of the park is the dunes. However, the variety of landscapes is low.

The score of 9 points for the biosphere is explained by the fact that the park presents a huge variety of species of animals, plants, fish, insects and, especially, birds. Unique woodlands with unique natural monuments, such as the "Dancing

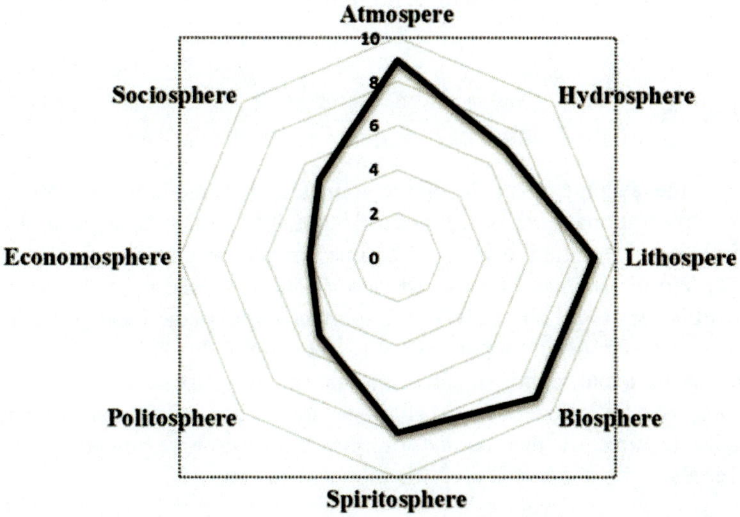

Fig. 3 Visualization of the results of valuating of the tourist and recreational potential of the national Park "Curonian Spit"

Forest" attract tourists. An ornithological tourism is in great interest for fans because of "Fringilla" biostation, which has been conducting research on migratory birds since the nineteenth century (Baranov et al. 2020, p. 145). The highest tourist attraction occurs in spring and autumn—the seasons of mass migration of birds (Tsvey and Sokolov 2014). However, there is also a core area of some parasites, such as ticks.

The spiritual sphere of society is estimated at 8 points, as on the one hand, there is a sense of peace, solitude with nature, where everyone can find a secluded place to enjoy the silence of nature, on the other hand, there is an insufficient amount of development of organized tourism in favor of amateur tourism. The latter circumstance somewhat reduces the quality of perception.

The political sphere of the society (Baranov and Asadov 2015, p. 169) is estimated at 5 points. This is due to the fact that there are still a number of complex administrative and managerial problems on the territory of the park. Some of them arise from the border status of the territories (Korneevets et al. 2018). Other problems arise from irrational nature management caused by amateur tourism.

The economic sphere of the society is estimated at 4 points due to the fact that entry price is getting more and more expensive every year, not every resident of Kaliningrad can afford to rest on this territory. The cost of accommodation and food, entertainment of tourists is several times higher than the cost in the city, which also affects the tourist's wallet. Souvenir products are mainly represented by amber products, which does not reflect the specifics of the national park. In general, the tourist infrastructure of the park is not well developed.

The social sphere is estimated at 5 points due to the fact that the park's resources are not available to all categories of Russian citizens, especially for pensioners and people with disabilities. The level and quality of accommodation, food and entertainment of tourists leaves much to be desired.

Conclusions

All in all, the average valuable of the tourist and recreational potential of the Curonian Spit National Park is 7.0 (out of 10 possible). The ratio of natural spheres was 8.5 against 5.5 cultural ones. At the same time, this object is represented in the UNESCO World Heritage List by criterion v—cultural landscape. In general, the assessment is very high, but it gives an important signal to the local authorities that there are very significant opportunities to increase it by solving the identified problems in the social, political, and economic spheres of society.

The proposed GIS model "World Heritage" makes it possible to analyze not only the spatial features of the studied objects, but also to estimate their tourist attractiveness.

References

Amelkina, D. V. (2020). Modern theoretical approaches to understanding and topical issues of research on the territorial organization of recreation and tourism. *Bulletin of Tver State University. Series: Geography and Geoecology, 2*(30), 43–59.

Asadov, B., Baranov, A., Baranova, S., Bobrova, A., & Philippova, I. (2021). Problems of underwater cultural heritage research in the context of the Baltic Sea drainage basin ecology. In *E3S Web of Conferences* (Vol. 291, p. 02020). EDP Sciences.

Asadov, B. R. O., Baranov, A. S., Bogdanov, E. I., Pogodina, V. L., & Filippova, I. G. (2017). Some aspects of the formation of the imaginary space of St. Petersburg as a UNESCO World Heritage Site in the youth environment. In *The Tenth International Congress on Social Sciences and Humanities* (pp. 65–73).

Asadov, B., Baranov, A., & Philippova, I. (2019, November). World natural heritage of the Baltic Drainage Sea Basin: Problems and solution. In *IOP Conference Series: Earth and Environmental Science* (Vol. 390, No. 1, p. 012025). IOP Publishing.

Baranov, A. S. (2005). *Methods of studying the World Heritage using computer technology in the geography course of the 10th grade* (Doctoral dissertation). Herzen State Pedagogical University, St. Petersburg, Russia.

Baranov, A. S. (2017). *The phenomenon of World Heritage in the systems of education, science and tourism: A retrospective analysis. Natural and cultural heritage: Interdisciplinary research, conservation and development: A collective monograph* (pp. 22–29). Publishing House of A.I. Herzen RSPU.

Baranov, A. S., & Asadov, B. R. (2015). On the influence of the youth environment on the process of tourist development of the political sphere of society. *Bulletin of Perm University. Political Science, 2*, 168–180.

Baranov, A. S., & Baranova, S. A. (2021). *Visualization of the results of valuating the tourist and recreational potential of the territory drawing on example of the World Heritage Site "Curonian spit"* (pp. 154–159).

Baranov, A. S., Bogdanov, E. I., Vereshchagina, N. O., & Filippova, I. G. (2020). *Monuments of the World Natural and cultural heritage of Russia in the tourism system: Textbook.* Directmedia Publishing LLC.

Baranov, A. S., & Fedorova, U. A. (2012). The study of the UNESCO World Heritage System as the basis for the development of cultural and educational tourism. *Russian Scientific Journal Bulletin of the National Academy of Tourism, 2*(22), 25–28.

Grishin, S. Y. (2014). *Tourist and recreational design.* SPbGEU.

Gumilev, L. N. (2008). *Ethnogenesis and the biosphere of the Earth.* Iris-Press.

Korneevets, V. S., Zaitseva, N. A., Dragileva, I. I., Dmitrieva, N. V., Silaeva, A. A., Boboshko, V. I., & Boboshko, N. M. (2018). Development problems and prospects of environmental tourism in the territory of an UNESCO World Heritage object "curonian spit" in cross-border cooperation projects. *Ekoloji, 27*(106), 1667–1673.

Rosentau, A., Klemann, V., Bennike, O., Steffen, H., Wehr, J., Latinović, M., et al. (2021). A Holocene relative sea-level database for the Baltic Sea. *Quaternary Science Reviews, 266*, 107071.

Tsvey, A. L., & Sokolov, L. V. (2014, March). Impact of climate change on the physiological condition of passerine birds during migration. In *Doklady Biological Sciences* (Vol. 455, No. 1, p. 99). Springer Nature BV.

Yavorska, V. V., Hevko, I. V., Sych, V. A., & Kolomiyets, K. V. (2018). Periodization of the studies of territorial organization of recreation and tourism. *Journal of Geology, Geography and Geoecology, 27*(3), 520–528.

World Labor Market: The Influence on State Economic Security

Natalia Nikolaevna Reshetnikova, Elena Sergeevna Zakharchenko, and Zhanna Viktorovna Gornostaeva

Introduction

In world science, economic security at the macro level is interpreted as financial stability. Modern foreign research in the field of ensuring and assessing financial stability has two directions:

1. analysis of the role of the central bank in ensuring financial stability and the onset of global financial crises. A description of this direction is given in the works of M. Bordo (2018), M. Bordo and C.M. Meissner (2016), M. Woodford (2016) and others.
2. development of indicators of financial shocks (stresses) designed to assess the instability of the state's financial system. A description of this direction is given in the works of S. Giglio et al. (2016), L. Sun & Y. Huang (2016), J.F. Sandahl et al. (2015), J. Huotari (2015) and others.

The second line of research, according to the authors, is of the greatest interest and is considered in the context of this application.

Financial stability indicators must be used to assess financial security (in particular, in the works of A.C. Bertay et al. (2016) and C. Kim (2016).

N. N. Reshetnikova (✉)
Rostov State University of Economics (RINH), Rostov-on-Don, Russian Federation

Don State Technical University, Rostov-on-Don, Russian Federation

E. S. Zakharchenko
Don State Technical University, Rostov-on-Don, Russian Federation

Rostov Branch, Russian Customs Academy, Rostov-on-Don, Russian Federation

Z. V. Gornostaeva
Kadyrov Chechen State University, Grozny, Russian Federation

A. Rumyantseva et al. (eds.), *Challenges and Solutions in the Digital Economy and Finance*, Springer Proceedings in Business and Economics,
https://doi.org/10.1007/978-3-031-14410-3_39

Separately, it should be noted the "financial secrecy index"—an indicator developed and calculated by the international organization Financial Secrecy Index (2021a, b). Based on this index, countries are ranked according to their level of security and the scale of their offshore financial activity.

To achieve the goal mentioned above (studying the mutual influence of the labor market and the economic security of the state), it seems necessary to solve the following tasks:

1. The Global Financial Safety Net, GFSN description.
2. International organizations' place determination.
3. The economic security indicators identification.
4. The world labor market tendencies description.

The working hypothesis of the study is that the state of the world labor market is one of the key indicators of the economic security of any state.

Materials and Methods

The Global Financial Safety Net (GFSN) was created after the global financial crisis of 2008–2009, to effectively prevent and resolve crises.

The researches of the authors Scheubel and Stracca (2016), Denbee et al. (2016), Weder di Mauro and Zettelmeyer (2017) describe the issues of the functioning of the Global Financial Safety Net, role, fragmentation, management in a multipolar system in the global financial architecture.

In particular, the work of the author, explores the prospects for the development of the Global Financial Safety Net, its structure, including the International Monetary Fund, regional financial agreements (RFAs), as well as bilateral or multilateral central bank swap agreements and relationship between them. In addition, the author emphasizes the heterogeneity of the GFSN, and the lack of much reason to expect significant progress in improving cooperation between the RFAs and the IMF, unless the latter's governance structure undergoes significant changes. The author concludes that the GFSN will become even more fragmented with the further development of the European Stability Mechanism and the advent of the BRICS Contingent Reserve Agreement. This requires significant reform of the IMF's governance.

These researches and their works Zakharchenko et al. (2021), Reshetnikova (2021), Reshetnikova et al. (2019, 2021, 2022), Gornostaeva et al. (2020), Gornostaeva (2021) are among the scientists of the Rostov research school who have accumulated significant scientific experience in the study of the global competitiveness of the economy and ensuring national economic interests in the system of world economic relations.

The activities of the International Monetary Fund (IMF) are mainly focusing on relatively short-term macroeconomic crises to promote international cooperation development in the monetary and financial sphere. The activities of the IMF are given a special place in the world and domestic literature.

It seems necessary, when working with the economic security concept, to use synthesis as a method of scientific knowledge and refer to the problem description. The economic security category definition variety, the state national security structure description is made with the help of the method of graphic images.

A systematic approach to possible division of workers into categories was reflected in the study.

Results

The economic security concept, its content and composition are widely represented in Russian and foreign researchers scientific works (Fig. 1). This figure was presented in work of authors Abdullaeva and Zakharchenko (2018).

However, regardless of the author's approach to this category definition, the main concepts are the possibility of realizing the interests of the state and leveling the danger, which, it seems, is due to the aggravation of contradictions between states, international organizations and network institutions.

Thereby, the state economic security is the most important characteristic of the state economic system, which determines its ability to maintain conditions for the normal life of the population, to consistently provide resources for economic entities. The state economic security has a complex structure, being at the same time an

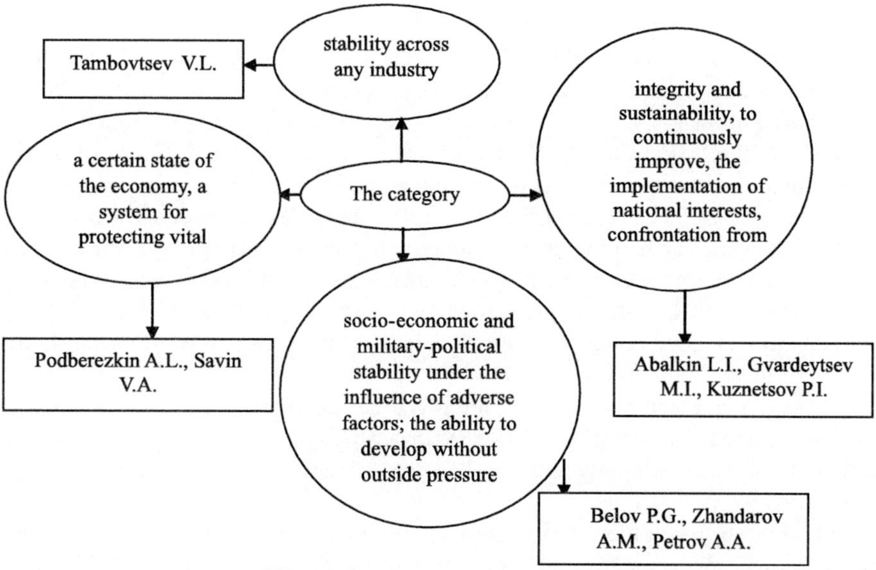

Fig. 1 The economic security category definition variety (Abdullaeva and Zakharchenko 2018)

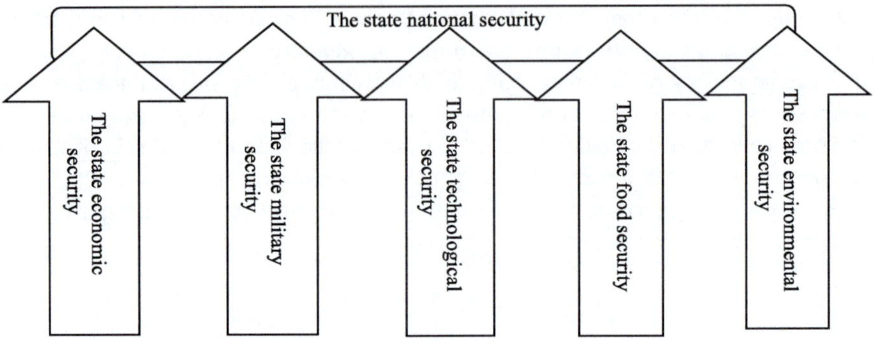

Fig. 2 The state national security structure (Abdullaeva and Zakharchenko 2018)

integral part of national security (see Fig. 2). This figure was presented in work of authors Abdullaeva and Zakharchenko (2018).

Thus, the financial system condition of the state is an indicator of its economic security, and the concept of "financial security".

The economy digitalization accelerated during the coronavirus pandemic (due to the most types of work removal) has led to new trends in the labor market. According to The Future of Jobs Report (2020) the job losses trend due to the massive spread of new technologies will be replaced by the economic crisis that has just begun. By 2025, 85 million jobs will be lost and replaced by 97 million new ones. But replacement process will be much slower, then the loss of old jobs, which will be observed in the short term, creating a significant distortion in the labor market.

In addition, the coronavirus pandemic has caused inequality in the labor market, for the first time since the crisis of 2008–2009 which stopped the growth of employment and accelerated the development of the fourth industrial revolution and universal digitalization, as provides The Future of Jobs Report (2020) by World Economic Forum.

Thus, in the IMF report (2017) on the adequacy of the Global Financial Safety Net, proposals were discussed to reform the Fund's lending tools in order to further strengthen the Global Financial Safety Net. In this context, a proposal was considered of a new mechanism for maintaining liquidity, as well as improving the existing tools of the Fund to prevent crises. The reforms emerging from these discussions are part of the Fund's broader program to strengthen the Global Network, through the introduction of new policy coordination tools and an expanded framework for cooperation with regional funding mechanisms.

For example, Forster et al. (2019) highlight the structural adjustment programs of the International Monetary Fund (IMF). In accordance with the authors, it is an important but internationally understudied determinant of inequality in the developing world.

It is obvious that in the medium term, approximately by 2025, it seems expedient for companies to change their business goals, revise the structure of jobs (taking into account softskills and hardskills of personnel).

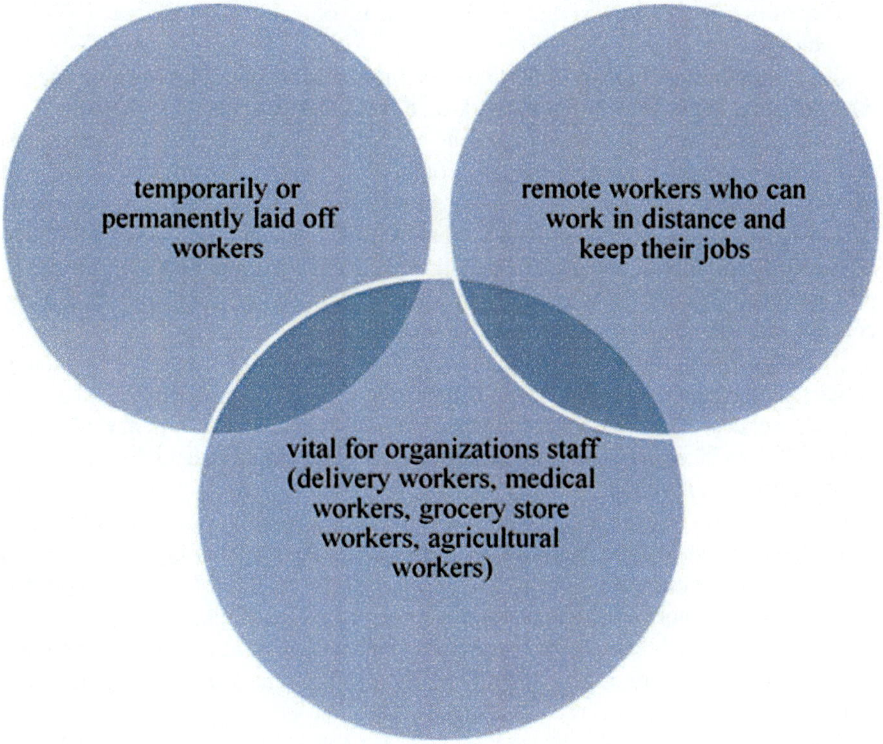

Fig. 3 The division of workers into categories (considering the adjustments of the coronavirus pandemic) (Abdullaeva and Zakharchenko 2018)

The Fig. 3 presents three categories of workers that form the labor market around the world today. This figure was presented in work of authors Abdullaeva and Zakharchenko (2018). What these categories have in common is that representatives of each of them are faced with a continuous change in working methods since 2020, which creates a need for additional education and ongoing training. Thus, companies will prefer a "hybrid" format of work.

Discussion

A study by the Boston Consulting Group (2021) says that the criterion for making a decision to dismiss an employee can be considered the readiness (or not) of additional training, changing skills, and improving existing qualifications The most vulnerable in terms of maintaining current employment are low-skilled employees and young people under 20 years of age (about 40% were fired). The largest number of layoffs in the world was observed in the areas of art, sales, media and light industry. The smallest number of cuts associated with the coronavirus pandemic and lockdowns as a result was noted in the areas of healthcare, science, IT and technology.

So, companies are highly likely to reduce their headcount in favor of freelancers, who are most likely to be hired to work on a particular project.

In addition, due to the spread of remote communication technologies, the need to hold offsite events, meetings, forums, conferences, and professional training will become irrelevant. Thus, it will not be necessary to plan the expenses for staff business trips. These funds are supposed to be directed to training new staff skills, advanced training, which seems to be a priority from our point of view in order to prevent an increase in the unemployment rate (Singapore's experience). So the concept of human capital seems to be viable.

The trigger for global changes that actualize the problem of ensuring global and national financial security in 2020–2021 was the COVID-19 pandemic.

These respectable researches and their works (Mühlich 2020; Gelpern et al. 2020; Stubbs 2021; Gallagher 2021) focus on expanding the global financial safety net in response to COVID-19.

Thus, in world science, the main focus is on the prevention of financial crises and macroeconomic shocks using indicators of financial security (stability), as well as on cross-country comparisons of financial security levels and the formation of a global financial security network and the contradictions of its development.

In the domestic literature, the problem of an indicative assessment of national financial security is being developed in a slightly different plane, among the main issues:

- selection of socio-economic indicators that are optimal for displaying the level of national financial security.
- determination of threshold values of financial security indicators to monitor its level.
- timely and effective response to changes in the level of financial security.

Rubtsov (2015, 2020) devotes his works to the development of financial and economic systems in the last two decades., which pay special attention to the consequences of the global financial crisis that erupted in 2007–2008 and the measures to overcome it.

By the way, Rubtsov B.B. in his work (2020) considers changes in the institutional architecture of financial markets after the global economic crisis, and assesses the role of the financial sector in the economies of leading countries, using the statistics of the Financial Stability Board as an empirical base.

A review of the above works allowed us to establish ourselves in the opinion that the global perspective of the study of national financial security, new threats and challenges associated with the global financial market, the imperfection of the structure and activities of the global financial security network require reflection.

The structural transformation of the financial market in the context of digitalization and "radical uncertainty" determined the formation of a new approach to solving these scientific and methodological problems through the prism of a global dimension.

This allowed the team of authors to form a conceptual approach to the study of financial security: a methodology for global measurement of national financial security and further development of the global financial security network and tools for embedding countries with emerging markets (including Russia) and developing countries into it.

An analysis of the current state of the studies known to us allows us to conclude that the problem statement in this aspect is original and there is no comprehensive study of national financial security in the perspective proposed in this project through the prism of structural changes in the development of the global financial market and integration into the global financial security network based on market instruments.

Analysis of the risks associated with this process is of particular importance in the study of the digital transformation of the financial market in Russia. The impact of digitalization on the risks of the financial sector, as well as the levels and goals of managing these risks, are presented in the work Maramygin (2019). Khalin and Chernov (2018) presented an analysis of the risks of digitalization for society and the Russian economy as a whole.

In recent decades, it is financial transformations in the aggregate of all systemic transformations that have played a special role (Khostinskaya and Chernikova 2013, 2018) investigate problematic issues of transformation of the Russian financial market, as well as possible ways to solve them.

Possible negative consequences of digitalization on the capital market on a global scale are investigated by E. Paraná (2018).

Since the project covers the problems of employment and income of the population, it is advisable to note the state of research on this problem.

Conclusion

Long-term perspectives of the process of transformation of labor as a source of livelihood in the so-called post-work world, the key drivers of which are the development of artificial intelligence and automation, shifts in the value system (from traditional to postmodern), a decrease in the role of work as a separate a measurable type of activity due to its combination with other aspects of people's daily life are investigated in the work of A. Hines (2019).

Medium and short-term effects of digitalization for the labor market and employment structure, their subdivision into destructive and transformational ones, as well as a model for mapping professions according to the degree of exposure to the effects of digitalization, are given in the article by Fossen and Sorgner (2019).

Analysis of the impact of digitalization on the Russian labor market is presented in studies of international organizations (such as McKinsey (Aptekman et al. 2017), Boston Consulting Group (Lisnyanskaya 2021), etc.), as well as in the works of Russian scientists, including M.A. Izmailova (2019), S.P. Zemtsov (2017), V.S. Osipov (2019), E. Sadovaya (2018) and others. Noteworthy is the project

Atlas of Emerging Jobs, which incorporates the results of the Foresight Competency 2030 study organized by the Moscow School of Management SKOLKOVO (2021) and the Agency for Strategic Initiatives.

The presented studies on the transformation of the Russian labour market in the context of digitalization emphasize the potential threat of an increase in socioeconomic inequality, as well as the risks of large-scale technological unemployment under certain conditions, but do not touch upon the corresponding consequences for the welfare of the population and its participation in the activities of the Russian financial market. These consequences and aspects are supposed to be investigated within the framework of the proposed project.

Acknowledgements Rostov State University of Economics, Russian Customs Academy, Don State Technical University, Institute of Service and Entrepreneurship (branch) of Don State Technical University in Shakhty.

References

Abdullaeva, V., & Zakharchenko, E. (2018). Tax crimes counteraction as the way to provide national financial security. *Finansovye issledovania, 1*(58), 31–39.

Aptekman, A., Kalabin, V., Klincov, V., Kuznetsova, E., Kulagin, V., & Yasenovec, I. (2017). *Digital Russia: A new reality: Report of McKinsey*. http://www.mckinsey.com/~/media/McKinsey/Locations/Europe%20and%20Middle%20East/Russia/Our%20Insights/Digital%20Russia/Digital-Russia-report.ashx

Bertay, A. C., Demirguc-Kunt, A., & Huizinga, H. (2016). Should cross-border banking benefit from the financial safety net? *Journal of Financial Intermediation, 27*, 51–67.

Bordo, M. D., & Meissner, C. M. (2016). Fiscal and financial crises. *Handbook of macroeconomics, 2a*, 355–412.

Bordo, M. D. (2018). An historical perspective on the quest for financial stability and the monetary policy regime. *The Journal of Economic History, 78*(2), 319–357.

Denbee, E., Jung, C., & Paterno, F. (2016, March 18). *Stitching together the global financial safety net*. Bank of Italy Occasional Paper No. 322. https://ssrn.com/abstract=2772513 or https://doi.org/10.2139/ssrn.2772513

Financial Secrecy Index. (2021a). *Method and concepts*. https://fsi.taxjustice.net/en/methodology/method-and-concepts

Financial Secrecy Index. (2021b). *What is the Financial Secrecy Index?* https://fsi.taxjustice.net/en/faq

Fossen, F., & Sorgner, A. (2019). Mapping the future of occupations: Transformative and destructive effects of new digital technologies on jobs. *Foresight and STI Governance, 13*(2), 10–18. https://doi.org/10.17323/2500-2597.2019.2.10.18

Forster, T., Kentikelenis, A. E., Reinsberg, B., Stubbs, T. H., & King, L. P. (2019). How structural adjustment programs affect inequality: A disaggregated analysis of IMF conditionality, 1980–2014. *Social Science Research, 80*, 83–113.

Gallagher, K. (2021). Safety first: Expanding the global financial safety net in response to COVID-19. *Global Policy, 12*(1), 140–148.

Gelpern, A., Sean, H., & Adnan, M. (2020, April). Debt standstills can help vulnerable governments manage the COVID-19 crisis. In M. Obstfeld & A. Pozen (Eds.), *How the G20 can hasten recovery from COVID-19*. Peterson International Institute for International Economics Briefing, 20-1.

Giglio, S., Kelly, B., & Pruitt, S. (2016). Systemic risk and the macroeconomy: An empirical evaluation. *Journal of Financial Economics, 119*(3), 457–471.

Gornostaeva, Z. (2021). The ratio of staff and artificial intelligence in modern entrepreneurship: The specifics of small and medium-sized businesses. *Studies in Systems, Decision and Control, 314*, 137–144.

Gornostaeva, Z., Gribova, O., Tregulova, N., & Vasenev, S. (2020). A new approach to company's HR management in the conditions of Russia's turning in a growth pole of the global economy: From management to marketing. *Lecture Notes in Networks and Systems, 73*, 159–166.

Hines, A. (2019). Getting ready for a post-work future. *Foresight and STI Governance, 13*(1), 9–30. https://doi.org/10.17323/2500-2597.2019.1.19.30

Huotari, J. (2015). Measuring financial stress a country specific stress index for Finland. *SSRN Electronic Journal, 7*, Bank of Finland Research Discussion Papers.

International Monetary Fund. (2017). *IMF policy paper adequacy of the Global Financial Safety Net—Considerations for Fund Toolkit Reform.* Press Release No. 17/507. http://www.imf.org/external/pp/ppindex.aspx

Izmailova, M. (2019). The impact of global trends on the transformation of the world labour market: A challenge for Russia. *Voprosy regionalnoj ekonomiki., 39*(2), 23–28.

Khalin, V., & Chernova, G. (2018). Digitalization and its impact on the Russian economy and society: Advantages, challenges, threats and risks. *Upravlencheskoe konsul'tirovanie – Management Consulting, 10*(118), 46–62.

Khostinskaya, G., & Chernikova, L. (2013). *Systemic transformations in macro- and microfinance.* Scientific Technologies.

Khostinskaya, G., & Chernikova, L. (2018). *Transformations in economic systems: A financier's view*. Knorus.

Kim, C. (2016). Strengthening the global financial safety net: Challenges and prospects. *Emerging Markets Finance and Trade, 52*(10), 2212–2220.

Lisnyanskaya, V. (2021). *Study or be fired: How the pandemic has changed the labor market.* Boston Consulting Group Survey (in Russian). https://trends.rbc.ru/trends/education/60ae308e9a7947daf82dcdd9

Maramygin, M., Chernova, G., & Reshetnikova, N. (2019). Digital transformation of the financial services market in Russia: Trends and specificity. Upravlenets. *The Manager, 10*(3), 70–82. https://doi.org/10.29141/2218-5003-2019-10-3-7

Mühlich, L. (2020). *The global financial safety net tracker: Lessons for the COVID-19 crisis from a new interactive dataset.* Boston University, Global Development Policy Center.

Osipov, V. (2019). Gap of the employment in the digital economy. *CITISE, 1*(18), 35.

Paraná, E. (2018). *Digitalized finance: Financial capitalism and informational revolution.* https://doi.org/10.1163/9789004383920.

Reshetnikova, N. (2021). Digital society and the information age: Challenges and threats for man, business and the state. Reality of politics. *Estimates-Comments-Forecasts, 17*, 145–154.

Reshetnikova, N., Magomedov, M., & Buklanov, D. (2021). Digital finance technologies: Threats and challenges to the global and national financial security. In *IOP Conference Series: Earth and Environmental Science* (Vol. 666, p. 062139).

Reshetnikova, N., Magomedov, M., Buklanov, D., & Zakharchenko, E. (2019). *The international business cooperation and its influence on enterprise financial security under globalization.* The future of the global financial system: Downfall or harmony. "Lecture Notes in Networks and Systems" Cham, Switzerland (pp. 294–308).

Reshetnikova, N., Magomedov, M., Zmiyak, S., & Chernysheva, Y. (2022). Digital technologies adoption in the agro-industrial complex as a priority of regional development in the conditions of global macroeconomic changes. In *XIV International Scientific Conference "INTERAGROMASH 2021"* (pp. 3–12).

Rubtsov, B. (2015). *The current state of development and anti-crisis regulation of the financial and economic system.* SRC INFRA.

Rubtsov, B. (2020). The shifts in the financial markets architecture after the global economic crisis of 2007–2009. *Banking Services, 10*, 9–21. https://doi.org/10.36992/2075-1915_2020_10_9

Sadovaya, E. (2018). Digital economy and a new paradigm of the labor market. *Mirovaya ekonomika i mezhdunarodnye otnosheniya, 62*(12), 35–45.

Sandahl, J. F., Holmfeldt, M., Ryden, A., & Stromqvist, M. (2015). An index of financial stress for Sweden. *Sveriges Riksbank Economic Review, 2*, 49–67.

Scheubel, B., & Stracca, L. (2016). *What do we know about the global financial safety net? Rationale, data and possible evolution*. ECB Occasional Paper. https://ssrn.com/abstract=2843608

Skolkovo. (2021). *Atlas of emerging jobs*. https://skolkovo.ru/public/media/documents/research/sedec/SKOLKOVO_SEDeC_Atlas.pdf.

Stubbs, T. (2021). Whatever it takes? The global financial safety net, Covid-19, and developing countries. *World Development, 137*, 105171.

Sun, L., & Huang, Y. (2016). Measuring the instability of China's financial system: Indices construction and an early warning system. *Economics, 10*(1), 1–39.

Weder Di Mauro, B., & Zettelmeyer, J. (2017, January 30). *The new global financial safety net: Struggling for coherent governance in a multipolar system CIGI essays in international finance* (Vol. 4). https://ssrn.com/abstract=2946452

Woodford, M. (2016). Quantitative easing and financial stability. *Journal Economía Chilena (The Chilean Economy), Central Bank of Chile, 19*(2), 4–77.

World Economic Forum. (2020). *The future of jobs report. OCTOBER 2020*. World Economic Forum https://www3.weforum.org/docs/WEF_Future_of_Jobs_2020.pdfv

Zakharchenko, E., Brichka, E., Reshetnikova, N., Magomedov, M., & Tenova, Z. (2021, March). The main global investment activity and digitalization trends in the context of the COVID-19 pandemic. In *IV International Scientific and Practical Conference* (pp. 1–5).

Zemtsov, S. (2017). Robots and potential technological unemployment in the Russian regions: Review and preliminary results. *Voprosy Ekonomiki, 7*, 142–157.

Financial Inclusion of Rural Areas in the Conditions of Digitalization: Studies in Europe and Russia

Dinar R. Baetova, Oleg A. Blinov, and Anastasia A. Zagorenko

Introduction

The financial services availability, the possibility to raise funds and the financial flows optimization contribute to the economy development. This aspect of the financial inclusion role is noted in many studies, both foreign (Milana and Ashta 2020; Saha and Dutta 2020) and Russian (Danilov 2019). The World Bank Group considers the financial inclusion development as a key factor in reducing poverty and improving the society welfare (Demirguc-Kunt et al. 2018). One of the effects of the financial inclusion development is engagement in the formal economy. The active use of financial services in rural areas by business and the population is more often considered as a source of financing the capital deficit and cash services for business. Thus, financial inclusion is considered as an actual direction in the rural areas economy development. In the pandemic conditions the processes of the financial sector digitalization and the partial change of financial services towards the digital field have accelerated. Given the disconnection and insufficient financial infrastructure of rural areas, it can accelerate the pace of financial inclusion development. However, with all the evidence of the ongoing processes, many aspects need additional study, the results of which must be considered in the financial inclusion development in rural areas in the context of digitalization. The purpose of the article is to study the topic of considering the financial inclusion of rural areas in the context of digitalization in published studies.

D. R. Baetova (✉) · O. A. Blinov · A. A. Zagorenko
Federal State-Funded Educational Institution of Higher Education Omsk State Agrarian University named after Pyotr A. Stolypin, Omsk, Russian Federation
e-mail: dr.baetova@omgau.org; oa.blinov@omgau.org; aa.zagorenko1817@omgau.org

A. Rumyantseva et al. (eds.), *Challenges and Solutions in the Digital Economy and Finance*, Springer Proceedings in Business and Economics,
https://doi.org/10.1007/978-3-031-14410-3_40

Materials and Methods

Separate methods of bibliometric analysis were used during the study. The study was carried out in accordance with the following stages: identification of search criteria and keywords, as well as the period of publication; selection of scientometric databases; export of publicly available papers; analysis of information and discussion of results. Thus, the keywords "financial inclusion", "rural areas", "digital financial services", "financial services digitalization", "digital infrastructure", "financial infrastructure" were chosen, but it is necessary to clarify that requests were stated by several keywords. The research area was focused on articles published in the period of 2017–2021. Even though the reference to the scientometric databases was carried out in January 2022, and there was not enough time for the publication of studies carried out in 2021, this year is not excluded from the study results. The search for previous publications within the framework of this study was carried out using the international scientometric databases Scopus and Web of Science. Despite the limitations of the text completeness of some papers, the use of these databases made it possible to collect enough information for the purposes of this study. The choice of Scopus and Web of Science is based on quality, the ability to search and filter search using several bibliographic parameters, the provision of easy access to full texts of documents, and the fact that these are the most used databases for obtaining useful information for researchers evaluating scientific activity. After the search results were obtained, both databases were searched to eliminate double counting of papers. The demand for the topic of financial inclusion of rural areas in the context of digitalization in Russian papers was considered using the RSCI Russian scientometric base. The main indicator of the method of counting the number of papers was their total volume. Publications of the authors of this work were excluded, when counting papers. After counting papers in scientometric databases for 2017–2021 in the context of the countries of publications, an extensive analysis of papers that are in the public access was carried out to group the topics of studying the financial inclusion of rural areas in the context of digitalization.

Results

The total number of papers in the international scientometric databases by the keywords "financial inclusion" amounted to 1822 publications for the analyzed period, 529 of them accrue to 2021 (even though not all studies of 2021 have been placed in scientometric databases yet). The diagram shown in Fig. 1 makes it possible to see the increase in the interest of researchers in the issues of financial inclusion by 2.2 times. There are less papers devoted to "digital financial services" and "financial services digitalization" without reference to the territorial indication

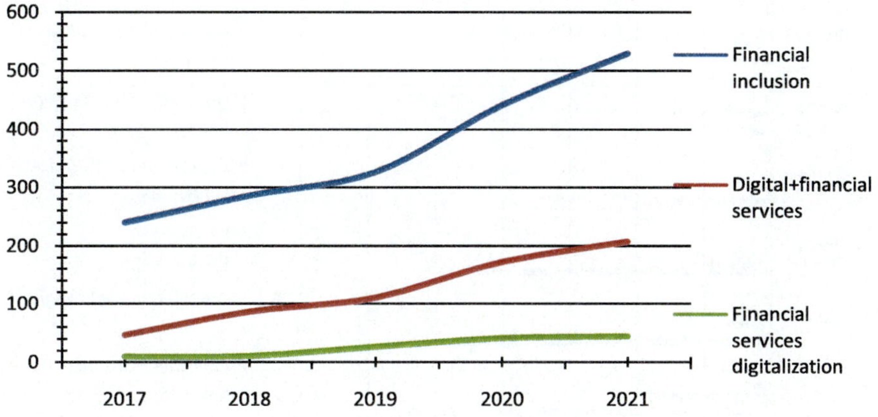

Fig. 1 Number of papers in Scopus and/or Web of Science on financial inclusion and financial services digitalization by year of publication

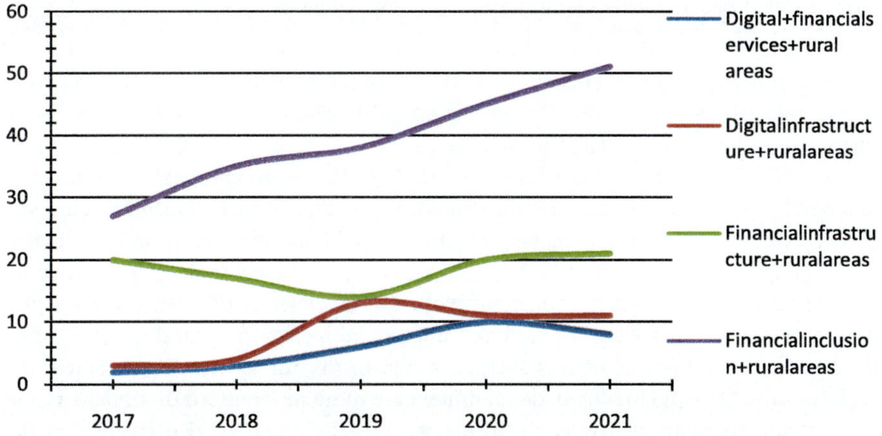

Fig. 2 Number of papers in Scopus and/or Web of Science on financial inclusion and financial services digitalization in rural areas by year of publication

"rural areas"—621 and 130 publications respectively. However, the increase in the growth of publications over 5 years is 4.5 and 4.8 times respectively.

The combination of these keywords in the search with the territorial indication "rural areas" gave a significantly lower result: from 29 to 196 publications totally over 5 years. The distribution of papers by year is shown in Fig. 2. The largest number of publications is associated with financial inclusion in rural areas (51 papers in 2021—there is an increase of 1.8 times over the analyzed period). Against the background of an increase in the number of publications on the digital infrastructure of rural areas, the number of papers on the financial infrastructure of rural areas is declining, due to the increasing role of remote access to financial services.

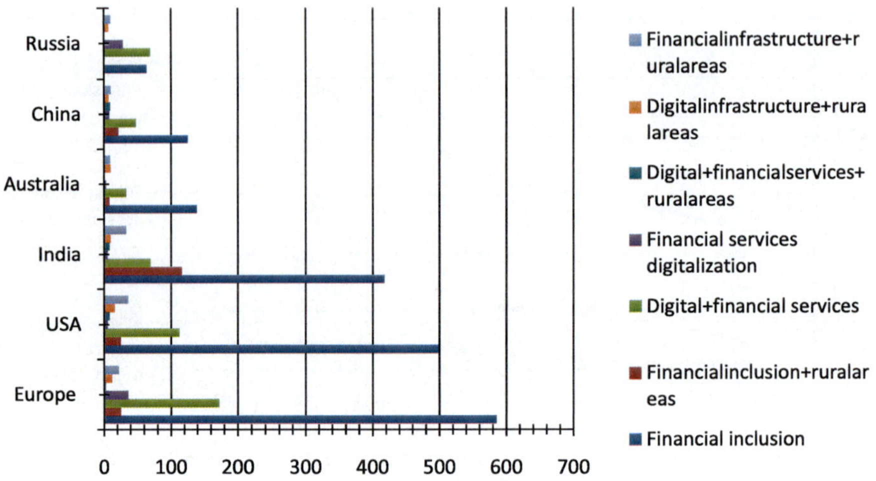

Fig. 3 Number of papers in Scopus and/or Web of Science on financial inclusion and financial services digitalization, including in rural areas, by country of publication

Researchers are less interested in studying digital financial services in rural areas (from 2 to 10 papers per year), which is probably due to the very fact that digital financial services are available in rural areas.

The distribution of publications for 2017–2021 in international scientometric databases by country of publication show a high activity of European researchers (Fig. 3). 32% of papers on financial inclusion are published by European researchers, 27% of papers are written in the USA, 22% of papers fall upon India. It is noteworthy that, in terms of research on financial inclusion in general, European studies predominate, while India leads in research on financial inclusion in rural areas. Despite the seeming strikingness of this fact, it is due to the following circumstance: the effects of the financial inclusion development are more pronounced in rural areas and are considered mainly in relation with poor population. It is also likely that the peculiarities of rural areas determine the fact that the issues of financial services digitalization in rural areas of India, Australia, and China are slightly presented in publications. Despite the fact that India is quite actively represented in papers on research topics, the most common language of publication is English. It is connected with the fact that the most relevant magazines on these issues are published in English. In addition, most publications have European or English-speaking co-authors.

It is noteworthy that, despite the significant share of rural areas and the rural population in the Russian Federation, the issues of financial inclusion of rural areas are practically not presented in Russian editions according to international scientometric databases. Only one Russian paper is devoted to digital financial services in rural areas.

It is assumed that the papers of Russian researchers may not be presented in the international databases, but exist in Russian scientific publications, so the RSCI

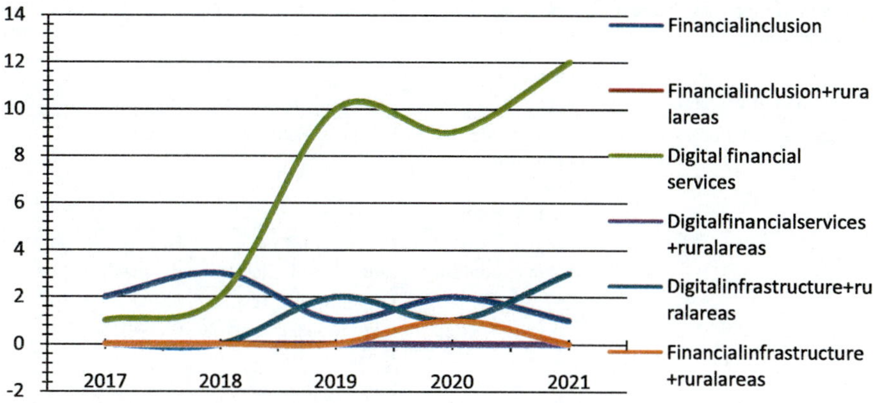

Fig. 4 Number of papers in the Russian Science Citation Index on financial inclusion and financial services digitalization, including in rural areas

Russian scientometric database was chosen (Fig. 4). When comparing the search results for Russian publications in Scopus, Web of Science and RSCI, it turned out that:

- there are more Russian publications on financial inclusion in the international databases than in Russian one (64 versus 9);
- there is only one publication on the financial inclusion of rural areas in Russian editions that is not presented in international databases;
- as in the international databases, publications on digital financial services prevail in Russian editions (69 and 34, respectively);
- there are no papers on digital financial services in rural areas in Russian editions, in the international databases there is only one work;
- a significant increase in publications for selected keywords is observed in 2019, while in the international databases the annual increase since 2017 is 1.2 times.

Thus, it can be stated that the topic of financial inclusion in rural areas in the context of digitalization has not been sufficiently studied in Russia.

Papers that are in the public access are analyzed by research areas. An attempt to carry out some systematization of research in order to understand the level of certain issues coverage in the financial inclusion development in rural areas in the context of digitalization is made (Table 1).

There are two aspects in the context of understanding the essence of financial inclusion:

- financial inclusion as an internal state of the individual, including theirselves in the system of financial services; the study is noteworthy (Kangwa et al. 2021), where respondents consider financial services to be something exclusive even with active possession of digital skills, as well as studies showing the lack of need for financial inclusion among the population of rural areas (Lopez and Winkler 2018; Urbano et al. 2021);

Table 1 Directions of research on the financial inclusion of rural areas in the context of digitalization

Directions of research	Issues of research	Topics of research	Including
Financial inclusion essence	Acceptance, inclusion in the financial services system	Financial behavior (Kangwa et al. 2021)	
	Financial services availability	Financial infrastructure (Ji et al. 2021)	New model of financial inclusion in the context of digitalization
		Digital infrastructure (Aldashev and Batkeyev 2021)	
		Suitability of the financial services offer to the needs (Baber 2021)	
Digital financial inclusion impact	Financial inclusion risks (Çera et al. 2021)	For population (Vörös et al. 2021; Lopez and Winkler 2018)	
		For financial institutions (Ozili 2021; Singh et al. 2021; Ahamed et al. 2021)	
	Positive impact	Economic growth (Papadimitri et al. 2021; Kling et al. 2020)	
	Negative impact	Poverty trap (Vörös et al. 2021)	
		Lack of awareness (Vörös et al. 2021)	
Factors for the development of financial inclusion in rural areas in the context of digitalization	Personal	Personal factors (Potocki 2019)	Education (Nawaz 2021)
			Type of activity
		Use of gadgets (Bayar et al. 2021)	Activity of online social contacts (He and Li 2020)
		Availability of financial contacts with other territories	migrant worker remittance (Berk Saydaliyev et al. 2020)
	FinTech achievements (Takeda and Ito 2021; Kanga et al. 2021)		
	Territorial	Share of black economy	Typical financial behavior (Urbano et al. 2021)
		Features of rural areas (Potocki and Cierpia-Wolan 2019)	Disconnection
			Remoteness
		Increase in the share of non-agricultural activities (Potocki 2019)	
		Digital infrastructure (Aldashev and Batkeyev 2021)	
		participation of state and municipal authorities in the digital financial inclusion development (Kabakova and Plaksenkov 2018)	

– financial inclusion as financial accessibility, so its formation is determined by the issues of infrastructure and the availability of financial products and services that consider the characteristics of business and the population of rural areas (Pakhnenko et al. 2021).

Most of the papers evaluating the positive impact of financial inclusion of rural areas, including the digital one, are devoted to the economic growth and agricultural organizations profitability rise. From the point of view of the population, with the seeming ease of obtaining digital financial services, studies on the risks and negative consequences of digital financial inclusion in rural areas are becoming more relevant. It is the rural population that is characterized as insufficiently financial literate that can lead both to the bankruptcy of individuals (if there is such procedure) or a debt trap, and to an increase in the risks of financial institutions.

A significant number of papers is devoted to the factors of financial inclusion development. FinTech achievements are an undoubted factor in the financial inclusion development in rural areas in the context of digitalization. On the one hand, the financial technologies development expands the possibilities of digital channels for promoting the services of traditional financial institutions. On the other hand, FinTech as an independent area of financial activity contributes to the development of alternative opportunities for financial services (crowdfunding platforms, p-2-p platform opportunities, etc.). Separate studies find a correlation between the level of financial inclusion of the rural population and the level of education, types of activity, activity of social online contacts and labor migration. Among the features of rural areas as a factor of financial inclusion, the disconnection of rural settlements, remoteness from megacities and financial centers are traditionally considered. The study that shows a link between the increase in financial inclusion and the growth of employment of the rural population in non-agricultural activities is notable.

With no doubt, financial inclusion is significantly influenced by the financial behavior typical for these territories: the habit of borrowing from informal sources, participation in the black economy.

Studies on various aspects of financial inclusion in rural areas recognize the importance of the policies of state and municipal authorities regarding the financial inclusion development. At the same time, it is pointed out that it is necessary to consider not only the positive effects, but also the risks of the financial inclusion development of rural areas, as well as to take into account the results of individual studies.

Discussion

An extensive analysis of publicly available papers show that the concept of financial inclusion has evolved in the context of the financial services digitalization, creating a new model of financial inclusion in recent years, where digital infrastructure and digital skills are becoming both an integral element of financial inclusion and necessary condition for its development. It is especially true for rural areas, where digitalization and the ubiquity of gadgets should create conditions for financial inclusion.

The significant activity of publications in the context of financial inclusion, digital financial services, and the growing interest in the financial inclusion of rural areas are clear evidence of this topic importance. At the same time, in the context of the digitalization development, there is a decrease in interest in studying the financial infrastructure of rural areas. There is a slight interest in digital financial services in rural areas, although this channel will be the main one for the financial inclusion development in rural areas. This is because even with insufficient digital infrastructure, the availability of mobile Internet is the necessary minimum for receiving digital financial services. On the other hand, the absence of significant differences in the mechanism for promoting and providing digital financial services between the city and the rural settlement reduces the circle of potential researchers on this issue.

The financial inclusion development in rural areas is associated with the fight against poverty, in this regard, India leads the research. Nevertheless, the available papers indicate that the financial inclusion development in rural areas is relevant in the context of the alternative employment development for the rural population.

It is noteworthy that Russia, as a country with a significant share of rural areas and the rural population, with achievements in the FinTech area and with an obvious potential interest in the financial inclusion of rural areas against the backdrop of ongoing digitalization, does not occupy a significant share in the number of publications on this topic. All studies recognize that the financial inclusion of rural areas in the context of digitalization has become a hot topic both at the level of state and municipal government, the management of commercial companies, and at the level of research activities.

The material in this article may be useful for research in the context of financial inclusion, as it provides a global perspective on this area of research. Thus, the article is a contribution to identifying the main trends in research on financial inclusion in rural areas in the context of digitalization.

Conclusion

This paper reveal that numerous studies of the financial inclusion of rural areas for the period of 2017–2021 are either synoptical or focused on local issues (such as the impact of the black economy share on financial inclusion). The issues of the impact of the financial inclusion of rural areas on their sustainable development, as well as issues of theoretical and methodological support are not reflected in the studies available in the public access. In addition, the publications practically do not reflect the experience of financial inclusion development in rural areas. Only one paper state that the breadth (population coverage) of financial inclusion activities in rural areas has a greater impact than the depth of translated financial knowledge and skills. Thus, stating the relevance of research on the financial inclusion of rural areas in the context of digitalization for both Europe and Russia, it is necessary to emphasize the significant prospects for further research.

Acknowledgements The project is being implemented with the financial support of the European Commission within the framework of the Erasmus + Jean Monnet project, module "European practice in the digital financial services development in rural areas: knowledge and inclusion", project No. 620032-EPP-1-2020-1-RU-EPPJMO-MODULE. The content of this material reflects the views of the authors, the European Commission is not responsible for the use of the information contained in it.

References

Ahamed, M. M., Ho, S. J., Mallick, S. K., & Matousek, R. (2021). Inclusive banking, financial regulation and bank performance: Cross-country evidence. *Journal of Banking & Finance, 124*, 106055. https://doi.org/10.1016/j.jbankfin.2021.106055

Aldashev, A., & Batkeyev, B. (2021). Broadband infrastructure and economic growth in rural areas. *Information Economics and Policy, 57*, 100936. https://doi.org/10.1016/j.infoecopol.2021.100936

Baber, H. (2021). Financial inclusion and crowdfunding-A study of European countries. *Review of Applied Socio-Economic Research, 22*(2), 37–48.

Bayar, Y., Gavriletea, M. D., & Păun, D. (2021). Impact of mobile phones and internet use on financial inclusion: Empirical evidence from the EU post-communist countries. *Technological and Economic Development of Economy, 27*(3), 722–741. https://doi.org/10.3846/tede.2021.14508

Berk Saydaliyev, H., Chin, L., & Oskenbayev, Y. (2020). The nexus of remittances, institutional quality, and financial inclusion. *Economic Research-Ekonomska istraživanja, 33*(1), 3528–3544. https://doi.org/10.1080/1331677X.2020.1774795

Çera, G., Khan, K. A., Rowland, Z., & Ribeiro, H. N. R. (2021). Financial advice, literacy, inclusion and risk tolerance: The moderating effect of uncertainty avoidance. *E&M Economics and Management, 24*(4), 105–123. https://doi.org/10.15240/tul/001/2021-4-007

Danilov, Y. A. (2019). The present state of global scientific debate in the field of financial development. *Voprosy ekonomiki, 3*, 29–47. https://doi.org/10.32609/0042-8736-2019-3-29-47

Demirguc-Kunt, A., Klapper, L., Singer, D., & Ansar, S. (2018). *The Global Findex Database 2017: Measuring financial inclusion and the fintech revolution*. World Bank Publications. https://doi.org/10.1596/978-1-4648-1259-0

He, J., & Li, Q. (2020). Can online social interaction improve the digital finance participation of rural households? *China Agricultural Economic Review, 12*(2), 295–313. https://doi.org/10.1108/CAER-11-2019-0213

Ji, X., Wang, K., Xu, H., & Li, M. (2021). Has digital financial inclusion narrowed the urban-rural income gap: The role of entrepreneurship in China. *Sustainability, 13*(15), 8292. https://doi.org/10.3390/su13158292

Kabakova, O., & Plaksenkov, E. (2018). Analysis of factors affecting financial inclusion: Ecosystem view. *Journal of Business Research, 89*, 198–205. https://doi.org/10.1016/j.jbusres.2018.01.066

Kanga, D., Oughton, C., Harris, L., & Murinde, V. (2021). The diffusion of fintech, financial inclusion and income per capita. *The European Journal of Finance*, 1–29. https://doi.org/10.1080/1351847X.2021.1945646

Kangwa, D., Mwale, J. T., & Shaikh, J. M. (2021). The social production of financial inclusion of generation Z in digital banking ecosystems. *Australasian Accounting, Business and Finance Journal, 15*(3), 95–118. https://doi.org/10.14453/aabfj.v15i3.6

Kling, G., Pesque-Cela, V., Tian, L., & Luo, D. (2020). A theory of financial inclusion and income inequality. *The European Journal of Finance, 1-21*. https://doi.org/10.1080/1351847X.2020.1792960

Lopez, T., & Winkler, A. (2018). The challenge of rural financial inclusion–evidence from microfinance. *Applied Economics, 50*(14), 1555–1577. https://doi.org/10.1080/00036846.2017.1368990

Milana, C., & Ashta, A. (2020). Microfinance and financial inclusion: Challenges and opportunities. *Strategic Change, 29*(3), 257–266. https://doi.org/10.1002/jsc.2339

Nawaz, T. (2021). What's in an education? Implications of CEO education for financial inclusion. *International Journal of Finance & Economics*. https://doi.org/10.1002/ijfe.2348

Ozili, P. K. (2021). Has financial inclusion made the financial sector riskier? *Journal of Financial Regulation and Compliance, 29*(3), 237–255. https://doi.org/10.1108/JFRC-08-2020-0074

Pakhnenko, O., Rubanov, P., Hacar, D., Yatsenko, V., & Vida, I. (2021). Digitalization of financial services in European countries: Evaluation and comparative analysis. *Journal of International Studies, 14*(2), 267–282. https://doi.org/10.14254/2071-8330.2021/14-2/17

Papadimitri, P., Pasiouras, F., & Tasiou, M. (2021). Financial leverage and performance: The case of financial technology firms. *Applied Economics, 53*(44), 5103–5121. https://doi.org/10.1080/00036846.2021.1915949

Potocki, T. (2019). Financial capability among low-income households in rural parts of Poland. *Argumenta Oeconomica, 2*(43), 85–114. https://doi.org/10.15611/aoe.2019.2.04

Potocki, T., & Cierpiał-Wolan, M. (2019). Factors shaping the financial capability of low-income consumers from rural regions of Poland. *International Journal of Consumer Studies, 43*(2), 187–198. https://doi.org/10.1111/ijcs.12498

Saha, M., & Dutta, K. D. (2020). Nexus of financial inclusion, competition, concentration and financial stability: Cross-country empirical evidence. *Competitiveness Review: An International Business Journal, 31*(4), 669–692. https://doi.org/10.1108/CR-12-2019-0136

Singh, K., Misra, M., & Yadav, J. (2021). Corporate social responsibility and financial inclusion: Evaluating the moderating effect of income. *Managerial and Decision Economics, 42*(5), 1263–1274. https://doi.org/10.1002/mde.3306

Takeda, A., & Ito, Y. (2021). A review of FinTech research. *International Journal of Technology Management, 86*(1), 67–88.

Urbano, D., Felix, C., & Aparicio, S. (2021). Informal institutions and leadership behavior in a developing country: A comparison between rural and urban areas. *Journal of Business Research, 132*, 544–556. https://doi.org/10.1016/j.jbusres.2021.04.073

Vörös, Z., Szabo, Z., Kehl, D., Kovacs, O. B., Papp, T., & Schepp, Z. (2021). The forms of financial literacy overconfidence and their role in financial well-being. *International Journal of Consumer Studies, 45*(6), 1292–1308. https://doi.org/10.1111/ijcs.12734

Evolution of Payment Instruments and Their Development in the Digital Economy

Natalia A. Lazareva, Irina G. Zaiceva, and Elena M. Zvyagina

Introduction

The widespread use of digital technologies in banking has significantly changed the way banks work and the content and procedure for their interaction with customers, competitors, and the regulator in a very short time. Bank payment services were the first to undergo digitalization and up to the present time they hold leadership positions in the use of FinTech tools. According to Ernst & Young, 96% of bank customers around the world are aware of FinTech solutions in the field of payments and transfers. According to this indicator, Russia occupies the leading positions in the world; the figure reaches 99.5% here (Khutko, 2019).

The payment industry in the Russian Federation is characterized by interconnection with the financial technology market, growing investments in FinTech startups, omnichannel interaction, instant payments, personal financial management and biometric identification (Korobeynikova et al., 2021).

The use of financial innovations, on the one hand, affects the development of payment instruments for cashless payments that in turn stimulates the demand for more and more innovations for the payment sector (Krivosheya, 2020). On the other hand, the use of modern payment instruments can have a positive impact on even informal retailers that play an important role in countries with emerging financial markets (Adhikary et al., 2021).

The digital technologies development is changing the preferences of payment services users. The key factors that they evaluate are the convenience, speed and

N. A. Lazareva (✉) · E. M. Zvyagina
St. Petersburg University of Management Technologies and Economics, St. Petersburg, Russian Federation

I. G. Zaiceva
St. Petersburg State University of Economics, St. Petersburg, Russian Federation

A. Rumyantseva et al. (eds.), *Challenges and Solutions in the Digital Economy and Finance*, Springer Proceedings in Business and Economics,
https://doi.org/10.1007/978-3-031-14410-3_41

reliability of the operations performed, as well as the possibility of choosing several alternative options for the transfer. The choice of mobile and other digital payment instruments provides consumers with significant agility, portability, and efficiency benefits (Sahi et al., 2021).

So far, digital payments have revolutionized the financial sector, providing it with features such as user friendliness, convenience, and faster transfer of funds compared to traditional payment systems (Chaveesuk et al., 2021).

Analyzing the main reasons preventing the spread of new digital payment instruments on the market Dimitrova, I., Öhman, P., & Yazdanfar, D. divided them into functional and socio-psychological ones. The first group included the risk of confidentiality loss, the risk of security loss and the risk of lack of access to money due to technical failures. The second group of factors relates to the risk of depersonalization, with the fear of not getting the necessary service due to the lack of direct contact with bank employees. Overcoming these dangers and competent management of emerging risks will contribute to a deeper penetration of digital tools into the market (Dimitrova et al., 2021).

Most of the papers devoted to the analysis of factors influencing the spread of digital payment instruments are based on stable crisis-free conditions. The COVID-19 pandemic has changed the behavior of many consumers globally. There has been an increase in the use of cashless and contactless payment methods (Huterska et al., 2021).

Considering the global transformations in the sphere of money payments under the influence of digitalization, the rate of payment instruments evolution is becoming faster; the diversity and functionality of the instruments used is increasing. Finding the factors contributing to the evolution of tools and patterns that characterize it will allow to determine the prospects for their further development. Therefore, the purpose of the study is to uncover the factors that determine the development of retail payment instruments in the Russian Federation, a country that is among the leaders in the use of financial innovations in the payment sector.

Materials and Methods

To achieve the goal and obtain primary data, there was a survey among users of retail payment instruments. The survey among students and teachers of economic and financial universities was conducted in St. Petersburg in 2021. The survey involved 156 people: 9 teachers and 147 students. There were some students from other cities, so the opinion of people from different parts of the country was considered.

Information was collected using a Google form for 2 weeks. All survey participants answered the questions of the form that included 17 questions, 5 of them allowed the possibility of submitting several answers and 12 suggested the choice of only one answer. The obtained data were analyzed using statistical methods, including methods of descriptive statistics, systematized, and interpreted. Conclusions that are important for the Russian payment sector on the eve of the introduction of a new digital payment instrument—the digital ruble—were drawn.

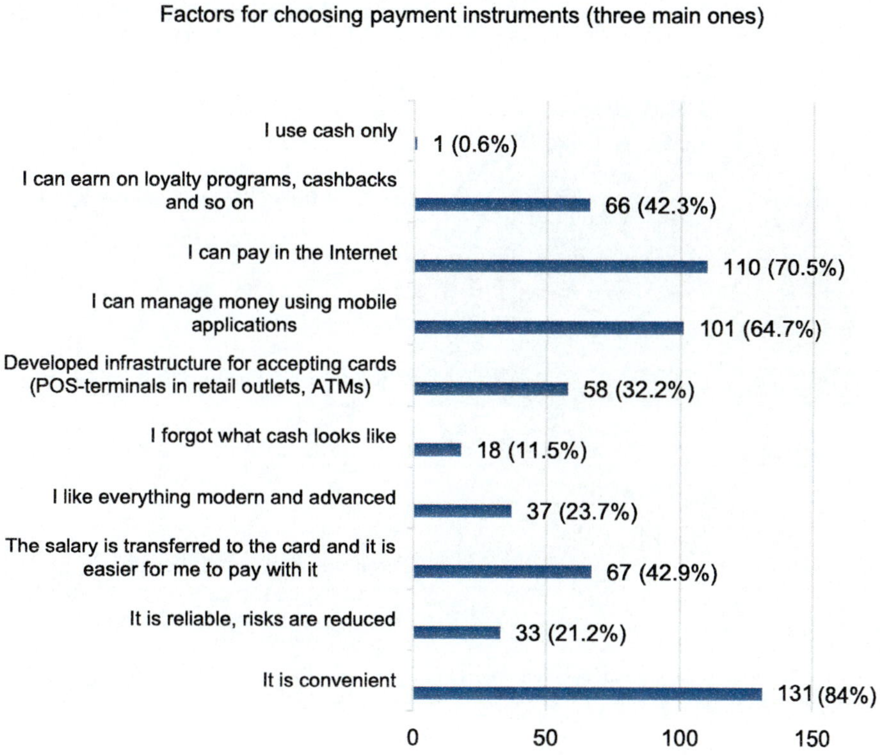

Fig. 1 Factors for choosing payment instruments

Results

The main participants in the survey were undergraduate students—48.7% and postgraduate students—45.5%. Teachers accounted for 5.9%. More than 82% were young people under the age of 24. This largely predetermined the positive attitude of the respondents to modern payment instruments and digitalization in general. The majority of participants—95.4%—prefer to use non-cash money in everyday life. Only 3% of survey participants prefer cash. 7.7% of participants noted that they had experience using cryptocurrencies in the last few years. Electronic money is little in demand by the target audience—63.8% of respondents do not use it. Users were divided approximately in equal shares between—Money and QIWI systems. The main demanded electronic means of payment for respondents were debit cards (82%), mobile applications of banks (70.5%), transfers through a personal account on the bank's website (19.9%), credit cards (15.4%). The key factor in the choice of cash was the absence of the technical failure danger—it was identified by 75% of respondents, as it can be seen from Fig. 1.

The study showed that the use of bank cards is very popular: 21.2% of respondents have one card, the majority have more than one and almost 43% have three or

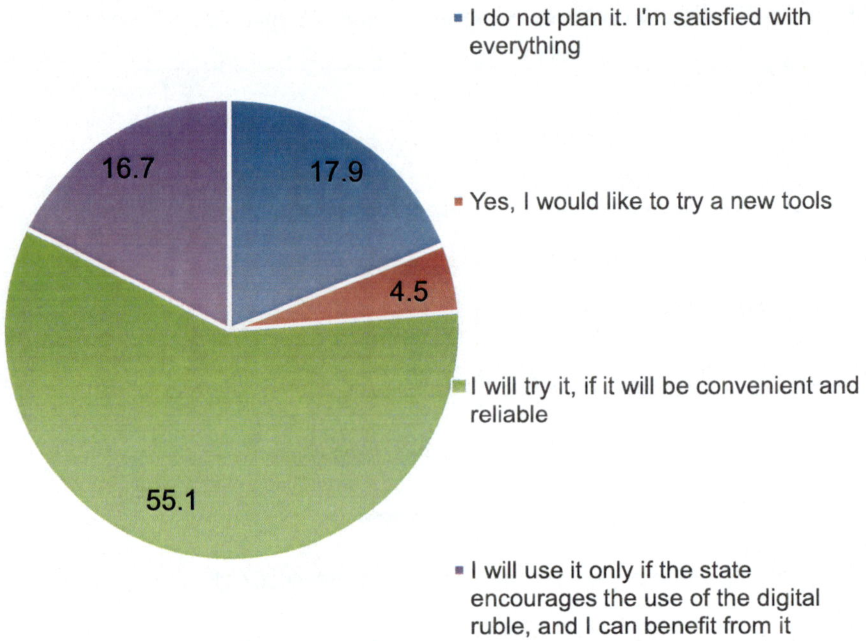

Fig. 2 The structure of respondents' answers to the question, "Do you plan to use the digital ruble in your everyday life?"

more cards in different banks. The priority payment technology is the use of a smartphone and a mobile banking application. Contactless payment with a plastic card occupies the second place. When asked about submitting their own biometric data, plans to introduce a digital ruble, 21.8% of respondents provided their biometric data to the bank and the same number are not going to do this at all. The rest are not ready to decide now, because they are not sure about the safety of personal information. 59% of respondents are aware of the introduction of a project to implement the digital ruble in the Russian Federation in 2022, and 76% are generally ready for its introduction if it is convenient, reliable, and profitable (Fig. 2).

Determining the factors that will be considered when evaluating new payment instruments, the respondents keep their characters and choose the following as the key ones (Fig. 3):

In addition, 81.4% of respondents cannot imagine their modern life without non-cash money. The data obtained during the survey generally allow to draw conclusions about the factors that will determine the evolution of retail payment instruments in Russia in the coming years and, accordingly, may be useful to regulators and commercial banks to further expand the use of digital payment instruments.

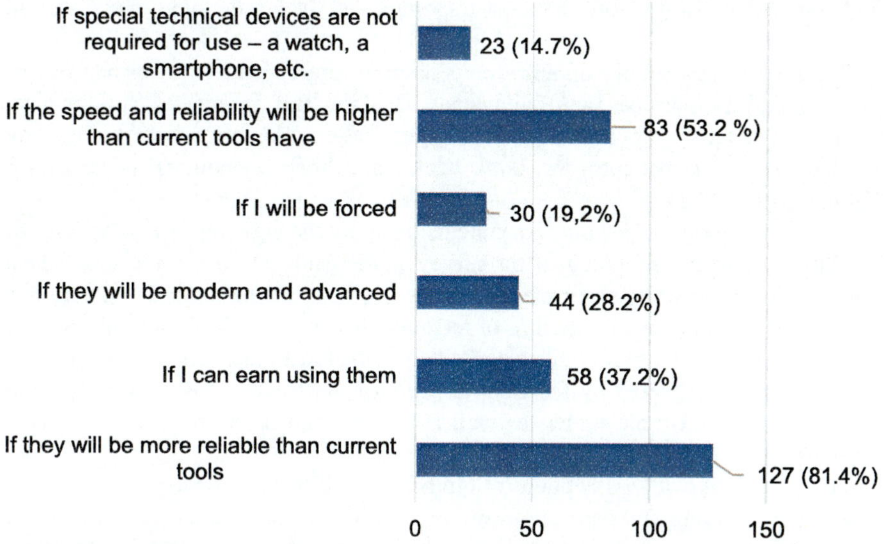

Fig. 3 Respondents' opinions: what will increase the activity of using the digital ruble or other new payment instruments

Discussion

Currently, central banks in many countries are actively planning the launch of centralized digital money—Central Bank Digital Currencies (CBDC). To date, no country has formally introduced a large-scale CBDC mode, but many central banks have launched pilot programs and research projects to determine the viability and usability of CBDCs. China is the undisputed leader in this matter—the digital Yuan has been tested since April 2020. Caribbean countries and Nigeria launched their CBDCs in 2021, Hong Kong, Thailand, Venezuela, and Ghana are preparing to launch. The United States also plan to introduce the digital dollar. Testing of the digital ruble in Russia should begin in 2022, so in the future it will be necessary to investigate the consequences of its implementation. It is very likely that the future of money will be a combination of centralized, decentralized, account and token based, with CBDC, stable coins and cryptocurrencies settlements acting alongside traditional digital and physical currencies.[1]

Eliminating the shortage of physical money, the need to replace worn-out ones, withdraw them from circulation, carry out the cost of issuing, reissuing paper money

[1] Central Bank Digital Currencies and the Future of Money [Electronic source]. URL: https://www.pwc.com/m1/en/media-centre/2021/documents/central-bank-digital-currencies-and-the-future-of-money-part1.pdf (access date 20.01.2022).

and coins have always been the most important task facing the governments of all states.[2]

The positive impact of digitalization is obvious: digital leaders in various sectors of the economy overtake non-digital ones. Banking and, in general, the financial sector is the most important leader in digitalization. Remote access to banking services provides benefits for both banks and their customers (Harasim & Klimontowicz, 2021).

This study allows to formulate a conclusion about the high level of acceptance of the latest payment technologies by survey participants. At the same time, most participants see insufficient protection from the risks (technological, operational, fraudulent) that accompany the use of technologies as the main limitation on these tools' distribution. This fact quite clearly defines the most important area of work for the regulator represented by the Central Bank of the Russian Federation that will become the digital ruble issuer, as well as commercial banks providing payment services.

As E. Semerikova (2019) points out in her paper "Payment strategies of Russian consumers: attitudes and perceptions of payment instruments", the personal range of payment instruments and methods has become more extensive, and the consumers' choice of which one to use for payment began to depend not only on socio-demographic factors, but also on their attitudes, perceptions and emotional reactions.

The widespread adoption of digital payment instruments is now common in many emerging markets. In India a ban on the use of large denomination cash notes at the end of 2016 was the catalyst for the growth of cashless payments. In the countries of Latin America money transfers are offered through the WhatsApp service. In the USA 27% of small businesses report an increase in mobile payment services (like Apple Pay) during the crisis. Thus, contactless payments and mobile transfers have become the main impetus for the cashless payments development and have become a competent alternative to cash payments, including under the influence of the COVID-19 pandemic (Teng & Khong, 2021).

Economically developed countries are also undergoing a transformation in the payment sector. In London, with a population of 8 million, cash has long been considered the "king" in retail, but by 2017 cash payments across the UK had fallen by 15%, and debit and credit cards had become the predominant payment method. Cash payments account for 40% of customer payments and are expected to drop to 21% by 2026 (Fleck & Ozlanski, 2021).

Another manifestation of the contactless payment's growth is the active use of wearable devices—smartphones, watches, rings, key fobs, and other devices. Wearable payments increased by 365% between 2017 and 2020, and a quarter of

[2]The History of Coins. Certified Coin Consultants Inc. Powered by Shopify. 2021. URL: https://certifiedcoinconsultants.com/pages/the-history-of-coins. (access date 23.12.2021).

Europeans plan to increase them further.[3] Allied Market Research predicts that the wearable payment market will reach $1.37 trillion by 2027.[4]

Wearable devices significantly expand the possibilities of making payments and provide access to a wider range of financial services. Payment instruments are becoming not just an applied means of payment, but also an image element, an attractive and convenient item that performs several functions. At the same time, it gives banks more opportunities to attract customers' attention and increase sales per customer. The list of banking payment services is expanding significantly, the customer has an ever-wider choice of tools and means of payment that can be integrated into his personal payment strategy.

Conclusion

The study was aimed at identifying the main factors that determine the development of retail payment instruments in modern conditions. During the study it was found that the key factors are:

- minimizing the risks of using payment instruments.
- ease of use.
- efficiency of calculations.
- financial benefit (availability of business partners of the bank, cash back, etc.).

Other factors are less significant from the users' point of view. Moreover, the factors that influence the consumers' choice when using one or another payment instrument are usually considered as a whole—the desire to reduce risks is taken into account along with convenience, the speed of money transfer—with the possibility of obtaining benefits. The results obtained can help financial institutions, especially banks, create higher quality products that solve the problems of users of certain payment instruments and develop strategies that consider the preferences and behavioral characteristics of Russian users.

References

Adhikary, A., Diatha, K. S., Borah, S. B., & Sharma, A. (2021). How does the adoption of digital payment technologies influence unorganized retailers' performance? An investigation in an

[3] Wearable Tech 2021: The Future Is Now. URL: https://smartpaymentassociation.com/index.php/liste-documents/public-resources/position-papers/867-21-04-27-spa-wearables-payment-final/file (access date: 21.12.2021).

[4] Just a tiny chip: how Mastercard envisions the future of wearable tech Money [Electronic source]. URL: https://www.thedrum.com/news/2020/12/15/just-tiny-chip-how-mastercard-envisions-the-future-wearable-tech (access date 20.01.2022).

emerging market. *Journal of the Academy of Marketing Science, 49*(5), 882–902. https://doi.org/10.1007/s11747-021-00778-y

Chaveesuk, S., Khalid, B., & Chaiyasoonthorn, W. (2021). Digital payment system innovations: A marketing perspective on intention and actual use in the retail sector. *Innovative Marketing, 17*(3), 109–123. https://doi.org/10.21511/im.17(3).2021.09

Dimitrova, I., Öhman, P., & Yazdanfar, D. (2021). Barriers to bank customers' intention to fully adopt digital payment methods. *International Journal of Quality and Service Sciences*. https://doi.org/10.1108/IJQSS-03-2021-0045

Fleck, E. M., & Ozlanski, M. E. (2021). Cash: Never leave home with it? *CASE Journal, 17*(2), 182–201. https://doi.org/10.1108/TCJ-06-2019-0055

Harasim, J., & Klimontowicz, M. (2021, September). Digitalisation as a determinant of new payment methods' development: The evidence from Eurozone and Poland. In *ECIE 2021 16th European Conference on Innovation and Entrepreneurship* (Vol. 1, p. 423). Academic Conferences. https://doi.org/10.34190/EIE.21.194.

Huterska, A., Piotrowska, A. I., & Szalacha-Jarmużek, J. (2021). Fear of the covid-19 pandemic and social distancing as factors determining the change in consumer payment behavior at retail and service outlets. *Energies, 14*(14), 4191. https://doi.org/10.3390/en14144191

Khutko, A. (2019). *Russia has become one of the world leaders in terms of FinTech services penetration*. Ernst & Young Global. https://www.ey.com/ru_ru/news/2019/11/news-ey-fintech-survey-2019.

Korobeynikova, O. M., Korobeynikov, D. A., Popova, L. V., Glazova, M. V., & Kuzmina, E. V. (2021). Innovations in the retail payment market of Russia. In *Socio-economic systems: Paradigms for the future* (pp. 421–430). Springer. https://doi.org/10.1007/978-3-030-56433-9_44

Krivosheya, E. (2020). The role of financial innovations in consumer behavior in the Russian retail payments market. *Technological Forecasting and Social Change, 161*, 120304. https://doi.org/10.1016/j.techfore.2020.120304

Sahi, A. M., Khalid, H., Abbas, A. F., & Khatib, S. F. (2021). The evolving research of customer adoption of digital payment: Learning from content and statistical analysis of the literature. *Journal of Open Innovation: Technology, Market, and Complexity, 7*(4), 230. https://doi.org/10.3390/joitmc7040230

Semerikova, E. (2019). *Payment strategies of Russian consumers: Attitude and perception of payment instruments*. 2nd International Conference on Applied Research in Management, Economics and Accounting. https://doi.org/10.33422/2nd.iarmea.2019.06.1047.

Teng, S., & Khong, K. W. (2021). Examining actual consumer usage of E-wallet: A case study of big data analytics. *Computers in Human Behavior, 121*, 106778. https://doi.org/10.1016/j.chb.2021.106778

Electric Power Industry of Russia in the Transition to a Low-carbon Economy

Natalia V. Vasilenko

Introduction

Over the past 150 years, there has been steady growth in electric energy consumption: it grew by a factor of 6.6 between 1860 and 1900, 3 between 1900 and 1950, 5 between 1950 and 1995, and 6.5 over the past 20 years (Konstantinov et al., 2020). The growth rate accelated in the last decade primarily due to two reasons: digitalization of business processes and a shift in the energy mix towards decarbonization.

Digitalization, which solves problems associated with collecting, processing, trans-mitting, storing, and using data by means of digital telecommunication channels, requires a lot of electricity. According to the Higher School of Economics, in 2020, Russia ranked 43rd out of 63 countries (between Italy and Turkey) in the World Digital Competitiveness Index and 42nd out of 79 countries (between Romania and Oman) in the Global Connectivity Index, with the United States and Singapore leading both rankings (https://issek.hse.ru/).

Decarbonization, being part of the fourth energy transition (Smil, 2016) and supported by the Paris Agreement, implies an increase in the share of clean energy in national energy mixes, which can be achieved, among other things, by using cars with electric motors in-stead of those with internal combustion engines and adapting renewable energy on a large scale (Nuno, 2021). Decarbonization is a means of solving environmental problems and reducing human impact on the environment be means of reducing the global carbon footprint (Ilinova et al., 2020). In the BP's ranking of countries by greenhouse gas emissions, Russia ranks fourth after China, the United States, and India, with its GDP having much higher carbon emission

N. V. Vasilenko (✉)
St. Petersburg University of Management Technologies and Economics, St. Petersburg, Russian Federation
e-mail: n.vasilenko@spbacu.ru

A. Rumyantseva et al. (eds.), *Challenges and Solutions in the Digital Economy and Finance*, Springer Proceedings in Business and Economics,
https://doi.org/10.1007/978-3-031-14410-3_42

intensity compared to the top ten countries and being very close to that of India (0.912 and 0.936 t/thousand US dollars, respectively) (Kharitonova et al., 2021).

The transition to a low-carbon economy that ensures sustainable development in such domains as ecology and economics will inevitably affect the key infrastructure industries of the economy, among which are transportation, telecommunications, and power generation.

As a result of the reform 1998–2008, the centralized state regulation of the Russian electric power industry was supplemented by market mechanisms where possible. However, the industry is still highly concentrated and is influenced a lot by the centralized government regulation. Also, there are external factors influencing its development, such as having to function in different climate conditions due to Russia's huge area. Also, a lot of plants and machinery are characterized by a high degree of wear and tear. All these and other factors make it necessary to study in detail whether Russia's electric power industry is ready to function successfully in the context of current low-carbon trends. Therefore, the study aims to identify the baseline conditions, prospects for, and barriers to the development of Russia's electric power industry during the transition to a low-carbon economy.

Materials and Methods

The research methodology is based on the following assumptions.

There are three vectors in the transformation of the electric power industry that correspond to the 3Ds concept (Decarbonization; Decentralization, Digitalization) (Kholkin & Chausov, 2020):

Energy system decentralization implies the inclusion of microgeneration facilities, with some of them operating on renewables, and an increase in the number of prosumers (power network users who both produce and consume) who can be flexible and independent in determining energy supply conditions and selling excess energy to other users of the network (Linder & Lisovsky, 2017). As the number of users grows, so do transaction costs, while the systems efficiency decreases.

A means of reducing transaction costs in distributed generation is to use business models based on the integration of digital solutions and services into a unified digital infrastructure. Developing such an infrastructure for the generation, trade, and consumption of distributed energy is the key task for the digitalization of the electric power industry (Zubakin, 2019). As a result of digitalization, production and transaction costs of electricity producers and consumers are optimized by means of integrated digital services.

The evolutionary nature of the institutional transformation of energy systems during the transition to a low-carbon economy can be described using the mechanism of the institutional market, according to which, if the probability of choosing a certain institutional norm exceeds 2/3 (in our case, the choice is in favor of a new energy source), no additional effort is required for the dominant norm to take over the remaining ones (Tambovtsev, 2001).

To ensure that the institutional transformation is successful, it is necessary to con-sider the initial conditions for such transformation of energy systems, which are formed by:

- the achieved level of development of a particular national economy, which is reflected in such indicators as GDP growth rates, its energy efficiency, and the structure of the primary energy mix;
- national guidelines and policies aimed at achieving sustainable development goals since the fourth energy transition agenda emphasizes the need to prevent climate change be means of decarbonization and switching to environmentally friendly energy sources;
- customer habits in electricity consumption and saving, whose importance is growing in the context of decentralized energy systems.

Results

- An analysis of the structure of Russia's electric power industry shows that during the transition to a low-carbon economy, «path dependence» is manifested in a major share of hydropower and nuclear power in the country's energy mix. Wind and solar energy does not play a big role but is supported by the government and its share is growing fast.
- The institutional environment that stimulates the transition of Russia to a low-carbon economy includes international, national, corporate levels and individual consumers, within which renewable energy sources and intelligent power systems are promoted.
- In Russia the implementation of the 3Ds concept involves changing the corporate, market and consume routines of energy market participants. In Russia's system services market, the most rapidly growing sector is energy demand management.
- Compared with other developed countries, Russia is less prepared for energy transformation. The principal barriers to the development of Russia's electric power industry within the framework of the 3Ds concept are technological and psychological.

Discussion

"Path Dependence" in the Transition of Russia's Electric Power Industry to Low-carbon Energy

Russia's current electricity generation mix is historically conditioned. In the course of the transition from a traditional to industrial economy, the Russian energy sector relied first on water engines, and then on steam engines at the beginning of the industrial revolution. The country lagged behind European nations in the use of the latter (17.4% versus 45% by 1913) and its level of industrialization was low (Alexeyev, 2018). To solve this problem, the first Soviet plan for the electrification of Russia (GOELRO) was implemented from 1920 to 1935, which resulted in the construction of forty modern power plants that mainly relied on local resources such as water and peat (Degtyarev et al., 2016). By integrating power plants into a single power system, a lot of advantages were gained in the field of resource and costs management due to a decrease in the total peak load, the opportunity to reduce the total capacity margin, more favorable conditions for automating the process of electricity transmission, and, consequently, an increase in the security of power supply, which also created more profitable operating modes for various types of power plants (Vagin, 2021).

Russia's transition to a market economy in the 1990s demanded a restructuring of the electric power industry. It took its form in the creation of private property as a share capital with the controlling share belonging to the government. Energy prices were state-controlled and often undercut at the level of regional commissions during this period. As a result, energy companies started experiencing financial problems, which led to a deterioration of their fixed assets and key performance indicators. In the period from 2002 to 2008, power generation decreased and energy quality in the national power grid worsened, which made it necessary to continue with reforms, introduce markets for electricity, capacity and systems services, and develop necessary engineering and commercial infra-structures in the Unified Power System of Russia (Urinson et al., 2020). Due to Russia's large area and unevenly distributed population, several individual power supply systems of various sizes developed in some regions (Siberia, the Far East, the Arctic, Crimea, etc.), which have prospects for replacing diesel power plants operating on fossil fuels with installations running on renewable energy sources due to the opportunity of load shifting in time (Zhukovskiy et al., 2021).

At the current stage of the development of Russia's electric power industry as an infrastructural metasystem, it is necessary to search for opportunities to integrate centralized and decentralized energy supply systems. Russia's energy industry combines the processes of generation, transmission, and distribution of electricity, supplying both companies and individuals. Most electricity in Russia is generated by thermal (62.92%), hydro (17.61%), nuclear (19.32%), wind (0.03%), and solar (0.12%) power plants (https://minenergo.gov.ru/, 2019).

Thermal power plants (TPPs), which generate the major share of electricity, operate on fossil fuels but tend to switch to gas to replace coal and peat. According to the Corporate Energy Institute (https://tesiaes.ru/), more than 350 thermal power plants of various configurations are operating in Russia. They are located in large cities and serve consumers within territorial entities. The largest of these are the Surgutskaya, Reftinskaya, and Kostromskaya TPPs with an installed capacity of 5.6 GW, 3.8 GW, and 3.6 GW, respectively. Most of Russia's TPPs are owned by seven OGKs (generating companies serving the wholesale power market) and fourteen TGKs (territorial generating companies), among which the leaders are Gazprom Energoholding LLC, Unipro PJSC, Quadra PJSC, Enel Russia PJSC (Russian subsidiary of the international Enel Group), and Fortum PJSC (Russian subsidiary of the European company of the same name), etc.

Almost a fifth of all the electricity produced in Russia is generated at eleven nuclear power plants. The largest among them are the Balakovskaya, Kalininskaya, Kurskaya, and Leningradskaya nuclear power plants (NPPs) with a capacity of 4.0 GW each. According to forecasts by the International Renewable Energy Agency (IRENA) and the World Energy Council (WEC), the share of nuclear power in the global energy mix may more than double as a result of the fourth energy transition. Russia's NPPs are operated by Rosenergoatom.

Hydropower plants (HPPs) are the third-biggest energy provider in Russia. The largest HPPs in the country are part of one group: Sayano-Shushenskaya (6.4 GW) and Krasnoyarskaya (6.0 GW) on the Yenisei River; Irkutskaya (0.66 GW), Bratskaya (4.3 GW), and UstIlimskaya (4.32 GW) on the Angara River. In the European part of Russia, a big group of HPPs is located on the Volga and includes nine power plants, of which the Volzhskaya HPP has the highest installed capacity (2426 GW). The total capacity of the group is 10.91 GW (Vagin, 2021). Most of the capacities in Russia's hydropower sector are operated by RusHydro PJSC.

Solar power plants (SPPs) and wind farms produced only 0.12 and 0.03% of all electricity generated in 2019. Solar energy is most developed in Russia's southern regions (the Republic of Crimea, Stavropol Krai, and Astrakhan Oblast) (https://recyclemag.ru/). In the wind power sector, Russia's largest wind farms are the Kochubeevskaya and the Bondarevskaya located in Stavropol Krai with an installed capacity of 210 and 120 MW, respectively, and the Adygea wind farm in the Republic of Adygea with an installed capacity of 150 MW. They are operated by NovaWind JSC, a high-tech division of Rosatom. Fortum PJSC is actively developing wind power (https://www.fortum.ru/razvitie-vetroenergetiki-v-rossii).

The integration of individual power plants into energy systems is based on harmonizing the processes of production, distribution, and consumption of electricity. At present, the Unified Power System of Russia (UPS) consists of seven interconnected power systems (IPSs) that service the territories outlined in the very first GOELRO plan. The power such regions of the country as North-West, Center, Middle Volga, Ural, South, Siberia, and East (https://minenergo.gov.ru/). About 90% of Russia's power transmission lines with voltages from 220 kV to 750 kV are operated by Rosseti FGC UES, a natural monopoly which is a subsidiary of Rosseti. As of the end of 2020, the assets of Rosseti FGC UES included 149.1 thousand km of high-voltage transmission lines, 889 substations with a total capacity of 352 GW, and 21.7 thousand employees (https://www.fsk-ees.ru). In the process of fulfilling contracts for the provision of electricity transmission services in 2020, the company interacted with 8408 suppliers and 669 consumers. In the power distribution sector, fourteen interregional grid operators service lines with a voltage of 0.4 to 110 kW and about 3000 territorial grid operators service mainly lines with a voltage of 0.4 to 10 kW. These types of operators share the market for power grid services in the proportion of 7 to 3.

Institutional Environment Stimulating the Transition to a Low-carbon Economy

In modern conditions, the institutional environment that stimulates the transition of countries to a low-carbon economy consists of various levels:

- international level, which includes international agreements initiated by the UN and creating responsibilities for individual countries.
- national level, which includes documents, strategies, and structures regulating the development of the electric power industry and related industries to achieve decarbonization goals.
- corporate level, which includes documents and strategies that steer company policies towards using renewable energy sources and fostering digitalization and decentralization as the most important areas of transformation of the electric power industry.
- consumer level, where it is important that customer habits should create demand for energy optimization technologies in the distributed network.

At the international level, the key elements for Russia are as follows: Kyoto Protocol (1997) that supplements the UN Framework Convention on Climate Change (1992), Paris Climate Agreement (2015), Sustainable Development Goals (SDGs) (2015), etc.). All of these initiatives imply specific measures for the implementation of a general decarbonization policy to achieve climate stabilization by 2050. We should also note the importance of scenario forecasts which have been is-sued in the last decade by the International Energy Agency (World Energy Outlook), the International Renewable Energy Agency (IRENA) (World Energy Transitions Outlook), the US Energy Information Administration (International Energy Outlook), OPEC (World Oil Outlook), BP (Energy Outlook), and other international organizations.

Institutional support for the development of Russia's electric power industry at the national level is made of the following:

- federal legislation that includes federal laws regarding the electric power industry ("On Electric Power Industry" (2003), "On Natural Monopolies" (1997), "On Protection of Competition" (2006), "On the Use of Atomic Energy" (1995)), those governing related areas ("On Technical Regulation" (2012), "On Environmental Protection" (2002), "On the Security of the Critical Information Infrastructure of the Russian Federation" (2017)), and others.
- documents regulating the operation of infrastructure organizations and energy markets approved by the President of the Russian Federation and the Government of the Russian Federation, including those in the fields of electric power systems operation and control, pricing policies, information disclosure by subjects of the wholesale and retail electricity markets, goods and services quality assessment, monitoring investment programs in the energy sector, antimonopoly regulation and control, and others.

- strategic documents for the development of the industry, which can be divided into three groups:
  - documents on environmental issues ("Fundamentals of State Policy in the Field of Environmental Development of the Russian Federation for the Period up to 2030" (2012), "Strategies for the Environmental Safety of the Russian Federation for the Period up to 2025 (2017)", etc.);
  - documents on research and development ("Strategy for Advancement in Research and Development of the Russian Federation" (2012, 2021), etc.);
  - documents regarding the energy sector itself ("The Doctrine of Energy Security of the Russian Federation" (2019), "Energy Strategy of the Russian Federation for the Period up to 2035" (2020), "Strategy for the Development of the Electric Power Transmission Systems of the Russian Federation" (2013, 2017), "General Layout of Power Facilities Until 2035" (2017), "Scheme and Program for the Development of the Unified Power System of Russia for 2021– 2027" (2021)).The documents of the third group note that the cost of electricity supplied by the Unified Power System of Russia and the cost of electrical energy generated locally by industrial consumers (including construction costs) are often almost the same, which, among other things, which stimulates distributed generation and encourages consumers to stop using the electricity provided by centralized power systems. Also, it is noted that the industry's fixed assets continue to deteriorate, which leads to the need for greater investments in the coming years to maintain the already achieved indicators of quality and supply security. Hence, the key tasks for the government in the field of power distribution include creating economic methods to stimulate the efficiency of grid operators, providing conditions for electricity price stabilization, and attracting new capital to the electric grid systems in an amount sufficient for upgrading and reconstructing power grids in order to ensure power supply security.

A special place in the institutional support of Russia's electric power industry is occupied by the trend towards developing the renewable energy sector. It paves the way for decarbonization as it contributes to an increase in the share of green energy in the energy mix and solves decentralization problems as renewable power installations can be used separately from the UPS, with such configurations enabling electricity flows in both directions.

The opportunity to deliver renewable power to the grid was first discussed in a decree of the Government of the Russian Federation in 2008. In 2011, by order of the Ministry of Energy of Russia, a plan was approved for constructing power generation facilities based on the use of renewable energy sources on the territory of the Russian Federation. As a result of competing for government support in 2013–2015, 93 solar, wind, and small hydropower projects were implemented (Ratner & Nizhegorodtsev, 2017).

An important direction in the institutional support of Russia's electric power industry is the improvement of information exchange systems between the subjects of the electric power industry, including the introduction of smart metering systems with a guarantee of access to a particular set of functions.

It should also be noted that Russia's strategic documents in the field of the development of the electric power industry build upon earlier documents. Regulations on the topic are constantly updated to take into account the need to improve the quality and security of energy provided to Russian companies and citizens and the trends in the transformation of energy systems all around the world.

Practices and Consumer Behavior in Energy Markets

What makes customer behavior particularly important is that it is the factor that spurs real institutional changes in the energy system and, consequently, in the entire economy of the country, while the legislation and strategic institutional initiatives only create the prerequisites for such changes.

Contracting in the energy market can be analyzed at three levels:

- corporate level, which can be analyzed by looking at whether the lines of development in energy companies correspond to the 3Ds concept discussed in Section "Results" .
- market level, which can be studied by studying the problems associated the markets for capacity, electricity, and systems services.
- consumer level, which can be studied by analyzing demand response.

Based on the results of a content analysis in which we studied reports and strategic documents issued by leading power generating and power distribution companies (Gazprom Energoholding, Unipro, Quadra, Enel Russia, Fortum, Rosenergoatom, RusHydro, and Rosseti), the following two changes can be identified in the behavior at the corporate level. The first one is investments in projects for constructing solar power plants and wind farms.

The second change manifests itself in projects for making electric power systems automated and intelligent on six different levels: control centers, system and equipment troubleshooting, maintenance and repairs, procurement and resource management, process control and analysis (including labor costs analysis), compliance with occupational safety and health rules and regulations. An example is the introduction of smart meters in Kaliningrad, Yaroslavl, and Tula in cooperation between Rosseti and the Russian Direct Investment Fund.

In 2011, Russia launched a market for systems services to maintain the required level of security and quality of the UPS's functioning. It is still developing and includes different kinds of services. Currently, the most actively developing sector in the market is the one for demand response services. In 2020, the total sum paid for such services amounted to 1,434,299 thousand rubles (https://www.so-ups.ru/).

As for consumer stimulation within the demand response framework, it is assumed that it will make the demand for electricity more elastic, with consumers reducing electricity consumption when prices grow. Electricity demand elasticity can be increased by the use of smart metering technologies that rely on state-of-the-art hardware, latest software, and interval meters. Smart metering facilitates data collection and processing and pro-vides for automatic and remote control.

Currently, only about 9% of Russia's electricity meters belong to the category of smart meters (7 out of 80 million), while in Europe this figure reaches almost 90% (Hovalova & Zholnerchik, 2018). About two-thirds of electricity meters in Russia are installed in households, which means that it is individuals who can greatly influence the state of electricity metering systems, but the figures show that the choice for smart systems is not a popular one.

Assessment of Russia's Readiness for the Energy Transition: The Position of the World Energy Forum

Countries demonstrate different levels of preparedness for transforming their energy systems. To measure them, such indicators are used as the Energy Transition Index (ETI), which forms a ranking of 115 countries. It allows for assessing the current efficiency of national energy systems and coordinating government policies with market factors on the way to transforming these systems.

As of 2021, ten developed European countries (led by Sweden, Norway, and Den-mark) topped the ETI ranking. Some other economies ranked as follows: the US—24, China—68, Russia—73, India—87 (WEF Fostering Effective Energy Transition 2021). The group of countries that includes Russia (Commonwealth of Independent States) improved their aggregate ETI by 5% over the last decade thanks to advances in environmental sustainability, access to energy and quality of electricity supply. A bigger improvement (6%) was only demonstrated by emerging and developing Asian countries.

With a Global Average of 59%, Russia's ETI is 56%, which is similar to the results demonstrated by Jordan, Indonesia, and Bolivia. It places Russia between the United Arab Emirates, Vietnam, Morocco, Philippines, China, and Sri Lanka as the upper bound, and Oman, Tajikistan, Egypt, and Guatemala at the lower bound.

Russia's system performance indicator (SP), which characterizes the efficiency of the energy system, is 66% (between the Philippines (66.5%) and Algeria (66.2%)). The following factors positively influence Russia's SP: quality of electricity supply—8 out of 8; reduction in CO2 intensity (kg CO2/GJ) for the period from 2010 to 2019–5%. At the same time, Russia's SP was negatively influenced by a slowdown in economic growth associated with the volatility of global oil and gas prices, including due to the COVID-19 pandemic.

The transition readiness indicator (TR), which reflects the stability of the political environment, the state of the investment climate, the level of consumer involvement, as well as the intensity of the development and implementation of new technologies, is 45.7% (between Honduras (46.1%) and Ukraine (45.5%)) with an average of 54.7%. The low value of this indicator is explained by the fact that at present Russia is not yet among the countries with a clearly defined and documented goal of achieving absolute zero CO2 emissions, and also has rather modest (in comparison with the top ten countries) investments in both mature renewable energy technologies and new ones.

Barriers to the Development of Russia's Electric Power Industry in the Transition to a Low-carbon Economy

For the Russian electricity industry to succeed in its institutional transformation during the transition to a low-carbon economy, progress is needed in the following areas concerning decarbonization: legislation; industrial infrastructure; financial support; consumer readiness. If there is no progress in these areas, it creates barriers for the whole industry.

The analysis presented above of the institutional environment in terms of whether it stimulates Russia's electric power industry towards the transition to a low-carbon shows that the regulatory framework of Russia undergoes discrete changes corresponding to the transformation trends witnessed in the global electric power industry. However, the state of the industrial infrastructure run by energy companies and grid operators is characterized by a high degree of depreciation of fixed assets cause by wear and tear and becoming obsolete. The wear and tear rate of equipment exceeds 40% at TPPs, is about 50% at HPPs, and reaches 63% at substations. The share of distribution networks with expired service life is 50%, and 7% of electrical grids have been operating twice as long as expected. Overall, the wear and tear rate of distribution networks has reached 70%. The wear and tear rate of the main power grids operated by Rosseti FGC UES is about 50% (https://minenergo.gov.ru/).

Energy system transformation should be supported by up-to-date technologies. For the transition to a low-carbon economy, such technologies are needed as photovoltaics, controlled thermonuclear fusion, high-capacity batteries that can be recycled, electric vehicles, etc. The digitalization of the electric power industry is supported by the development of the Internet of Energy technology and by products and services such as smart turbines, digital power plants, virtual demand aggregators, energy hedging, and others. It is as a result of the development of technologies that the costs of power generation using renew-able sources have been steadily decreasing. From 2011 to 2019, they de-creased by 35% and 20% for solar power and wind power, respectively (http: /www.ren21.net), and the prices of electricity generated with their help decreased by 73% and 22% for the period from 2010 to 2018 (IRENA 2019). Currently, it is considered promising to develop and test mini-grids (Belsky & Dobush, 2019) as well as autonomous hybrid energy systems consisting of photovoltaic, wind, diesel, and battery energy storage systems that can have different configurations depending on the payback period and CO2 emission limits (Lavrik et al., 2021).

Institutional transformation of Russia's electric power industry is currently financially supported by the government program titled "Energy Sector Development".

Studying the experience of carrying out institutional changes, including those in the field of digital transformation of energy companies, shows that a great barrier in such a situation is not costs associated with the implementation of a technological or digital solution but the readiness of the company's employees and clients to welcome innovations (Afanasev & Vorontsov, 2019), which is influenced by three key factors: general awareness; being sure in job security, career prospects, or comfort in energy consumption; willingness to incur costs associated with change in anticipation of future benefits that exceed costs.

The analysis presented above allows for concluding that the most successful adaptation to the transition to low-carbon energy is demonstrated at the corporate level in the context of hierarchical relations and management transactions. Changing behaviors at the market level is hindered by the imperfection of the mechanisms of contractual interaction inherent in the legal and regulatory framework. At the consumer level, habits will be changed only when Russian consumers accept the "choice-responsibility" dichotomy of the market in the field of demand response and become aware of new digital ways. All this taken together creates a rather high psychological barrier.

Conclusion

The success of institutional transformation of electric power industry in the transition to a low-carbon economy depends on the initial conditions and is measured by how close the evolutionary trajectory of the institutional environment is to the desired directions, which are framed in the presented research by the 3Ds concept.

The initial conditions include the fact that the country's electric power industry used to develop according to a unified national plan before being restructured to promote competition among energy producers and on the electricity market within the Unified Power System. The evolution of legislation and the regulatory framework takes the form of discrete changes and moves towards improving the availability of energy markets. All this is due to Russia's path dependence that has its roots in the first Soviet plans and the reforms of the period from 1998 to 2008.

In general, the evolution of Russian's institutional environment goes in line with the current climate agenda. Decarbonization of Russia's energy sector is manifested in stimulating the renewable energy sector and making renewable power more available to the wholesale and retail electricity and capacity markets. The decentralization of Russia's energy system is facilitated by the development of demand response services and the introduction of automated and intelligent electricity systems in the course of implementing digitalization projects.

At the same time, Russia's overall readiness for energy transformation remains be-low average compared to other countries. First of all, this is explained by technological and psychological barriers to the development of the electric power industry in Russia during the transition to a low-carbon economy.

References

Afanasev, V., & Vorontsov, N. (2019). Intelligent digital solutions for increasing operational efficiency and labor productivity in electric power industry. *Vestnik Universiteta, 9*, 39–47. https://doi.org/10.26425/1816-4277-2019-9-39-47

Alexeyev, V. (2018). The energy challenge for imperial Russia. *Ural Historical Journal, 2*(59), 6–12. https://doi.org/10.30759/1728-9718-2018-2(59)-6-12

Belsky, A., & Dobush, V. (2019). Small wind-driven power plant operating experience. *IOP Conference Series: Materials Science and Engineering, 489*, 1–6. Accessed August 24, 2021, from. https://doi.org/10.1088/1757-899X/489/1/012013/meta

Degtyarev, K., Zalikhanov, A., & Solovyev, A. (2016). GOELRO plan and renewable energy sources. *Energy Policy, 3*, 55–65.

Hovalova, T., & Zholnerchik, S. (2018). The effects of the introduction of smart grids. *Strategic Decisions and Risk Management, 2*(105), 92–101.

Ilinova, A., Romasheva, N., & Stroykov, G. (2020). Prospects and social effects of carbon dioxide sequestration and utilization projects. *Journal of Mining Institute, 244*, 493–502. https://doi.org/10.31897/pmi.2020.4.12

Kharitonova, N., Kharitonova, E., & Pulyaeva, V. (2021). Carbon footprint of Russia: Realities and prospects of economic development. *Russian Journal of Industrial Economics, 14*(1), 50–62.

Kholkin, D., & Chausov, I. (2020). The new formula of energy transition. *Energy Policy, 12*(154), 40–53.

Konstantinov, A., Konstantinova, Y., Kvak, D., Koltsov, V., & Stanishevsky, S. (2020). Energy transition. Past and future. *Transport of the Asia-Pacific Region, 4*(25), 84–89.

Lavrik, A., Zhukovskiy, Y., & Tcvetkov, P. (2021). Optimizing the size of autonomous hybrid microgrids with regard to load shifting. *Energies, 14*, 5059–5059. https://doi.org/10.3390/en14165059

Linder, N., & Lisovsky, A. (2017). The development of the electricity market in Russia: Main trends and prospects. *Business Strategies, 2*, 48–54.

Nuno, C. (2021). The effects of corruption, renewable energy, trade and CO2 emissions. *Economies, 9*(2), 62. https://doi.org/10.3390/economies9020062

Ratner, S., & Nizhegorodtsev, R. (2017). Barriers to energy efficiency: An empirical study. *Economics of Contemporary Russia, 4*(79), 103–117.

Smil, V. (2016). *Energy transitions: Global and national perspectives* (Vol. 2, 2nd ed.). Praeger, ABC-CLIO. publish-er.abc-clio.com/9781440853258

Tambovtsev, V. (2001). Institutional market as a mechanism of institutional changes. *Social Sciences and Contemporary World, 5*, 25–38.

Urinson, Y., Kozhukhovski, I., & Sorokin, I. (2020). The Russian electricity reform: Achievements and unresolved issues. *HSE Economic Journal, 24*(3), 323–339.

Vagin, G. (2021). The state and prospects of development of the electric power industry in Russia. *Intelligent Electrical Engineering, 2*(14), 4–14.

Zhukovskiy, Y., Lavrik, A., Vasilkov, O., & Semenyuk, A. (2021). Potential for electric consumption management in the conditions of an isolated energy system in a remote population. *Sustainable Development of Mountain Territories, 4*(12), 583–591.

Zubakin, V. (2019). State stimulation of transformation of power industry. *Strategic Decisions and Risk Management, 10*(4), 320–329. https://doi.org/10.17747/2618-947X-2019-4-320-329

Cryptocurrency Development in Russia

Natalya Vitalievna Usova, Mikhail Pavlovich Loginov, and Elizaveta Igorevna Makovkina

Introduction

The legitimacy of the circulation of cryptocurrencies in national financial markets is in the center of attention of various states. However, in Russia until recently there has been debate regarding the feasibility of recognizing cryptocurrencies and their legal circulation in the country. As part of this study, the authors are interested in a very debatable issue regarding the resolution of cryptocurrency in the territory of the Russian Federation and the possible consequences of making this decision.

Materials and Methods

The authors have studied both the scientific works of several researchers and analytical materials on cryptocurrencies. During the study, the authors used methods such as analysis, comparison, and grouping.

Rodina (2021) presented a political economy approach to the emergence of cryptocurrency. Gorda (2018) in his work analyzed the global cryptocurrency market in terms of risks and regulatory mechanisms. Tkach and Lazarchuk (2022) also contemplated cryptocurrencies on a global scale, namely, they considered the world practice of taxation of operations with cryptocurrency. Eshkov (2020) recognized in his article cryptocurrency as a digital financial asset. Khadzhiev et al. (2020) explore cryptocurrencies as a form of private money. The potential of cryptocurrency in the context of the evolution of technological structures is presented in the work of Safiullin et al. (2020).

N. V. Usova (✉) · M. P. Loginov · E. I. Makovkina
Ural State University of Economics, Yekaterinburg, Russian Federation

A. Rumyantseva et al. (eds.), *Challenges and Solutions in the Digital Economy and Finance*, Springer Proceedings in Business and Economics,
https://doi.org/10.1007/978-3-031-14410-3_43

Results

Before examining various points of view on cryptocurrency, we will review various approaches to the category of "cryptocurrency." This is necessary due to the transition to a digital model of the national economy and the formation of digital segments in various areas of activity.

Khadzhiev et al. (2020) adhere to the point of view that "cryptocurrency is a financial non-state banknote used in circulation by individual entities". In turn, Rodina (2021) notes that "modern cryptocurrency is a harbinger of a full-fledged electronic instrument of the highest liquidity for the universal measurement of value". Mikhailov (2020) presented the following approach in his work that "cryptocurrency is a type of digital currency, the creation and control of which is based on cryptographic methods". Thus, cryptocurrency is a digital currency issued by indefinite individuals or private companies that are not secured and not having a physical expression.

The issue of transition to digital currencies is a logical continuation of the formation of a digital model of the economy and the transformation of economic processes into digital ones. It cannot be said that cryptocurrency is an obligatory element of the financial system of any country, and it should be used as a means of payment. Nevertheless, cryptocurrency will be a participant in financial relations, in any case, including the shadow economy. It should be noted that the pandemic and the transition to a digital model of the national economy have increased the importance of developing a digital financial market. There are several areas of development that are considered in the Russian Federation: development of digital technologies and digital services in the financial market; introduction of a national digital currency issued by the Central Bank of the Russian Federation and secured by the state; regulation of the cryptocurrency's circulation.

Within the framework of this study, it is the third issue that is of interest to the authors. As Mikhailov (2020) notes. "A monetary system based on a digital currency may seem attractive, since it provides an opportunity to limit the role of the central bank". In addition, the refusal or limited use of the possibilities of digital currencies can lead to a number of negative consequences. For example, it can lead to an outflow of financial resources and the growth of the shadow economy. Safiullin et al. (2020) note that "cryptocurrencies have the potential to eventually become one of the world or regional currencies, given the unique characteristics that this technology has (limited emission, crypto-protection, decentralization of settlements, smart contracts, tokens)".

Given that at the global level there are different approaches to the possibility of using cryptocurrencies, in Russia this issue has become a priority relatively recently. The authors present a classification of countries according to the status of cryptocurrencies based on the work of Gorda (2018) (Table 1).

Thus, we see that in most cases, cryptocurrency is recognized as an asset. In turn, discussions regarding the eligibility of using cryptocurrencies in the territory of the Russian Federation began in January 2022. We will examine the points of view that

Table 1 Classification of countries based on the status of cryptocurrencies

Country	Cryptocurrency status
Cryptocurrency is considered an asset	
Australia	Asset for capital gains
Brazil	Asset
Great Britain	Asset
Israel	Asset
Norway	Asset
Cryptocurrency is considered a product	
Canada	Product
USA	Cryptocurrency is a product. ICO tokens—securities
Cryptocurrency is considered a financial service	
Finland	Financial service
Other approaches	
Belarus	Activities in the field of high technologies
Denmark	Not considered a currency
Netherlands	Barter operation
Germany	Private money. Settlement currency
Singapore	Not considered a currency or a product
Switzerland	Foreign currency
Japan	Asset-like values used in payments

existed at that time. The Central Bank of the Russian Federation acted as an opponent of cryptocurrencies. The mega-regulator presented his point of view in the report for public discussions "Cryptocurrencies: trends, risks, measures". As risks, high volatility and a significant prevalence of fraud, restriction of the sovereignty of monetary policy, the withdrawal of citizens' savings from the country, and the growth of illegal activities were identified. To minimize the identified risks, the Central Bank of the Russian Federation proposed to ban the use of cryptocurrency. This includes not only the use of cryptocurrencies as a means of payment, but also the organization of the issuance or circulation, as well as the possibility of investing financial institutions in cryptocurrencies. A different point of view is shared, for example, by the Ministry of Digital Development, Communications and Mass Media of the Russian Federation, as well as the Ministry of Finance of the Russian Federation. So the Ministry of Digital Development adheres to the position that the ban on the circulation of cryptocurrencies in the national market will have a negative impact. The ban will find its expression in a slowdown in the development of the IT sector, as a result, in a decrease in the level of the country's competitiveness in the international arena. In this case, it is advisable to exercise regulatory influence on the cryptocurrency market, but not prohibit it. A similar position regarding the need to ban cryptocurrency is shared by the Ministry of Finance of the Russian Federation. In particular, it is noted that the ban on the high-tech industry is inappropriate. Recently, the world has become very virtualized and the importance of forming a legal framework for the possibility of protecting citizens when using cryptocurrencies has increased.

Discussion

Based on the results of the review and using analytical materials, the authors structured and supplemented the risks and opportunities for the national economy from cryptocurrencies (Table 2).

Thus, we see that there are a fairly large number of risks and opportunities associated with the functioning of cryptocurrencies in the national market.

The authors agree with the opinion of the Ministry of Digital Development and the Ministry of Finance of the Russian Federation. However, at the same time, the fears of the Central Bank of the Russian Federation and their position are understandable. Thus, in our opinion, it is necessary to take into account all the pros and cons that were presented above developing regulatory legal acts, as well as determining the strategic directions for the development of the national financial market in the medium term.

Thus, the authors propose the following activities:

- to determine at the legislative level the possibilities of using cryptocurrencies and responsibility for the illegal circulation of cryptocurrencies in the country.

Table 2 Risks and opportunities of using cryptocurrencies for the national economy

Risks	Opportunities
Growth of the shadow economy	Development of digital infrastructure
Decrease in the level of economic security	Development of the national financial market digital segment
Depreciation of the national currency	Attraction of foreign investments in various projects
Increasing the influence of foreign capital	Protection of citizens in the cryptocurrency market, subject to the formation of an appropriate regulatory framework
Destabilization of the national economy	Verification the legality of transactions in order to prevent illegal actions (money laundering, financing of terrorism, purchases of prohibited goods in the country)
Increasing incidence of fraud and cybercrime	Increase in the number of jobs due to the legalization of mining as an entrepreneurial activity
Creation of a financial pyramid	Growth of tax revenues to the federal budget
Reorientation of the population and legal entities from national currencies, including digital ones, to cryptocurrencies	Avoiding anonymity when conducting transactions due to the creation of system organizers and a digital currency exchange operator registered in Russian jurisdiction

- to conduct training programs in educational institutions on cryptocurrencies, as well as post training programs/games, explanatory materials on the internet.
- in the media, to conduct programs on the risks and opportunities of cryptocurrency in order to increase the level of financial literacy of the population that does not use the digital space.
- to identify priority areas where the authorities welcome the influx of foreign investment, including in cryptocurrency.

Conclusion

Based on the results of the analysis, the authors concluded that the cryptocurrency is becoming a full-fledged participant in economic relations. It is necessary to review existing financial instruments and the possibility and expediency of using cryptocurrencies in Russia. We have grouped and supplemented the risks and opportunities of cryptocurrencies and proposed several measures.

References

Central Bank of the Russian Federation. (2022). *Cryptocurrencies: Trends, risks, and regulation*. Consultation paper, Central Bank of the Russian Federation. Retrieved from https://www.cbr.ru/Content/Document/File/132242/Consultation_Paper_20012022_eng.pdf

Eshkov, A. V. (2020). Cryptocurrency as a digital financial asset. *Accounting, 12*, 131–134.

Gorda, A. S. (2018). World market cryptocurrency: risksand mechanisms of regulation Мировой рынок криптовалют: риски и механизмы регулирования. *Scientific Bulletin: Finance, Banking, Investment, 4*(45), 96–104.

Khadzhiev, M. R., Batukaeva, L. S. E., & Mezhieva, H. B. (2020). Cryptocurrency as a form of private money in the modern economy. *Herald of GSTOU. Humanitarian, Social and Economical Sciences, 16*(3), 25–32.

Mikhailov, A. Y. (2020). Cryptocurrency market development: Hurst method. *Finance: Theory and Practice, 24*(3), 81–91.

Rodina, G. A. (2021). Politeconomic approach to the appearance of cryptocurrency. *Theoretical Economics, 6*(78), 52–59.

Safiullin, M. R., Savelichev, M. V., & El'shin, L. A. (2020). Cryptocurrency potential in the context of transition of global economy to the sixth technological paradigm. *Finance and Management, 3*, 74–85.

Tkach, Y. B., & Lazarchuk, A. A. (2022). World practice of taxation of cryptocurrency's transactions. *Finance, 1*, 58–64.

Mechanism for the Development of the National Digital Services Market

Natalya Vitalievna Usova

Introduction

The service sector is currently undergoing transformational processes. Moreover, the transformations concern all types of services such as retail trade, catering, education, and transport services. The Russian Federation is steadily included in the list of the first 50 countries that are actively introducing digital technologies. The introduction of digital technologies has become a prerequisite for ensuring the competitiveness of the national economy, improving macroeconomic indicators, and ensuring leadership positions in international ratings.

To ensure the competitiveness of the national economy at the international level and improve the quality of life of the population, it is necessary to determine the mechanism for developing the national digital services market.

Materials and Methods

The study was based on the author's own works, as well as works that were written in co-authorship (Loginov et al., 2021, 2021). The study is also based on the works by Abalkin (2000), Izmalkov et al. (2008), Kulman (1993), Slepov et al. (2011), Chalenko (2010), Shitukhina (2018) and others.

During the study, such methods as analysis, grouping, and comparison were used.

N. V. Usova (✉)
Ural State University of Economics, Ekaterinburg, Russian Federation

A. Rumyantseva et al. (eds.), *Challenges and Solutions in the Digital Economy and Finance*, Springer Proceedings in Business and Economics,
https://doi.org/10.1007/978-3-031-14410-3_44

Results

The development of digital services at the present stage is a certain challenge. The need to modernize the existing model of the national economy is a priority development.

According to the author, the national market for digital services is an organizational and economic relationship related to the process of interaction among the manufacturer and the consumer of the service through digital information communications, digital intermediaries within the economic space of the state.

Thus, there is a need to form the mechanism for managing the development of the digital services market. At the same time, it is impossible to consider the mechanism without considering the specifics of the applied model of market development management, as well as a possible set of management methods, in view of the specificity and multidimensional nature of the mechanism of managing the development of the studied market.

Figure 1 shows four models for managing the development of the digital services market, each of them has certain specifics. According to the author, the Russian

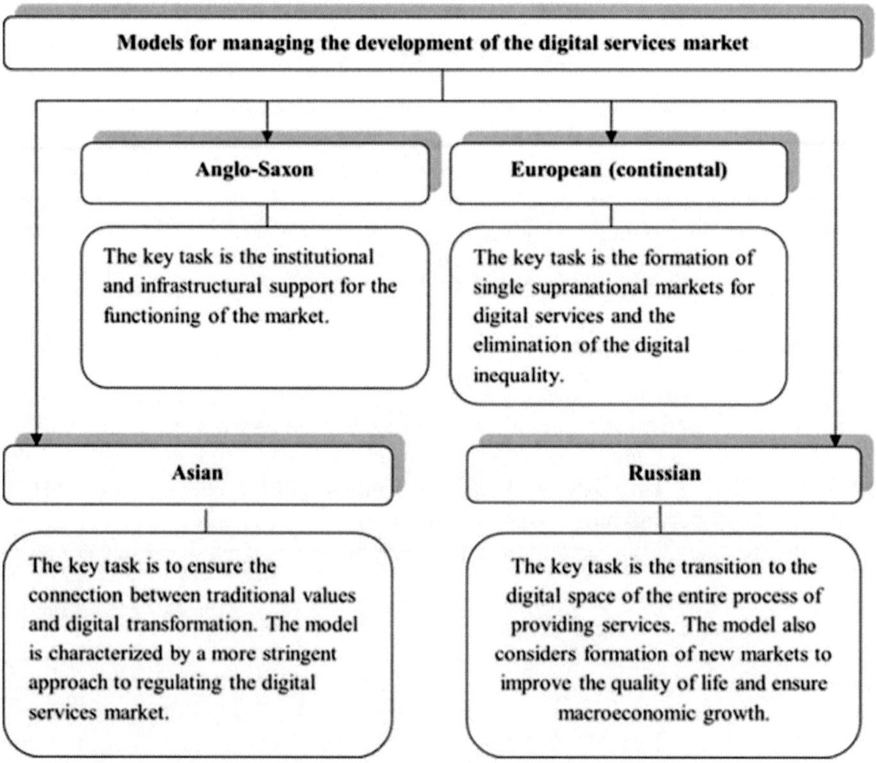

Fig. 1 Models for managing the development of the digital services market (Compiled by the author using (Shitukhina, 2018))

Table 1 Approaches to the category "mechanism" (Compiled by (Abalkin, 2000; Chalenko, 2010; Izmalkov et al., 2008; Kulman, 1993; Loginov, 2015; Slepov et al., 2011))

Approach, authors	Specifics of the approach
Structural (system-organizational) Abalkin (2000)	Using the category "mechanism" to describe the functioning of any economic system. At the same time, there is no hierarchy and structured mechanisms, as well as a generally recognized methodology for the formation of mechanisms and management.
Informational (Gurvits L., Maskin E., Myerson R.)	The mechanism is based on the use of methods of mathematical game theory. The priority direction is to take into account the subjective preferences and strategies of economic entities. This approach does not take into account the effectiveness of the market regulator, as well as the processes taking place in the economic sphere and the objective interests of market entities.
Functional Kulman (1993)	The mechanism is the interconnection that occurs between various economic phenomena under certain conditions under the influence of the initial impulse in a natural way.
Combined Slepov et al. (2011), Chalenko (2010)	The mechanism is a set of interrelated elements and resources of the economic process and ways to connect them to ensure the performance of the functions of the process. Moreover, within the framework of this approach, the mechanism becomes an executive element that provides resources for the implementation of managerial influence.
Cybernetic Loginov (2015)	The hierarchy of mechanisms is distinguished, namely economic and management mechanisms.

Federation has its own model of managing the development of the digital services market. This model is based on both the development of existing markets in the digital space and the formation of innovative digital services markets that allow increasing the growth rate of macroeconomic indicators and the population's quality of life. Thus, territorial boundaries are being erased and the likelihood of national companies operating in the digital space entering the international market increases.

Each of the management models is based on management methods and management mechanism. For example, all methods of management can be divided into economic method, ensuring the coordination of market processes through taxation, influencing prices, providing benefits, budget lending and investment; organizational method, expressed in providing participants in the digital services market with up-to-date and reliable information, implementing measures of program documents, developing the material, technical and digital base of service enterprises, ensuring the proper level of employee qualifications, and forming a system of state expertise; regulatory and legal method, which is the streamlining of market processes by working with legislative acts, instructions, orders, instructions, rules, as well as the application of control measures.

After examining the models and methods of management, it is necessary to move on to the mechanism for managing the development of the national digital services market (Table 1).

As can be seen from the table above, approaches to the mechanism category have evolved. About the digital services market, the author follows a cybernetic approach and proposes the following definition. The mechanism for the development of the digital services market is a system of tools for managing the aggregate of economic relations between market participants regarding the production, distribution, exchange, and consumption of goods provided in intangible form.

The use of the cybernetic approach is due to several reasons. Firstly, it allows building direct connections as well as feedback in the management process. Secondly, there is the division of the entire managed system into separate subsystems with subsequent management. Thirdly, there is the ability to monitor environmental indicators and correlate them with standard indicators. Fourthly, there is the permanent process of collecting, transmitting, and transforming information about the control object.

All these make it possible to obtain a certain effect from the development of the digital services market, associated with a reduction in costs (time, organizational, financial) for organizing, providing, and receiving digital services, increasing the transparency of the provision of services, as well as leveling administrative barriers.

Figure 2 presents the mechanism for the development of the digital services market.

Each of the participants selected in the mechanism performs certain functions.

The relevant state government body develops strategic documents, determines the priority directions for the development of the digital services market and monitors the achievement of its goals. In addition, the relevant body is in close cooperation with the Ministry of Digital Development, Communications and Mass Media of the Russian Federation, which ensures the development of digital infrastructure. This digital infrastructure is a mandatory component for the development of the digital segment of the national economy.

The Federal Antimonopoly Service supervises the activities of participants in the national digital services market, ensuring a competitive environment and market development.

The rating agencies form the ratings of digital platforms (mobile applications, marketplaces, online stores).

The interaction between the seller and the buyer of digital services is carried out on digital platforms. In addition, logistics companies that ensure the delivery of goods to the end user can take part in the process of providing digital services. In particular, the digital retail services market is characterized by the presence of logistics companies or their own logistics services that provide services for transporting goods to the buyer.

In general, the provision of digital services is based on several principles:

- accessibility is the ability to receive a digital service at any time, regardless of the territorial location.
- adaptability is the ability to adapt a digital service to the needs of a particular consumer.
- permanence of quality is the quality of the service provided does not change; all processes are algorithmized and digitized.

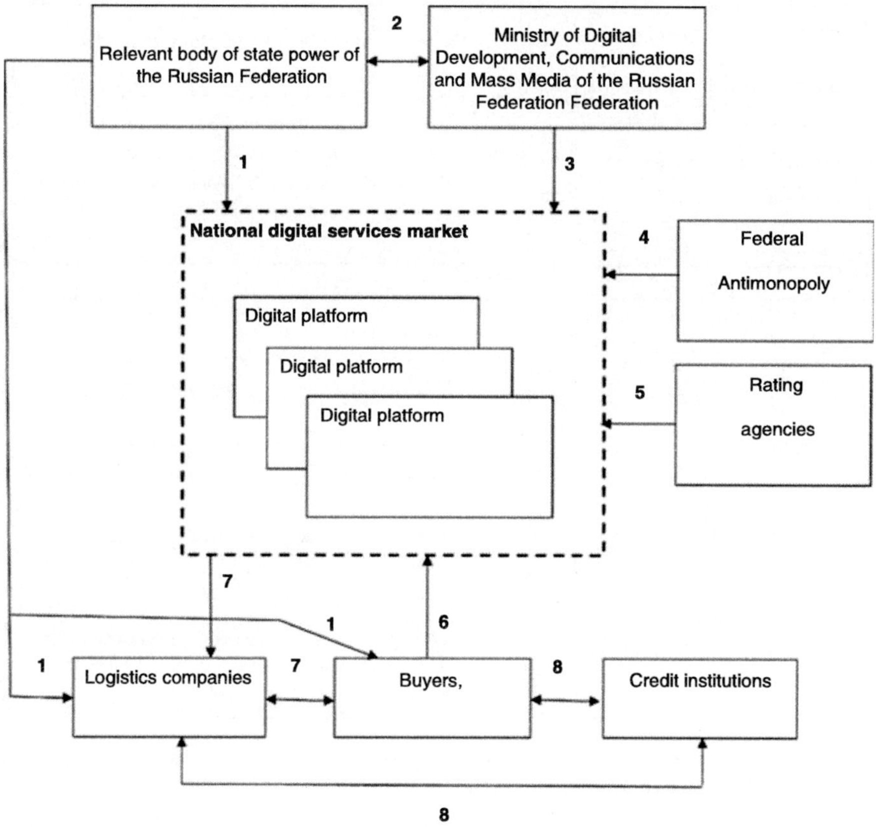

Fig. 2 Mechanism for the development of the digital services market: Designations: 1—regulation of interaction between participants in the process of purchase and sale; 2—interaction; 3—development of digital infrastructure; 4—supervision; 5—formation of ratings; 6—interaction of sellers and buyers on digital platforms; 7—ensuring the logistics of providing services for transactions; 8—ensuring transactions

- efficiency is the process of interaction between participants in the digital services market is aimed at achieving the ultimate goal, namely, meeting the needs and needs of consumers by providing a digital service that meets the expectations of consumers and achieving a target level of economic or social efficiency.
- customer orientation is the development of the digital services market depending on external trends and consumer needs.
- differentiation is the possibility of providing various digital services, as well as a set of services, depending on the needs and specifics of the target audience.

The tools of influence are also very diverse and vary depending on the stage of management of the digital services market (Fig. 3).

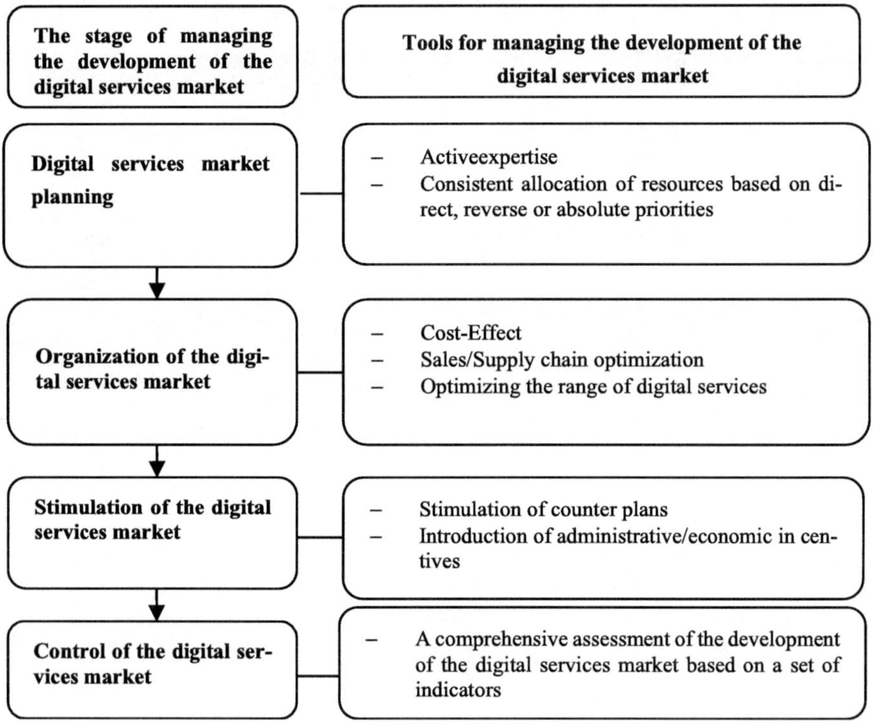

Fig. 3 Tools of the cycle of managing the development of the digital services market (Compiled by the author based on the study of various works)

Cost-effect tool is increasing the efficiency of resource use, reducing the subjectivity of decisions made, encouraging digital service market participants to use resources more efficiently.

The tool for optimizing the sales/supply chain is aimed at increasing the economic efficiency of the activities of participants in the digital services market by optimizing the distribution channel or the chain of digital service/resource providers to create digital services; increasing the accuracy of planning and the efficiency of decisions made in conditions of instability and risk.

The tool for optimizing the range of digital services is aimed at improving the provided digital services, implementing innovative policies, taking into account the transformation of the external environment and changing preferences of the target audience.

The tools of the stage of stimulating the digital services market provide an increase in the economic efficiency of the activities of participants in the digital services market, the fulfillment or over fulfillment of the targets set out in strategic documents, improving planning accuracy.

The integrated assessment tool is aimed at regularly monitoring the situation in the digital services market and timely responding to identified trends in order to make changes to strategic documents and adjust regulations that ensure the functioning of the digital services market within the national legal framework.

Applying various management tools and taking into account the specifics of the mechanism for the development of the digital services market, the results of managerial influence on the development of the national digital services market can be a reduction in the level of digital inequality; improving the quality of life of the population; expanding the range of digital services offered; growth of macroeconomic indicators; development of the economy of mobile applications and platform economy; formation of digital ecosystems; active introduction of modern digital technologies to improve the efficiency of the digital services market; the emergence of new segments of the digital services market.

Discussion

Based on the results of the study, the author proposes an approach to the categories "national digital services market" and "mechanism for the development of the digital services market" and develops a mechanism for the development of the digital services market. As a result, the author's approach to identifying tools to influence the national digital services market, depending on the stage of management development of the digital services market is proposed.

Conclusion

Thus, the mechanism for the development of the digital services market is aimed primarily at ensuring the accessibility of digital services for the population, as well as the growth of key macroeconomic indicators. It is especially important in the current conditions of strengthening global transformation processes.

References

Abalkin, L. (2000). Selected works. On the way to reform. The economic mechanism of a developed socialist society. A new type of economic thinking. Perestroika: Ways and problems. *Moscow: Economics, 2*, 911.

Chalenko, A. (2010). On the conceptual uncertainty of the term "mechanism" in economic research. *Industrial Economics, 3*(51), 26–33.

Izmalkov, S., Sonin, K., & Yudkevich, M. (2008). The theory of economic mechanisms (Nobel Prize in economics 2007). *Questions of Economics, 1*, 4–26.

Kulman, A. (1993). *Economic mechanisms*. Progress.

Loginov, M. (2015). Economic mechanisms: Essence, classification, cybernetic approach. *Problems of Theory and Practice of Management, 9*, 94–102.

Loginov, M., Usova, N., & Baigotanova, A. (2021). Ensuring sustainable development of the national financial market based on digital financial services. In *SHS web of conferences*, vol. 93. EDP Sciences.

Loginov, M., Usova, N., & Drevalev, A. (2021). Models of digital services development during an economic crisis. In *Proceeding of the international science and technology conference FarEastCon 2020* (pp. 555–567).

Shitukhina, N. (2018). Analysis of foreign experience in regulating the sphere of information services in the context of digitalization. *New Technologies, 3*, 145–152.

Slepov, V., Burlachkov, V., & Ordov, K. (2011). On the theory of economic mechanisms. *Finance and Credit, 24*(456), 2–8.

Silver Economy of a Megapolis: St. Petersburg Case Study

Natalia Nikolaevna Shestakova, Mikhail Georgievich Djanelidze, and Margarita Borisovna Skvortsova

Introduction

In modern conditions, the problem of population aging is becoming relevant in the world, due to a number of circumstances (the main ones are a decrease in the birth rate and an increase in life expectancy) and predetermining many consequences. According to international organizations such as the UN and WHO, the number of people over 60 years of age will increase to 2 billion people by 2050, which will increase the share of this cohort in the total population to 22% (World Health Organization, 2018). The number of people in the 80+ age group will triple by 2050 compared to 2015 and will amount to 434 million people.

Such trends are typical for the Russian Federation as a whole, and for St. Petersburg, due to the age structure of its population.

Materials and Methods

According to Rosstat data (The population of the Russian Federation, 2021), today every fourth (26.7%) resident of St. Petersburg belongs to the category of the population older than the working age.

As of January 01, 2021, 94,3009 people aged 65+ lived in St. Petersburg, which was 17.5% of the total population of the city (The population of the Russian Federation 2021). For comparison: on average in the Northwestern Federal District, this indicator was at the level of 16.8%, and in Moscow it was equal to 17.9%.

N. N. Shestakova (✉) · M. G. Djanelidze · M. B. Skvortsova
Institute for Regional Economy Studies Russian Academy of Science, St. Petersburg, Russian Federation

A. Rumyantseva et al. (eds.), *Challenges and Solutions in the Digital Economy and Finance*, Springer Proceedings in Business and Economics,
https://doi.org/10.1007/978-3-031-14410-3_45

According to the information of the Committee on Social Policy of St. Petersburg (Government of St. Petersburg 2021), as of October 01, 2021, 1228.6 thousand people over 60 years of age lived in the city (22.8% of the total population of the city—5384.3 thousand people), including:

- 644.6 thousand people aged 60 to 70 years.
- from 70 to 80 years—360.6 thousand people.
- from 80 to 90 years—194.2 thousand people.
- from 90 to 100 years—28.8 thousand people.
- over 100 years—343 people.

The age of the oldest resident of the city was 108 years old at that moment.

And this is quite consistent with the state guidelines for increasing life expectancy by 2030 to 78 years and increasing healthy life expectancy by 2024 to 67 years, announced within the framework of the project "Older Generation" of the National Project "Demography" (2019–2024) (Ministry of Labor of the Russian Federation 2018).

That is, in the foreseeable future, every fourth resident of St. Petersburg will still fall into the older age category, the border of which can be conditionally designated as exceeding the working age.

Thus, it is obvious that a new rather large, significant group of the population is gradually being formed in society, and in fact there has already been formed in St. Petersburg, which has its own very specific features, interests, priorities, needs, requests. Representatives of the silver generation are becoming one of the main consumers of services provided by the fields of medicine, tourism, hotel business, health, and beauty industries, etc.

Results

The transformation of the age structure of the population towards the stabilization of the share of the population of older age groups imposes both economic and social obligations on the economy and the social sector of the city.

The economic and social consequences of population ageing are:

- an increase (or, in the conditions of the launched pension reform, at least, not a decrease) of the share of pensioners by age in the total population and, accordingly, the corresponding burden on society.
- an increase in the share of elderly and old people, which exacerbates the problem of their financial support for society, especially since the proportion of people over 80 years of age is growing faster than the share of elderly people.
- the increasing need for medical care, pension, and social security for the elderly, which, accordingly, requires both additional funds and the expansion of the network of medical, social, gerontological institutions, a fundamental restructuring of the healthcare system.

- reduction in the number and proportion of people of working age in total population.
- necessity in providing opportunities for social (including labor) activity of the elderly population, and first "young elderly" who want to work.

Current economic policy should respond to emerging challenges in one way or another. And this is already happening in the most economically advanced states: objectively emerging "age trends" are beginning to dictate their own conditions for the functioning of the economy.

Now more complicated approaches appear that consider the aging of the population as new opportunities for economic development, too. We are talking about the so-called longevity dividend, which can be realized through the launch and promotion of the so-called silver economy model.

The scientific community has proposed many definitions of the category "silver economy". Among the numerous attempts to determine this phenomenon, we have chosen two. The first is I. P. Tsapenko's definition, it narrows the re- search object: "In a narrow sense, the Silver economy is defined as a set of sectors focused on the production of goods and services for the elderly population, in a broad sense—as a set of types of national economic activities that directly produce products for these cohorts of the population, as well as economic activity in other areas generated by the functioning of the relevant sectors" (Tsapenko 2016).

The second, expanded definition of the silver economy, is given by the Belgian economist David Eatock for the Report of European Commission (Eatock 2015). According to him, the silver economy is not only the satisfaction of a growing number of needs and the development of a new market of services for the elderly, but also the use of human capabilities during the period called "aging". With this approach, the silver economy covers changes in the labor market to expand employment opportunities for older age groups and measures to promote active and healthy aging, which allow increasing the labor potential of older people.

However, at the initial stage of the study, we decided to be guided by the first, narrower version of the interpretation of the category.

Today, the phenomenon of the silver economy has already become a reality in several countries. Moreover, these are not only the officially recognized developed countries of Western Europe (Germany, France, Great Britain), North America, Asia (Japan. South Korea), but also Poland, China and some others. For example, in 2014, the production of goods and services for the elderly brought China $652 billion, or 8% of GDP. By 2050, according to the forecast of the National Committee on Aging, the silver hair market will make up a third of the entire Chinese economy (Zotov, 2017). In Europe, the European Commission estimates that consumer spending by people aged 60 and older has grown 50% faster over the past two decades than spending by people under the age of 30. According to forecasts, the volume of the silver economy will steadily and steadily grow in the EU (an increase of about 5% per year) and by 2025 will amount to 5.7 billion euros. It is expected that by 2025, the silver economy of Europe will provide 88 million jobs. This will amount to 38%

of the employment of the European Union (Growing the European Silver Economy, 2015).

According to the World Data Lab, the total purchasing power of the elderly population worldwide in 2020 was about \$ 8.4 billion. This figure is expected to grow to \$ 14 billion over the next decade (Mookerjee & Pham, 2021).

According to their forecasts, the world's exit from the coronavirus pandemic will provoke an increase in spending by older consumers and open new opportunities for investors in the silver economy industry. This will happen due to the realization of the deferred demand for medical services, tourism and travel, luxury goods, and the growth of demand for insurance services. Experts also expect that the forced access of the elderly to the Internet during the period of isolation "will forever open this demographic group to e-commerce companies and social networks" (Mookerjee & Pham, 2021).

Moreover, World Bank experts believe that older people will play an important role in the economic recovery after COVID-19. According to international estimates, it is expected that the share of 65+ consumers in the 2020s in the total consumer spending will be 66%.

Researchers from World Data Lab put Russia on the sixth (after the US, China, Japan, Germany, and France) place for long-term (2030) and the volume of the silver markets of the economy (Fengler, 2021).

Discussion

Meanwhile, domestic experts do not yet give certain estimates of the volume of such a market. "It is not easy to assess the market of goods and services for the elderly—it includes many industries: from cosmetics, food additives, equipment for recreation and sports to specialized medical equipment, medicines and literature," the authoritative publication Delovoy Peterburg noted in 2015 (Vasilyeva, 2015).

Our analysis shows that local St. Petersburg manufacturers of goods and services for the elderly mainly targeted at needs of sick and infirm, not to produce goods and services for active older people. This is evidenced by the motto "Care, Help, Mercy", under which an exhibition of social, medical, charitable assistance, goods and services for the elderly is held within the framework of the International Forum "The Older Generation" held annually (since 2005) in St. Petersburg.

According to the estimates of entrepreneurs, in the general structure of the market for the population of older age groups, local manufacturers occupy about 50–60% of the market of non-technical goods, hygiene products for the elderly. But in rehabilitation and medical equipment, this share is no more than 7–8%.

As of 2015, the following problems were inherent in the market segment of manufacturers of equipment for rehabilitation after illness (considered as a segment of the market of goods and services for the elderly) (Vasilyeva, 2015):

- the market is focused on the public procurement system, the share of sales to consumers from the business sector and households is low and amounted to about 15%.
- the factor of priority of domestic producers over foreign ones set "from above" during state orders and tenders limits the range of purchased goods.
- in addition, it was noted the low level of development of the relevant sectors ("it is possible to buy only consumables and the simplest equipment (crutches, mattresses, walkers) from Russian manufacturers, and machinery and equipment from foreign manufacturers", G. Lyutykh, owner of the private nursing home "Century").

At the same time, there are also positive trends in the development of certain segments of the silver economy. This is evidenced by a wide range of organizations interacting within the framework of the already mentioned International Forum "The Older Generation". So, among them:

- State authorities (Social Policy Committee of St. Petersburg, Health Committee of St. Petersburg, Employment Service of St. Petersburg, Department for Horticulture Development of St. Petersburg, Commit-tee for Social Protection of the Population, Pension Fund of the Russian Federation—Department for St. Petersburg and Leningrad Region, GU—St. Petersburg Regional Department of the Russian Federation Social Insurance Fund).
- Organizations implementing compulsory health insurance programs.
- Specialized public organizations (Union of Pensioners of Russia; St. Petersburg Public Organization of Veterans (Pensioners, Invalids) of War, Labor, Armed Forces and Law Enforcement Agencies)
- Companies producing rehabilitation equipment, providing goods, social and medical services for the disabled and elderly.
- Companies providing educational services.
- Banks, law firms, consumer cooperatives.
- Food retailers.
- Clothing, shoes, and accessories retailers.
- Cosmetic products, health products, medical optics, dietary supplements retailers.
- Organizations providing Spa treatment.
- Travel companies.
- Religious organizations.
- Libraries, mass media.
- Charitable organizations and some others.

Moreover, within the framework of the Forum, together with the Union of Industrialists and Entrepreneurs, a competition among enterprises "The best manufacturer of goods and services for the elderly" was established. Four nominations are proposed for evaluation: industrial and technical products; products for the population; food products and services.

Thus, the executive director of the Association of Companies of the Industry for the Older Generation (St. Petersburg), V. Barausov, analyzing services for out-ofhospital (home) care for the elderly, believes that " The global trend that should reach Russia is the orientation to a longer aging of a person at home (aging in place) due to adapted real estate, services and technologies. This market is many times larger than the market of private boarding houses and judging by the interest of our developers who are engaged, for example, in apartments, we can also expect a breakthrough here" (Zarubina, 2019).

Almost all "market participants agree that it has a huge growth potential and work in this area can become very profitable, but so far investors consider the market of assistance to elderly people unattractive" (Vasilyeva, 2015).

It is important to note that our research being launched logically fits into the framework of the international project OSIRIS (participants: Latvia, Lithuania, Estonia, Denmark, Russia and Finland, start—2019), aimed at exploring the possibilities of applying an intelligent approach to use the opportunities of the silver economy in the Baltic Sea region (Osiris project 2021).

It should also be noted that a systematic approach to launching and purposefully promoting the Silver Economy model on a national scale at any level of the territorial division of the Russian Federation, based on an analysis of available resources and socio-economic reserves, has not been carried out so far. And this opens vast prospects for the activities of both the authorities of each of the regions, including St. Petersburg, and for a wide range of economic entities located within them.

Conclusion

In general, the phenomenon of the silver economy can be considered as a window of opportunities for economic growth, the use of which at the city level will contribute to solving its pressing problems, and in future will expand the directions of its socio-economic development and give St. Petersburg additional competitive advantages at the interregional level.

As it was demonstrated above, the trend of population aging is beginning to dictate its conditions for the functioning of the economy. The current economic policy is forced to respond to emerging challenges, and this is already happening in the most economically developed countries and regions. In the Russian Federation, the share of the population of older age groups is significantly differentiated from region to region, but according to our study, the silver economy elements are more characteristic for megapolis such as Moscow and St. Petersburg. And it was shown in this article on the example of St. Petersburg.

References

Eatock, D. (2015). *The silver economy: Opportunities from ageing. European Parliament*, Retrieved from Accessed January 3, 2022, from https://epthinktank.eu/2015/07/16/the-silver-economy-opportunities-from-ageing/

Fengler, W. (2021). *The silver economy is coming of age: A look at the growing spending power of seniors*. Available at: https://www.brookings.edu/blog/future-development/2021/01/14/the-silver-economy-is-coming-of-agea-look-at-the-growing-spending-power-of-seniors/. Accessed 7 Jan 2022.

Government of St. Petersburg. (2021), *October 1 - International day of the elderly people*. Retrieved from https://www.gov.spb.ru/gov/otrasl/trud/news/222346

Growing the European Silver Economy. (2015). European Commission. Background Paper. Available at: https://digital-strategy.ec.europa.eu/en/library/growing-silver-economy-background-paper. Accessed 21 Jan 2022.

Ministry of Labor of the Russian Federation. (2018). Federal project passport "*Development and implementation of systemic support and improvement of older citizens life quality program*". Retrieved from https://mintrud.gov.ru/ministry/programms/demography/3

Mookerjee, I., & Pham, L. (2021). *After the pandemic, a wave of spending by older consumers*. Available at: https://www.bloomberg.com/news/articles/2021-06-12/pandemic-turbocharges-aging-theme-for-global-equitiesinvestors?fbclid=IwAR3p5JWZuOKXH5ISaLTGdwBdSmhaKQvxI0TGu3BqkXvPWDmLvIbDo0F2QVg. Accessed 18 Jan 2022.

Rasstat. (2021). The population of the Russian Federation for gender and age on January 1, 2021. *Statistical Bulletin. Moscow, 8–9*, 443.

Tsapenko, I. (2016). *The social context of economic development in the XXI century*. Institute of World Economy and International Relations RAS, Moscow.

Vasilyeva, M. (2015). *The Russian market of goods and services for the elderly may soon become very profitable*. Available at: https://www.dp.ru/a/2015/03/11/Raschet_na_paqensii. Accessed 18 Jan 2022.

World Health Organization. (2018). *Aging and health*. Retrieved from https://www.who.int/news-room/fact-sheets/detail/ageing-and-health

Zarubina, O. (2019). *Adult industry*. Available at: https://spb.plus.rbc.ru/news/5c65464f7a8aa9145cbf36d2. Accessed 18 Jan 2022.

Zotov, G. (2017). *The market of "silver hair". Why are old people in China the engine of the economy*? Available at: http://www.aif.ru/money/economy/rynok_serebryanyh_volos_pochemu_stariki_v_kitae_-_dvigatel. Accessed 7 Jan 2022.

Digitalization of Electricity Suppliers' Activities in the Arctic Zone

L. N. Babkina, O. V. Skotarenko, E. S. Kuznetsova, and E. S. Khatsenko

Introduction

Creating a digital economy ecosystem in Russia, where digitalized data are the core production factor in all areas of socioeconomic activity (President of Russian Federation, 2016, 2018).

The need to integrate digitalization into management systems of production, distribution, and consumption of any resource, including electricity, in any area, especially in the Arctic, depends both on the natural and climate conditions there and on the need to improve the quality of life, foster a more sustainable social, environmental, and economic development, and increase the efficiency of energy pro-duction and consumption. Therefore, integrating digital technologies into all stages of electricity production, distribution, and consumption in all target segments of customer markets as well as into all management functions, such as recording, monitoring, forecasting, programming, and planning, will help achieve the stated objectives step by step and pre-vent possible future problems. For instance, energy

L. N. Babkina (✉)
North-Western Institute of Management, Branch of the Russian Presidential Academy of National Economy and Public Administration attached to the President of the Russian Federation, St. Petersburg, Russian Federation

O. V. Skotarenko
Murmansk Arctic State University, Murmansk, Russian Federation

Military Academy of Logistical Support named after General of the Army A. V. Khrulyov, St. Petersburg, Russian Federation

E. S. Kuznetsova
Murmansk State Technical University, Murmansk, Russian Federation

E. S. Khatsenko
Chair of Youth Department Murmansk Region, Murmansk, Russian Federation

A. Rumyantseva et al. (eds.), *Challenges and Solutions in the Digital Economy and Finance*, Springer Proceedings in Business and Economics,
https://doi.org/10.1007/978-3-031-14410-3_46

consumption is one of the most important indicators in international statistics. It is used to measure the degree of industrial development and the general socioeconomic development and living standards (UNCTAD, 2019).

The goal of the study is to explain the need for digitalization of communication between suppliers of electricity, an important resource, and customer markets using a case study of a big company operating in the Russian Arctic Zone.

Materials and Methods

Some Russian scholars believe that digitalization and industrialization are not mutually exclusive but complementary, which means using the experience in scenario modelling widely implemented in the USSR (Bodrunov et al., 2017).

Scholarly papers also attempt to solve the issue of digitalizing economic processes based on software for identifying structural components of economic potential in Russian regions (Skotarenko et al., 2019).

A "development model for regional clusters during transition to a digital economy and an interaction mechanism for cluster participants in management of a regional socioeconomic environment" were developed for management of cluster interaction (Charykova & Markova, 2019).

An adequate interpretation of determinants indentified with a multifactor human resource efficiency model for a digital economy helps choose the right tools for management of human resources as a factor of economic growth (Kelchevskaya & Shirinkina, 2019).

Given the ongoing changes, it is also recommended to use digital technologies in the banking sector (Belous & Lyalkov, 2017).

A lot of attention is also paid to the problems hindering digitalization of company-based economic processes (Pasandideh et al., 2013).

Changes in European and international legislation in a digital economy are described in studies by international scholars (Ting & Gray, 2019; Trenta, 2019; Willems & Kamau, 2019).

To carry out this study, we used data from annual government statistics reports of Russia, a World Bank statistics reference book, and annual financial and statistical reports of energy suppliers.

The study is based on main principles of the system-based, comprehensive, program-goal, standards-based, program-orientated, and process-based methodological approaches. They laid the groundwork for using the theoretical principles of consistency, complexity, program goals, comparability, adequacy, informative value, program orientation, and proportion in assessing communication between suppliers of electricity, an important resource, with customer markets. The study also uses the methods of statistical and economic analysis.

Results

A study carried out by the authors showed that the division of countries into groups by per-capita gross national product (GNP) in USD used in World Bank statistics reference books could also be done using another indicator: per-capita electricity consumption (kWh). Several groups of countries by per-capita GNP are identified: low-income, medium-income, below-average income, above-average income, the Euro area, and high-income (UNCTAD, 2019).

In accordance with the World Bank classification, Russia belongs to the group of above-average income countries by per-capita gross national product. However, by per-capita electricity consumption in 1990, it was 2.18 times above the indicator value of its group and 1.24 times that of the Euro area, being behind only high-income countries by 15%. Those proportions indicate an important level of production and consumption at the time, but later, the indicator did not grow but fell by 15.35%, making the difference in electricity consumption from the Euro area equal to 20% and from high-income countries 69% despite the fact that the value was higher than that of its own group of countries by 84.77%.

In Arctic regions, which include federal subjects, such as Murmansk Oblast and three autonomous Okrugs: Nenets, Yamalo-Nenets, and Chukotka, electricity production and consumption, both overall and per capita, determines both the level of regional socioeconomic development and the conditions for living, and even survival, of the people, which has a direct impact of national security and further exploration of those areas.

For instance, an analysis of statistical data (Rosstat, 2021) by the indicator of power plant capacity (million kW) in 2017, 2018, and 2019, showed that the cumulative capacity of all power plants in all the four regions was 7.9 m kW in 2017. In 2018, the cumulative capacity grew to 8.3 m kW or by 10.5%. In 2019, the cumulative capacity was 9.0 m kW, which accounted for approximately 9% growth compared with the previous year. Therefore, the growth slightly slowed down. In the total cumulative capacity of power plants, the shares of the regions in 2017 were as follows: 45.6% in Murmansk Oblast, 40.5% in Yamalo-Nenets Autonomous Okrug (AO), 10.1% and 3.8% in Nenets and Chukotka Autonomous Okrugs, respectively. In 2018, the share of Murmansk Oblast decreased to 43.37%, and the share of Yamalo-Nenets AO grew to the same value of 43.37%. The shares of Nenets and Chukotka Autonomous Okrugs slightly decreased to 9.6% and 3.6%, respectively. In 2019, the shares of Murmansk Oblast and Chukotka AO kept getting smaller, to 40.0% and 3.3%, respectively. In the other two regions, Yamalo-Nenets and Nenets Autonomous Okrugs, the shares grew to 46.7% and 10.0%, respectively.

By the second indicator, electricity production in billion kWh, Murmansk Oblast had a share of 61.4% of the total amount of 28.5bn kWh produced in the four regions in 2017. The share of Yamalo-Nenets AO was 29.47%, Nenets AO 6.67%, and Chukotka AO 2.46%. In 2018 and 2019, the share of Murmansk Oblast started to decrease, similarly to that of the power plant capacity, making up 59.0% and 54%, respectively. Yamalo-Nenets AO demonstrated an opposite trend, with its share in the total electricity production growing to 32.4% in 2018 and 37.2% in 2019. In Nenets and Chukotka AO, no definite trend was observed. In 2018, the shares got slightly smaller, to 6.14% and 2.39%. In 2019, they grew to 6.15% and 2.59%. In general, both in 2018 and 2019, electricity production in all of the regions grew by 3% and 5.5%, respectively, compared with the previous year.

The next indicator is per-capita electricity production. It is a specific indicator that had to be calculated based on the statistical data on annual average population and regional electricity production.

For example, over the latest reported five years from 2015 to 2019, the largest per-capita production of electricity (kWh) was observed in Nenets Autonomous Okrug, which can be shown as the following sequence of figures: 15.91, 13.64, 18.18, 18.18, and 20.45. After a small drop in consumption in 2016, growth was observed in 2017, followed by stagnation in 2018 and more growth in 2019.

An analogous situation was observed in Murmansk Oblast, which can be shown as the following sequence of figures: 4.86, 4.76, 4.77, 4.81, and 4.86. Therefore, after an insignificant reduction in production in 2016, stagnation was observed in 2016 and 2017, followed by slight growth in 2018. In 2019, the figure grew back to the value of the first year of observation, 2015.

In Yamalo-Nenets AO, the calculated data show that the indicator value grew, then stagnated, and then grew again: 5.62, 5.97, 5.95, 6.65, and 7.72.

Chukotka AO was different from all the other regions by the fact that the energy production did not change over the five years, remaining at the level of 6.0 kWh.

Nenets AO was an obvious leader in the indicator, whereas Yamalo-Nenets AO was behind by a minimum factor of 2.28 or 128% in 2016 and a maximum factor of 3.06 (206%) in 2018.

A similar lag the leading region in indicator values was also observed in Chukotka AO, varying from a factor of 2.27 (127%) in 2016 to 3.41 (241%) in 2019.

Murmansk Oblast held the worst position because the minimum lag Nenets AO was by a factor of 2.87 (187%) in 2016, and the maximum lag was by a factor of 4.21 (321%).

Discussion

Thus, improvement of Murmansk Oblast electricity suppliers is a priority problem to be solved in the Arctic.

The Kola Branch of the Atomenergosbyt Corporation is one of the biggest companies in Murmansk Oblast, a region located in the Russian Arctic Zone. The company supplies electricity to two customer markets: industrial enterprises and the public. The main goal of the company is to provide electricity in full and on time to the two customer markets, i.e., individuals and households, and to the sectoral markets, i.e., legal entities, and ensure efficient energy distribution in each target segment of those customer markets in the Russian Arctic Zone.

The subject is proven to be relevant because improvement is needed for information support to the company as a whole and its branches, including the Kola Branch. Digitalising the document flow will help make contractual relations with customer markets more efficient. The main goal of digitalization is to provide continuous availability of the required information stored in the database, reduce the time it takes to sign a contract or make summaries and reports, and create a state-of-the-art database for record-keeping of supervisory activities.

The Atomenergosbyt Corporation was established in 2001 to provide electricity to nuclear facilities. The main activity of the company (according to OKVED, the National Classification of Economic Activity Types) is 35.14 «Selling electricity». Some of the additional activities include 33.12 «Repairing machinery and equipment», 35.11 «Power generation», and 35.13 «Power distribution».

In 2015, by Order of the Russian Ministry of Energy No 14 of 23 January 2015, the company was assigned the status of the default supplier of electricity in Murmansk Oblast (Russian Ministry of Energy, 2015).

The main strategic economic goal of the company and its branches is to raise the profitability of its supply business. It has the following strategic vectors or means to achieve the goal: reduce the scope and growth of debt due from the customers, ensure that payment for the energy supplied exceeds the average for the energy market in the Russian Arctic Zone, and develop and implement new products, services, and work on the respective customer markets.

The goals are achievable by reaching the following programme goals:

- standardise the customer support system.
- promote remote and interactive services on the customer markets.
- raise the competence of the customer service employees.
- standardise the work with the debtors.
- advance the activities aimed at optimisation and reduction of the debt due from the customers.
- unify and automate contractual activities with consumers on the customer markets by digitalising the document flow.
- automate the claim and legal action unit in the billing system for debt collection.
- prioritise goals and criteria for the new Smart Home activity.

A dynamic for the total amount of electricity consumers on the consumer market serviced by the Kola Branch is shown in the figure. As seen from the figure, the number of legal entities serviced by the Branch insignificantly decreased over the last three years. However, the number of individuals steadily grew over the studied period. In 2021, the company reached 9,9 thousand legal entities and 342 individuals in Murmansk Oblast.

The Kola Branch revenue figures for selling electricity and capacities in Murmansk Oblast in 2019–2021 show that it was 11,865 m RUB in 2019 and 12,544 m RUB in 2020. The revenue growth was 105.72% compared with 2019. In 2021, the revenue was 13,948 m RUB, showing a growth of 111.19% (1404 m RUB) compared with 2020.

The rate of growth had therefore grown from 5.72% to 11.19%, i.e., almost doubled.

Experience has shown that digitalisation and introduction of advanced information technologies is beneficial for any company and its competitiveness on customer markets. The current competition within the sector is medium level. The Kola Branch is the biggest electricity supplier in Murmansk Oblast, but there are competing companies on the market of electricity supply and distribution. Those are two corporations: Murmanskaya Gorodskaya Elektricheskaya Set and Oboronenergo, as well as Murmanskaya Oblastnaya Elektrosetevaya Kompaniya.

Experts consider it unlikely that new competitors may appear at this sectoral market of electricity resources, since the entry costs are extremely high. Apart from that, the government management system sets restrictions and imposes regulations on companies.

The rate of customers' leaving the company's reach is assessed by experts as medium because the Kola Branch has a considerable capacity to improve its services. It is highly trusted by the target segments of the sectoral and consumer markets because it is the default provider of electricity and capacities.

Experts also assess the strength (degree) of influence from various resource providers on the Kola Branch as medium. However, the company has a high priority for resource providers.

To prevent new competitors from producing replacement products, it is necessary to use new service management technologies and digitalisation, raise the quality of service, reduce costs as the power generation increases, set and achieve customer loyalty goals, and encourage activities of intermediaries. It is possible to make up for a certain loss of customers following a marketing study of their needs and reactions to a change in the cost, quality, or properties of a service. It is also necessary to make sure that the Kola Branch is prepared for a timely and adequate response to new needs of the customer markets and fast-emerging demand for services. Despite the high priority level with the existing resource suppliers, the Branch management should analyse their competitive position on the resource and service markets and look for new partners providing the Branch with new technological, information, labour, financial, and commodity opportunities.

We can therefore conclude that the current competitiveness of the Branch helps it expand its services and increase its share on the customer markets.

An expert analysis of the degree the strengths and weaknesses of the Branch affect its activities has shown that the following opportunities, goals, and activities should be currently prioritised: digitalisation of underperforming contractual relation processes; introduction of state-of-the-art service technologies; increased automation to improve major activities and several actions and functions of the staff working with end users.

Importance of the Branch's further development strategies have been analysed to determine their importance factor. The strategies are aimed at adding value and include the following: reducing costs and time (importance factor: 0.36) and increased customer service efficiency (importance factor: 0.16). The analysis has shown that the strategies can be achieved by improving the information support of

Table 1 List of documents for contractual relations

Name	Flow of information
1) Application for power supply services	Input
2) Statement of application review	Output
3) Power supply contract	Output
4) Reporting documentation	Output
5) Claims	Output
6) Notifications	Output
7) Writ of execution	Output

core business processes (importance factor: 0.39), which includes digitalisation of contractual relations.

The entire process can be divided into four sub-process stages done by the Sales Department and the Contracts Department: «Processing and accounting of application for power supply», «Signing and record-keeping of contracts», «Claims and legal action», «Routine reporting».

The first stage starts at registering applications for power supply services received be the Contracts Department. The applications are assigned a number in the registration logbook to track the follow-up work with them.

At the second stage, the registered application serves as a basis for a power supply contract detailing the main parameters and the customer account number.

The Sales Department calculates and sends the invoices. It also collects customer debt data based on the payment information received from the Accounting.

The debt data are sent to the Legal Department, which uses them to draft claims to be filed to court.

Having studied all the process stages, we have identified a list of the documents present in the contractual relations (Table 1).

The analysis has shown major time and labour constraints during preparation and submittal of the documents, e.g., most of the claims and legal action activities are done manually, which both extends the procedure and leads to failure to meet deadlines.

To determine the sequence of digitalisation of documents and automation of contractual management in the Kola Branch, we have used the ABC (Pareto) analysis method. The analysis has shown that four (67%) documents should be digitalised and automated. They take up 80% of the total document processing.

The quantitative indicators of strategic goals and specific activities in the information support system include the average time it takes to process an application, produce a decision, prepare the supporting documents, and make reports as well as the number of signed contracts.

The projected distributed information support system aimed at more time-efficient contractual relations, reliability and timeliness of data must conform to the following requirements:

- flexible and personalised adjustment for each employee of the respective department.

- unification of data processing.
- restoration of modified data.
- active information exchange with the existing information system (STEK-Energo calculation and billing system).
- support for reporting to the management.
- information and technical support for creating reports to the management
- standard software modules as elements of a system that makes it possible to add new modules without changing the operation algorithms and integration of the existing elements and the database.
- duplication and implementation of both individual software modules and the system in other similar facilities or in other regions.
- a standardised audit procedure to control the automated functions of information collection, processing, storage, and display as well as to detect irregularities.
- a united information environment and database for the data now contained in different databases, for the purposes of easy availability of information and eliminating redundancy.

In accordance with the united information environment concept, all elements of the information system must function under the same rules and principles and unite all the contractual relations processes. Each task must start with a certain input document requiring specific actions: registration, analysis, and decision. Each new action must follow a strictly defined sequence, without exclusion of other actions, and cause the next operation to start.

A suggested information system design must provide for multi-user operation. The client-server technology with a two-tier and three-tier architecture, like the existing STEK-Energo architecture, is therefore chosen for implementation.

That architecture conforms to the requirements of reliability, scalability, performance, flexibility, and security.

The main components of the information system to be designed include the following:

- a STEK-Energo database containing debtor data (first level).
- digitised paper or electronic documents containing court and financial data (first level).
- a united database (second level).
- an ASP application receiving user requests for information from employees to be forwarded to the united database and returned with the search results to the users (third level).

Despite the increase in the information system software tasks, the functions of the contractual support staff will not be reorganised. Besides, the expanded software capacity includes storage of full-text documents (contracts, claims, and writs of execution).

The automated contractual relations system of the Kola Branch of the energy supplier will be an information and reference system and will perform the following tasks: collection and registration of input data, verification and correction of input data, storage of full-text documents, information processing with predefined algorithms, search using search queries, and display of electronic documents.

Input data will be manually collected, registered, and entered the STEK-Energo software complex, which will make it possible, if needed, to adjust the user personal data or reference information resulting from recalculation.

If there is a file write or read error, the error code and description is displayed. Writing and reading is done on the all-or-nothing basis: the data are considered transmitted/received when all the objects have been written without any error.

To ensure integrity and authenticity of the data stored in the unified management system, the project suggests using the following ways of verifying the manually added data:

- routine visual data check for unacceptable characters.
- checking the quantitative data range or variation for amounts.
- checking the input data for accurate syntax structures (syntax check).
- checking the presence and meaningful interrelation of the indicators to be calculated (semantic check).

The suggested ways of input data checks will contribute to a reduction in the number of malfunctions and errors in the management system.

All the documents on the input and output documents list related to the automated management system for contractual relations can be divided into two groups by document form:

1. Documents with a rigid structure and guidelines, incapable of modification.

This group of documents includes power supply contracts of the company and court documents (claims, court rulings, writs of execution, enforcement procedure orders, etc.).

2. Documents with a flexible structure and guidelines for internal needs of branch management. This group of documents includes various reports, e.g., company debtors report.

However, one of the main tasks for the information system design was automation of court claims, frequently filed and rigidly structured and designed documents.

The project does not aim at complete replacement of hard copies with electronic input and output documents because some input documents will be stored as hard copies, with the required data to be entered into the database manually.

Output information will be displayed on-screen electronically and as a hard copy at the employee's discretion provided that such a report (table, chart, diagram, etc.) is available in the system.

Introduction of the suggested information system is projected to have a management, social, and economic effect.

The expected management effect is expressed in reduced time loss, faster availability of analytical information, fewer errors in record-keeping, higher reliability of information, lower workload of department staff, reduced paperwork, and, consequently, an improvement of the general management culture.

Table 2 Labour input

Process	Labour input, hours/month		Labour input reduction, hours
	Base option	Suggested option	
Processing input documents	231	198	33
Reporting	120	94	26
Making court claims	265	204	61
Total	616	496	120

The social effect can be expressed in increased stability of the Branch's operation, which results in job security and higher quality of service.

The economic effect is expressed in its respective monetary equivalent (roubles, equivalent units). To calculate the indicator, first, we must assess the amount of time reduction in managing the Branch's contractual relations (in particular, claims and courts), using the following formula for the time effect (TE) (1):

$$\mathrm{TE} = \mathrm{T}_0 - \mathrm{T}_1, \quad (1)$$

where

T_0 is the time consumption prior to the introduction of the automated management system.

T_1 is the time consumption after the introduction of the automated management system, hours.

Then we need to calculate the cost saving resulting from the time effect.

The current time consumption by department staff can be reduced by expanding the functions of the STEK-Energo software complex. The project solution will therefore help find additional time for different, more complex operations. The labour efficiency is projected to increase because of the reduced labour input for document processing, reporting, and compliance.

Apart from that, it results in more time available for the employees to deal with more complex, nonstandard operations.

Table 2 shows a comparison of the labour input for the existing (base) and suggested (automated) option and a calculation of the indicator of labour input reduction for each process task. It demonstrates that introduction of the contractual management system will reduce the labour input by 120 hours a month, with the hourly wage of a legal action department employee being about 170 RUB.

Thus, the yearly cost-saving (CS year) from the operation using the designed information subsystem will be (2):

$$\text{CS year} = 120 \text{ hours} \times 170 \text{ RUB} \times 12 \text{ months} = 24.800 \text{ RUB} \tag{2}$$

No staff will be laid off because the extra time saved by the labour input reduction will be used for analytical tasks.

We can therefore calculate the expected annual cost efficiency (CE year) of the project implementation, given the annual cost-saving (CS year) of 244,800 RUB, the annual information system support costs of 50,000 RUB (SC year), and the lump one-off cost (LC) of 157,600 RUB, using the formula (3):

$$\text{CE year} = (\text{CS year} - \text{SC year}) : \text{LC} \tag{3}$$

$$\text{CE year} = (244.800 - 50.000) : 157{,}600 = 1.236 \text{ RUB/RUB}$$

The expected breakeven period (BP) of the project as the inverse value of the cost efficiency is calculated using the formula (4):

$$\text{BP} = 157.600 : (244.800 - 50.000) \approx 0.809 \text{ yr} \tag{4}$$

Thus, the project will break even in about ten (9.71) months.

Practical relevance of the findings is that improvement of information support of the company will help solve the following tasks: automated accounting of contracts and required and implemented supply of electricity to each customer market as a whole and to each customer separately; monitoring of ongoing contractual performance; and timely and efficient management of contractual responsibilities.

Development and integration of an automated information system can achieve the following results: cut the costs of document creation, processing, storage, search, and retrieval; formalise and expedite preparation of relevant contractual documents; improve the quality of customer service; improve the efficiency of energy consumption; reduce the economic risk of delays in obtaining information from consumers on customer markets and maintenance and repair units of the company; reduce the probability of energy leakages and loss in distribution networks; and provide the management with access to current data on contractual performance by customers.

Main requirements to improvement of information support for company's communication with customer markets include the following: consistency control of input, stored, and output data on plans and reports; integrity and accessibility of the data stored in the database; and a multi-user mode of the system.

Conclusion

In the context of the study, we should note that timely contractual relations help develop the company, create product turnover, build industrial, financial, economic, and social capacity, and satisfy a growing resource demand on customer markets. The growing demand is due to the strategic goal of accelerated development of the

Russian Arctic Zone by building new industrial facilities and infrastructure and raising investment for the existing tradition-al economic sectors.

A digitalized contractual relations system gives the company the opportunity to avoid threats, such as economic losses and opportunity costs, avoid the risks caused by errors and ambiguities in relevant documents, significantly reduce contract default penalties and manage the debts due from the customers and payment deadlines for the energy pro-vided.

That is the reason why digitalization of the contractual management process in the existing market economy is a priority goal of the management system as a whole. Using technical means, databases, and software for timely and efficient information processing will also help make contractual relations more accurate and less time-consuming.

References

Belous, A. P., & Lyalkov, S. Y. (2017). A vector of business development in the flow of digital revolution. *Bankovskoye delo, 10*, 16–19.

Bodrunov, S., Plotnikov, V., & Vertakova, Y. (2017). Technological development as a factor of ensuring the national security. In *Vision 2020: Sustainable economic development, innovation management, and global growth* (pp. 2666–2674).

Charykova, O. G., & Markova, E. S. (2019). Regional clustering in the digital economy. *Ekonomika Regiona, 2*, 409.

Kelchevskaya, N. R., & Shirinkina, E. V. (2019). Regional determinants of effective use of human capital in the digital economy. *Economy of Region, 15*(2), 465–482.

Pasandideh, S. H. R., Niaki, S. T. A., & Mousavi, S. M. (2013). Two metaheuristics to solve a multi-item multiperiod inventory control problem under storage constraint and discounts. *The International Journal of Advanced Manufacturing Technology, 69*(5), 1671–1684.

President of Russian Federation. (2016). *On the strategy for scientific and technical development of the Russian Federation*. Retrieved from http://www.kremlin.ru/acts/bank/41449/print

President of Russian Federation. (2018). *On national development goals and strategic tasks of the Russian Federation until 2024*. Retrieved from http://kremlin.ru/events/president/news/57425

Rosstat. (2021). *Regions of Russia. Socioeconomic indicators – 2021*. Retrieved from https://rosstat.gov.ru

Russian Ministry of Energy. (2015). *On assigning the default supplier status: Order of the Russian Ministry for Energy No 14 of 23 January 2015*.

Skotarenko, O., Babkin, A., Senetskaya, L., & Bespalova, S. (2019). Tools for digitalization of economic processes for supporting management decision-making in the region. In IOP Conference Series: Earth and Environmental Science (302, 1, 012147). IOP Publishing https://doi.org/10.1088/1755-1315/302/1/012147.

Ting, A., & Gray, S. J. (2019). The rise of the digital economy: Rethinking the taxation of multinational enterprises. *Journal of International Business Studies, 50*(9), 1656–1667.

Trenta, C. (2019). European VAT and the digital economy: Recent developments. *eJTR, 17*, 118.

UNCTAD. (2019). *Handbook of statistics*. Retrieved from http://unctadstat.unctad.org/

Willems, A., & Kamau, M. (2019). Of binding provisions and trust marks; roadmap to a global legal framework for the digital economy. *Legal Issues of Economic Integration, 46*(3), 225–246.

The System of Mechanisms for Improving Russia's Industrial Policy While Expanding Industrial Network Interaction with the Republic of Belarus

Gregory V. Lepesh, Olga Dmitrovna Ugolnikova, and Irina V. Makarova

Introduction

Among the classical foreign researchers of territorial aspects of industrial policy, such classics as D. Rodrik (2004), Pack and Saggi (2006), Wilson (1990) and others can be distinguished. P. R. Krugman (1983), S. Ramelli et al. (2011) distinguish three levels of industrial policy, interpret the possibilities of its implementation at each level differently.

The content of the industrial policy category was considered in several works by Russian economists: Sukharev (2018), Tatarkin and Romanova (2014), etc. Sukharev (2018) wrote: «Industrial policy is a set of methods, tools, institutions subordinate to the task of developing specific production activities, sectors, subordinating the available resource capabilities to this».

In his works, Tatarkin and Romanova (2014) defined industrial policy as «a functional multi- subject industrial policy, that is, a system of relations between state and municipal authorities, economic entities, scientific organizations and civil institutions regarding the formation of a structurally balanced, competitive industry, the intellectual core of which is represented by the latest technological framework».

The works of Belarusian researchers are devoted to determining the peculiarities of industrial cycles in the post-Soviet space, they show the role of innovations in the modernization of Belarusian industry and developed training mechanisms for Belarusian industry. So, Gursky (2014, p. 53–59) wrote that «industry is the cornerstone of sustained economic growth and, accordingly, the goal of our industrial policy will be to create a super-industrial economy». Considering the conceptual foundations of

G. V. Lepesh (✉) · O. D. Ugolnikova
St. Petersburg State Economic University, St. Petersburg, Russian Federation

I. V. Makarova
Administration of the Governor of Perm Krai, Perm, Russian Federation

A. Rumyantseva et al. (eds.), *Challenges and Solutions in the Digital Economy and Finance*, Springer Proceedings in Business and Economics,
https://doi.org/10.1007/978-3-031-14410-3_47

the industry of Belarus, economists of the republic, as a rule, pointed out that its feature is the strengthening of the role of the state. The concept of a network society as a special quality of a post-industrial society is considered by Castells (1996, p. 151–200).

The program for the development of the industrial complex of the Republic of Belarus for the period up to 2020 defined industrial policy as a system of legal and economic measures and actions of industrial policy entities based on priority ensuring the competitiveness of the national economy, stable and innovative socio- economic development of the Republic of Belarus, and priorities for the development of the industrial complex—the creation of fundamentally new high—tech and knowledge-intensive industries corresponding to the V and VI technological frameworks.

The size of the Belarusian economy makes it possible to attribute it to economies that do not have a significant impact on the global market (small), and its integration into the mega-economy—to an open economy.

So far, there is no work on the peculiarities of the formation of industrial policy of the territories of neighboring countries, which are the Russian Federation and the Republic of Belarus, focused on the development of network interaction. To date, positive network effects have been established: they arise due to transaction costs, the conditions for the development of network structures are indicated. Industrial policy issues within the framework of the concept of network development have not yet been considered.

The technical and technological modernization of the Belarusian and Russian economy should be carried out in unity with the organizational and managerial aspects of the transformation of the global economy. The superindustrial economy requires the development of an interstate industrial policy, it should be aimed at expanding the network interaction between the industrialized regions of Russia and the Republic of Belarus.

Materials and Methods

The materials for the study were works on the industrial partnership of the border states of the post-Soviet space, the prospects for their cooperation, industrial integration in the economic era 4.0. The methodology is represented by theoretical provisions on identity in the economic, economic, and other spheres, a stable system of relationships based on network solutions to the problems of modernization of the industrial complex and ensuring the economic security of territories in the conditions of digitalization. General scientific methods were used: analysis, synthesis, modeling, generalization, systemic approach; among empirical—a survey, a method of expert assessments.

Results

This section analyses the state and development of production of Russian and Belarusian industrial products. Industrial capacity of Republic of Belarus is defined by generally assembly enterprises which remained after the collapse of the USSR, providing in the world market large deliveries of heavy-load dump trucks, trucks, tractors and agricultural machinery, products of electronics, potash fertilizers, oil products, products of chemical, food and light industry. Currently, Belarus is a country with an economy focused mainly on exports but working on imported energy resources. Most of the volumes of raw materials and materials for industrial production are exported from the Russian Federation and Belarus supplies most manufactured products for export (Russia accounts for 42% of the country's exports—more food products and mechanical engineering products). The ratio of exports of goods and services to GDP is 66% (Russia—28%, Ukraine—41%, Poland—56%). The unique geographical position of Belarus between the EU and the EAEU in the center of Europe has long provided it with the place of the main transit of goods passing through the borders of the EU and the EAEU.

The basis of the country's economic potential is industry: it accounts for about 40% of the main assets of enterprises in all sectors of the economy and created almost 30% of GDP. According to the National Statistical Committee of Belarus, in 2021 the volume of industrial production in current prices was at the level of 154.4 billion rubles or increased by 6.5% compared to 2020. In 2019, the growth of industrial production in Belarus amounted to 12% compared to 2018, and in 2020 the decrease in industrial production amounted to only 0.7%. Thus, despite the protests and pandemic, economic indicators in Belarus are among the best in Europe. According to the forecast, in 2022 industrial production is expected to grow by 5.3%. Positive dynamics by economic activity is shown in Fig. 1.

Data from the World Bank and Belstat show that Belarus "real GDP has doubled since 1990, while industrial production has tripled, while Russia's GDP has grown by about 20% over the past 30 years. G.V. Lepesh (2021) noted that even though in Belarus the volume of GDP of the manufacturing sector is dominated by low- tech

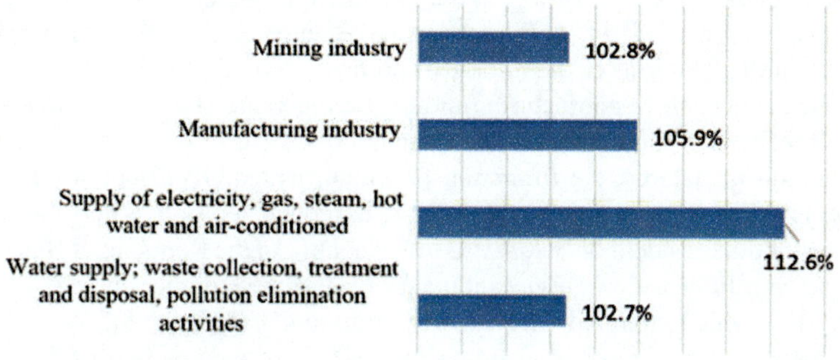

Fig. 1 Belarus industrial production dynamics in 2021 by economic activity, Belarus today (2022)

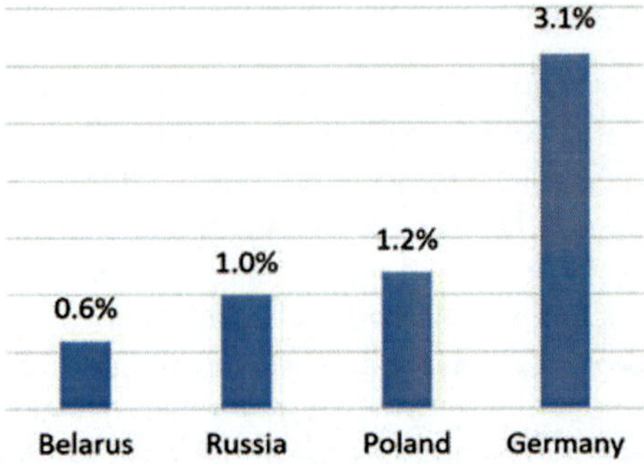

Fig. 2 Share of GDP of national enterprises spent on domestic research and development

and medium-tech enterprises, the state creates incentives and actively finances innovations in the development of enterprises themselves and the production of high-tech products. The share of GDP of Belarusian enterprises spent on domestic research and development is significantly lower than, for example, in Russia and the leading Western countries (Fig. 2). About half of the funds are provided by the state, three quarters of these funds are received by enterprises in the commercial segment of industry. The share of innovative products in Belarus' GDP is greater than in Russia. The share of GDP of national enterprises spent on domestic research and development is shown in Fig. 2. In the same work, G.V. Lepesh emphasized that the state gives special attention to the knowledge-intensive industry sector, which leads to an increase in its competitiveness index (Competitive Industrial Performance Index (CIP—annually calculated by the United Nations Industrial Development Organization (UNIDO) to assess the level of competitiveness of manufacturing industries in the world), which reflects the ability of countries to produce.

The increase in Belarus position is mainly due to an increase in the production and export of manufactured goods. So, A. G. Shumilin (2020) pointed out that in Belarus in 2019 there was a significant increase in indicators: y manufacturing exports per capita, dollars. US (in 2010 prices)—from 2079.2 to 2641.4 dollars. United States (27.0% increase); y the country's share in the world value added of the manufacturing industry—from 0.17% to 0.21% (an increase of 23.1%). The decline in national values was observed only for one indicator «The share of medium and high-tech goods in the total export of the manufacturing industry»—from 41.9% to 39.8% (a decrease of 5.0%). At the same time, Belarus ranks 35th in the world (out of 150 countries) in terms of the «Share of production of high technological level in the added value of the manufacturing industry», the value of which was 38.8%.

The information and communication technologies (ICT) sector occupies particular attention in Belarus, the formation of which, writes G.V. Lepesh (2021), has become possible with serious state support based on legislative and educational initiatives of the President of Belarus A. Lukashenko. In the Republic of Belarus, the Computer and Information Services industry was formed, numbering 1894 companies in 2021, which provided GDP growth of up to 4% (Belstat, 2022).

Currently, Belarus closely cooperates with 80 constituent entities of the Russian Federation in trade, economic, scientific, technical, and cultural areas of activity.

Within the framework of interregional cooperation, more than 300 agreements and agreements were concluded between Belarusian regions and Russian municipalities. Belarus employs 2.5 thousand enterprises with Russian capital, more than 60% of large and medium-sized Belarusian industrial enterprises cooperate with Russian partners. The basis for the development of Russian-Belarusian relations is the harmonization of legislation in the most pressing areas of cooperation, such as transport, energy, security, agriculture, as well as the adoption of decisions on the implementation of existing extremely important infrastructure and investment projects, knowledge-intensive industries in the space industry, petrochemicals, energy, pharmaceuticals, agriculture, and other sectors.

Discussion

We indicate the spatial factors and strategic priorities of the regional industrial policy in the context of the network industrial interaction with the Republic of Belarus.

The most complete essence of industrial policy is defined as... "a system of relations between the state, its territorial entities and economic entities regarding the formation of competitive industry based on modern technological frameworks for its development, adequate to the laws of cyclical dynamics. The System of State Support for Industrial Development is a mechanism for implementing industrial policy, promoting a competitive industrial complex, ensuring its effective functioning, and solving social problems of the population" (Tatarkin et al., 2012, river 46). In this definition, the term "industrial policy implementation mechanism" defines a system of legal, organizational, and other measures that promote the formation of a competitive industrial complex and ensure its effective functioning. The definition extends the list of subjects of industrial policy, including state structures. The subject of industrial policy of Russia is the Government in the person of the Ministry of Industry and Trade, which does not fulfill an imperative function. The role of the state is mostly coordinating and is mandatory only for several measures that support the most significant projects in the development of industry, through financing national projects, concessional lending, etc. The subjects of industrial policy today are state corporations, systemically important companies, the public sector, and regions with their own development strategies. The mechanism for implementing the Russian industrial policy has undergone significant changes since the 90s of the last century, when its implementation was carried out through the development of knowledge-intensive industries, the restructuring of traditional industries and a compensation mechanism that provides a solution to the problem of structural unemployment, infrastructure support for industrial structural adjustment, etc. The new mechanisms being developed are based on complementary mechanisms for the implementation of private-state partnership, federal and regional development institutions, on project and cluster approaches in the development of industry in the Russian regions.

The Republic of Belarus began the transformation of industry at the stage of the country's transition to new socio-economic relations, in conditions of significant technological backwardness and low competitiveness of the products produced, due to the non-compliance of its world standards and excessively high production costs. In part, these realities determine the predominance in the Belarusian industry of products of the low-tech sector. In 2021, Belarus adopted the Program for Socio-Economic Development of the Republic of Belarus for 2021–2025 (hereinafter referred to as the Program), which defines the main provisions of the industrial development strategy. In accordance with it, the industrial development strategy provides for the transformation of domestic production into a competitive complex that quickly and flexibly responds to the global situation and the needs of the domestic market. The bet is made on the accelerated development of high-tech industries in pharmaceuticals, optics and electronics: electronic components, dual-use optical and electronic systems, medical equipment, diagnostic equipment and safety systems. The strategy for the development of the industrial sector of Belarus is aimed at deepening the integration taking place in the economy and politics, primarily with Russia within the framework of the Union State, respecting national interests. The main criterion for integration is ensuring equal economic conditions, energy consumption, market access, participation in public procurement, and the use of financial instruments.

The process of economic integration presents some difficulties arising from the compromise, where each side defends the fundamental issues of the functioning of a future economic association. In this regard, Russian-Belarusian integration has certain difficulties related to the state ownership of large machine-building enterprises in Belarus. The implementation of the Program of Social and Economic Development of Belarus includes the solution of the key task of the five-year period—the work of conducting an inventory and involving inefficiently used resources and national wealth in the economic turnover, "which practically eliminates barriers to investment due to the state ownership of some key industrial facilities of the Republic of Belarus. The improvement of the mechanism for implementing the industrial policy of the Republic of Belarus is carried out by the State by supporting the most knowledge-intensive industries in the areas of creation: high-precision automated equipment, measurement systems and tools, technical diagnostics, optical-mechanical and optical-electronic products, as well as production of new substances and materials of various functional purposes, including nanomaterials and nanotechnologies, new forming technologies, where integration institutes between Russian and Belarusian enterprises and scientific and technical complexes are already functioning. A significant increase in integration is observed in the framework of the joint production of military equipment and weapons of both countries. Lepesh (2021) concluded that according to the Program", joint action programs on specific areas of activity, deepening of production cooperation, development of regional cooperation will act as effective tools."

An effective form of interaction between countries in the field of science, industry, construction, innovation, ICT is allied programs, further improvement of national innovation legislation, their consistent harmonization and unification. The development of science and innovation requires an integrated approach. First, there is a need for coordinated development of effective financing mechanisms and tax incentives, which will contribute to increasing the innovative potential of the economies of both countries. To do this, we need unified regulatory legal acts. The key in the legal system should be the level of allied decisions of direct effect with legal force. The legal regulation of similar legal relations should be unified, based on the same principles and approaches.

The development and implementation of a coordinated industrial policy aimed at creating conditions for sustainable economic growth in Russia and Belarus is one of the main areas of activity of the Union State. Currently, more than twenty intergovernmental agreements are in force in the field of industrial policy of the Union State. Since 2021, the intensity of integration processes taking place between Russia and Belarus has increased significantly. Countries have begun to form a unified industrial policy. Already, there is an agreement between the countries to stimulate the development of joint ventures, as well as the implementation of a single policy to support production and sales. An important solution is the introduction of uniform rules for access to government orders and public procurement.

Thus, there are all prerequisites for synergy between the achievements of Russian and Belarusian scientific and production complexes in a single economic space on common platforms and in equal conditions for economic entities, including when the parties implement national programs. The prospect of expanding investment cooperation, the transition from simple trade to more effective long-term and high- tech forms of work was determined.

Let us further present the model of industrial policy of the Union State. The formation and implementation of industrial policy is based on regulation and regulatory and legal support. Suppose there are three interrelated elements governing industrial policy, as in the proposed "three-link model" Bashkirtsev (2013). Adding a fourth link—an integrated structure, we will get a model of integration industrial policy based on the harmonization of industrial legislation, the concept of industrial development and agreed enterprise development programs in the context of the integration of industrial production.

The spatial development strategies of Russia and Belarus provide for increasing the competitiveness of economies by providing conditions for the development of the production of goods and services in the sectors of promising economic specializations of subjects of states, strengthening interregional cooperation and developing trade and economic cooperation. In these countries, national security requires the development of certain geostrategic territories at a rate that exceeds the average rate in the country.

The Republic of Belarus is building its industrial policy in the direction of improving the sectoral structure of the industrial complex based on the implementation of projects of a knowledge-intensive, high-technology orientation in regions with high scientific and technical potential. At the same time, the development of industries and industries operating on local raw materials (Brest, Grodno and Minsk regions) is a priority. The introduction of energy-saving and resource-saving technologies is especially significant in regions focused mainly on imported raw materials (Vitebsk, Gomel and Mogilev regions). The main industrial production facilities of Belarus are concentrated in urbanized zones belonging to the zones of influence of large cities characterized by a high concentration of urban population and its high mobility.

For Russia and Belarus, the development of industrial production today is the only way to strengthen its position on the world stage. At the same time, in shaping the industrial policy of both countries, priority is given to innovative development as the only possible form of economic progress. The most promising method of cooperation between the states is to increase the volume and synchronize the production of joint high-tech products through the implementation of a coordinated scientific, technical, and industrial policy of Belarus and Russia, based on the harmonization and unification of the legislation of both countries. A significant example of successful cooperation is the implementation of allied joint development programs aimed at increasing the efficiency of economic sectors and creating new state-corporate integration structures. Examples of already functioning structures are Bryanskselmash, LLC MTZ-ElAZ Trading House (Elabuga, Tatarstan). Examples are successful scientific and technical projects of Belarus and Tatarstan in the field of machine tool building, the organization of assembly industries of Belarusian agricultural equipment in the Altai Territory, Krasnoyarsk Territory, etc., the supply of innovative products of Russian enterprises for the Minsk Automobile Plant, to the management company of the holding "Belkommunmash," to the joint Belarusian-Swiss enterprise "Stadler Minsk" and others, supplies of Belarusian components for the Russian (MTZ) has cooperative ties with 157 enterprises of Russia supplying it with raw materials, materials and components.

The Russian Federal Nuclear Center of the All-Russian Research Institute of Experimental Physics (RFNC VNIIEF), together with several Russian and Belarusian enterprises and organizations, is implementing an innovative project for the modernization of motor vehicles in terms of the use of natural gas as motor fuel in diesel engines—the Motor Synthesis Gas project. The dynamics of Russian- Belarusian industrial cooperation is also observed in the military-industrial sphere.

Zalessky et al. (2017, p. 46) wrote that the most important area of Russian-Belarusian technological and innovative cooperation, both in the industrial sector and in other areas, is the transfer of improved production methods, the creation of production chains of demanded products with high added value with the active participation of small and medium-sized businesses of the two countries in this process.

Of particular importance is the further development of the Union State based on the forecast of scientific and technological development, the development on its basis of a coherent industrial policy and its implementation in accordance with the joint programs and projects of the Union State.

Conclusion

The first task in the Union State is being solved today—the creation of a material basis for implementation, 28 union programs approved by the Presidents of Belarus and Russia, ensuring the economic integration of the republics in the short term until 2023. (medium-term) in the Russian Federation, in accordance with the draft Regional Industrial Policy Strategy until 2024 and for the period until 2035, "the industrial potential of the regions will be preserved and the conditions for their long-term economic growth will be laid down (including the implementation of individual programs has been completed. Phase II (long-term) involves the expansion of the modern competitive industrial sector in regions, accompanied by accelerated growth in fixed capital investments and an increase in exports of non- primary non-energy products, "will already take place within the framework of the implementation of regional industrial policy, considering the integration of industrial enterprises of Belarus.

Acknowledgements The study was carried out with the financial support of RFFI and BRFFI within the framework of scientific project №20-510-00002.

References

Bashkirtsev, A. S. (2013). Regional legal basis for the development of industrial complex: Structure and content. *Problems of the Modern Economy, 3*(47), 290–296.
Castells, M. (1996). *The information age: Economy, society and culture*. The Rise of the Network Society (Vol. 1, pp. 151–200). Blackwell Publisher.
Gursky, V. L. (2014). Prerequisites for the formation of the industrial policy of Belarus in the process of Eurasian integration. *Bulletin of BarSU, 2*, 53–59.
Krugman, P. R. (1983). Targeted industrial policies: Theory and evidence. *Industrial Change and Public Policy* (pp. 123–155). Kansas City: Federal Reserve Bank of Kansas City
Lepesh, G. V. (2021). Analysis of the state of the industrial complex of the Republic of Belarus in terms of prospects for digitalization of production, services and business models. *Technical and Technological Problems of the Service, 4*(58), 3–11.
Pack, H., & Saggi, K. (2006). Is there a case for industrial policy? A critical survey. *World Bank Research Observer, 21*(2), 267–297.
Ramelli, S., Christensen, P., & Allen, C. (2011). Local economies in a globalizing world: The role of European industrial policy. *Local Economy, 26*(6–7), 588–593.
Rodrik, D. (2004). *Industrial policy for the twenty-first century*. Available at SSRN 666808. Centre for Economic Policy Research.
SB.BY. (2022). *Belarus today: 17.01.2022*. Retrieved from www.sb.by/articles/vvp-vyros-na-2-3-a-obem-promproizvodstva-na-6-5-belstat-opublikoval-pervye-itogi-2021-goda.html
Shumilin, A. G. (2020). Problems and prospects of innovative development of the Republic of Belarus. *Bulletin of the Institute of Economics of the NAS of Belarus*, 50–57.
Sukharev, O. S. (2018). Industrial policy: The problem of technological update. *Scientific Works of the Free Economic Society of Russia, 211*(3), 630–659.

Tatarkin, A. I., & Romanova, O. A. (2014). Industrial policy: Genesis, regional features, and legislative provision. *Economy of Region, 2*, 9–21.

Tatarkin, A. I., Romanova, O. A., Chenenova, R. I. & Makarova, I. V. (2012). *Regional industrial policy. From macroeconomic conditions of formation to new institutes for development. Yekaterinburg*.

Wilson, G. K. (1990). *Business and politics. A comparative introduction*. Chatham House Publishers.

Zalessky, B. L., Valkovsky, M., & Greshnikov, A. (2017). *Technologies and innovations in the cooperation of the regions of the union state as an impetus for the development of Eurasian integration. Information and integration project*. Businessofset.

Adaptation of the Terminological and Lexical System in the Field of Customs to the Modern Conditions of Digitalization and Sustainable Development

Maria Viktorovna Grudina

Introduction

The sphere of foreign economic activity, foreign trade operations, customs control of passengers and cargo is becoming increasingly important and is largely focused on the use of an extensive vocabulary borrowed from other languages. As you know, due to historical and social reasons, English is recognized as the most common language in the designated professional area (Chernykh et al., 2021).

The modern world has been going through a range of significant changes that have an influence on the most spheres of people's operations. The modern world has faced a new reality due to a range of trends on the global arena and all the fields involved must adjust to the modern situation (Kostin et al., 2021a).

One of the most topical trends is in the sphere of digitalization and further globalization of the international space. All the countries all over the world are focused on developing and adapting their trade systems as well as the practical instrument for facilitating and serving the international trade—customs procedure and the fields of customs control. The most recent international agreements as well as countries' attempts to unify the global trade and transportation space are aimed at creating one single platform for simplification and standardization in the mentioned areas to save time and efforts and that way contribute to sustainable development goals in terms of developing international partnership.

There is certain sphere that requires the development of the standardized language in the field of global trade and customs that would simplify signing international agreements and unification of documentation and finding the common language in

M. V. Grudina (✉)
Russian Customs Academy St.-Petersburg branch named after Vladimir Bobkov, St. Petersburg, Russian Federation

A. Rumyantseva et al. (eds.), *Challenges and Solutions in the Digital Economy and Finance*, Springer Proceedings in Business and Economics,
https://doi.org/10.1007/978-3-031-14410-3_48

the area. For this purpose, it is required to adapt the language so that it could serve the goals of international communication (Kostin et al., 2021b).

Materials and Methods

Currently, in linguistics there is a growing interest in the analysis of the specialized vocabulary of certain areas of knowledge. Considering the fact that the specifics of the professional terminological system of the customs sphere it still not analyzed very carefully, it is necessary to consider specialized vocabulary with a focus on a comparative analysis of the terminological systems of the Russian and English languages in order to identify patterns of formation and functioning in modern English and Russian.

The sphere of foreign economic activity, foreign trade operations, customs control of passengers and cargo is becoming increasingly important and is largely focused on the use of an extensive vocabulary borrowed from other languages. The international nature of the spheres is historically related to the intercultural goal of the formation of these terms (Kostin et al., 2021c). The customs affairs started with the developing international trade, so that historically traces back to adjusting to different languages and different cultures. That logically means that there is a wide range of the words taken from the other languages and, as well, historically from the languages of those nations that Russia was trading with. And as well, due to historical and social reasons, English is recognized as the most common language in the designated professional area.

Results

If we consider the specification of the terminological system of a specific professionally oriented sphere—in our situation, in the sphere of customs affairs and international trade—it has several features that are primarily associated with the following specifics:

1. The vocabulary of any sphere in its development is based on professional activity. It in its turn is reflected in the functions of lexical units, which fix the main patterns, stages, and problematic aspects of development. So, the language can absorb and store the main features of the development of the sphere and the development of trade and nation, as well as it can fix both positive and problematic fields in the development and the level of the language at the current stage can fix, as well, the connotations and senses, ideas, and beliefs that this or that word or specific expression is associated with at the current moment. When time passes by, some words can modify their meaning or get modified themselves, however

some of the words or terms will keep their historical background throughout their further use in the language.

2. Terminological units in the language are distinguished by a special feature—the consciousness of the formation of a lexeme and the category associated with it, and a certain kind of artificiality in the formation of a conceptual apparatus that is easily amenable to conscious regulation and ordering.

   This fact can be easily explained by the applied nature of the terminological system that arises and develops as a means of expressing certain professional interactions and needs. And it can easily explain the situation with the development of the terminological system and its adaptation to modern needs.
3. Special vocabulary easily reveals the relationship with the material and spiritual culture of the people.
4. The term system reflects the processes of mutual influence of languages, primarily the processes of borrowing and tracing. This aspect is especially evident in the situation of customs terms since this professional field since its formation has been aimed at internationalization and cross-border cooperation.
5. It should be noted that the terminological system of professional activity (in particular, in the customs sphere) is extremely dependent on the development of scientific and technical knowledge, and all achievements in the field of computer science and technology, as well as factors of automation and globalization of processes, primarily have a direct impact on the formation of a whole layer of vocabulary (Grinev-Grinevich & Sorokina, 2018).

As well, it is important to bear in mind that when teaching a foreign language to students of the customs university we can provide them with the general understanding of the processes and links and connections in the languages as the educational process becomes more professionally oriented when students realize the significance of this or that term in the conditions when it was stored in the language or developed in the system.

Discussion

Given this need, it is impossible not to wonder whether a non-linguistic university should aim to master or develop foreign language skills. However, the position of a foreign language as a discipline in a non-linguistic university is associated with certain kinds of difficulties. Among the cycles of disciplines for a non-linguistic university, according to federal state educational standards, a foreign language, as a rule, does not fall into the cycle of general professional or professional disciplines, and according to the standards of a specialty, it often turns out to be among optional disciplines. It should also be considered that the teaching of a foreign language in Russia, which inherited a lot of stereotypes in this sense, also developed in rather difficult conditions. The communicative approach, which managed to become a classic abroad, has taken root in Russia relatively recently, and at the current stage, the methodology of teaching a foreign language continues to rapidly adapt and use the latest developments and trends that are increasingly appearing in this field (Internet laboratories, programs for "devices").", electronic textbooks, etc.).

Since the professional activity of a customs officer belongs to the "Person-to-Person" type (as indicated in the Professiogram of labor and employment agencies on the territory of the Russian Federation), that is, it is focused on communication and interaction with people, among the formed competencies, a group of linguistic, speech and communicative competencies: the ability to apply modern communication technologies, including in a foreign language(s), for academic and professional interaction, as well as analyze and take into account the diversity of cultures in the process of intercultural interaction.

Currently, terminological vocabulary is the object of numerous scientific studies.

According to some researchers, relationships are the central element of the situation. Based on this theoretical basis, A.A. Aldokhina (2015) classifies social relationships in which a customs officer communicates and identifies the following situations: a state customs officer—a civil servant of the law enforcement system, a state customs officer—a civil servant in the field of passport control, a state customs officer—a resident of Russia, a state customs officer—a resident of a foreign state.

The criteria for evaluating the terminological system in the professional field of customs can be presented within the following categories:

1. The functional criterion of analysis is non-one-dimensional. It involves the identification of several varieties of its targets.
2. Thematic criterion is related to the scope of the topics presented.
3. The coherent criterion is represented by different methods of conjugation.
4. The delimitation criterion is associated with the establishment of boundaries. This is a very difficult research task related to the definition of the term boundary.
5. Structural criterion in its most general form involves the allocation of elementary and non-elementary units.

The development of terminology in professional training involves the use of a lexical and terminological base, considering the functional features of the units of the system and their contextual tasks. By function, the units used are formed depending on the context in which they operate (Ehlich, 1999). So, for the inspection of passengers or cargo, customs specialists enter a dialogue and use the lexical and terminological system in a format adapted for the understanding of the interlocutors. When working with colleagues, a purely professional format will be used to a sufficient extent, available for communication in the professional sphere "in the same language".

The thematic criterion is associated with the formation of thematic lexical fields and the distribution of vocabulary and terminology in separate semantic groups. Terminological units in the language are distinguished by a specific feature—the consciousness of the formation of a lexeme and the category associated with it and a certain kind of artificiality in the formation of a conceptual apparatus that is easily amenable to conscious regulation and ordering (Vinokur, 2007).

Conclusion

The coherent criterion is associated with various ways of conjugating vocabulary and terminology in the context of a communicative situation, in this case a professional one. There are two clearest ways:

(a) pairing with a whole, expanded dialogue.
(b) pairing with its individual replicas.

This is the contextual interaction of lexical units in the language system, as well as their work in the context of use.

The delimitation characteristic relates to the establishment of the boundaries of the dialogue and context. In most communicative situations, the delimitation criterion involves defining the boundary between the use of contextual units in the framework of a dialogue with one or more interlocutors.

Structural characteristics in the most general form involves the allocation of elementary and non-elementary terminological units. They can be different in volume, starting with a set of two constituent words and ending with fixed phrases, reaching a dozen words included in one unit.

Thus, the importance of a deep study of terminological units is due to the increasing role that these units play during the period of rapid development of modern economic relations. Successful professional activity of a customs officer requires the development of both highly specialized vocabulary, which includes customs terminology, customs economic terminology, customs legal terminology, and terminology related to customs areas of activity: legal terminology, socio-political terminology, educational and pedagogical terminology, vocabulary denoting the names of goods, information and computer terminology, vocabulary of the transport sector of activity, etc.

References

Aldohina, A. A. (2015). The main theoretical aspects of the formation of the communicative competence of specialists in the field of "customs". *Lecturer XXI Century, 1*(4), 69–74.

Chernykh, A., Voskresenskaya, E., & Martynkevich, M. (2021). Social policy governance mechanisms as a tool to ensure stability of the state and society. In *SHS web of conferences* (Vol. 94). EDP Sciences. https://doi.org/10.1051/shsconf/20219403016

Ehlich, K. N. (1999). *Permanent source of scientific terminology* (pp. 135–137). CUP.

Grinev-Grinevich, S. V., & Sorokina, E. A. (2018). Perspective directions of development of terminological research. *Vestnik MGOU, 5*, 18–28.

Kostin, G. A., Chernykh, A. B., Andronov, I. S., & Pryakhin, N. G. (2021a). Perception and interpretation of general humanistic values through communication Technologies in the Digital Transformation of Society. In *2021 communication strategies in digital society seminar (ComSDS)* (pp. 162–165). IEEE. https://doi.org/10.1109/ComSDS52473.2021.9422839

Kostin, G. A., Chernykh, A. B., Andronov, I. S., & Pryakhin, N. G. (2021b). Perception and interpretation of general humanistic values through communication technologies in the digital transformation of society. In *2021 communication strategies in digital society seminar (ComSDS)* (pp. 162–165). IEEE. EpSBS. https://doi.org/10.15405/epsbs.2021.05.245

Kostin, G. A., Chernykh, A. B., Andronov, I. S., & Pryakhin, N. G. (2021c). The phenomenon of tolerance and non-violence in the development of the individual in the digital transformation of society. In *2021 communication strategies in digital society seminar (ComSDS)* (pp. 213–216). IEEE. https://doi.org/10.1109/ComSDS52473.2021.9422873

Vinokur, T. G. (2007). *Speaking and listening: Variants of speech behavior*. LKI.

New Technologies of Financial Support of Municipalities

G. Morunova

Introduction

The scientific problem of developing new technologies for attracting investment in the development of the territory based on municipal target bonds is the lack of scientific research by Russian scientists on the methodology and practice for implementing target loans in the financial and economic activities of local governments.

The search for additional options for financial support of local issues and the organization of economic activities of municipalities is particularly relevant in the context of the economic crisis and budget constraints. At the same time, in the winter and spring of 2021, the Bank of Russia recorded a steady interest of Russian citizens in investing in securities against the background of a decline in deposit yields, which underlines the availability of a resource for issuing subnational bonds, including municipal ones. It is obvious that additional sources can be the funds of citizens and businesses, which once again emphasizes the need to develop a mechanism that enables the accumulation of these funds. The role of such a mechanism can be carried out by a targeted municipal loan using bonds. At the same time, the models built based on inter-municipal cooperation are also interesting. However, the implementation of such proposals will require amendments to the budget legislation of the Russian Federation.

Thus, the scientific novelty of the research lies in the development and justification of methodological provisions, methodological recommendations, and practical proposals for attracting additional investments in the socio-economic development

G. Morunova (✉)
St. Petersburg State University of Economics, St. Petersburg, Russian Federation
e-mail: morunova.g@unecon.ru

A. Rumyantseva et al. (eds.), *Challenges and Solutions in the Digital Economy and Finance*, Springer Proceedings in Business and Economics,
https://doi.org/10.1007/978-3-031-14410-3_49

of territories on the basis of municipal target bonds, taking into account positive foreign experience, as well as historical domestic experience.

Practice in Applying Targeted Municipal Bonds

In the context of the current economic crisis, the search and scientific justification of new directions for attracting investment in the socio-economic development of the territory are becoming particularly relevant. The relevance of this topic is also confirmed by the statement of the Chairman of the Government of the Russian Federation, Mikhail Mishustin, made in the framework of the Report on the Work of the Government of the Russian Federation on May 12, 2021, in the State Duma of the Russian Federation. It says that infrastructure bonds are another tool that will attract additional extra-budgetary funds for housing construction, the creation of engineering, transport, and social facilities. In addition, in 2020, the Bank of Russia prepared a new version of the Regulation on Securities Issuance Standards No 706-p dated December 19, 2019, which sets out the specifics of issuing bonds with the intended use of funds received from the placement, including issues of "green", social and infrastructure bonds. According to this document, "green" bonds are such bonds, the funds from the placement of which are aimed at the preservation and protection of the environment, a positive impact on the environment and comply with international principles and standards in the field of ecology and (or) "green" financing; social bonds are such bonds, the funds from the placement of which are aimed at the development of public life and comply with international principles and standards in the field of social financing and (or) sustainable development; infrastructure bonds are defined as bonds, the funds from the placement of which are used for purposes related to the financing of a project for the creation and (or) reconstruction of property that is the object of a concession agreement, a public-private partnership agreement or a municipal-private partnership agreement.

It is assumed that special interest in the issue of targeted bonds will arise from state authorities and local governments in the context of budget deficits caused by the economic crisis. However, there are several issues on the target bonds that remain unresolved. So, for the decision on the issue of "green" and social bonds, the projects for which it is planned to attract funds must be registered. There must be verification that these projects comply with Russian and international standards. The decision to issue the bonds should include the issuer's obligation to disclose on a regular basis information on the direction and use of funds raised for these projects, and to provide information on the development of these projects. The requirement to disclose the verification institution is also introduced. An important and difficult issue is the creation of motivations for the issuer.

Russian practice shows occasional cases of issuing "green" bonds, including those issued by regional state authorities. While foreign practice on this issue is very representative and is developing rapidly.

At the municipal level, the situation is complicated by the low level of financial preparedness of specialists of local administrations, who are obliged to create conditions that meet the daily needs of citizens to address local issues, including to develop and maintain the proper level of infrastructure and social security of the territory.

By and large, there is no market for targeted municipal bond loans in our country (Morunova et al., 2018). However, this tool is successfully used in developed countries. For example, in the United States, municipal infrastructure bonds are issued by municipalities as "special" purpose bonds for specific infrastructure projects—income-bearing bond loans. The peculiarity of the mechanism is the requirement to lay payments for servicing the debt on municipal bonds in the local budget when it is formed. A distinctive feature of the mechanism of municipal infrastructure bonds in the Netherlands is a legally established system of guarantees of the solvency of municipalities, which reduces the risks for investors to zero.

The experience of using target bonds of foreign countries has a rich positive history and is represented by several studies (Combes et al., 2021; Gross, 2021; Hattori, 2018; Herrera & Rangaraju, 2019; Van Hecke, 2013).

Among developing countries, the experience of China is noteworthy, where the interest of individuals in financial investments is rapidly growing. Research on this topic is also covered in scientific publications of foreign authors (Chen et al., 2020; Ding et al., 2014; Gao et al., 2021; Walker et al., 2021; Wang et al., 2008; Wang & Yang, 2021).

The share of savings in the GDP structure of the People's Republic of China is quite high, which reflects the traditional approach of the country's population to ensuring financial stability. In this regard, municipal bond funds that finance infrastructure projects are increasingly in demand. The leaders in the volume of such operations are large cities with a high level of economic development, such as Beijing and Shanghai. The maximum volume of municipal infrastructure bond issuance occurred in the post-crisis period (2008–2012), when the Government decided to change the development model from export-oriented to growth due to domestic factors and domestic consumption to maintain economic growth. Assessing the high contribution of infrastructure to GDP growth, local governments began to issue appropriate bonds, which on the one hand gave an impetus to economic activity, and on the other hand reflected a heavy debt burden on the economy. To reduce the risks of issuing unsecured securities, the country is carefully honing the decision-making mechanism for issuing local bonds. It includes an analysis of the balance and debt sustainability of the municipal budget; an analysis of the macroeconomic and social indicators of the municipality; an assessment of the cost and effectiveness of alternative sources of financing for infrastructure projects; an assessment of the qualifications of local government employees (to ensure a sufficient level of competence for the implementation of relevant projects); an assessment of the transparency and information openness of the activities and financial condition of municipalities. In addition, as in other countries, it is important to monitor the implementation of the turnover of municipal bonds, which in China is carried out by the National Audit Office. Periodic inspections of compliance with the law have a positive impact on ensuring financial discipline. This experience can be useful for reducing the risks of uncontrolled emissions and subsequent financial and economic problems.

Thus, the international experience of the structure of subnational target bonds, the mechanism of their issuance and regulation has a rich history and can be applied considering its adaptation to the existing realities of the Russian economy.

The problem of finding investments for infrastructure development is quite acute in Russia, especially about the financing of urban infrastructure. The Russian infrastructure mainly functions due to the exploitation of the assets of the Soviet Union, the level of depreciation of which is extremely high. The instruments for financing new infrastructure that have become widespread include income from the sale of housing, and only a small part is covered by budget financing. Thus, in the context of financial constraints, social significance and low marginal income, the development of long-term financing tools for urban infrastructure is particularly relevant. One of the mechanisms that can solve the problems of urban infrastructure development is municipal infrastructure bonds, which have shown their effectiveness abroad.

Theoretical and methodological aspects of the formation and development of the market of "green" bonds, including municipal bonds, are reflected in the works of foreign authors (Naeem et al., 2021; Zerbib, 2016). In the Russian literature, this topic is also actively studied by such authors as: N. Ivanova, I. Katsyuba, E. Firsova. "Green" finance (Febriani et al., 2020); A. Ramazanov, K. Grigorian. Municipal Loans as a Tool for Solving the Financial Problems of the Territory (Ramazanov & Grigorian, 2015); I. Polyakova, E. Vasilyeva. Benefits of Public-and-private Partnership for the Creation of the Infrastructure of the Urbanized Territories in Russia (Polyakova & Vasilyeva, 2016). However, the issue of the practical application of municipal target bonds in domestic practice requires additional research. In this regard, it is necessary to solve the following tasks:

1. Study of the use of subnational target bonds to attract investment to address issues of subnational significance in foreign practice (study of legislation, practice of application, positive experience), including considering the experience of the Zemstvo reform in Russia in the XIX.
2. Development of general principles for the application of subnational target bonds in modern Russia based on the current legislation and proposals for its improvement, considering specific types of these bonds: "green", infrastructure, social.
3. Development of a new model for attracting investment in the development of the territory based on subnational target bonds.
4. Development of organizational and methodological approaches to the introduction of a new model for attracting investment in the development of the territory based on subnational target bonds.
5. Creation of specific conditions for the introduction of subnational target bonds at the municipal level.

Conclusion

In our opinion, the introduction of targeted bonds for the population at the municipal level will allow the implementation of "green", infrastructure and social development projects. This tool can be used to finance key projects of the territory. The issued municipal target bonds will attract funds from citizens and businesses for the reconstruction of housing and utilities facilities and social facilities. By purchasing these bonds, public legal entities will be able to participate in the co-financing of these projects.

In conclusion, it should be noted that in the context of the economic crisis, when financial support is reduced, municipalities are increasingly using new financial technologies to support local issues, such as public and municipal private partnerships, initiative budgeting, and self-taxation of citizens. It is obvious that only cooperation and interaction at all levels will be able to ensure the solution of several tasks, and as a result—territorial development. An instrument of such cooperation (inter-municipal, inter-level, within the territory with citizens and businesses) can be a target bond loan for the implementation of a specific project—municipal target bonds issued on behalf of a public legal entity for the population to finance a specific project. For the implementation of this tool, it is necessary to provide for corresponding changes in the budget legislation.

References

Chen, Z., He, Z., & Liu, C. (2020). The financing of local government in China: Stimulus loan wanes and shadow banking waxes. *Journal of Financial Economics, 137*(1), 42–71. https://doi.org/10.1016/j.jfineco.2019.07.009

Combes, J. L., Minea, A., & Sawadogo, P. N. (2021). Does the composition of government spending matter for government bond spreads? *Economic Modelling, 96*, 409–420. https://doi.org/10.1016/j.econmod.2020.03.025

Ding, C., Niu, Y., & Lichtenberg, E. (2014). Spending preferences of local officials with off-budget land revenues of Chinese cities. *China Economic Review, 31*, 265–276. https://doi.org/10.1016/j.chieco.2014.10.001

Febriani, A., Syafriana, V., Afriyando, H., & Djuhariah, Y. S. (2020). The utilization of oil palm leaves (Elaeis guineensis Jacq.) waste as an antibacterial solid bar soap. IOP conference series: Earth and environmental science 572, 1, 012038 IOP Publishing. doi:https://doi.org/10.1088/issn.1755-1315.

Gao, H., Ru, H., & Tang, D. Y. (2021). Subnational debt of China: The politics-finance nexus. *Journal of Financial Economics, 141*(3), 881–895. https://doi.org/10.1016/j.jfineco.2021.05.028

Gross, T. (2021). Dynamic optimal fiscal policy in a transfer union. *Review of Economic Dynamics, 42*, 194–238. https://doi.org/10.1016/j.red.2020.10.010

Hattori, T. (2018). Decomposing Japanese municipal bond spreads: Default and liquidity premiums in times of crisis. *Journal of Asian Economics, 59*, 16–28. https://doi.org/10.1016/j.asieco.2018.09.002

Herrera, A. M., & Rangaraju, S. K. (2019). The quantitative effects of tax foresight: Not all states are equal. *Journal of Economic Dynamics and Control, 107*, 103726. https://doi.org/10.1016/j.jedc.2019.103726

Morunova, G., Kuzmina, S., & Sorvina, T. (2018). Options for financial support of local issues. In *MATEC web of conferences* (Vol. 193, pp. 05058). EDP Sciences. https://doi.org/10.1051/matecconf/201819305058

Naeem, M. A., Farid, S., Ferrer, R., & Shahzad, S. J. H. (2021). Comparative efficiency of green and conventional bonds pre-and during COVID-19: An asymmetric multifractal detrended fluctuation analysis. *Energy Policy, 153*, 112285. https://doi.org/10.1016/j.enpol.2021.112285

Polyakova, I., & Vasilyeva, E. (2016). Benefits of public-and-private partnership for the creation of the infrastructure of the urbanized territories in Russia. *Procedia Engineering, 165*, 1380–1387. https://doi.org/10.1016/j.proeng.2016.11.868

Ramazanov, A. V., & Grigorian, K. A. (2015). Municipal loans as a tool for solving the financial problems of the territory. *Procedia Economics and Finance, 24*, 537–542. https://doi.org/10.1016/S2212-5671(15)00627-9

Van Hecke, A. (2013). Vertical debt spillovers in EMU countries. *Journal of International Money and Finance, 37*, 468–492. https://doi.org/10.1016/j.jimonfin.2013.06.011

Walker, T., Zhang, X., Zhang, A., & Wang, Y. (2021). Fact or fiction: Implicit government guarantees in China's corporate bond market. *Journal of International Money and Finance, 116*, 102414. https://doi.org/10.1016/j.jimonfin.2021.102414

Wang, J., Wu, C., & Zhang, F. X. (2008). Liquidity, default, taxes, and yields on municipal bonds. *Journal of Banking & Finance, 32*(6), 1133–1149. https://doi.org/10.1016/j.jbankfin.2007.09.019

Wang, Z. J., & Yang, J. (2021). Cross-trading and liquidity management: Evidence from municipal bond funds. *Pacific-Basin Finance Journal, 67*, 101564. https://doi.org/10.1016/j.pacfin.2021.101564

Zerbib, O. D. (2016). Is there a green bond premium? The yield differential between green and conventional bonds. *Published in the Journal of Banking and Finance, 98*, 39–60. https://doi.org/10.1016/j.jbankfin.2018.10.012

Digitalization of Modern Educational Processes and Improvement of Inclusive Programs

Vladimir V. Okrepilov and Alexey D. Shmatko

Introduction

The world is rapidly undergoing changes in all spheres of human life. These changes are especially visible in the field of education. It is necessary to note the current trends in the development of education. Currently, Russian educational organizations of higher education pay great attention to the automation and digitalization of educational and extracurricular activities. Digital transformation is becoming the main reason for the radical changes taking place in almost all areas of scientific, educational, and professional activities. Technologies and principles for creating new products and services are fundamentally changing, as well as the ideology and culture of their use (Uvarov et al., 2019).

Modern education is subject to globalization processes. The processes of internationalization and integration of education inevitably lead to the universalization of national education systems. The academic mobility of students, teachers and researchers is increasing. The demand for higher education is growing rapidly. In recent years, expenses for higher education have increased significantly. Now they are much higher than for other levels of education. This is a factor in increasing investment in human capital. The growth of competition and offers for additional education programs affect changes in the market of educational services. This is due to the implementation of the concept of continuous education. It involves the constant updating of knowledge, skills, and abilities throughout life in a dynamically changing environment.

V. V. Okrepilov (✉) · A. D. Shmatko
Institute for Regional Economic Studies, Russian Academy of Sciences, St. Petersburg, Russian Federation
e-mail: pokid@vmail.ras.ru

A. Rumyantseva et al. (eds.), *Challenges and Solutions in the Digital Economy and Finance*, Springer Proceedings in Business and Economics,
https://doi.org/10.1007/978-3-031-14410-3_50

Materials and Methods

The purpose of this study is to examine the impact of digitalization on the educational process. The process of technological transformation generates qualitatively new socio-economic and social relations that highly affect modern education.

To achieve the goal of the study, the following research tasks were formulated. It is necessary to:

- analyze the results of scientific research and statistical data in the field of education.
- assess the advantages and disadvantages of digitalization.
- analyze statistics that ensure the right to affordable education.
- prove that the use of a unified electronic information and educational environment improve the workflow and the education process.
- highlight the exact features of digitalization that can affect the abilities of modern students.
- find the relation between digitalization and inclusiveness of education.

Statistical analysis serves as a methodological basis for the research. This research method is necessary to identify dependencies in the dynamics of different indicators in different time periods. The study reveals the connection between the education development and the digitalization. It allows to detect a positive effect of digitalization in the context of the universalization of national education systems and dynamically changing environment.

This article is based upon official data of Federal State Statistics Service, Constitution of the Russian Federation and Federal Law "On Education in the Russian Federation" (The State Duma, 2022). The results of selective statistical observation of people's participation in continuing education is analyzed.

In the scientific research literature, there are many arguments about the positive relationship between digitalization and the abilities of modern students. The article examines different points of view on the impact of digitalization of modern education, that are discussed in various scientific works.

The technological transformation and its effect on modern education is analyzed in the article on "Difficulties and prospects for the digital transformation of education" (Uvarov et al., 2019). The changes in the field of digital technologies and the educational process, which will become dominant in the development of the education sector in the next decade, are considered.

The exact features of digitalization that can affect the cognitive and other abilities of modern students are considered in the research on "Positive and negative aspects of using computer technologies by children and adolescents" (Maslova, 2013) and "Cognitive abilities: phenomenology, diagnostics, development" (Solovieva, 2010). The importance of computer technologies for the education, development, and communication of people with serious illnesses and disabilities is also shown.

The Federal State Statistics Service was used to define the relation between digitalization and inclusiveness of education. The implementation of inclusive education is also analyzed in the article on "Inclusive education: transformation of the education system" (Kotov, 2016) and "Inclusive education: what needs to be done for its successful implementation" (Kulakova et al., 2016).

This article refers to previous scientific research of the authors, that includes analysis of the concept of scientific and technological development of St. Petersburg for the period up to 2030. It examines the genesis of qualitatively new socio-economic and social relations in the process of technological transformations (Okrepilov & Shmatko, A.D. 2021).

The patented database for building a model of the quality of life (The Quality-of-Life Database) is also discussed (Okrepilov et al., 2021).

Result

To ensure the right to affordable education, the requirement to create special conditions is enshrined at the legislative level. Their absence can make the development of educational programs more difficult. This applies to all citizens, regardless of their physical, intellectual, social, emotional, linguistic and other characteristics.

Obtaining education by persons with disabilities is guaranteed by the Constitution of the Russian Federation (Official Legal Information Internet Portal, 2022), the Law of the Russian Federation of December 29, 2012 No 273-FZ "On Education in the Russian Federation" (The State Duma, 2022) and other federal laws. The legislation of the Russian Federation and the main international documents in the field of education assumes the principle of equality in obtaining education by persons with disabilities. Definitions relating to education for socially vulnerable groups of the population (persons with disabilities and persons with disabilities) are defined by law. The essence of education at the university for both healthy students and students with disabilities is determined. In other words, it is an inclusive education that excludes any kind of discrimination. Inclusive education is recognized as the most humane, and therefore has become one of the leading forms of education in Russian educational policy (Kotov, 2016; Kulakova et al., 2016). The growth of the global market for online education is a condition for providing accessible education to people with disabilities and increasing the level of participation in lifelong education.

Federal State Statistics Service (2022) sample observation of the participation of the population in lifelong education in 2020 in the Russian Federation is presented in Table 1.

The level of participation in lifelong education of these socially vulnerable groups of the population is low and was about 15% in 2020 (Table 2).

The process of technological transformation generates qualitatively new socio-economic and social relations. This requires flexibility and the ability to quickly adapt to changing conditions when implementing an education development strategy. This can be done through the development of the online education market. It

Table 1 Population aged 15 and over participating in continuing education, with a disability (thousand people)

	Population participating in continuing education with disabilities	Participation forms		
		General or professional education	Further education or professional training	Self-education
Population				
Total	930.7	76.0	194.1	785.3
Including by disability groups:				
I group	95.7	5.2	24.8	79.4
II group	378.0	19.1	69.9	321.5
III group	416.0	12.8	93.6	363.9
Disabled children	41.0	38.9	5.8	20.5
Men				
Total	444.8	48.3	106.5	358.8
Including by disability groups:				
I group	51.4	3.8	14.6	42.3
II group	175.8	14.0	40.0	141.3
III group	192.8	7.2	47.3	162.4
Disabled children	24.8	23.3	4.5	12.8
Women				
Total	485.9	27.7	87.7	426.5
Including by disability groups:				
I group	44.3	1.4	10.2	37.1
II group	202.2	5.1	29.8	180.3
III group	223.2	5.6	46.3	201.5
Disabled children	16.2	15.6	1.3	7.6
Urban population				
Total	684.2	45.6	127.0	594.6
Including by disability groups:				
I group	72.6	3.7	15.2	64.3
II group	273.5	13.3	46.3	234.5
III group	315.4	7.3	61.9	282.8
	22.8	21.3	3.7	13.1

(continued)

Table 1 (continued)

	Population participating in continuing education with disabilities	Participation forms		
		General or professional education	Further education or professional training	Self-education
Disabled children				
Rural population				
Total	246.5	30.4	67.1	190.6
Including by disability groups:				
I group	23.1	1.5	9.6	15.1
II group	104.6	5.8	23.6	87.0
III group	100.6	5.5	31.8	81.1
Disabled children	18.2	17.6	2.1	7.4

includes the development, implementation and distribution of new platforms and distance learning technologies, as well as increasing the role of technology in higher education (Okrepilov & Shmatko, 2021). In the context of the pandemic in 2020, distance learning was massively introduced at universities. This allowed the participants in the educational process to quickly assess the advantages and disadvantages of digitalization.

The digitalization of education and the widespread use of digital educational technologies had a significant positive impact on the abilities of modern students. The use of information technology in the learning process contributes to the development of intelligence and creativity. High cognitive activity of students, interest in obtaining and consolidating knowledge remains for a longer period (Maslova, 2013). Through the use of distance learning, students develop digital competence. The subjects of the educational process master the use of information, practical tools, specialized and general technologies. Students expand their digital literacy by working with different operating systems, programs, platforms, and devices (Solovieva, 2010).

The use of a unified electronic information and educational environment ensures the information openness of the university. It complies with the requirements of the current legislation of the Russian Federation and increases the efficiency of the educational process. This system is an organized set of information and educational resources, computer equipment, information and telecommunication technologies, hardware and software, organizational and methodological support. The system is focused on meeting the needs of users in information services and educational resources.

Table 2 Level of participation in lifelong education of the population aged 15 years and older with a disability, %

	Population participating in continuing education with disabilities	Participation forms		
		General or professional education	Further education or professional training	Self-education
Population				
Total	15.0	1.2	3.1	12.6
Including by disability groups:				
I group	11.4	0.6	3.0	9.5
II group	12.5	0.6	2.3	10.7
III group	18.0	0.6	4.0	15.7
Disabled children	80.6	76.5	11.5	40.3
Men				
Total	15.9	1.7	3.8	12.9
Including by disability groups:				
I group	12.4	0.9	3.5	10.2
II group	13.5	1.1	3.1	10.8
III group	18.5	0.7	4.5	15.6
Disabled children	79.5	74.7	14.4	41.2
Women				
Total	14.2	0.8	2.6	12.4
Including by disability groups:				
I group	10.4	0.3	2.4	8.7
II group	11.8	0.3	1.7	10.5
III group	17.5	0.4	3.6	15.8
Disabled children	82.4	79.4	6.8	38.9
Urban population				
Total	15.8	1.1	2.9	13.7
Including by disability groups:				
I group	12.5	0.6	2.6	11.1
II group	12.9	0.6	2.2	11.1
III group	19.7	0.5	3.9	17.6
	79.5	74.3	12.9	45.8

(continued)

Table 2 (continued)

	Population participating in continuing education with disabilities	Participation forms		
		General or professional education	Further education or professional training	Self-education
Disabled children				
Rural population				
Total	13.0	1.6	3.5	10.1
Including by disability groups:				
I group	8.9	0.6	3.7	5.8
II group	11.6	0.6	2.6	9.6
III group	14.1	0.8	4.5	11.4
Disabled children	82.1	79.4	9.7	33.2

Automation of educational processes optimizes the workflow. The use of information technology expands the possibilities of higher education. Digitalization can diversify learning, increase the speed of obtaining information and the effectiveness of cognitive activity, develop independence and creativity.

Technology for student learning is changing at an incredible rate, and it is necessary to make the most of old and new learning formats at the same time. It is also important to note the disadvantages of education digitalization. These are such social problems as the difficulty of adaptation in the digital environment, the difficulty of socialization, the possible development of asociality and the lack of formation of communication skills. Education risks losing its personal development function, including the transmission of general cultural values.

Conclusion

The COVID-19 lockdown has influenced more active use of automated control systems. Now educational standards and programs requirements are constantly being reformed. Under these conditions, it is impossible to imagine a modern university without a single technological digital platform or ecosystem. This system provides the necessary level of information support for business processes.

A database has been created and registered, which can serve as an information and analytical basis for the formation of state administration systems in regions. It considers the typological features of different management objects, including the education sector. The quality-of-life measurement model is a strategic management tool. It can be used throughout the entire period of implementation of the strategy. It is also useful for the implementation of regional policy in the field of education. The model provides a combination of requirements for the quality of life and the possibility of choosing the best ways to meet these requirements. This model has broad prospects for the use of the database. It can be used to select strategic guidelines, as well as to form criteria and scales for assessing the quality of life, including indicators of the education sector (Okrepilov et al., 2021).

The advantages of using distance learning technologies should be noted. They can increase the efficiency of managing the educational process, reduce costs and speed up the process of interaction between teachers and students. They also ensure the availability of educational resources and influence the expansion of the export of educational services. Digitalization contributes to the development of inclusiveness in education and the right to affordable education for socially vulnerable groups of the population. At the same time, there is a risk of decreasing the quality of education, because the learning process depends too much on the reliability and stability of digital systems.

References

Federal State Statistics Service. (2022). *Results of selective statistical observation of people's participation in continuing education*. Retrieved March 15, 2022, from https://gks.ru/free_doc/new_site/population/trud/inobr2020/index.html

Kotov, S. V. (2016). Inclusive education: Transformation of the education system. *Proceedings of the Southern Federal University. Pedagogical Sciences, 12*, 30–36. Retrieved from https://elibrary.ru/item.asp?id=28140896

Kulakova, E., Lyubimova, M., & Lyubimov, M. (2016). Inclusive education: What needs to be done for its successful implementation. *Social Pedagogy in Russia, 6*, 14–23. Retrieved from https://elibrary.ru/item.asp?id=28784168

Maslova, Y. V. (2013). Positive and negative aspects of using computer technologies by children and adolescents. *Educational Technologies and Society, 16*(4), 493–503. https://readera.org/pozitivnye-i-negativnye-aspekty-ispolzovanija-kompjuternyh-tehnologij-u-detej-i-14062522

Official Legal Information Internet Portal. (2022). *Constitution of the Russian Federation* (adopted by the nationwide vote on December 12, 1993 with amendments approved during the nationwide vote on July 01, 2020). Retrieved from http://publication.pravo.gov.ru/Document/View/0001202007040001

Okrepilov, V. V., & Shmatko, A. D. (2021). Genesis of qualitatively new socio-economic and social relations in the process of technological transformations (on the example of the analysis of the concept of scientific and technological development of St. Petersburg for the period up to 2030). *Economic Revival of Russia, 1*(67), 81–88. https://doi.org/10.37930/190-9780-2021-1-67-81-88

Okrepilov, V. V., Shmatko, A. D., & Gagulina N. L. (2021). *Quality of life database (database for building a model of the quality of life)*. Patent RF, no. 2021622426, 2021.

Solovieva, O. V. (2010). *Cognitive abilities: Phenomenology, diagnostics, development*. Stavropol State University.

The State Duma. (2022). *Federal Law "On Education in the Russian Federation"* dated December 29, 2012 No 273-FZ (last edition). Retrieved from http://pravo.gov.ru/proxy/ips/?docbody=&nd=102162745

Uvarov, A. Y., Gable, E., Dvoretskaya, I. V., Zaslavsky, I. M., Karpov, I. A., Mertsalova, T. A., et al. (2019). *Difficulties and prospects for the digital transformation of education*. HSE Publishing House.

Analysis of Instruments and Restrictions of Foreign Trade Before and During the Pandemic

S. E. Demidova and M. M. Balog

Introduction

Over the past decade, the world economy has been developing in the projection of globalization and the World Trade Organization (WTO) has become the leading international institution in this area, which has adopted the powers of the General Agreement on Tariffs and Trade (GATT). The GATT/WTO ideology is based on the theories of comparative advantage and free trade. According to the first theory, foreign trade is beneficial for the participating country (Ricardo, 2007). However, the international division of labor determines the rejection of the development of some types of production in favor of others, which causes criticism of the theory regarding the predominance of the interests of developed countries and industrial discrimination of developing countries. Significant effects of WTO membership are received by countries with developed industry (Gowa & Kim, 2005), which is reflected in the growth of their foreign trade. The profitability of mutual trade between the two countries (even with the same productivity) explains the economies of scale and providing more choice of products for consumers (Krugman, 1979). According to the concept of free trade, the reduction of customs tariffs leads to an increase in trade turnover. (Fernandes et al., 2020) At the same time, geographically discriminatory tariffs (preferences) lead to economic inefficiency (Baldwin et al., 2009; Mattoo et al., 2019).

There are three key economic aspects of tariff preferences. (1) Preferential tariffs. They can be so large that even countries receiving preferences may be in a worse position—this situation is called the «Wiener ambiguity» (Viner, 1950). Third

S. E. Demidova (✉)
Financial University Under the Government of the Russian Federation, Moscow, Russian Federation

M. M. Balog
Pskov State University, Pskov, Russian Federation

A. Rumyantseva et al. (eds.), *Challenges and Solutions in the Digital Economy and Finance*, Springer Proceedings in Business and Economics,
https://doi.org/10.1007/978-3-031-14410-3_51

parties are likely to suffer from such preferences. If the customs tariff in a country is zero or almost zero, regional trade agreements will not be able to create significant discrimination. (2) Rules of origin. Preferential tariffs always include "rules of origin". The rules of origin act as "fine-tuning" of trade barriers, often restrict the freedom of companies to choose the most efficient international supply chain, as there is a possibility of losing the status of "country of origin" and the preferences provided in this regard. Thus, the "rules of origin" increase export costs. (3) Cumulation rules. The rules of origin always include rules about where export value can be added. The most common is bilateral cumulation, which allows companies to calculate value added in either of the two countries. Permissive cumulation rules can mitigate the protectionist content of the rule of origin by expanding the choice when making decisions in the international supply chain.

An analysis of the literature assessing the impact of WTO membership on the trade structure of developing member countries in comparison with developed countries did not reveal a noticeable conditionality. This is due to the high level of protectionism on the part of developing countries. (Mokgethi, 2016; Larch et al., 2019). At the same time, calculations were obtained confirming an increase in bilateral trade after joining the WTO in different growth ranges: by 86%, by 72%, 60% (Herz & Wagner, 2011; Liu, 2009; Tomz et al., 2007). On the other hand, tariff liberalization is based on global value chains, which can be supported on the Governments' own initiative. According to World Bank estimates, a 10% reduction in tariff protection increases the country's international trade volumes by almost 2%; a reduction in non–tariff measures by another 2% (Hoekman & Nicita, 2008). The consequence of a multilateral tariff reduction may be an increase in prosperity around the world (Bureau et al., 2019).

Both institutional embeddedness and informal participation contribute to the development of foreign trade. For example, the Russian Federation's compliance with the Partnership and Cooperation Agreement with the European Union before joining the WTO; the Republic of Belarus, Kazakhstan are EAEU member states and are forced to comply with WTO rules.

The development of regional integration, the formation of economic communities does not reduce the role of multilateral global regulation. In international trade, WTO rules and principles continue to dominate, since any preferential trade agreement (PTA) has territorial restrictions, although their number is constantly increasing. According to the WTO, by the end of 2020, 305 RTAs were in operation, including 44 in the European Union, 31 in the European Free Trade Association, 31 in Chile, 25 in Singapore, 22 in Mexico and Turkey, 19 in Peru, 18 in South Korea and Ukraine.

The WTO implements its activities through a multilateral package of agreements, the rules and regulations of which regulate about 95–97% of international trade relations. In addition to a single package of documents, there are so-called «plurilateral agreements», the versatility of which is limited to the participation of individual WTO members in them.

Developed countries support the integration of developing countries through unilateral preferential measures in the form of mutual trade preferences. A number of researchers believe that developing countries should abandon unilateral trade preferences in favor of mutual agreements. (Gil-Pareja et al., 2019) The least developed countries export more as a result of obtaining this status, which is associated with higher aggregate exports (Klasen et al., 2021). Harmonization of regulations in the field of economic integration or harmonization of regulations between developing and developed countries, net exporters and importers stimulate trade flows (Zongo, 2021).

The WTO's institutional structure has a mechanism for resolving trade disputes recognized by all participants. Decisions of the WTO Court are approved automatically, the procedure can be suspended only by the unanimous opinion of all participating countries on the refusal to approve it. Thus, if one of the countries wants to block a court decision, it is necessary to enlist the support of all WTO member countries without exception in the need for such a decision. It is possible to withdraw from the WTO or change the mandatory conditions only after 3 years and at the same time pay individually determined compensation.

If the respondent country does not comply with the recommendations on compensation or an agreement on them has not been reached, the WTO may authorize the suspension of concessions and obligations on the part of the accuser country in favor of the defendant.

The key principle of the WTO «non-discrimination» is ensured by the systemic interaction of two regimes: the most favored nation (MFN) regime and the national regime (NR). If a country introduces or maintains a quantitative restriction, then this application should be non-discriminatory. At the same time, WTO «exception clauses» apply both to developed countries and (to a greater extent) to developing countries.

There is an opinion that multilateral trade agreements, in addition to liberalizing trade in goods and services, strengthen monopolies at the pre-production stage by controlling knowledge in the form of intellectual property rights (patents, industrial designs), and at the post-production stage, as well as by strengthening control over branding and marketing (Siddiqui, 2016; Ariu et al., 2018). As a result, the added value of such trade is concentrated in developed countries, while developing countries compete for production in low-cost segments.

Extreme events such as international conflicts and disputes, as well as political tensions, are likely harmful to international trade, as they affect export and import decisions (16), consumption, exchange rate, tax and monetary policy (Gulotty, 2022). In addition, there are disagreements between developed countries (USA, European Union) and developing WTO members (Brazil, India, China, etc.). The lack of consensus leads to the fact that negotiation tracks turn into plurilateral initiatives (Balcilar et al., 2018).

The purpose of the work is to study the restrictions established by international acts regarding the application of various fiscal and financial measures to stimulate trade, to identify trends in international trade between Russia and foreign countries in the period before and during the pandemic.

Materials and Methods

The analysis was carried out based on WTO regulations, foreign and domestic scientific publications, opinions of international and domestic expert organizations, legal and regulatory documents, considering statistical and empirical data. At the first stage, a review of international regulatory sources was conducted for the established requirements and restrictions. At the second stage, the statistical indicators of export-import operations of the Russian Federation and foreign countries, the restrictions imposed are analyzed. At the third stage, conclusions were drawn about the limitations for stimulating production.

Results

The rules of international trade are related to the movement of goods and services. Mandatory rules in the field of tariffs and non-tariff barriers were agreed by the GATT/WTO. The rules of direct taxation are usually adopted unilaterally, so taxation disputes quite often arise between WTO member countries (more often it concerns excise taxes on alcohol, sales taxes, VAT). International rules concerning measures affecting trade and rules concerning direct taxation have common goals (removing obstacles to the cross-border movement of goods, services, capital, labor, technology).

The analysis of the legal framework of the World Trade Organization is presented in Table 1 and reflects the main requirements and restrictions applicable to the countries of the WTO jurisdiction.

WTO member countries are prohibited from imposing or maintaining any form of prohibitions or restrictions other than duties, taxes, or other charges. At the same time, WTO members are not prohibited from applying export taxes, and about a third of countries use this mechanism. Export regulation is also carried out through quotas and licenses, the latter limit not the volume of exports, but exporters of goods and services. Export restrictions have several consequences: bans have a negative impact on countries that do not have production facilities; for a country imposing restrictions, negative consequences arise if it is necessary to import raw materials. Tariffs or export licenses may constrain trade in addition to price increases. Export restrictions can create uncertainty in the investment strategies of enterprises and undermine confidence in the international rules of the game.

Governments, using fiscal and other incentives to attract foreign investment, adopt conditions to ensure that investments correspond to national priorities (in terms of content, industries, productivity). Such conditions can distort trade relations through the establishment of import tariffs and export subsidies. In the field of production regulation, the fundamental principles of the WTO provide for a reduction in state subsidies, a reduction in import duties, and the abolition of import quotas.

Table 1 Analysis of the legal framework (key documents) of the World Trade Organization

Document	Restriction
General Agreement on Tariffs and Trade of 1994	Measures to restrict imports are being taken for balance of payments reasons solely for the purpose of limiting the overall level of imports and cannot go beyond the necessary, due to the state of the balance of payments.
General Agreement on Trade and Services	Subsidies can have a distorting effect on trade in services. Solutions: multilateral rules that reduce distortions; acceptability of procedures compensatory measures. The GATS Dispute Settlement Commission has the right to veto government decisions any country in terms of trade regulation.
Agreement on Trade-Related Aspects of Intellectual Property Rights	The national regime should be extended for foreign participants who have joined to TRIPS. Postponement opportunities for developing countries up to 4–5 years. For the least developed countries −10 years.
Agreement on Trade-Related Investment Measures	Prohibition to establish restrictive measures with a negative impact on foreign investment in the manufacturing sector. Partially exempt from the requirements developing countries.
Agreement on Agriculture	Subject to compliance with internal support obligations, the participant does not provide support to national producers beyond the levels stipulated by the obligations. Developing member countries are provided with differentiated and more favorable treatment for fulfilling obligations to reduce domestic support (up to 10 years). The least developed member countries are not required to make commitments by reduction.
Agreement on Sanitary and Phytosanitary Measures	The application of sanitary, veterinary and phytosanitary measures should be based on based on international standards, scientific data. Protection measures should be no more restrictive than required, be applied without discrimination. In cases where substantial investments are required to meet the SPS conditions by exporting developing countries, the possibility of providing technical assistance to exporters is being considered, allowing them to maintain and expand market access opportunities for a particular product.
Agreement on Technical Barriers to Trade	Sanitary and phytosanitary measures do not fall under the provisions of the Agreement. Special and differential treatment for developing WTO member countries. The provisions apply only to agricultural subsidies.
Agreements on subsidies and countervailing measures	Displacement or serious infringement interests do not arise if: • a ban/restriction on the export of a similar product or import to the market of a third country is implemented; • the decision of the importing country to reorient

(continued)

Table 1 (continued)

Document	Restriction
	imports to another country for non-commercial reasons; • natural disasters, strikes, transport disruptions or other force majeure affecting the production, quality, quantity or prices of goods available for export; • the existence of agreements restricting exports; • voluntary reduction of the possibility of exporting goods; • non-compliance with standards and other administrative requirements in the importing country. Subsidies that do not give grounds to take measures: • for research (no more than 75% of the cost of industrial research or 50% of the cost of development at the pre-competitive stage); • support of disadvantaged territories allocated according to a single methodology; • assistance in the adaptation of existing production facilities to the new requirements for environmental protection
Agreements on special protective measures	Special protective measures are not applied against goods originating from a developing member country until, so far, its share in the import of this product any importing member does not exceed 3%, provided that the total share developing member countries with less than 3% share do not exceed 9% of the total volume of imports of this product. A developing member country has the right to extend the period of application of a special protective measure to 2 years beyond the maximum period
Application Agreement Articles VI of GATT 1994	The anti-dumping duty remains in force only in time and in the amount necessary to counter dumping. It is desirable that the introduction of the duty should be moderate in the territory of all WTO member countries and that the duty should be less than the dumping margin (and sufficient to eliminate the damage to the industry).

Currently, WTO rules are hindered by shifts in world trade volumes away from geopolitically oriented rich states and changes in global supply chains. The problems have intensified due to the increasing role of China in world trade, geopolitical tensions in relations with the United States, the use of economic sanctions and counter-sanctions as a tool of foreign policy (Kudrin et al., 2021; Gupta et al., 2019; Hongyuran & Wang, 2019).

The COVID-19 pandemic has made significant adjustments to the rules of international trade. Since the outbreak of the pandemic, 140 trade and trade-related measures have been implemented by the economies of the G20 countries, of which 101 (72%) have facilitated trade, and 39 (28%) can be considered trade restrictions.

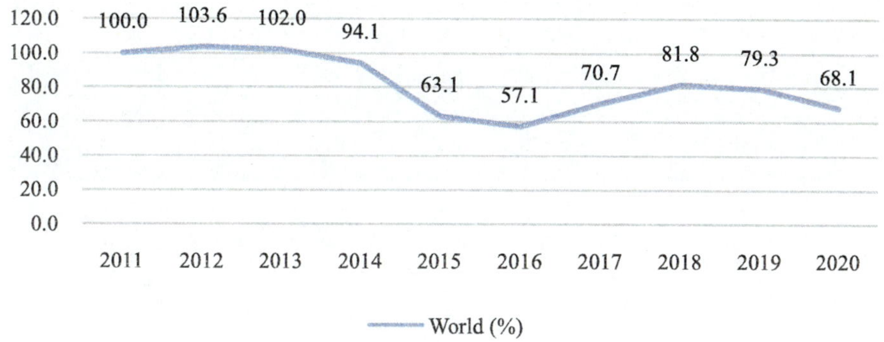

Fig. 1 Dynamics of world trade volumes of the Russian Federation in the period 2011–2020 with an estimated change in % to the base year 2011, in current prices, million US dollars, (compiled by the author, Bloomberg data, https://bba.bloomberg.net)

Export bans accounted for more than 90% of all registered restrictive measures. The reduction or cancellation of import tariffs and import taxes account for 60% of the trade facilitation measures taken, several countries have reduced their tariffs on various goods, such as disinfectants, medical equipment and medicines. Three countries have temporarily lifted their import tariffs on vaccines. In addition, new trade-restrictive measures not related to the pandemic were introduced, mainly due to tariff increases, import bans, stricter customs procedures and export duties. [1]

In 2020 (the year of the spread of the COVID-19 pandemic), the volume of international trade in the Russian Federation decreased and the volume of Russian exports decreased by 20.6% (Fig. 1).

In 2020, Russia was the eleventh economy in the world in terms of GDP (the seventieth in terms of GDP per capita), the thirteenth economy in terms of total exports, the twenty-first in terms of total imports, ranked 43rd in the ECI economic complexity index. [2] The main export items of Russia are oil and refined petroleum products, natural gas, gold, coal briquettes (exported mainly to China). Russia is also the world's largest exporter of wheat, frozen fish, nickel, and cast iron. Russia imports cars, broadcasting equipment, medicines, computers, hydraulic turbines, aluminum oxide, copper, etc.

As a result of the import substitution policy, the number of technological operations carried out by foreign investors in the country has been increasing over the last 10 years. The most technologically advanced Russian industries include automotive, transport engineering, agricultural engineering, heavy machinery, metallurgy, construction materials industry, civil aircraft, pharmaceutical and medical industries and the military-industrial complex. In general, import substitution provided a reduction

[1] Report on G20 trade measures (mid-october 2019 to mid-may 2020). URL:https://www.wto.org/english/news_e/news20_e/report_trdev_jun20_e.pdf

[2] Economic Complexity by Region Level. URL: https://oec.world/en/profile/country/rus/#economic-complexity

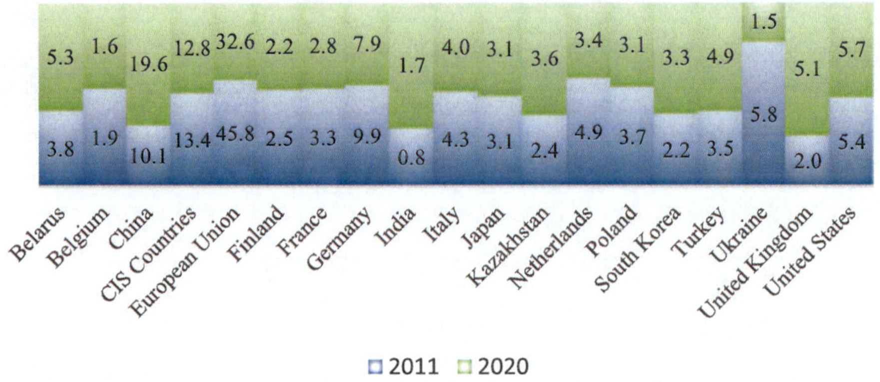

Fig. 2 Changes in foreign trade relations of the Russian Federation with foreign countries in 2011 and 2020 (total trade), % of global indicators (compiled by the author, Bloomberg data, https://bba.bloomberg.net)

in dependence on imported industrial goods, their share was 40% in 2020 against 49% in 2014.

In 2011–2020, the balance between the priority directions of the development of trade relations of the Russian Federation in the perimeter of foreign partners has changed (Figs. 2 and 3).

The data presented in Figs. 2 and 3 show a decrease in the share of the European Union countries in Russia's foreign trade (from 45.8% to 32.6%), the share of CIS countries—from 13.4% to 12.8%, the largest decrease is observed in Ukraine (from 5.8% to 1.5%). There was a significant increase in the share of China as a trading partner (from 10.1% to 19.6%), India (from 0.8% to 1.7%), South Korea (from 2.2% to 3.3%), Turkey (from 3.5% to 4.9%), Kazakhstan (from 2.4% to 3.6%), the Republic of Belarus (from 3.8% to 5.3%) in the period from 2011 to 2020. One of the reasons for such changes is the attempts of companies to diversify supply chains, violations of which were caused in 2020 by the waves of the COVID-19 pandemic, including through strategies for regionalization of supplies, strategies from the «near shore» (Crozet & Hinz 2020) (Fig. 4).

Figure 3 shows the dynamics of net exports of the Russian Federation in the period 2011–2020. As shown in Fig. 3, the volume of net exports of the Russian Federation for the analyzed period 2011–2020 decreased in a sample of countries, except for China (from −7.0 to 2.2 billion dollars), India (from 1.3 to 3.8 billion dollars), South Korea (from 2.9 to 11.5 billion dollars), Turkey (with 18.6 to 21.1 billion dollars), Italy (from 10.9 to 12.7 billion dollars). The decrease in net exports with the countries of the European Union amounted to $146.16 billion (from 176.0 to 29.8), with CIS countries—from $46.9 billion to $22.5 billion. Thus, Russia's main partners are China, Belarus, and South Korea. At the same time, until 2022, for the European Union (EU), Russia was the fifth largest partner in exports and the third in imports of goods, and the largest partner from the EU for Russia was Germany.

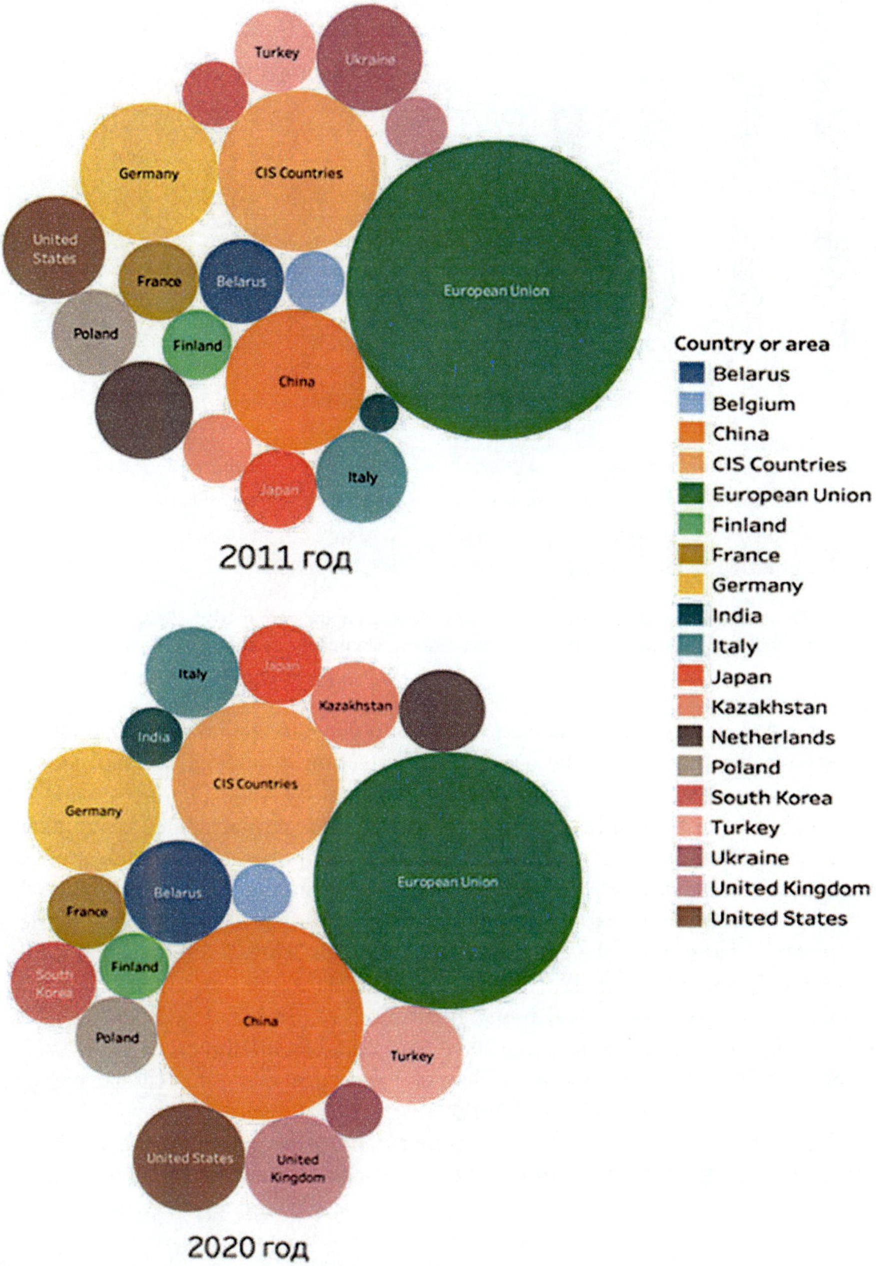

Fig. 3 Dynamic of the volume of foreign trade turnover of the Russian Federation with foreign countries in 2011 and 2020 (total trade), USD (compiled by the author, Bloomberg data, https://bba.bloomberg.net)

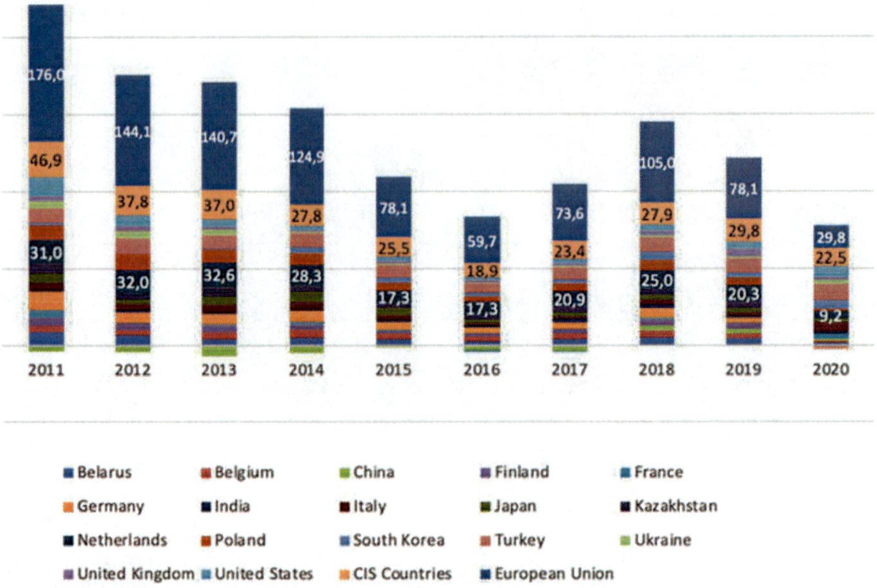

Fig. 4 Dynamics of net exports of the Russian Federation in the period 2011–2020 (in a sample of countries), billion US dollars (compiled by the author, Bloomberg data, https://bba.bloomberg.net)

Foreign trade in 2020 was influenced by the policy of trade restrictions, including on exports. Russian raw materials exports are seriously affected by sanctions on petroleum products, reducing the value of exports by about 36.56%, while the consequences of sanctions for exports of non-oil products are insignificant. Counter-sanctions cause a decrease in the cost of importing agricultural products by 54.52% and a decrease in the cost of importing non-agricultural products by 20.86%. The decline in exports and imports indicates that these non-tariff barriers are mutually harmful for the country responding with counter-sanctions and for the countries imposing sanctions. (Nguyen & Do, 2021) According to the Ministry of Economic Development of the Russian Federation, 34 restrictive measures on access to foreign markets of Russian goods have been curtailed or eliminated in 2020. The result of the WTO policy was the leveling of 7 measures restricting Russian exports, the main results are presented in Table 2.

Russia, through the WTO dispute resolution mechanism, proved the validity of claims against the European Union to take into account the costs of purchasing raw materials by enterprises in anti–dumping processes; to Ukraine—to remove restrictions on access to railway equipment. No less significant contribution was made by the activities of intergovernmental commissions on trade, economic, scientific, and technical cooperation, of which there are more than 100 with the participation of Russia, as well as bilateral strategies for the development of cooperation with foreign countries.

Table 2 The use of WTO instruments to support Russian export potential in 2020 by leveling the restrictions of foreign countries

Country	Export goods	Eliminated/prevented/suspended measures
Brazil	Rolled steel	Anti-dumping measures
Ukraine	Fertilizers	Special protective investigation
Ukraine	Passenger cars	Compensatory measures
Turkey	Steel products	Increased duties
Australia	Ammonium nitrate	Anti-dumping measures
Vietnam	Conformity assessment of cars	Discriminatory procedures
Bosnia and Herzegovina	Petroleum	Margin Limitation
Egypt	Products registration of enterprises	Discriminatory procedures
Bangladesh	Waste	Discriminatory rules
Saudi Arabia	Confectionery	Product composition requirements (amount of sugar)

In 2020, amid the spread of the pandemic, governments of different countries imposed restrictions, trade barriers, including those contrary to the principles of the WTO. In relation to Russian goods, eight measures violating WTO rules were established by the following countries: Bangladesh, Great Britain, USA, India, Tajikistan. By the end of 2020, about 195 restrictive measures have been implemented in relation to Russian goods.[3]

In 2019, among the countries that applied the most significant restrictions to Russian goods were (in descending order of restrictive measures): the European Union, Ukraine, India, Turkey, Belarus, the USA, China, Vietnam, Kazakhstan.

In 2020, the geography has changed somewhat: the European Union, Ukraine, Turkey, Belarus, the USA, India, Kazakhstan, Morocco, Australia. The most common type of restrictions are anti-dumping duties, administrative measures, sanitary, phytosanitary, and veterinary measures.

Discussion

Summarizing the research findings on the role of the WTO, it can be stated that the WTO principles are designed to develop trade relations between participants with economic intentions. The WTO legal framework includes a differentiated approach to developed, developing, and least developed countries. At the same time, developed and large-scale economies that are ready for radical reductions in tariff restrictions benefit more from trade liberalization. Tariffs continue to form the

[3] Results of foreign economic activity of the Russian Federation in 2020 and the first half of 2021/ Ministry of Economic Development of the Russian Federation. 2021. 98 p.

bulk of trade facilitation measures, followed by decisions to abolish import taxes and simplify customs procedures. The reduction of tariffs on imported goods displaces domestic producers from the domestic market, reduces the volume of production.

About a third of WTO members impose export duties in various forms. All types of export taxes inevitably reduce export volumes, being a form of export restriction. Export regulation through quotas is associated with the maximum restriction of export volumes, through licenses—with the restriction of exporters of goods and services.

Export restrictions have several consequences: bans have a negative impact on countries that do not have production facilities; for a country imposing restrictions, negative consequences arise if it is necessary to import raw materials. Tariffs or export licenses may constrain trade in addition to price increases. Export restrictions can create uncertainty that will affect firms' investment strategies and, over time, reduce confidence in the international rules of the game.

Supply chains remain particularly vulnerable to shocks and disruptions, almost 90% of respondents surveyed by Mc&Kinsey see a way out to a certain degree of regionalization soon (Alicke et al., 2021).

The long-term future of world trade is being transformed by technological shifts, trade wars, and «black swans» in the form of the COVID-19 pandemic.

Conclusion

The study concluded that tariffs still constitute the bulk of trade facilitation measures, followed by decisions to abolish import taxes, and simplify customs procedures. The reduction of tariffs on imported goods displaces domestic producers from the domestic market, reduces the volume of production. Developed and large-scale economies that are ready for radical reductions in tariff restrictions benefit from trade liberalization since their scale allows them to influence world prices.

The implementation of WTO rules is hindered by the displacement of world trade leaders, geopolitical tensions, uncertainty shocks (pandemic), sanctions entailing changes in global supply chains.

Foreign trade in 2020 was influenced by the policy of trade barriers, including on exports, some of them contradicted the principles of the WTO.

References

Alicke, K., Barriball, E., & Trautwein, V. (2021). *How COVID-19 is reshaping supply chains*. McKinsey and Company, 2011–2020. Retrieved from https://www.mckinsey.com/business-functions/operations/our-insights/how-covid-19-is-reshaping-supply-chains. Accessed: 03/21/2022

Ariu, A., Mayneris, F., & Parenti, M. (2018). *One way to the top: How services boost the demand for goods*. National Bank of Belgium Working Paper No. 340.

Balcilar, M., Bonato, M., Demirer, R., & Gupta, R. (2018). Geopolitical risks and stock market dynamics of the BRICS. *Economic Systems, 42*(2), 295–306.

Baldwin, R., Evenett, S. S., & Low, P. (2009) Beyond tariffs: Multilaterising non-tariff RTA commitments. January. pp. 79–141. https://doi.org/10.1017/CBO9781139162111.005

Bureau, J. C., Guimbard, H., & Jean, S. (2019). Competing liberalizations: Tariffs and trade in the twenty-first century. *Review of World Economics, 155*(4), 707–753. https://doi.org/10.1007/s10290-019-00346-1

Crozet, M., & Hinz, J. (2020). Friendly fire: The trade impact of the Russia sanctions and counter-sanctions. *Economic Policy, 35*(101), 97–146.

Fernandes, A., Kee, H. L., & Ozden, C. (2020). Free trade now: A case for tariff reductions and non-tariff measures simplifications to fight COVID-19 (coronavirus). *World Bank Blogs*, 11. Retrieved from https://blogs.worldbank.org/developmenttalk/free-trade-now-case-tariff-reductions-and-non-tariff-measures-simplifications-fight

Gil-Pareja, S., Llorca-Vivero, R., & Martínez-Serrano, J. A. (2019). Reciprocal vs nonreciprocal trade agreements: Which have been best to promote exports? *PLoS One, 14*(2), e0210446. https://doi.org/10.1371/journal.pone.0210446

Gowa, J., & Kim, S. Y. (2005). An exclusive country club: The effects of the GATT on trade, 1950–94. *World Politics, 57*(4), 453–478.

Gulotty, R. (2022). WTO subsidy disciplines. *World Trade Review* (pp. 1–12). doi:https://doi.org/10.1017/S1474745622000118.

Gupta, R., Gozgor, G., Kaya, H., & Demir, E. (2019). Effects of geopolitical risks on trade flows: Evidence from the gravity model. *Eurasian Economic Review, 9*(4), 515–530.

Herz, B., & Wagner, M. (2011). The 'real' impact of GATT/WTO–a generalised approach. *The World Economy, 34*(6), 1014–1041.

Hoekman, B., & Nicita, A. (2008). Trade policy, trade costs, and developing country trade. *World Development, 39*(12), 2069–2079. Retrieved from https://openknowledge.worldbank.org/bitstream/handle/10986/6341/WPS4797.pdf?sequence=1&isAllowed=y

Hongyuran, W., & Wang, F. (2019). *Central bank Chief says China will guard against Belt and Road debt risks, Caixin Global*. Retrieved from https://www.caixinglobal.com/2019-04-25/central-bank-chief-says-china-will-guard-against-belt-and-road-debt-risks-101408561.html

Klasen, S., Martínez-Zarzoso, I., Nowak-Lehmann, F., & Bruckner, M. (2021). Does the designation of least developed country status promote exports?. *The Journal of International Trade and Economic Development, 30*(2), 157–177. Retrieved from https://EconPapers.repec.org/RePEc:taf:jitecd:v:30:y:2021:i:2:p:157-177

Krugman, P. R. (1979). Increasing returns, monopolistic competition, and international trade. *Journal of International Economics, 9*(4), 469–479. Retrieved from https://seinst.ru/files/vehi_6_034_krugman_vozrastayushchaya-otdacha.pdf

Kudrin, A. L., Avksentiev, N. A., Arlashkin, I. Y., Bakvenye, T. A., Barbashova, N. E., & Belov, S. G. (2021). *Economics and economic policy in the context of a pandemic*. Publishing House of the Gaidar Institute.

Larch, M., Monteiro, J. A., Piermartini, R., & Yotov, Y. (2019). *On the effects of GATT/WTO membership on trade: They are positive and large after all*. Retrieved from https://www.wto.org/english/res_e/reser_e/ersd201909_e.pdf

Liu, X. (2009). GATT/WTO promotes trade strongly: Sample selection and model specification. *Review of International Economics, 17*(3), 428–446.

Mattoo, A., Mulabdic, A., & Ruta, M. (2019). *Trade creation and trade diversion in deep agreements*. https://thedocs.worldbank.org/en/doc/328491559591517896-0050022019/original/TradeCreationanfTradeDiversion.pdf

Mokgethi, T. (2016). *Effects of WTO membership on developing members' trade: A case for South Africa*. Retrieved from https://www.diva-portal.org/smash/get/diva2:937035/FULLTEXT01.pdf

Nguyen, T. T., & Do, M. H. (2021). Impact of economic sanctions and counter-sanctions on the Russian Federation's trade. *Economic Analysis and Policy, 71*, 267–278.

Ricardo, D. (2007). *Principles of political economy and taxation: Selected works*. Eksmo.

Siddiqui, K. (2016). International Trade, WTO and Economic Development. *World Review of Political Economy. Winter, 7*(4), 424–450.

Tomz, M., Goldstein, J. L., & Rivers, D. (2007). Do we really know that the WTO increases trade? Comment. *American Economic Review, 97*(5), 2005–2018.

Viner, J. (1950). *The customs union issue*. Carnegie Endowment for International Peace. VIII.

Zongo, A. (2021). The impact of services trade restrictiveness on food trade. *International Economics, 166*, 71–94. Retrieved from https://hal.archives-ouvertes.fr/hal-03258290

Transformation of Logistics Systems in the Context of Digitalization

Natalia Taskaeva, Olga Boyarskaya, and Natalia Meshkova

Introduction

The modern theories and world practice in the era of digitalization confirm the objective necessity of applying the logistics principles and logistics management mechanisms in the management of material flows. Theoretical and methodological foundations and conceptual approaches to logistics management of the domestic enterprises, considering the socio-economic peculiarities, are just being formed (Shepherd & Hamanaka, 2015). Without the proper scientific basis, structurally deformed and functionally limited logistics formations are created in practice, which cannot ensure efficient organization and management of commodity flows based on consistency, comprehensiveness, integration, manufacturability, efficiency (Démare et al., 2017). The organization of integrated logistics entities in the context of the creation of logistics systems of the enterprise and the introduction of management tools with full digitalization of the life cycle. The creation of the organizationally, technologically perfect, and cost-effective system of delivering commodity flows from the producers to the consumers of the various goods, increasing the competitiveness of the supply chain participants (Langley et al., 2020).

The works of such leading scientists as Zine Benotmane, Ghalem Belalem, Abdelkader Neki (Benotmane et al., 2018) became the theoretical basis for the

N. Taskaeva (✉)
National Research University Moscow State University of Civil Engineering, Moscow, Russian Federation
e-mail: TaskayevaNN@mgsu.ru

O. Boyarskaya
Gzhel State University, Electroizolyator, Russian Federation
e-mail: boyarskayaov@internet.ru

N. Meshkova
Lugansk Vladimir Dahl State University, Lugansk, Ukraine

A. Rumyantseva et al. (eds.), *Challenges and Solutions in the Digital Economy and Finance*, Springer Proceedings in Business and Economics,
https://doi.org/10.1007/978-3-031-14410-3_52

study of the features of creating enterprise logistics systems (Langley et al., 2020). Appreciating the scientific achievements of the mentioned scientists, it should be noted that there is a lack of comprehensive consideration of the problems of organization of the enterprises logistics activities in the conditions of formation of the global information and telecommunication environment and integration of economic processes (Bondar et al., 2019).

The issues of the network organization of logistics, the directions of development of the system form of logistics, do not allow the participants of the market relations to choose the development strategy reasonably, and they remain insufficiently elaborated. There is an objective need to create the theoretical and methodological basis and substantiation of practical recommendations to enhance the formation and to ensure the effective functioning of the enterprise of digital logistics systems (Botayeva et al., 2018). The justification of the theoretical foundations, conceptual provisions, and development of the scientific and methodological recommendations for the formation and development of the enterprise of digital logistics systems determine the purpose of the research (Galkin & Sysoyev, 2020).

The active reform of the production sector implies the need to develop integrated logistics services, expand the network of multimodal transport and logistics infrastructure, deepen the customer orientation of transport and logistics services, develop intermodal and multimodal transportation, digitalize transport and logistics systems (artificial intelligence, blockchain technology, international information platforms, communication tools, server systems) (Mamuladze et al., 2021). The introduction of the logistics approach into the activities of enterprises is considered as a tool to improve the efficiency of their functioning, to ensure competitive advantages and economic growth (Nekrasov & Sinitsyna, 2021).

Materials and Methods

The set of management tools for logistics activities and logistics formations, except for logistics management in practice also includes mechanisms, methods, algorithms, techniques, methods of logistics processes organization, borrowed from the other types of activity and branches of knowledge. They are focused on the use of ideas and principles different from the logistics approach. As a result, the practice of logistics management can lead to systemic problems in the processes of development of the real sector of the economy. Therefore, it is important to form logistics management processes on the high quality theoretical and methodological basis, focused on improving the activities and increasing the efficiency of enterprises in the modern creative and digital industry (Kalyakina et al., 2020).

There are different approaches to the identification and systematization of the allocation of total logistics costs in economic practice:

1. to the individual elements depending on the specific logistics operation, the performance of which gives rising to the corresponding costs.

2. for the aggregated cost groups (inventory formation and storage costs, costs of digitalization and management functions, transportation, and procurement costs).
3. on the certain types of logistics costs (costs of logistics operations and functions, losses from immobilization of funds in inventories, losses from insufficient level of logistics management and service, costs of logistics digitalization) (Efimova et al., 2021).

Considering the logistics processes in the complex of logistics activities, it is advisable to combine all the logistics costs into the consolidated groups related to:

1. performing a set of transportation (forwarding) operations.
2. performance of a set of warehousing operations and storage of cargoes.
3. management of logistics processes and flows.

The costs of the first two groups reflect the volume of logistics costs for logistics operations and processes of logistics activities. It is processing of the material flow, transformation of its elements. The costs combined in the third group characterize the volume of costs for the performance of logistics operations and processes related to the sphere of logistics management.

In the modern economy, the share of logistics costs in the gross domestic product is growing due to the development of logistics and the expansion of its implementation in the activities of enterprises, as well as the irrationality of the organization of logistics flows. Logistics costs are one of the fixed components of total costs, and, in some cases, they can be one of the most important factors in the formation of product prices. The conducted studies show that the logistics costs in sales are, for example, in Germany—12.05%, in France—8.71%, in the Netherlands—6.74%, in the UK—7.74%, in the USA—7.6%. Considering that the share of logistics costs is 5–35% of sales depending on the commercial activities of enterprises, and their share, according to expert estimates, in total production costs is very significant. The amounts are up to 30% in some industrial enterprises, it is therefore advisable to pay special attention to the process of their determination (Sremac et al., 2018).

With the development of digital technologies in logistics, the concept of costs has also changed. Costing is done in the result-oriented way: first, the scope and nature of the logistics system's work is determined, and then the costs associated with its implementation are determined. If the goal is to improve the quality of the system, then in its implementation, up to a certain level, logistics costs increase linearly, and then exponentially. The introduction of the digital logistics system requires the development of the new approach to the exchange of costs, the so-called logistics costing, focused on the result (Gviliya et al., 2019).

The basis for decision-making in this case is the concept of total logistics management costs. The costs associated with procurement, transportation, storage, inventory management and other logistics operations at the stages of material (commodity) supply and goods movement depend on each other. The attempts to minimize them only within a single activity may cause an increase in total costs in the end-to-end logistics chain. The concept of minimizing overall costs means a comprehensive assessment of the costs incurred by all the enterprises participating in the logistics chain to regulate certain types of costs in such a way as to achieve a reduction of their overall size. The objective of logistics management is to find ways to minimize costs to the appropriate level of service and customer satisfaction (Maire, 2017).

It is obvious that it is necessary to formulate and to agree on a common goal for all the participants of the logistics chain and to determine the composition of the functions, operations and types of work that is achieved in the digital environment. Subsequently, the simultaneous analysis of the costs of logistics activities of each of the participants of the logistics chain and the redistribution of costs by reviewing the methods, methods of logistics operations and the digital technologies involved. The search for the economic trade-off for the entire set of costs can be carried out by assessing and comparing different options for their redistribution using descriptive models (horizontal analysis, analytical coefficient system) and predicative models (factor and situational analysis) of financial analysis. It should be noted that within the framework of logistics chains such links should be connected in the single whole. They include the interrelated logistic functions of logistic operations over the commodity flow, as the procurement of commodity resources, their delivery to warehouses of enterprises, storage of goods and performance of internal warehouse operations in warehouses, physical distribution of the sold mass of goods. It also includes transportation to facilities, as well as sales of goods in trade with the provision of the appropriate set of aftersales. In this case, the implementation of these logistics operations is subject to the task of sequential execution of functional blocks to determine the needs for goods, order formation, organization of order execution and monitoring its implementation, as well as the receipt of goods at the warehouse of the enterprise with the subsequent performance of the function of maintaining inventories (Korchagin & Rizayeva, 2021).

Logistic activity of the enterprise, focused on the fullest satisfaction of customer demand within the complex of its activities, requires the effective organization of goods transportation processes and their subprocesses. The main ones are determining the need for goods, making decisions on procurement, procurement, and control of its implementation, creating inventory, its regulation and sale of inventory goods), each of which is in one way, or another related to the inventory management system. Inventory management should be subordinated to the objectives of the higher-level system, the goods management system. Respectively, the current objectives of the inventory management system should be subordinated to the objectives of the goods flow management in the logistics chain, but the implementation of goods management objectives is not possible if the inventory management problems are not properly addressed (Brah & Lim, 2006).

The analysis of the costs of the industrial enterprise has shown that the material costs occupy the significant part in their composition—more than 85%. The high level of the material intensity of the enterprise production indicates the need to search for reserves to reduce material costs using the digitalization process (Gerzanych & Kyrlyk, 2019).

The assessment of the consequences of digitalization of the enterprise, which is considered because of improving the management system, should be comprehensive. Specific manifestations of a comprehensive assessment of the consequences of digital transformation include their multidimensional nature, the interconnectedness of elements and conclusions, as well as multivariance (Kochegura, 2021). The effectiveness of improving the management system in the conditions of digital transformation is determined by the formula:

$$E = \frac{P}{I} \quad (1)$$

where

P—expected revenue growth potential due to digital transformation.
I—investments in digitalization.

Results

The important construction of priorities to simultaneous optimization of the various components requires more resources, time and has a higher probability to lead to unsatisfactory results (Jabbour et al., 2019). Therefore, it should be carried out before the immediate start of optimization processes. This prioritization can be formed after analysis of company strategy, available budget, and desired level of logistics services, i.e., simpler and less costly optimization processes receive a higher priority for further development than those that require a radical restructuring of business or logistics processes. Despite the fact there is a risk of constant changes in the conditions of a particular product or service that in the competitive market environment (it may be due to the entry of a new competitor or alternative service), priorities are a dynamic indicator. It can constantly change, helping to adapt to new factors, and can be adjusted by the experts. They will be involved in the formation of logistics systems.

The analysis of the industrial enterprise's activity showed that there are no logistical links, and it leads to unreasonable downtime (Table 1).

The analysis has shown that production downtime due to the lack of materials accounts is for almost 15.6% of total downtime. This is the rationale for the formation of the logistics chain, which makes it possible to improve the management of material flows of production connected between the individual shops and to ensure optimal costs.

The rearrangement of flow diagrams includes the improvement of the management system of transport service and eliminates delays in the movement of flows, creates a real basis for the implementation of logistics schedules based on optimal time standards of the transport operations (Jovčić et al., 2019).

The implementation of the logistics strategy for cost management of the industrial enterprise is recommended to be carried out by the logistics center. The main responsibilities of the logistics center are as follows: implementation of digital management system of the enterprise with allocation of subsystems of cost

Table 1 Downtime analysis

Reasons for downtime	Downtime, hour	Relative deviation from the plan, %			
	Plan	Specific weight, %	Fact	Specific weight, %	
Total downtime	2800	100	3360	100	20
Repair of production equipment	1925	68.75	2005	59.7	4.15
Repair of power equipment	300	10.71	280	8.33	−6.7
Repair of lifting and handling equipment	325	11.6	350	10.4	7.7
Repair of auxiliary equipment	250	8.9	200	5.9	−20
Lack of materials	–	–	525	15.6	–

management, activity profiles, production, finance, supply, sales, which contributes to optimization of enterprise cost management. The system of digitalization in the field of logistics can be represented in the form of a community formed based on the use of digital platforms and their applications, developers, suppliers and users of logistics services, agents with experience in the logistics field of competencies acquired through the use of digital information technologies. In the context of the fact that the digital transformation of logistics systems is potentially focused on improving the results of their activities and requires significant financial costs, it becomes advisable to determine the effect of the measures taken to justify and make appropriate management decisions. The complexity of determining the economic result of the proposed measures of digital transformation of logistics enterprises is since it is necessary to consider such a feature that it is necessary to consider the indicators of the enterprise that are not directly related to digital transformation (Asaul et al., 2017).

The purpose of determining the effect of the digital transformation of an enterprise is to conclude in the form of calculated and justified indicators on the results of its implementation to justify the management decisions made and evaluate the effectiveness of improving the enterprise management system. This conclusion allows us to assess the condition and possibilities of further development of the object of assessment, which is of particular importance in modern conditions.

According to the data of the German consulting company Roland Berger, which, based on the analysis of data from the existing experience of digitalization of enterprises, calculated a possible reduction in operating, capital costs and income growth of the enterprise, because of digital transformation, significant changes in economic performance results are achieved, presented in Table 2.

It is possible to assess the economic effect of the implementation of measures to digitalize the enterprise in two scenarios: optimistic and pessimistic and draw conclusions about the result. An assessment of the projected economic effect of the digital transformation of an industrial enterprise is presented in Table 3. In addition to assessing the economic effect of digital transformation, it is necessary to evaluate the effectiveness of improving the enterprise management system, which

Table 2 Expected changes in indicators because of management system improvements based on digital transformation based on foreign experience

The economic result	The economic effect, %
The revenue	+2...+3
Material costs	−3. . . -4
Fuel	−2. . . −4.5
Labor costs	−4,5. . . −6
Repair costs	−5. . . −7
Administrative costs	−6. . . −7

Table 3 Projected economic effect of digitalization of an industrial enterprise

The economic result	Million rubles	The economic effect	
		Pessimistic forecast	Optimistic forecast
The revenue	1,662,574	33,251	49,877
Material costs	99,832	2995	3993
Fuel	32,489	650	1462
Labor costs	462,579	20,816	27,755
repair costs	98,589	4929	6901
administrative costs	171,139	10,268	11,980
the amount		72,909	101,968
Economic efficiency (49,877–33,251)/ 5366.47 = 3.098			

allows you to summarize the result of the funds spent on improvement and get a conclusion about the achievement of improvement goals. The assessment of the economic efficiency of the digital transformation of industrial enterprises is carried out considering the expected potential of the digital transformation of the enterprise, because of which the net income of the enterprise and the costs of implementing and developing information technologies change. The calculation results are presented in Table 3.

Thus, the projected economic effect from the introduction and implementation of digital solutions in the work of an industrial enterprise in a pessimistic scenario will amount to 72.909 thousand rubles, in an optimistic scenario—101.968 thousand rubles. The economic efficiency of improving the management system based on digital transformation is 3.098. That is, 1 ruble of investment in digitalization brings the enterprise 3.098 rubles of income.

Discussion

A high level of organization of the logistics activity of the enterprise is achieved because of the formation of the logistics system of the enterprise as a complex production, economic, technical and technological system. The set of subsystems and elements forms an integrated unity and interacts with the purpose of organizing, implementing, and managing logistics flows and processes based on the introduction of new technologies and digital solutions.

Logistics has turned from an auxiliary industry into a key factor in ensuring business efficiency and competitiveness. Today logistics is actively developing and changing under the influence of global trends and digital transformation of the market.

The logistics activity of the enterprise should be considered as a complex of logistics processes aimed at continuously improving the characteristics of the parameters and aspects of the logistics management system of the enterprise, focused on achieving a result that allows the enterprise to compete in the conditions of modern development, in particular digitalization. Based on practical recommendations for the introduction of digital technologies and the formation of an appropriate environment for the implementation of the necessary digital changes, the implementation of which will achieve the expected effect of digital transformation.

Conclusion

The applied tasks of digitalization are to reduce the time, labor and financial costs associated with data retrieval, as well as IT applications for the formation of optimal business partnership schemes based on effective modeling of horizontal production, economic, trade and economic relations between various organizations involved in the logistics process. The introduction of digital technologies will increase the efficiency of interaction between participants in the logistics process, create organizational and technological conditions not only for concluding mutually beneficial contracts for transportation, but also for automating the processes of monitoring the movement of vehicles and operations, improving the quality of work of employees, simplifying the procedure for obtaining processing documents for settlements with all participants in the logistics chain.

The impact of digital transformation on the improvement of the logistics system of the enterprise is assessed, the effect of digital changes is determined. Expected improvement of operational characteristics: increase of operational efficiency, productivity, increase of efficiency of interaction with customers, development of new areas of activity, increase of cybersecurity. The necessity of integrating domestic logistics into the global ecosystem of digital logistics is substantiated.

Digital transformation of an enterprise allows you to get additional opportunities and improve the results of its activities, but it should not be carried out in fragments, but consider a systematic approach, be organically integrated into the enterprise management system and stimulate its continuous improvement.

References

Asaul, A., Malygin, I., & Komashinskiy, V. (2017). The project of intellectual multimodal transport system. *Transportation Research Procedia, 20*, 25–30.

Benotmane, Z., Belalem, G., & Neki, A. (2018). A cost measurement system of logistics process. *International Journal of Information Engineering and Electronic Business (IJIEEB), 5*, 23–29.

Bondar, A., Andrievska, V., & Onyshchenko, S. (2019). Identification of creation and development projects of logistic systems. *Development of Transport Management and Management Methods, 69*(4), 26–37.

Botayeva, S. B., Anarbayev, B. Y., & Botayev, B. B. (2018). Main types and functions of logistic systems. In *Industrial Technologies and Engineering (ICITE-2018)* (pp. 188–191).

Brah, S. A., & Lim, H. Y. (2006). The effects of technology and TQM on the performance of logistics companies. *International Journal of Physical Distribution and Logistics Management, 36*(3), 192–209.

Démare, T., Bertelle, C., Dutot, A., & Lévêque, L. (2017). Modeling logistic systems with an agent-based model and dynamic graphs. *Journal of Transport Geography, 62*, 51–65.

Efimova, O., Vakulenko, S., Kurenkov, P., Magomedova, N., & Rudakova, E. (2021). Selection of schemes for delivery of foreign trade cargoes using digital technologies. In *MATEC Web of Conferences* (Vol. 334, p. 01018). EDP Sciences.

Galkin, A., & Sysoyev, A. (2020). Formalizing criteria of intelligent transportation and logistic systems functioning. *Transportation Research Procedia, 45*, 514–521.

Gerzanych, V. M., & Kyrlyk, N. Y. (2019). Logistic Systems – The Latest Tool of Integration Processes. *Journal of Automation and Information Sciences, 51*(6), 66–81.

Gviliya, N. A., Parfyonov, A. V., & Shulzhenko, T. G. (2019). Managing integrated interorganizational logistic systems in the digital economy. *Upravlenets, 10*(1), 40–51.

Jabbour, C. J. C., Sobreiro, V. A., de Sousa Jabbour, A. B. L., de Souza Campos, L. M., Mariano, E. B., & Renwick, D. W. S. (2019). An analysis of the literature on humanitarian logistics and supply chain management: paving the way for future studies. *Annals of Operations Research, 283*(1), 289–307.

Jovčić, S., Průša, P., Fedorko, G., Večeřová, A., & Dobrodolac, M. (2019). Creating a simulation model for an automated logistics system. In *MATEC Web of Conferences* (Vol. 263, p. 01010). EDP Sciences.

Kalyakina, I. M., Sakharova, O. N., & Dmitrieva, I. A. (2020, May). Digital technologies in management of logistic systems on the transport service market. In *2nd International Scientific and Practical Conference "Modern Management Trends and the Digital Economy: from Regional Development to Global Economic Growth"(MTDE 2020)* (pp. 650–655). Atlantis Press.

Kochegura, D. (2021). Efficiency of vehicle operation in the context of digital transformation of transportation technologies. In *MATEC Web of Conferences* (Vol. 341, p. 00005). EDP Sciences.

Korchagin, V., & Rizayeva, Y. (2021). The model of functioning of cargo transportation subsystem. In *MATEC Web of Conferences* (Vol. 334). EDP Sciences.

Langley, C. J., Novack, R. A., Gibson, B., & Coyle, J. J. (2020). *Supply chain management: A logistics perspective*. Cengage Learning.

Maire, L. (2017). Large solutions for cooperative logistic systems: Existence and uniqueness in star-shaped domains. *Applied Mathematics and Nonlinear Sciences, 2*(1), 249–258.

Mamuladze, G., Katamadze, D., Katamadze, G., Mamuladze, G., & Bakuradze, I. (2021). Problems of the formation of marketing communications in the field of management of logistics systems. In *MATEC Web of Conferences* (Vol. 339, p. 01005). EDP Sciences.

Nekrasov, A., & Sinitsyna, A. (2021). Proactive management transportation logistics infrastructure life-cycle (Digital transformation). In *MATEC Web of Conferences* (Vol. 334, p. 02019). EDP Sciences.

Shepherd, B., & Hamanaka, S. (2015). Overcoming trade logistics challenges: Asia-Pacific experiences. *Asia Pacific Journal of Marketing and Logistics, 27*(3), 444–466.

Sremac, S., Tanackov, I., Kopić, M., & Radović, D. (2018). ANFIS model for determining the economic order quantity. *Decision Making: Applications in Management and Engineering, 1*(2), 81–92.

Efficiency of Business Models Based on Innovations (Additive Technologies) in a Circular Economy

Anna Starodubova, Dinara Iskhakova, and Nailya Gareeva

Introduction

Consumer goods are giving way to specialty products. This is a global trend. Industries will have to move from a business model that was before product-driven to a new business model. This new business model is solution-oriented. In other words, it is possible to move away from mass production to piece production. Enterprises need to select not a consumer for an existing product, but a product for an existing consumer. They will have to build their business on a customization model. Additive technologies allow you to move to piece development, production.

Only 5% of the world's consumer goods are based on additive technology. According to the forecast of K. Schwab, consumer goods are based on the additive technologies should grow by 2025.

Additive technology is one of the elements of smart manufacturing. Jaspert et al. (2021), Yang et al. (2018), Geissbauer et al. (2017), Nedelcu et al. (2018) and Serrano-Ruiz et al. (2022) analyzed the efficiency indicators of smart manufacturing.

Al-Khatib and Al-ghanem (2022), Haricha et al. (2021), Zenisek et al. (2021), Rossmann et al. (2017) and Castelo-Branco et al. (2022) suggested the possible benefits of a smart manufacturing process, its potential.

There are studies (Maresova et al. 2018; Herrmann, 2018; Arjoni et al. 2017; Reynolds 2017) that consider the risks (disadvantages) of smart manufacturing. For example, they include articles on investment risks in smart factories.

The sources (Mittal et al., 2019; Zeba et al., 2021) define smart manufacturing, intelligent manufacturing.

A. Starodubova (✉) · D. Iskhakova
Kazan National Research Technological University, Kazan, Russian Federation

N. Gareeva
Kazan Innovative University Named After V.G. Timiryasov, Nizhnekamsk, Russian Federation

A. Rumyantseva et al. (eds.), *Challenges and Solutions in the Digital Economy and Finance*, Springer Proceedings in Business and Economics,
https://doi.org/10.1007/978-3-031-14410-3_53

The authors of Ben-Ner and Siemsen (2017), Braziotis et al. (2019), Caviggioli and Ughetto (2019) and Chong et al. (2018) noted the weaknesses and strengths of the application of additive technologies.

In most of the articles, the authors Delic and Eyers (2020), Beltagui et al. (2020) researched the impact of additive manufacturing on one model of the circulation economy. This is the distributed supply chain model. The other authors Hettiarachchi et al. (2022), Angioletti et al. (2016), Dev et al. (2020), Hamid et al. (2022) looked at the impact of Industry 4.0, including additive technologies, on the circular economy.

Researchers Colorado et al. (2020), Cruz Sanchez et al. (2020) and Afshari et al. (2020) investigated the relationship of additive technology to the circular economy and sustainable development.

All these scientific papers highlight the relevance of the topic.

The purpose of the study is to identify the efficiency indicators of business models based on additive technologies and compare them with traditional technologies.

The object of research is a business model based on innovation (additive technologies) and traditional technologies.

The subject of the study is an indicator characterizing the effectiveness of innovations (additive technologies) in manufacturing.

The hypothesis is that additive technologies should be combined with circular technologies if they moved to a later stage of the life cycle. This is necessary to enhance the efficiency of production. The efficiency of the business model is higher from the simultaneous use of these technologies than separately. If only one additive technology used in production, then they have insufficient efficiency in production.

Materials and Methods

This study consisted of the following stages.

In the first stage of the study identified the stages of the life cycle of additive technologies in the world for 2010–2022. And, the authors defined the time to adapt additive technology in industry. The authors applied the expert method for this purpose. The study was based on data from Gartner (USA).

In the second phase of the study, comparative analysis was used as a method. The authors identified the composition of additive technologies in national smart manufacturing strategies for the 2022.

In the third phase of the study, the efficiency indicators of the additive technologies found and selected in the business model. The authors applied the expert method for this purpose.

In the fourth stage, the authors collected data to calculate production efficiency indicators based on additive and traditional technologies. Business modeling was carried out to produce polymer products in the Russian Federation 2021.

These indicators were calculated for additive and traditional technologies. Authors used the international method of UNIDO for this purpose.

In the fifth phase of the study, the authors compared production efficiency indicators between additive and traditional technologies. This was done based on data at the fourth stage of the study.

Results

In the first stage of the study identified the stages of the life cycle of additive technologies in the world for 2010–2022. And, the authors defined the time to adapt additive technology in industry.

Every year, Gartner (USA) compiles the Hype cycle to predict the impact of new technologies in information technology markets (Gartner Corp 2022).

The Hype cycle identifies five successive stages that a technology can go through.

Stage 1 is the technological trigger. This stage includes the first publications about the new technology.

Stage 2 is the peak of excessive expectations. In this stage there is widespread public discussion of the new technology.

Stage 3 is the getting rid of illusions. This stage includes the discussion of the first drawbacks of the new technology.

Stage 4 is the elimination of disadvantages. This stage is the elimination of the shortcomings of the new technology, commercialization.

Stage 5 is a plateau of productivity. This stage means that society takes the new technology as a given.

For each technology, Gartner specifies the time it takes for experts to adapt the technology. "Gartner" uses the following time intervals for this: more than 10 years, 5–10 years, 2–5 years, less than 2 years.

Additive technology is the result of the emergence of stereolithography in 1948.

However, the first publications on 3D printing and 3D bioprinting appeared in 2010. Additive technologies in 2010 were in the first stage of Gartner's life cycle. This was the technological trigger stage. In 2010, the period of technology adaptation was for 3D printing (5–10 years) and for 3D bioprinting was (more than 10 years).

In 2015, Gartner introduced 3D printing to enterprises in Stage 4 on the lifecycle. It is the elimination of the disadvantages of new technology, commercialization. The period of technology adaptation was specified as 2–5 years. In 2015, 3D printing by consumers was presented in 2 stages of the life cycle. The period of technology adaptation was specified as 5–10 years. This stage meant the peak of excessive expectations. In 2015, the community began a wide-ranging discussion of additive technologies.

3D technologies are not listed as promising technologies as of 2017. 3D has been replaced by 4D technology. These technologies were planned to be widely used in more than 10 years in Smart Manufacturing. The 4D technology was in Stage 1 of its life cycle. It was a technology trigger in 2015.

As of 2019, 4D technologies are not listed as promising technologies. They have been replaced by Nanoscale 3D Printing. This technology has been Innovation triggered in 2019. The period of adaptation of the technology was specified as more than 10 years. For 2020–2022 "Gartner" did not list additive technologies as promising lifecycle technologies.

In the second phase of the study, comparative analysis was used as a method. The authors identified the composition of additive technologies in national smart manufacturing strategies for the 2022.

22 countries have created national associations for additive technologies. They have been united in the GARPA (Global Alliance of Rapid Prototyping Associations) since 1998. The Alliance approves regulatory documents and ensures cooperation of Alliance members in the development of 3D models. GARPA member countries include: Australia, USA, Canada, China, Hong Kong, Japan, India, Denmark, Egypt, Finland, France, Germany, Italy, Netherlands, Slovenia, Sweden, Spain, UK, Switzerland, Portugal, South Africa. GARPA publishes an annual worldwide review and analysis of 3D printing. Each year, the leading 3D printing countries of the US, EU and China spend 3–4% of GDP on their research and development (GARPA 2022).

The adoption of additive technology in Europe and the Middle East has been slower than in the United States. Among the EU countries, Germany stands out. It creates international alliances between 3D printer manufacturers and businesses with different economic activities. Since 2015, companies that use additive technology have received investment in leading countries for 3D technology. The goal of this investment is to expand the range of industries that use additive technology. Stimulating demand among consumers will accelerate the development of 3D printing.

The authors studied the national strategies developed in the regions in the field of smart manufacturing for 2022 (European Commission 2021a, b, c).

We have identified three possible response options for 7 regions (Table 1). The maximum number of national strategies developed ranged from 3.5 to 7 strategies. The minimum number of national strategies developed ranged from 1 to 3.5 strategies. A lack of national strategies has been identified as an option for regions.

According to Table 1, it is possible to distinguish differences in such countries as USA, Canada and Great Britain. Among the other 4 countries, only these countries have incorporated additive technologies into their national strategy.

In the third phase of the study, the efficiency indicators of the additive technologies found and selected in the business model.

Structure of efficiency indicators of the additive and circular technologies in the business model:

- Energy efficiency per 1 unit of finished product.
- Material intensity per 1 unit of finished product.
- Operative time of production per 1 unit of finished product.
- Cost of packaging (transport and consumer) per 1 unit of finished product.
- Payment of environmental taxes (from packaging and finished product) per 1 unit of finished product.

Table 1 The number of national strategies developed in the regions in the field of smart manufacturing for 2022

	The number of national strategies						
Type of patents	China	Japan	South Korea	Russia	Canada	United Kingdom	United States
Robotics	1	1	1	1	1	–	1
Artificial intelligence	1	1	1	1	1	1	1
Internet of things	–	–	1	1	–	–	1
Big data	1	–	–	–	–	–	1
Advanced manufacturing technology	1	–	1	–	1	–	1
Additive technologies	–	–	–	–	1	1	1
Advanced materials	–	–	1	–	1	–	1
Photonics	–	–	–	–	1	1	–
Security	–	–	–	–	1	1	–
Total	4	2	5	3	7	4	7

- The volume of exhaust gases produced by transports.
- Net present value.
- Profitability index.
- Internal rate of return.
- Discounted payback period.
- Degree of product customization.
- Degree of experience marketing implementation.
- Creation of new jobs.
- Diversity of labor and attraction of qualified personnel.

In the fourth stage, the authors collected data to calculate production efficiency indicators based on additive and traditional technologies. Business modeling was carried out to produce polymer products in the Russian Federation 2021.

The business models were made for 3 years. The polymer product chosen was a polypropylene flowerpot. As the traditional production technology of this product was chosen the method of molding with the help of injection molding machine.

These indicators were calculated for additive and traditional technologies. The authors presented the results of efficiency indicators of business models per 1 unit of finished product (Table 2).

The efficiency of investments in the business models were shown in Table 3. Authors used the international method of UNIDO for this purpose.

In the fifth phase of the study, the authors compared production efficiency indicators between additive and traditional technologies. This was done on the basis of data at the fourth stage of the study.

The difference between the models was presented by the authors (Table 3).

Table 2 Efficiency indicators of business models per 1 unit of finished product

Indicator	Model based on additive circular technology	Traditional model	The difference between the models
Energy efficiency, rubles	0.78	0.15	0.63
Material intensity, rubles	11	20.6	−9.6
Operational time of production, sec.	40	6.43	33.57
The cost of packaging (transport and consumer), rubles	–	2.88	−2.88
Payment of environmental taxes (from packaging and finished products), rubles	–	0.40	−0.40

Table 3 Efficiency of investment in business models over 3 years

Indicator	Model based on additive circular technology	Traditional model	The difference between the models
Cash flow from current operations, rubles	4,987,707.69	3,511,000.41	1,476,707.69
Cash flow from investment operations, rubles	−9,422,120.00	−1,264,088.85	-8,158,031.15
Discounting coefficient, relative units	0.71	0.71	–
Net present value cumulative income, rubles	356,744.67	2,620,348.10	-2,263,603.43
Profitability index	1.04	3.07	-2.03
Internal rate of return	14%	79%	-65%
Discounted payback period	2 years 11 months	1 year 8 months	1 year 3 months

Discussion

In the business models, the researchers calculated the time of manufacture of the product, without the time of its transportation and implementation to the consumer (Table 2). In the new business model, the authors have scheduled production closer to the customer than in the traditional model. Therefore, the new business model reduced the time required to transport and sell products. In the business model based on traditional technology, storage costs were present in a significant amount. But they have become minimal in the business model based on additive and circular technologies. In the new business model, vehicle exhaust volumes have been reduced by introducing point-of-sale production. The sales channel was a trade network of 28 stores in 14 cities.

Most authors (Ben-Ner & Siemsen, 2017; Braziotis et al., 2019; Caviggioli & Ughetto, 2019; Chong et al., 2018) cited increased energy efficiency as one of the advantages of additive technology. According to our calculations, this advantage was not confirmed by the example of this project (Table 2). Energy costs have increased compared to the traditional model. This was because the new business model used recycling equipment. The authors planned that customer would turn in used products.

Material intensity has decreased in the business model based on additive and circular technology (Table 2). The reason for this was the use of waste recycling and its use in production.

In the traditional technology-based business model, the manufacturer paid environmental fees on finished products and packaging. These costs were not present in the business model based on additive and circular technology (Table 2). The reasons are to reduce packaging and organize collection points for recycling.

Packaging costs (consumer and shipping) have decreased in the new business model (Table 2).

The new business model had product customization, unlike the traditional model. Consumers could choose the shape, color, size, and style of the product. Compared to similar individual products, this gives a competitive advantage, in addition to the low cost. Also, customization and personalization increase the emotion of the consumer's purchase. Customer loyalty becomes greater since the customer can see the production of the product in real time and participate in the process of its creation.

Some authors (Maresova et al., 2018; Herrmann, 2018) pointed out a disadvantage from the introduction of additive technology. That disadvantage was the reduction of jobs. Our study refutes this assumption. So, in the business model with traditional production 7 employees worked 8 h a day for a 5-day work week. In the additive and circular business model, 18 employees worked 8 h a day for a 5-day work week. These employees have changed job functions. Labor became more diverse and creative. Intellectual labor replaced physical labor. This business model required more professional qualifications from employees.

The hypothesis proposed by the authors has been confirmed. Additive technologies are no longer reflected in the lifecycle of advanced technologies from 2020 to 2022. Additive technology has become a traditional technology according to Gartner. Despite this, the U.S., Canada, and the United Kingdom have developed national strategies for smart manufacturing that incorporate additive technologies. The authors' calculations have shown insufficient efficiency of production based on additive technologies. This fact restrains enterprises from introducing additive technology on a large scale. Therefore, the authors suggested using circular technology to enhance the effect of additive technology as well.

Consider the novelty of this study. The authors determined the relationship of efficiency indicators of business models based on additive technologies with stage of their life cycle. For the first time, the authors combined business model efficiency indicators for additive technologies with circular technologies. Before this, the effectiveness of additive technology was considered without circular technology. Other authors (Colorado et al. 2020; Cruz Sanchez et al. 2020; Afshari et al. 2020) have not considered the dynamism of the stages of the life cycle of additive technologies.

Conclusion

Despite the wide popularity of additive technology in the world, the experience of implementing additive technology is not fully disclosed in enterprises. Many manufacturers are afraid to switch to additive technology. They need proof of their effectiveness to switch to these technologies. Industrial companies have a fear of starting to change their traditional business model.

Consider the prospects of the research. In the future, the authors plan to look at more examples of existing business models based on additive technologies in the countries. It is necessary to develop new business models of additive technologies with circular technologies. Researchers can consider for which types of economic activity the hypothesis is fulfilled in this article. Since the authors have tested on the example of the production of polymer products. It will be necessary to answer the question. What is the optimal size of enterprises? This is necessary for the enterprise to have efficiency from additive technologies.

References

Afshari, H., Searcy, C., & Jaber, M. Y. (2020). The role of eco-innovation drivers in promoting additive manufacturing in supply chains. *International Journal of Production Economics, 223*, 107538. https://doi.org/10.1016/j.ijpe.2019.107538

Al-Khatib, A. W., & Al-ghanem, E. M. (2022). Radical innovation, incremental innovation, and competitive advantage, the moderating role of technological intensity: Evidence from the manufacturing sector in Jordan. *European Business Review., 34*(3), 344–369.

Angioletti, C. M., Sisca, F. G., Luglietti, R., Taisch, M., & Rocca, R. (2016). Additive manufacturing as an opportunity for supporting sustainability through the implementation of circular economies. In *21st Summer School Francesco Turco 2016* (25–25). AIDI-Italian Association of Industrial Operations Professors. Retrieved from http://www.summerschool-aidi.it/

Arjoni, D., Madani, F., Ikeda, G., Carvalho, G., Cobianchi, L., Ferreira, L., & Villani, E. (2017). Manufacture equipment retrofit to allow usage in the industry 4.0. In *Proceedings 2nd International Conference on Cybernetics, Robotics and Control* (pp. 155–161). https://doi.org/10.1109/CRC.2017.46

Beltagui, A., Kunz, N., & Gold, S. (2020). The role of 3D printing and open design on adoption of socially sustainable supply chain innovation. *International Journal of Production Economics, 221*, 107462. https://doi.org/10.1016/j.ijpe.2019.07.035

Ben-Ner, A., & Siemsen, E. (2017). Decentralization and localization of production: The organizational and economic consequences of additive manufacturing (3D printing). *California Management Review, 59*, 5–23.

Braziotis, C., Rogers, H., & Jimo, A. (2019). 3D printing strategic deployment: The supply chain perspective. *Supply Chain Management, 24*(3), 397–404. https://doi.org/10.1108/SCM-09-2017-0305

Castelo-Branco, I., Oliveira, T., Simões-Coelho, P., Portugal, J., & Filipe, I. (2022). Measuring the fourth industrial revolution through the Industry 4.0 lens: The relevance of resources, capabilities and the value chain. *Computers in Industry, 138*, 103639.

Caviggioli, F., & Ughetto, E. (2019). A bibliometric analysis of the research dealing with the impact of additive manufacturing on industry, business and society. *International Journal of Production Economics, 208*, 254–268. https://doi.org/10.1016/j.ijpe.2018.11.022

Chong, L. C., Ramakrishna, S., & Singh, S. (2018). A review of digital manufacturing-based hybrid additive manufacturing processes. *International Journal of Advanced Manufacturing Technology, 95*(5–8), 2281–2300. https://doi.org/10.1007/s00170-017-1345-3

Colorado, H. A., Velásquez, E. I. G., & Monteiro, S. N. (2020). Sustainability of additive manufacturing: The circular economy of materials and environmental perspectives. *Journal of Materials Research and Technology, 9*(4), 8221–8234. https://doi.org/10.1016/j.jmrt.2020.04.062

Cruz Sanchez, F., Boudaoud, H., Camargo, M., & Pearce, J. M. (2020). Plastic recycling in additive manufacturing: A systematic literature review and opportunities for the circular economy. *Journal of Cleaner Production, 264*, 121602. https://doi.org/10.1016/j.jclepro.2020.121602

Delic, M., & Eyers, D. R. (2020). The effect of additive manufacturing adoption on supply chain flexibility and performance: An empirical analysis from the automotive industry. *International Journal of Production Economics, 228*, 107689. https://doi.org/10.1016/j.ijpe.2020.107689

Dev, N. K., Shankar, R., & Qaiser, F. H. (2020). Industry 4.0 and circular economy: Operational excellence for sustainable reverse supply chain performance. *Resources, Conservation and Recycling, 153*, 104583. https://doi.org/10.1016/j.resconrec.2019.104583

European Commission. (2021a). *International report USA*. Retrieved from https://ati.ec.europa.eu/reports/international-reports/report-united-states-america-technological-capacities-and-key-policy

European Commission. (2021b). *Report on Canada - Technological capacities and key policy measures*. Retrieved from https://ati.ec.europa.eu/reports/international-reports/report-canada-technological-capacities-and-key-policy-measures

European Commission. (2021c). *Report on the United Kingdom - technological capacities and key policy measures*. Retrieved from https://ati.ec.europa.eu/reports/international-reports/report-united-kingdom-technological-capacities-and-key-policy

GARPA. (2022). *Global Alliance of Rapid Prototyping Associations*. Retrieved from https://wohlersassociates.com/GARPA.html/

Gartner Corp. (2022). Retrieved from http://www.gartner.com/

Geissbauer, R., Schrauf, S., Berttram, P., & Cheraghi, F. (2017). *Digital Factories 2020: Shaping the future of manufacturing*. Retrieved from https://www.pwc.de/de/digitale-transformation/digital-factories-2020-shaping-the-future-of-manufacturing.pdf

Hamid, M. S. R. A., Masrom, N. R., & Mazlan, N. A. B. (2022). The key factors of the industrial revolution 4.0 in the Malaysian smart manufacturing context. *International Journal of Asian Business and Information Management, 13*(2).

Haricha, K., Khiat, A., Bahnasse, A., Issaoui, Y., & Hassan, O. (2021). Towards smart manufacturing: Implementation and benefits. *Journal of Ubiquitous Systems and Pervasive Networks, 15*(02), 25–31. https://doi.org/10.5383/JUSPN.15.02.004

Herrmann, F. (2018). The smart factory and its risks. *Systems, 6*(4), 38. https://doi.org/10.3390/systems6040038

Hettiarachchi, B. D., Brandenburg, M., & Seuring, S. (2022). Connecting additive manufacturing to circular economy implementation strategies: Links, contingencies and causal loops. *International Journal of Production Economics, 246*, 108414. https://doi.org/10.1016/j.ijpe.2022.108414

Jaspert, D., Ebel, M., Eckhardt, A., & Poeppelbuss, J. (2021). Smart retrofitting in manufacturing: A systematic review. *Journal of Cleaner Production, 312*(2), 127555. https://doi.org/10.1016/j.jclepro.2021.127555

Maresova, P., Soukal, I., Svobodova, L., Selamat, A., & Krejcar, O. (2018). Consequences of industry 4.0 in business and economics. *Economies, 6*(3), 46. https://doi.org/10.3390/economies6030046

Mittal, S., Khan, M., Romero, D., & Wuest, T. (2019). Smart manufacturing: Characteristics, technologies and enabling factors. *Proceedings of the Institution of Mechanical Engineers, Part B, Journal of Engineering Manufacture, 233*(5), 1342–1361. https://doi.org/10.1177/0954405417736547

Nedelcu, M., Dima, A., Dinulescu, R., (2018). Digital factory – A prerequisite for revitalizing the production sector. In *Proceedings of the 12th International management conference: Management Perspectives in the Digital Era* (pp. 520–529). https://ideas.repec.org/a/rom/mancon/v12y2018i1p520-529.html

Reynolds, E. (2017). Innovation and production: Advanced manufacturing technologies, trends and implications for US cities and regions. *Built Environment, 43*(1), 25–43. https://doi.org/10.2148/benv.63.3.25

Rossmann, M., Khadikar, A., Le Franc, P., Perea, L., Schneider-Maul, R., Buvat, J., & Ghosh, A. (2017). Smart Factories: How can manufacturers realize the potential of digital industrial revolution. *Capgemini*, 5–18. Retrieved from https://www.capgemini.com/consulting/wp-content/uploads/sites/30/2017/07/dti-smart-factory-research.pdf

Serrano-Ruiz, J. C., Mula, J., & Poler, R. (2022). Development of a multidimensional conceptual model for job shop smart manufacturing scheduling from the Industry 4.0 perspective. *Journal of Manufacturing Systems, 63*, 185–202.

Yang, S., Raghavendra, M., Kaminski, J., & Pepin, H. (2018). Opportunities for industry 4.0 to support remanufacturing. *Applied Sciences (Switzerland), 8*(7), 1177. https://doi.org/10.3390/app8071177

Zeba, G., Dabic, M., Cicak, M., Daim, T., & Yalcin, H. (2021). Technology mining: Artificial intelligence in manufacturing. *Technological Forecasting and Social Change, 171*, 1–18. https://doi.org/10.1016/j.techfore.2021.120971

Zenisek, J., Wild, N., & Wolfartsberger, J. (2021). Investigating the potential of smart manufacturing technologies. *Procedia Computer Science, 180*, 507–516. https://doi.org/10.1016/j.procs.2021.01.269

Auditors in Russia: Entry into the Profession

Iurii Nikolaevich Guzov, Nikolay Aleksandrovich Polyakov, and Yaroslav Iurievich Guzov

Introduction

Compared to the extensive European and American literature on the historiography of auditing—see, for example, Mattessich (2011), Baker (2014), Flesher, Previts, & Samson (2005), Ashton (2013)—the historiography of auditing in Russia is at an early stage of its development. Work in this field to date has been led by two scholars, Sokolov and Bychkova (1999, 2005a, b). Apart from three joint papers published in 1999 and 2005, they have also published individually (e.g. Bychkova, 1996) or in collaboration with others (e.g. Sokolov & Terekhov, 2004; Guzov et al., 2008). Some of these studies look back to the seventeenth century, when rules of public control were first established (e.g. Bychkova, 1996).

In Russia profession named 'auditor' was born in the army. Tsar Peter the Great (reigned 1682–1725) introduced the rank (office) of an auditor together with other military ranks—first, in the military regulations of 1716, and subsequently in the Table of Ranks of Imperial Russia. Allegedly Peter borrowed this title from the Polish rank system, where it designated first a judge, then a person who attended or participated in the hearing of legal acts, and later an investigator (examinant). Auditors were selected among junior and senior warrant officers; if for any reason it proved impossible to find a person suitable for such military duty among middle-ranking officers, the selectors proceeded to sub-officers and wachtmeisters. Under the military regulations, an auditor-general was placed in command of the body of

I. N. Guzov (✉) · N. A. Polyakov
St. Petersburg State University, St. Petersburg, Russian Federation
e-mail: y.guzov@spbu.ru; n.polyakov@spbu.ru

Y. I. Guzov
LLC "Geological Center of St. Petersburg State University", St. Petersburg, Russian Federation
e-mail: guzov@geospbu.ru

A. Rumyantseva et al. (eds.), *Challenges and Solutions in the Digital Economy and Finance*, Springer Proceedings in Business and Economics,
https://doi.org/10.1007/978-3-031-14410-3_54

auditors, aided by an auditor-lieutenant-general and several chief auditors as his aides de camp. By nature, the position of an auditor in the army did not differ in unity: he combined the features of a clerk, secretary, and prosecutor. In 1797 the auditors' titles were rebranded as civil ranks and during the military and judicial reform of 1867, the rank of an auditor was abolished in the Russian military.[1] In addition to the name, army auditors had little in common with the modern profession.

Another root of the auditors' profession is drawn to the history of professional associations. The idea of their creating in Russia received public recognition in the 1870s, when conditions arose for borrowing European models of professional organization. The first wave of bankruptcies that followed at that time convinced many that the audit of the financial reports by the shareholders itself was insufficient, and the work of independent accountants needed to be streamlined. Well-known accounting experts in Russia were aware of the activities of the UK independent corporate organizations—institutes of chartered accountants (Lvova, 2005). They saw the solution of urgent problems in the institutionalization of the profession on the foreign model. However, before the revolution, all attempts to establish the institution of independent auditors failed. The developers of various projects did not agree on the educational and qualification requirements of the future members of the institute. The level of accountants' education at that time was too different and overwhelming too low. Establishing the qualifications desired by the initiators of institutionalization turned the profession into an elitist.

In the Soviet times the fledgling administrative economy was not interested in carrying out independent expertise of its economic entities and their state. Audit was squeezed out by inspection that catered to the needs of the Soviet establishment. Subsequently, with the fall of the Soviet Union in the early 1990s the vagaries of state monopoly on entrepreneurship were overcame, the private property legalized and new legal forms of business entities—stock companies and other organizational types—established. Thus, the prerequisites for the emergence and development of an independent audit were created.

The modern history of auditing in Russia is approaching its 30-year anniversary. During this time, many events have occurred, and some of them are reflected in the articles by Enthoven et al. (1998) about first sEACesses of Russian audit, Sucher and Bychkova (2001) on auditor independence rule requirements, McGee & Preobragenskaya (2005), Samsonova (2009), and Samsonova-Taddei (2013) on evolution of methodological and legislative framework and finally Mennicken (2008, 2010) and Alon & Dwyer (2012) regarding activities of Russian audit firms in the context of international standards adoption, audit market creation and globalization.

Unfortunately, the sEACessful 1990s and early 2000s gave way to the recession of the Russian audit market in the 2010s. In the first section of the paper will be presented numerical data confirming this trend accompanied by a brief overview of

[1] The Small Brockhaus and Efron Encyclopedic Dictionary: in 4 volumes, 1907–1909.

modern audit history in Russia. The second section represents a consecutive description of entry into the profession of an auditor, which has changed significantly in the last 30 years. The following section analyzes the Russian rules for obtaining auditor qualifications. The fourth and the final section introduces new qualification examination model—2020, in the development of which participated the author of this study. This section connects the history of Russia's audit with its present and future, where the problems of education and qualification of auditors are usually recognized as a key element in the sEACess of the profession institutionalization.

The purpose of the work is to compare trends in the development of the audit market and the education system of auditors in Russia in the direction of compliance with international standards of audit and education of auditors.

Materials and Methods

Content analysis of dominant development of the audit market and education systems and qualification of auditors in Russia.

Result and Discussions

Brief History of Russian Audit in the Modern Era: From 1991 to 2019

The collapse of the Soviet Union marked the initial stage of the audit development and its regulation in Russia. During 1991–1993 audit established in a regulatory free space and audit companies based their activities on a variety of different principles, ranging from the soviet rules of state revision to international auditing standards (which were used by the then Big Six, now the Big Four).

Presidential Decree № 2296 of 12 December 1993, introduced the *"Provisional Rules of Audit Activity"*. Due to this state regulation audit was identified, first time ever in Russia, as an independent business activity. These rules together with subsequent Resolutions adopted by the Government of the Russian Federation were approved by the Commission on Audit Activity under the President of the Russian Federation. They established a system of auditor certification and the licensing of audit activity. Four kinds of auditor certificates and licenses were introduced: "general audit", "bank audit", "insurance audit," and "audit of commodity and stock exchanges", off-budgetary funds and investment institutions". As Guzov (2016, p. 653) notes, that provisions gave a definition of auditing and its types, asserted the principles of auditing as well as the rights and responsibilities of the auditor and the client, and defined types of audit reports.

Table 1 Russian auditing market in 2004–2017

Year	Income of Audit firms (bln. rubles)	Number of audit firms (thous.)	Number of auditors (thous.)	Number of clients (thous.)	Level of modified audit reports issuance (%)
2004	20.5				
2005	28.9	4.7		60.6	48.9
2006	34.3	6.1		80.3	43.1
2007	41.7	7.1		84.9	44.9
2008	50.1	6.4		93.0	45.1
2009	49.6	6.9	38.8	92.7	44.2
2010	49.1	6.3	26.3	60.6	41.4
2011	50.0	6.2	26.8 (1.1)	75.6	33.8
2012	51.0	5.7	24.1 (3.2)	70.0	29.0
2013	52.2	5.5	23.0 (3.2)	68.4	25.9
2014	53.6	5.3	22.2 (3.4)	67.9	23.5
2015	56.1	5.1	21.5 (3.5)	71.8	22.7
2016	57.1	5.0	19.6 (3.6)	74.5	23.3
2017	55.4	4.8	19.6 (4.0)	78.1	21.1

Source: Author's processing, based on the main indicators of the audit services market in the Russian Federation introduced by the Ministry of Finance (URL: https://www.minfin.ru/ru/perfomance/audit/audit_stat/MainIndex/)

The first Federal Law 'On Auditing' (2001) gave the audit activity a higher status of a legally regulated business, for which an independent branch of law was allocated. The Law introduced Federal Standards of Auditing Activity (FPSAD) that greatly facilitated the rapid growth of the auditing market. In 2008 by the new version of Law the concept of self-regulation of audit activity was introduced. It replaced the previous four types of auditor certification with a single auditor qualification certificate. Six self-regulatory organizations (SROs) of auditors were established and began to work: Audit Chamber of Russia, Institute of Professional Auditors, Moscow Audit Chamber, Russian Collegium of Auditors, and Auditor Association *Sodruzhestvo*, Guild of Auditors of Regional Institutes of Professional Accountants (of business since 2013).

The main quantitative indicators of the Russian auditing market development for the period from 2004 when statistics of audit activities appeared till the last data of 2017 are given below (Table 1).

The data for the reference period demonstrates the rapid growth of the market for audit services in 2008–2009, immediately after the adoption of the Law on Auditing and auditing standards. Then came a 10-year stagnation, briefly interrupted by the outbreak of the economic crisis of 2014, accompanied by a sharp fall in the ruble exchange rate. The stagnation first of all is indicated by such measures as revenue and the number of clients. The number of audit firms has remained almost unchanged since the beginning of observations, but the number of auditors since 2009 has steadily decreased. The reason for this trend could be found in introduction of self-regulation of auditing activities, coupled with a new unified qualification certificate.

The tendencies of the stagnation of audit firms' revenues and reduction in the number of auditors that obtained qualification certificates before 2011 are likely to continue in the near future. The number of auditors with new qualification certificates is growing extremely slowly due to the relatively stringent examination requirements. Statistics provides another interesting observation: revenues vary significantly among small-scale and large-scale audit firms. Wage levels in some audit firms increase other dozens of times over. Moreover, opportunities for career growth, as well as in-house methodological assistance, can also vary dramatically across the companies.

At present, a new edition of the Law on Audit is being drafted, where the Central Bank of Russia will act as an auditing activity regulator, in conjunction with the Ministry of Finance.

Modern Russian audit market.

The modern audit market in Russia includes 2 self-regulatory organizations "Commonwealth" and "Russian Union of auditors", 19 (nineteen) thousand auditors working in 4 (four) thousand audit firms.

The turnover of audit firms exceeds $ 1 billion. More than 78 (seventy eight) thousand firms are audited according to international auditing standards.

Entry into the Profession from the Evolutionary Perspective

Access to the profession of an auditor in Russia is currently determined by the following provisions:

1. Provisions pertaining to the qualification examination for the auditor qualification certificate:
   (a) educational requirements for candidates seeking to obtain an auditor qualification certificate;
   (b) requirements for areas of knowledge forming the basis for the list of questions to be offered to a candidate at the qualification examination (the list of knowledge areas is approved by the Council for Auditing Activities on 11 July 2011, protocol № 1);
   (c) requirements for the hands-on experience for candidates seeking to obtain an auditor qualification certificate;
   (d) qualification examination procedure and questions related thereto (Order of the Ministry of Finance of the Russian Federation No. 32, Order of the Ministry of Finance of the Russian Federation No. 51, Orders of the Autonomous Non-Commercial Organization *Unified attestation Commission* (hereinafter—the EAC)).

2. Requirements for the continuous professional development (Federal Law on Auditing Activity, Chap. 11).
3. Qualification requirements for performing certain job functions set in the professional standard *'Auditor'* (Ministry of Labour and Social Affairs of the Russian Federation, 2015) individual qualification requirements set in the ISAs and written explanations of the regulator (Ministry of Finance of the Russian Federation).

Therefore, initial requirements for auditors in Russia are set at the legislative level through provisions on the auditor qualification certificate. A person having no qualification certificate cannot be recognized as an auditor. The Council for Auditing Activities approves knowledge areas which serve as a basis for the list of questions offered to a candidate at the auditor qualification examination, but it does not set requirements for the auditor's competencies (qualifications) that are subject to assessment, as it is the case in the International Standards on Auditing—hereinafter ISA (IAESB, 2014–2015). Minimum requirements for the candidate's education level and work are also set via admission to the qualification exam.

At the same time, requirements for auditors' competencies with respect to their role in the auditing process are stipulated within the scope of the labour relations system (professional standard *'Auditor'* was approved by Order No. 728n of the Ministry of Labour and Social Affairs of the Russian Federation). Professional standards are designed in such way so as to make provisions for requirements for education level and work experience, eligibility criteria, a list detailing requisite knowledge and skills, needed for performing employment duties in line with a certain job function. Therefore, questions represented in the ISAs as a unified system are regulated in Russia predominantly at the level of auditing legislation and regulations, whereas the questions pertaining to competencies—at the level of employment and labour legislation by-laws. As of today, no unified framework for the auditors' education system exists in Russia.

At present, requirements for auditors' education level and competencies in the Russian Federation are initially set at the legislative level of laws and regulations, and most issued are addressed at the level of regulators of auditing activity. Russian governmental bodies that establish the frameworks and provisions of the requirements for the qualification examination are as follows:

- an authorized federal government body which designs directive principles of state policy and exercises regulatory control of auditing activities (the Ministry of Finance of the Russian Federation)—determines the qualification examination procedure, as well as the procedure of establishing a unified certification commission that holds the qualification examination and issues the auditor qualification certificate in the Russian Federation;
- the Council for Auditing Activities—determines the areas of knowledge that serve as a basis for the list of questions offered to the candidate at the qualification examination;
- the Unified Certification Commission—holds the qualification examination in line with the established procedure.

Table 2 Qualification levels and functions of the auditor

Job position	Qualification level	Employment functions
Assistant auditor	4.5	1. Performing specific assignments from the superior and rendering other services concomitant with auditing activities 2. Conducting auditing procedures (activities), performing individual transactions in connection with the provision of associated auditing and other services.
Lead auditor	6.7	1. Supervising the due performance of the auditing assignment and provision of other services in connection with auditing activities; 2. Performing quality assurance reviews to ascertain the quality of executed auditing assignments, in which the person concerned did not participate.
Methodologist	7	1. Designing relevant internal organizational and administrative documentation that regulates auditing activities in the company;2. Providing methodological guidance to audit firm employees with regards to the auditing legislation and its practical application;3. Audit firm personnel training and professional development.
Audit firm director	7	1. Planning and arrangement of activities of an audit firm subdivision; 2. Ongoing management and control of activities of an audit firm subdivision.

Source: Ministry of Labour and Social Affairs of the Russian Federation, 2015

Self-regulatory organizations of auditors and the professional community are not entitled to direct involvement in issues regarding the requirements for the candidates.

Presently, the requirements for the qualifications, job duties and employment functions, designed and approved by the Ministry of Labour and Social Affairs of the Russian Federation, are contained in the professional standard *'Auditor'*. The principal activity of the auditor is conducting audit and rendering associated services (Sukhanov, 2015). The auditor's career path does not have many ranks. Every stage along this path represents a specific qualification level (Table 2).

Job functions of a technical specialist performing auditing activities in the capacity of an assistant to the lead auditor are subject to the least number of requirements for professional standard.

As far as methodologists are concerned, such position requires higher education diploma: either a specialist's or a master's degree. Besides, further auditing education is also mandatory, being a prerequisite for obtaining the auditor qualification certificate. An employee should have at least 5 years' work experience at a company engaged in auditing activities or accounting records maintenance and accounting (financial) reporting; of those 5 years—no less than 3 years at an audit firm.

About a head of department (subdivision, division), the requirements are the same as stated above. Specials conditions for work authorization apply—a valid auditor qualification certificate and proven membership in one of self-regulatory organizations of auditors. To the extent provided for by the applicable Russian legislation on auditing activities—a valid auditor qualification certificate issued after the first of January 2011.

Qualification Certificates

Stages of development of examinations for auditor's qualification certificate in Russia.

- Period 1 (one). 1994–2001 (nineteen ninety-four till two thousand one), 4 types of qualification certificates on "General audit", "audit of exchanges and investment institutions", "Bank audit", "audit of insurance organizations" in the form of 2 exams (testing and written work on 5 topics: law, accounting, taxation, Finance and audit);
- Period 2 (two). 2002–2010 (two thousand two till two thousand ten), 3 types of qualification certificates on "General audit", "audit of exchanges and investment institutions", "Bank audit" are left, in the form of 2 exams (testing and written work);
- Period 3 (three). 2011–2019 (two thousand eleven—two thousand nineteen), introduced a unified qualification certificate (3 exams (test and 2 written works);
- Period 4 (four). From year 2020—... new qualification exam in the form of 12 (twelve) exams (5 tests, 6 written works, 1 audit qualification work).

As it follows from the ISAs, setting up requirements for technical competencies, professional skills, determining professional values, ethical and interpersonal principles that prospective professional accountants ought to demonstrate and encourage among themselves, serves several purposes: it protects public interests, enhances professional accountants' performance, and increases overall confidence in the accounting profession.

Within the Russian system, the purposes of qualification requirements as such are not specified. An 'entry point' into the profession is stipulated at the legislative level, referring to the auditor qualification certificate issued upon sEACessful professional examination and proof of appropriate work experience.

An auditor is a private entity that has obtained an audit qualification certificate and is a member of one of self-regulatory organizations of auditors. The auditor qualification certificate is issued by a self-regulatory organization provided that the candidate has met the following conditions:

1. sEACessfully passed the qualification examination (admission to the qualification examination is available to candidates with a higher education degree in one of state-accredited educational programs)
2. has, as of the date of submitting the application for the auditor qualification certificate, no less than 3 years' work experience in auditing activities or accountancy and accounting (financial) reporting. Of the last 3 years of the stated work experience the candidate should have worked at least 2 years at an audit firm.

Consequently, all requirements for prospective auditors in Russia are contingent upon conditions for obtaining the auditor qualification certificate. The applicable legislation does not make any provisions for the essentiality of professional training as a prerequisite for entry into the profession. The decisive criterion is the ability to sEACessfully pass the professional examination.

In Russia, requirements for auditors' training (initial professional training) are not specified. Requirements for requisite knowledge and skills are set up through the state program of professional examination for the auditor qualification certificate. Upon sEACessful passing of the examination and proof or relevant work experience the candidate is awarded with the auditor qualification certificate. The system of competencies detailing the requisite skills set and expertise for prospective auditors (candidates taking the professional examination) does not exist in Russia.

Presently, in connection to the above, there has been a de-facto substitution of the federal standards of auditing activity with the International Standards on Auditing (resulting from the adoption of the ISAs in Russia on the first January 2017).

On examining the current program of the qualification examination for the auditor qualification certificate and the list of questions offered to the candidate (EAC, 2016) we find that:

- individual questions on professional ethics and independence are included in the 'Audit' section;
- questions of information technologies and computer systems are not featured among examination questions;
- individual questions on management accounting are considered in the 'Finances' section;
- questions on business and organization environment, economics, business and business management strategies are not featured in the list of examination questions;
- professional skills as such are not examined (International Education Standard (IES) 3).

The model of the Russian qualification examination is aimed predominantly at assessment of knowledge in relevant areas independently from professional skills (individual skills can be demonstrated while performing examination tasks). Therefore, in their preparation for the assessment candidates are mainly focused on memorizing requisite material.

In Russia, all requirements for prospective auditors' work experience are established by the Federal Law on Auditing Activity through conditions for obtaining the auditor qualification certificate. The candidate should, as of the date of submitting the application for the auditor qualification certificate, have no less than 3 years' work experience in auditing activities or accountancy and accounting (financial) reporting. Russian laws and regulations provide for no further elaborations regarding work experience. Consequently, the specified requirement appears to be purely formal, can only be proved by the record of employment and is not subject to efficient monitoring, unlike the IES.

Table 3 Results of certification exams for 2012–2018

Year	Number of examinees	Number of persons who passed exam	The Percentage of those who passed the exam %
2012	156	56	36.0
2013	309	122	39.0
2014	429	123	29.0
2015	520	160	31.0
2016	995	268	27.0
2017	1056	297	28.1
2018	875	273	31.2
Всего	4340	1299	29.9

According to the official website of the EAC http://eak-rus.ru/about/godovaya_otchetnost

In the Russian legislation, the assessment procedure for auditors' professional competencies is effectively determined by the Federal Law on Auditing Activities. Such assessment (or 'qualification check', according to the term used in the Federal Law on Auditing Activities) is conducted in the form of a qualification examination held by the EAC.

Autonomous non-profit organization "unified attestation Commission" (ANO "EAC") was established by six self-Regulatory organizations of auditors (currently there are only two) on August 31, 2010. The task of ANO "EAC" is to conduct qualification examinations to verify the qualifications of persons applying for a qualification certificate. The first exam organized by the UAC was held on March 17, 2011. The first group consisted of 13 applicants.

The EAC holds the qualification examination, the procedure for which is established by the Ministry of Finance of the Russian Federation. The list of questions offered to the candidate at the qualification exam is compiled by the EAC, based on the knowledge areas approved by the Council for Auditing Activities.

The EAC's activities are controlled by the regulator of auditing activities—the Ministry of Finance of the Russian Federation (EAC, 2016). The EAC is set up together with self-regulatory organizations of auditors. A EAC's incorporation documents, along with any modifications introduced thereto, are subject to agreement with an authorized federal agency prior to their approval. The appointment of an individual executive body of the EAC and termination of such appointment are subject to prior approval by the Council for Auditing Activities. The EAC activities are based on the principles of independence, impartiality, openness, transparency and self-financing. The examination procedure, as well as the list of questions, are determined by the EAC (Table 3).

Following the results of the assessment, an auditor qualification certificate is issued; issuance procedure and format are approved by the Ministry of Finance of the Russian Federation. No self-regulatory organization of auditors is entitled put forward any demands or conditions upon issuance of the auditor qualification certificate.

These data indicate a constant ratio of the total number of candidates who passed the exam. The relatively high results of the first 2 years are since the small number of candidates compared to subsequent years were represented by experienced practicing auditors who did not have time to obtain a certificate under the simplified scheme and decided to obtain it in connection with the new more universal and prestigious status of the new certificate.

The EAC actively cooperates with international organizations responsible for professional qualifications. Primarily with ACCA ICAEW.

ACCA and EAC signed a Memorandum of understanding to join forces in promoting the sustainable development of the accounting profession. This partnership will help to maintain and improve professional, educational and ethical standards, as well as to provide interested people with the opportunity to receive qualifications that are in demand all over the world.

The EAC determines the list of questions to be offered to candidates at the qualification examination; these include multiple-choice tests, examination papers and practical tasks. That said, Russian laws and regulations do not stipulate exactly which skills are subject to assessment at the qualification examination for the auditor qualification certificate, and do not specify the assessment method. The same goes for the EAC documents, for instance, the Program of qualification examination for the auditor qualification certificate does not provide detailed interpretation of the skills assessment process, representing merely a list of topics covered in the examination questions in each knowledge area. As can be seen from the above, the question of skills assessment in Russia largely remains open, contrasted by the IES system where the results of training in each domain of competence (skills) are detailed and subject to targeted assessment. It appears that this issue will be addressed upon transition to the new qualification examination model, which is currently being designed by the EAC, as the assessment of knowledge will be substituted with the assessment of competencies.

Summing up what has been said, this sphere is overwhelmingly subordinated to and controlled by the state. Self-regulatory organizations of auditors are not mandated to participate in activities pertaining to the qualification examination. A difference in the IES is that organizations, members of the International Federation of Accountants (the IFAC), are initially responsible for the assessment of professional qualifications, while other interested parties can be involved in the assessment procedure to a considerable degree.

New Qualification Examination Model

In 2015, the development of a new model of qualification examination for auditors began at EAK. The main objective of this model is to enable transition to the competence-based assessment of candidates for the auditor qualification certificate.

The qualification examination model is based on the following principles:

1. level-based, hierarchical approach.
2. modular, competence-based structure of qualification examination.
3. practice-oriented examination base.
4. regular update of examination base and its sources.
5. integration of various examination formats.
6. a balance between assurance of examination quality and its financial affordability for candidates.

Requirements stipulated by the IES and the practice of their implementation in the internationally recognized accreditations are used for the purposes of describing the auditor's competencies and their assessment. Accounted for are also characteristics and specifics of current auditing practices in the Russian Federation, as well as employers' requirements and expectations with regard to employees' competencies in the auditing services market (provisions of the professional standard 'Audit' on the employment functions and associated knowledge and skills).

The new competence-oriented examination model is based on the premise stipulated in the IES that an auditor is, first and foremost, a professional accountant who possesses relevant competencies and is responsible for forming an opinion on the authenticity of financial records, grounded by his professional judgement. Therefore, alongside professional accounting competencies, the auditing profession requires additional competencies pertaining to the level of responsibility.

The model advocates a level-based approach to the qualification examination, ensuring a gradual transition from one competence level to another, more complicated level, and represents a system that integrates an algorithm whereby a candidate for an auditor qualification certificate acquires professional competencies, maintains them at a requisite level, and an assessment of these competencies is performed.

The new examination has a modular structure where competence assessment areas of various levels are divided into individual examination modules. Each level has a set of modules for assessing and examining auditors' technical competencies, professional skills and values, as well as methods of their assessment and distribution of examined competencies (knowledge, skills and expertise). Questions on professional values, ethics and relationships are integrated into all assessment areas, at every level.

The qualification examination model includes six competence assessment areas:

- Accounting and accounting reports,
- Auditing activities,
- Financial analysis and business analysis,
- Management accounting, risk management, internal control,
- Law,
- Taxation.

The designation of competence assessment areas is based on their correspondence to knowledge areas specified by the Council for Auditing Activities on 23.03.2017 (Civil Aviation Authority, 2018), IES recommendations, and economic feasibility of organizing and conducting the qualification examination along with its financial affordability for candidates.

The model of competence-based assessment of auditors—2020, designed by the EAC (2019), consists of three levels.

1. Base level

   Assessment of candidates' competencies across five examination modules, administered as computer-based testing. Test questions are aimed at examining the knowledge and understanding of regulatory and theoretical frameworks of auditing, ability to apply basic knowledge from each assessment area in resolving auditing-related model tasks, within individual examination modules. The list of questions, offered to candidates in the form of multiple-choice tests, should include no less than 2000 tests in five examination modules. Each individual test set offered to a candidate in every examination module, is assumed to contain 40 test questions.
2. Foundation level

   Includes six modules organized as a computerized written exam that assesses the ability of candidates to apply knowledge across all modules of specified assessment areas in the process of resolving several written assignments that contain practice-based questions and mini-tasks. Each module is aimed at evaluating competencies only from a given assessment area. With regard to the above, all candidates receive the same test assignment during their examination, which considerably facilitates and expedites test validation. Multiple assignment variants are designed to ensure information protection.
3. Qualification level

   Represents a resolution of a cross-cutting case assignment that allows to assess the candidate's ability to combine and balance competencies from different assessment areas, to evaluate the ability to form professional judgements in line with the MSA. Qualification level is termed a qualification task, as it completes the qualification assessment procedure, after which the candidate is considered eligible for the auditor qualification certificate. All candidates receive the same cross-cutting case assignment at the qualification level.

Computer-based tests for base level assessment modules are held throughout the year at accredited examination centers, according to the EAC's schedule, with an option of taking the test on an individual basis.

Foundation level modules are tested twice a year during the transition period. It is possible to increase the frequency of examinations upon attainment of financial sustainability. At the examination session, assessments across all six modules are held in the 'one day—two modules' format.

The minimum duration period of the qualification examination procedure, from the candidate's registration to receiving a favorable decision, could amount to 1 year. The maximum period depends on existing limitations on the number of attempts and validity periods of the results of assessment levels passed by the candidate.

The time allowed for answering all questions of an individual test set from each base level module is no more than 2 h, time allowed for the written assignment is no more than 3 h, for the cross-cutting case assignment—no more than 4 h.

The results of each examination module are assessed independently based on the grading scale, which evaluates the candidate's performance on competencies that are specified in the Qualification Examination Programme and correspond to a certain assessment module. A criterion for the passing grade is determined, equal for all modules of the same level.

Conclusion

Now, the assessment of professional competencies for 'entry' into the profession is regulated in Russia by the Federal Law on Auditing Activities and represents a qualification examination for the auditor qualification certificate. All matters pertaining to the qualification examination fall under the jurisdictions of the Ministry of Finance of the Russian Federation and the EAC. The assessment of competencies is based on professional knowledge areas. Specific professional skills (knowledge, competencies) that are subject to assessment have not yet been articulated in Russia. The absence of such requirements presents a problem in view of transition to the new qualification examination model. Provisions on employment functions and requisite skills contained in the professional standard *'Auditor'* does not entirely resolve this problem.

The model of competence-based assessment of auditors—2020 introduced by EAK is based on the verification of requisite competencies in candidates in compliance with the International Education Standards and the European Council for Business Education (ECBE) Standards. The proposed system of modules is based on the following core premises: full scope of IES requirements; compliance with the most authoritative, internationally recognized practices of auditing assessment; integration of provisions contained in the professional standard *'Auditor'*; continuity of the existing qualification examination with regard to its general guidelines and the scope of knowledge areas subject to assessment, with further detalization and elaboration of each area through integration of modules and determining specific competencies to be assessed at the examination, with command of competencies increasing in proficiency from the base level through the foundation level to the qualification level. The new model is aimed at enhancing the prestige of the auditing profession.

Summary. Implementation of ISAS into audit practice is faster in Russia than the transition to the use of International standards of auditors' education. First introduced in 2017, the new format of qualification exams will be introduced in 2020, which is based on international standards of education of auditors. The Soviet period had no trend of development of audit activity in the direction of international standards.

References

Alon, A., & Dwyer, P. D. (2012). Globalization and multinational auditing: The case of Gazprom and PwC in Russia. *Behavioral Research in Accounting, 24*(1), 135–160.

Ashton, R. H. (2013). *Some early contributions to the study of audit judgment (RLE Accounting)*. Routledge.

Baker, C. R. (2014). A comparative analysis of the development of the auditing profession in the United Kingdom and France. *Accounting History, 19*(1–2), 97–114.

Bychkova, S. (1996). The development and status of auditing in Russia. *European Accounting Review, 5*(1), 77–90. https://doi.org/10.1080/09638189600000004

Civil Aviation Authority. (2018). *Annual report & accounts 2017/18*. Civil Aviation Authority. Retrieved from https://publicapps.caa.co.uk/docs/33/CAP1671_Annual%20Report_201718(WEB).pdf

EAC. (2016). *The programme of the qualification examination for the auditor qualification certificate, comes into force on 14 November 2016 (approved by the Order of the EAC of 11 March 2014, No. 21 with amendments and additions)*. Retrieved from http://eak-rus.ru/about_attestation/podgotovka_k_sdache_examena1/programma_ekzamenov_na_poluchenie_attestata_auditora

EAC. (2019). *Orders of the Autonomous Non-Commercial Organization Unified Certification Commission*. Retrieved from http://eak-rus.ru/normativnye_dokumenty/prikazy_ano_eak

Enthoven, A., Sokolov, Y., Bychkova, S., Kovalev, V., & Semenova, M. (1998). *Accounting, auditing and taxation in the Russian Federation*. The University of Texas Press.

Flesher, D. L., Previts, G. J., & Samson, W. D. (2005). Auditing in the United States: A historical perspective. *Abacus, 41*(1), 21–39. https://doi.org/10.1111/j.1467-6281.2005.00167.x

Guzov, I. (2016). History of auditing in Russia: periodization and challenges of development. In *Paradigma contabilității și auditului realități naționale, tendințe regionale și internaționale* (pp. 257–263).

Guzov, I. N., Strelnikova, O. V., Bykov, V. A., Soboleva, G. V., Sokolov, Y. V., & Terenteva, T. O. (2008). *Auditing*. St. Petersburg State University.

Lvova, D. A. (2005). *Professional Societies of Accountants: A history of the Founding and Activity*. The Institute of Professional Accountants of Russia: Informational Agency IPBR-BINFA.

Mennicken, A. (2008). Connecting worlds: The translation of international auditing standards into post-Soviet audit practice. *Accounting, Organizations and Society, 33*(4-5), 384–414. https://doi.org/10.1016/j.aos.2007.06.001

Mennicken, A. (2010). From inspection to auditing: Audit and markets as linked ecologies. *Accounting, Organizations and Society, 35*(3), 334–359. https://doi.org/10.1016/j.aos.2009.07.007

Mattessich, R. (2011). *Two-hundred years of accounting research: An international survey of personalities, ideas, and publications from about 1800 to 2000*. Routledge.

McGee, R., & Preobragenskaya, G. (2005). *Accounting and financial system reform in a transition economy: A case study of Russia*. Springer.

Samsonova, A. (2009). Local sites of globalisation: A look at the development of a legislative framework for auditing in Russia. *Critical Perspectives on Accounting, 20*(4), 528–552.

Samsonova-Taddei, A. (2013). Social relations and the differential local impact of global standards: The case of international standards on auditing. *Abacus, 49*(4), 506–538. https://doi.org/10.1111/abac.12017

Sokolov, Y. V., & Bychkova, S. M. (1999). About independent financial control in Russia at the turn of the century. *Auditing, 8*, 52–60.

Sokolov, Y. V., & Bychkova, S. M. (2005a). Audit in the 20th century Russia. *Audit Bulletine, 3*, 74–83.

Sokolov, Y. V., & Bychkova, S. M. (2005b). Audit in the 20th-century Russia. *Audit Bulletine, 4*, 15.

Sokolov, Y. V., & Terekhov, A. A. (2004). *Essays on the development of auditing*. FBK Press.

Sucher, P., & Bychkova, S. (2001). Auditor independence in economies in transition: A Study of Russia. *European Accounting Review, 10*(4), 817–841. https://doi.org/10.1080/09638180120069142

Sukhanov, S. S. (2015). Professional standard 'Auditor' has come into force. *Audit Bulletin, 12*, 34–35.

Analysis and Forecast of Labor Immigration to Japan in 2009–2019

Rinas V. Kashbrasiev, Artur A. Fomin, Yuliya S. Valeeva, and Ilgiz I. Nurtdinov

Introduction

Nowadays international migration is a widespread process intensified by the impact of globalization and regional integration. According to the UN, there were 272 million migrants worldwide in 2019 (excluding internal migration within borders of states), which amounted to about 3.5% of the world population, which is much higher than in 2000, when 2.8% of the world population was involved in international migration (UN, 2019).

The modern theories of migration suggest that the main flow of international migration is the movement of the population from less developed countries to more developed countries, which are attractive for migrants with the more favorable socio-economic situation (Abylkalikov and Vinnik, 2012). In practice, the developed countries, acknowledging the important role of talented and highly educated human resources in economic development, are actively attracting highly skilled labor.

In addition, as it is shown in the collective work on the economy of aging population (Piggott and Woodland, 2016), in recent decades in many developed countries there has been a phenomenon of aging population due to declining birth rates, which also contributes to international migration around the world.

R. V. Kashbrasiev (✉)
Kazan (Volga Region) Federal University, Kazan, Russian Federation

A. A. Fomin
National Research University "Higher School of Economics", Moscow, Russian Federation

Y. S. Valeeva
Kazan State Power Engineering University, Kazan, Russian Federation

I. I. Nurtdinov
Kazan State Medical University, Kazan, Russian Federation

A. Rumyantseva et al. (eds.), *Challenges and Solutions in the Digital Economy and Finance*, Springer Proceedings in Business and Economics,
https://doi.org/10.1007/978-3-031-14410-3_55

Japan belongs to the group of countries that experiencing this phenomenon. This country is known all over the world not only for the high level of its economic development and well-being of its population but also for its demographic crisis. This problem, which threatens the Japanese economy and society, is regularly reported in both academia and the media. The terms "aged society" (*Koreishakai*, 高齢社会in Japanese) and "declining birthrate and aging" (*Shoshikoreika*, 少子高齢化in Japanese) have already become common in the country in regard to its demographic situation. According to the World Bank, Japan's population declined by 1.5 million from 2010 to 2018 (World Bank, 2022). The situation in the country is characterized by an increase in the share of elderly citizens and a decrease in the number of children whereas the overall population is decreasing. The population of Japan was 126.5 million in 2017, according to the Japanese National Institute for Population and Social Security Research. People aged 65 years and older accounted for 27.8% of the country's population, children under aged 14 years and younger—12.3%, people of working age from 15 to 64 years old—59.9%.

According to the forecast of this organization, by the year 2065, the population of the country would be around 88.07 million, and the ratio of these age groups would be 38.4%, 10.1%, and 51.4% respectively (National Institute of Population ..., 2017). At the same time, Japan is one of little number of the developed countries where the share of immigrants in the population remains relatively low. As the 2019 UN Human Development Report stated, the average share of immigrants in the population of the developed countries was 11.6% in 2017, whereas in Japan that belongs to this category of world countries the share of immigrants in the population is only 1.8% (UNDP, 2019).

According to the statistics bureau of Japan, in June of 2019, there were 2,829,416 registered foreign citizens (excluding foreigners staying on short-term visas that do not allow working, which are mainly tourists, businessmen, and illegally staying foreigners). Those mid- and long-term foreign residents accounted for 2.24% of Japan's population in 2019 (Statistics of Japan, 2020).

The purpose of this paper is to study the position and peculiarities of Japan in the system of international labor migration, as well as to analyze and forecast the number and structure of migrants in terms of their participation in Japan's labor market.

To achieve this goal, the present paper attempts to accomplish a number of tasks. Firstly, the brief review of Japanese migration legislation and, particularly, the Immigration Control and Refugee Recognition Act of 1981 with subsequent amendments (Immigration Control and Refugee Recognition ..., 2020) to understand the existing legal basis for the acceptance of foreign nationals in Japan and find out the categories of entry visas that allows judging the occupation of migrants. Materials from the website of the Japan Migration Bureau provided the information on the latest measures for attracting foreign laborers (Acceptance of New Foreign Human Resources ..., 2020) and highly qualified specialists (Points-based System for Highly-Skilled ..., 2020). Secondly, the review of approaches to the study of demographic processes. Along with the use of traditional methods and modeling the demographic situation, geoinformation monitoring of migration processes, as well as the processes of immigrant adaptation and integration into host communities seems to be promising (Tikunov et al., 2015). Thirdly, the study of international and Japanese demographic statistics and mathematical processing of the data. Fourthly, for discussing the results obtained from the research and determining the place of Japan in international labor migration, the literature on this topic played a significant role. For example, G. Vogt describes the evolution of the state of affairs regarding foreign nationals working in Japan (Vogt, 2015); Fomin A.A. and Kashbrasiev R.V. identify the main categories of migrants arriving in Japan and characterize them (Fomin and Kashbrasiev, 2020); B. Brody analyzes the history of immigration to Japan, in particular, examines in detail the repatriation of the descendants of the Japanese who had migrated to South America in the early twentieth century (Brody, 2002).

Materials and Methods

This study uses sources such as the UN Human Development Report 2019, which presents the current situation on international migration; the 2017 report of the Japanese National Institute of Population and Social Security Research, which includes a composition of the population and the forecast of demographic indicators; data from the Japanese Ministry of Justice (Ministry of Justice, 2020) and the Statistics Bureau of Japan, which have up-to-date information on the number of migrants, their distribution by category of entry visas and countries of origin.

Based on the approaches of regression analysis, the dynamics of the total number of immigrants to Japan in 2000-2019 was studied (Table 1). Pairwise linear regression was chosen at the specification stage. Its parameters were estimated with the least squares' method. The statistical significance of the equation was tested using the determination coefficient and Fisher's test. It was found that in the studied case

Table 1 Foreign nationals in Japan (2000–2019)

Year	Migrants, thousand people	Year	Migrants, thousand people
2000	1594	2010	2087
2001	1680	2011	2047
2002	1746	2012	2034
2003	1805	2013	2066
2004	1864	2014	2122
2005	1907	2015	2232
2006	1990	2016	2383
2007	2069	2017	2562
2008	2145	2018	2731
2009	2126	2019	2829

Source: Compiled by the authors using data from the Japanese Ministry of Justice. http://www.moj.go.jp/housei/toukei/toukei_ichiran_touroku.html

84.5% of the total variability in Y was explained by the change in X. It was also found that the parameters of the model were statistically significant.

Labor migration trends to Japan (Table 2) are presented by the multiple regression equation, the main purpose of which is to build a model with a large number of factors as well as their cumulative impact on the modeled indicator while determining the influence of each of the factors separately.

Next, the cluster classification of five objects (*X1–X5*) indicated in Table 2 was carried out, each of them is characterized by two features, the average value and the β_i coefficient, these were obtained in the standardized form of the multiple regression equation (Table 3).

The simplex method was used to identify the relationships between different categories of immigrants arriving in Japan. Simplex optimization with visualization of the results in the form of an isoline map with support of plans of 2nd, 3rd, 3.5th, and 4th order were implemented with Delphi 5.0. Scheffe's formula was taken as an approximating polynomial, which has the following general form:

$$\widetilde{y} = b_0 + \sum_{1 \le i \le q} b_i x_i + \sum_{1 \le i \le j \le k \le q} b_{ijk} x_i x_j x_k$$

The number of indexes for the responses and the degree of the polynomial must be equal (Fig. 1). Responses and design based on simplex 3:3 are shown in Table 4.

Considering that simplex lattices are full, resubstituting was applied to determine the coefficients of the polynomial. This means that the coordinates of the points were put one after another in the polynomial in order to derive formulas for the calculation.

Results

The economic interpretation of the parameters of the pairwise regression model, compiled based on the data in Table 1, shows that over the studied period, the number of migrants increased by an average of 50,446 people every year (Fig. 2). The decline in the number of migrants from 2008 to 2012 could be explained by the global financial crisis of 2008-2009 and the 2011 earthquake that led to the accident at the Fukushima Daiichi Nuclear Power Plant.

The multiple regression equation based on data of Table 2 (the estimation of the regression equation) is as follows:

$$\mathrm{Y} = -1013.0089 + 2.5928\mathrm{X1} + 8.3622\mathrm{X2} - 0.8716\mathrm{X3} + 1.1221\mathrm{X4} + 5.3041\mathrm{X5}$$

The standardized form of this regression equation is written as:

Table 2 Classification of migrants arriving in Japan

Year	Total migrants, thousand people	Number of migrants by main categories, thousand people				
		Permanent residents (Highly skilled professionals (scientists, engineers, top managers, etc.—holders of visas for the categories "technology and humanities, international work", "researcher", "managers and managers", "highly skilled professionals").)	Long-term residents (Highly skilled professionals (scientists, engineers, top managers, etc.—holders of visas for the categories "technology and humanities, international work", "researcher", "managers and managers", "highly skilled professionals").)	Highly skilled professionals (Highly skilled professionals (scientists, engineers, top managers, etc.—holders of visas for the categories "technology and humanities, international work", "researcher", "managers and managers", "highly skilled professionals").)	Trainees	Nursing and elderly care specialists (In 2018, a program was adopted according to which migrants would be accepted to work in 14 areas, where the lack of workers is especially acute: construction, automotive and shipbuilding, agriculture, etc.)
	Y	X1	X2	X3	X4	X5
2009	2153	533	222	152	65	0
2010	2105	565	195	147	109	0
2011	2055	598	178	144	145	0
2012	2034	625	165	146	153	1
2013	2067	655	160	151	157	1
2014	2122	677	160	161	169	1
2015	2232	701	162	182	194	2
2016	2383	727	169	211	230	3
2017	2562	749	180	245	276	3
2018	2731	772	192	287	330	4
2019	2829	784	198	321	369	5
Среднее	2298	671	180	195	200	2

Compiled by the authors using data from the Japanese Ministry of Justice http://www.moj.go.jp/housei/toukei/toukei_ichiran_touroku.html

Table 3 Data for cluster analysis of labor force migration to Japan in 2009–2019

Index	X1	X2	X3	X4	X5
Mean	671	180	195	200	2
βi	0.763	0.578	0.191	0.364	0.0318

Source: authoring

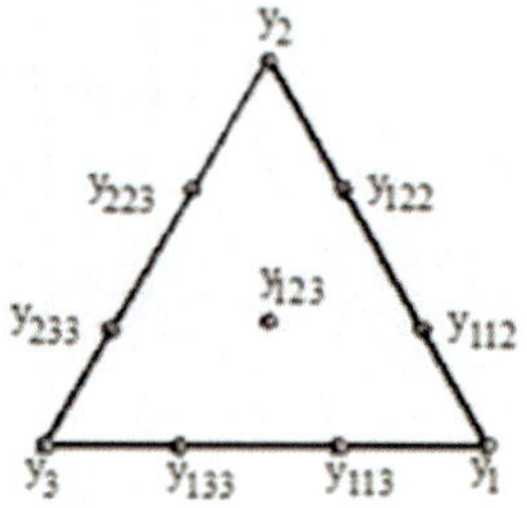

Fig. 1 Designation of responses at points of a simplex lattice

Table 4 Responses and design based on simplex 3: 3

	Lattice position	The values	№	Lattice position					
No	x_1	x_2	x_3		The values	x_1	x_2	x_3	
1	1	0	0	y_1	6	0	1/	2/	y_1
2	0	1	0	y_2	7	2/	3	3	y_2
3	0	0	1	y_3	8	3	1/	0	y_3
4	1/3	2/3	0	y_{122}	9	2/	3	1/	y_{122}
5	1/3	0	2/	y_{133}	10	3	0	3	y_{133}
			3			0	2/	1/	
						1/	3	3	
						3	1/	1/	
							3	3	

Source: authoring

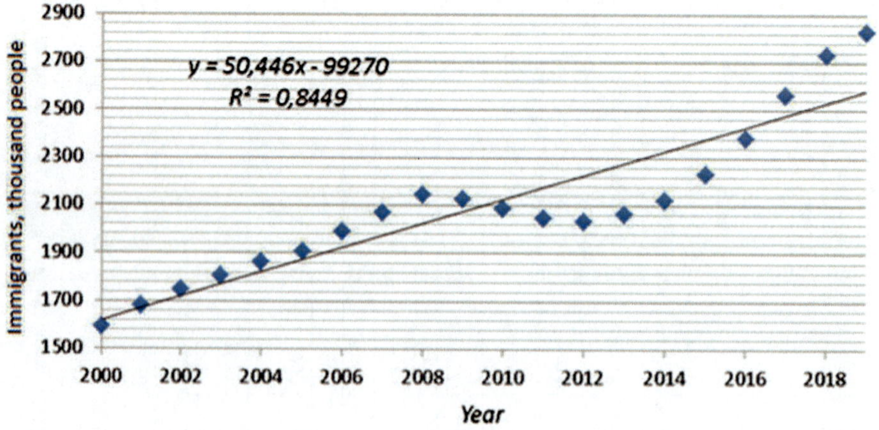

Fig. 2 Equation of pairwise regression for the number of immigrants in Japan dynamics (2000-2019)

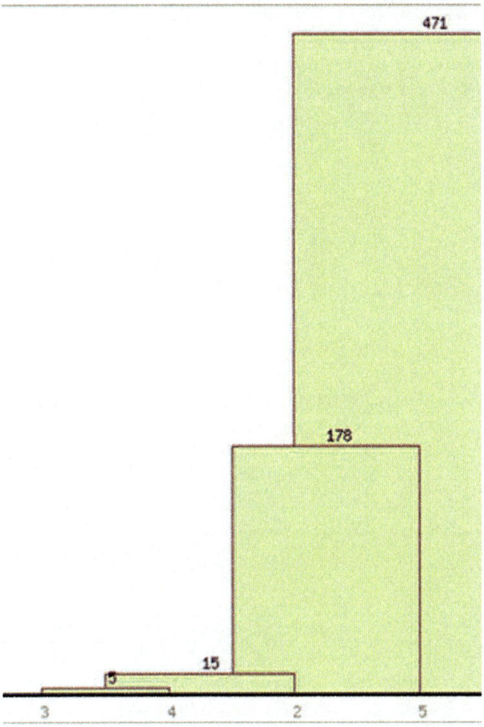

Fig. 3 Cluster classification of migrants results according to the principle of "nearest neighbor"

$$t_y = 0.763x_1 + 0.578x_2 0.191x_3 + 0.364x_4 + 0.0318x_5$$

With the maximum coefficient $\beta_1 = 0.763$ it might be concluded, that the factor x_1 has the greatest influence on the result Y. The statistical significance of the equation was tested using the determination coefficient and Fisher's test. It was found that in the studied case 99.8% of the total variability of Y is explained by changes in the factors X_j.

As a result of the cluster analysis based on the "nearest neighbor" principle, a cluster was formed: $S_{(2,3,4)}$ (Fig. 3), and the result based on the "far" neighbor principle (Fig. 4) was very close to the first result, which allows concluding that it is inexpedient to continue solving this problem with other clustering methods.

The closest in terms of dynamics categories of immigrants arrived in Japan in 2009-2019 were long-term residents (residence status: Long-term Residents 定住者—*teijusha*), highly skilled professionals (residence statuses: Professor (教授—*kyoju*), Artist (芸実—*geino*), Journalist (報道—*hodo*), Highly Skilled Professional (高度専門職–*kodosenmonshoku*), Medical Services (医療—*iryo*), Business Manager (経営・ 管理—*keiei-kanri*), Legal/Accounting Services (法律・ 会計業務—*horitsu-kaikei gyomu*), Researcher (研究–*kenkyu*), Instructor (教育—*kyoiku*), Engineer/Specialist in Humanities/International Services (技術・ 人文知識・ 国際業務 – *gijutsu-jinbunnchishoku-kokusaigyomu*)) and trainees (residence statuses:

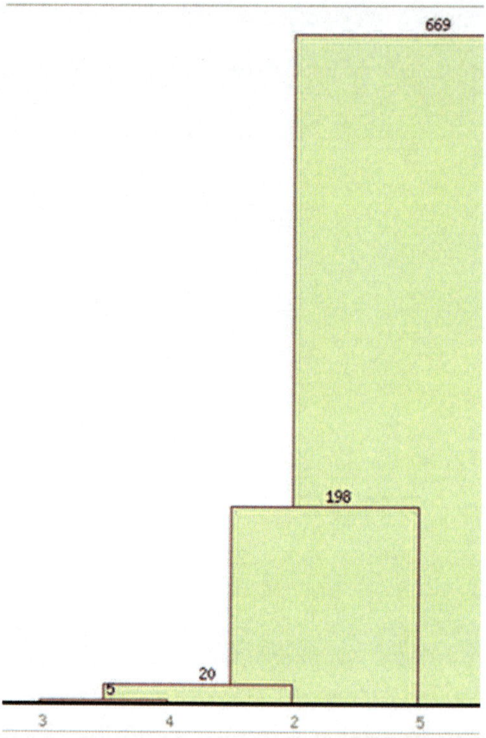

Fig. 4 Cluster classification of migrants results according to the principle of "distant neighbor"

Technical Intern Trainee (技能実習—*ginojisshu*) and Trainee (研修—*kenshu*)). Their distinctive ratio over the last decade could be determined on the basis of the simplex method, which is an algorithm for solving an optimization problem of linear programming through iterating over the vertices of a convex polyhedron in a multidimensional space (Fig. 5). In the computer program used to calculate the isoline map, the first variable (long-term residents) is assigned the value *x1*, the second (highly skilled professionals) *x2,* and the third (trainees) *x3* respectively.

As a result of using the simplex optimization method, it was shown that the most distinctive ratio of long-term residents (*x1*), highly skilled professionals (*x2*), and trainees (*x3*) to the total number of migrants to Japan (up to 56%) was found in the area of 45%:35%:20% respectively.

The forecast on the found most significant categories of migrants to Japan could be presented using a pairwise regression equation. However, in the case of long-term residents, it was found that only 3.1% of the total variability in *Y* is explained by a change in *X*, this makes it inappropriate to represent this forecast as a linear trend line. However, the polynomial form of the trend line shows a fairly high value of the approximation reliability, which amounted to 95.2% (Fig. 6).

Studying the results presented in this figure, it is necessary to consider that the significant decline in the number of immigrants in the long-term residents category, which began in 2009 and lasted until 2015, could be due to the consequences of the

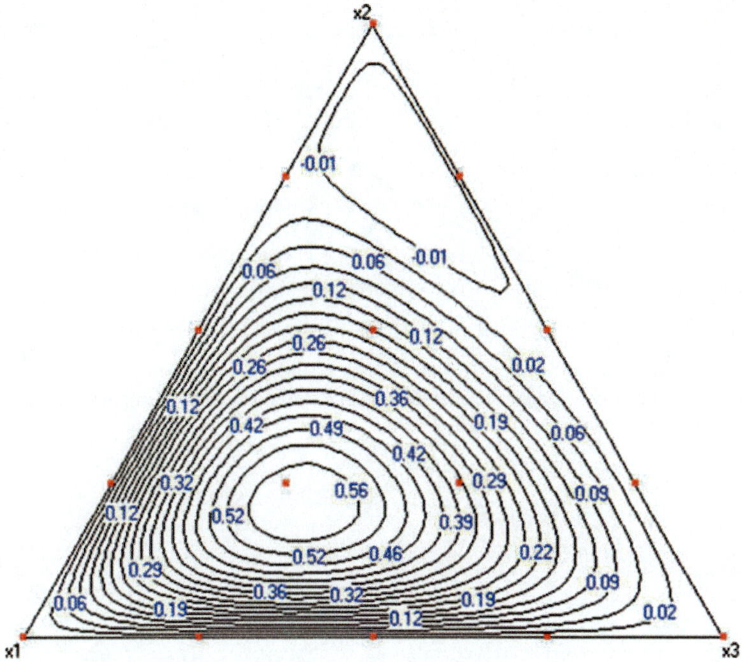

Fig. 5 Isoline map of the ratio of the number of long-term residents (x1), highly skilled professionals (x2), and trainees (x3) in the total number of migrants to Japan, %

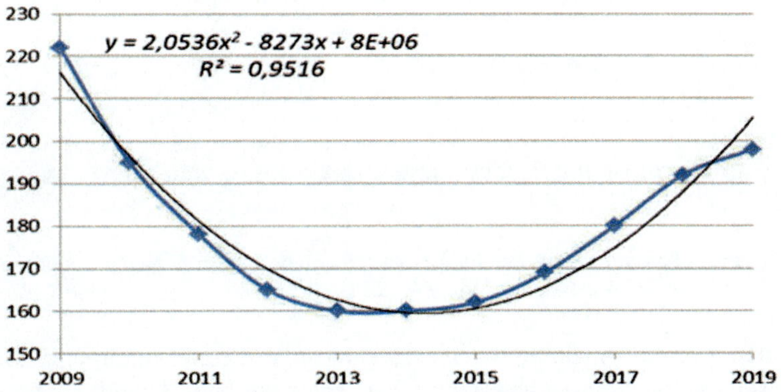

Fig. 6 Equation of pairwise regression for the dynamics of the number of migrants in Japan for the category "Long-term residents" in 2009–2019. Polynomial trend line included

global financial crisis of 2008–2009, which is probably could not influence the trend line after 2015. In this case, it seemed more expedient to construct a linear trend line for the dynamics of the number of immigrants in the long-term residents category for the period 2015–2019, the value of the approximation reliability here was 98.9% (Fig. 7).

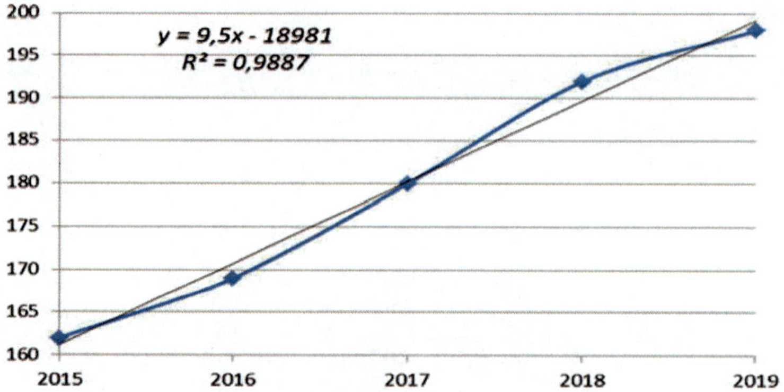

Fig. 7 Equation of pairwise regression for the dynamics of the number of migrants in Japan for the category "Long-term residents" in 2015–2019

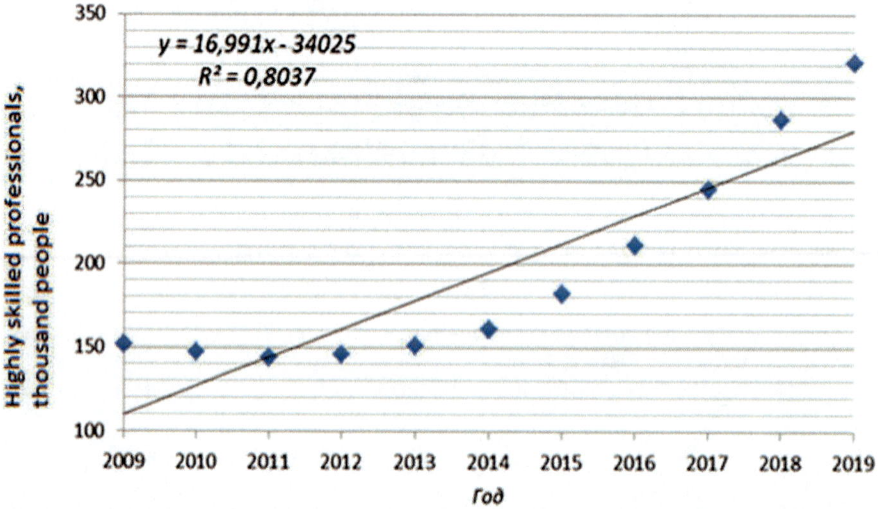

Fig. 8 Equation of pairwise regression for the dynamics of the number of migrants in Japan for the category "Highly skilled professionals" in 2009–2019

The trends of the remaining two most significant categories of migrants to Japan (highly skilled professionals and trainees) are shown in Figs. 8 and 9.

According to the calculations, the greatest dynamics of the increase in the number of migrants to Japan should be noted in the category of trainees. This could be explained by one of the highest in the world level of scientific and technological development of this country and also the high level of international cooperation with many countries (especially Asian ones), so internships in Japan must be demanded by from all over the world.

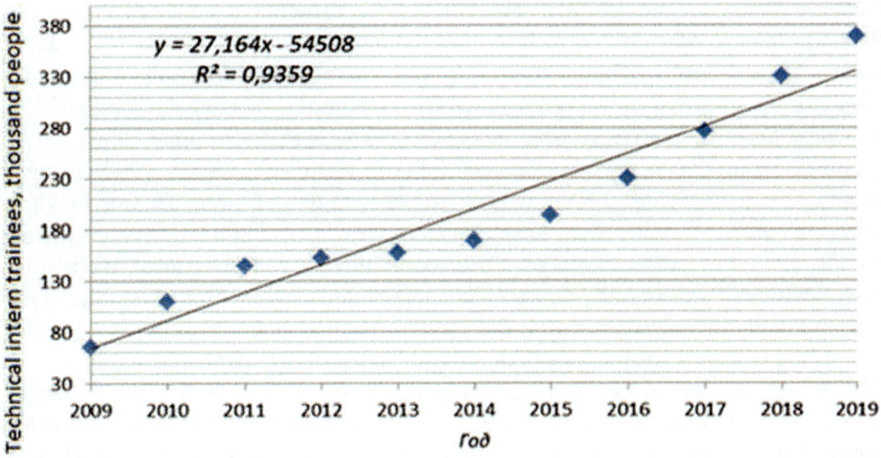

Fig. 9 Equation of pairwise regression for the dynamics of the number of migrants in Japan for the category "Trainees" in 2009–2019

Discussion

Table 1 shows that the number of migrants in Japan is gradually increasing and their share in the total population is growing noticeably. In 2000, there were 1.594 million foreigners on mid- and long-term visas in Japan, their share in the population was 1.26%, in 2010, there were 2.097 million of such people with a share of 1.63%. In June 2019, the number reached 2.829 million people (2.24% of the population). As can be seen from Fig. 2, until 2008 the number of foreigners was growing (annual growth for mid-term and long-term visas holders ranged from 2.3% to 8%). However, in 2008–2012, their number decreased, this might be explained by the global financial crisis and the 2011 earthquake, which led to the disaster at the Fukushima Daiichi nuclear power plant. Since 2013, growth has been continuing again with a rate from 1.3% to 7.5% per year.

Analysis of the legislative framework of Japan's migration policy allows having an insight into the types of activity, in which foreigners are engaged. Their distribution by visa categories makes it possible to highlight the main groups of migrants. Data on migrants provided by Japanese governmental organizations were summarized in Table 2. It reflects the problem of labor shortage in many sectors of the economy associated with the current demographic situation in Japan that is being solved by changing the directions of the state migration policy. Even during the studied period, the dynamics of various migrant groups changed significantly, and a new category of immigrants emerged such as the "Nursing Care" (介護–*kaigo*).

Nevertheless, as the review of the relevant literature indicates, this problem arose in the 80–90s of the twentieth century. The obvious need for Japan to attract new workers led to the fact that in 1990 the 1981 Immigration Control and Refugee Recognition Law was amended. Based on changes in this law, it became possible to come to Japan and immediately obtain a residence permit for the descendants of Japanese people who immigrated to other countries (*Nikkeijin*, 日系人 in Japanese—“a person of Japanese origin”) up to the third generation. This opportunity was mainly used by the citizens of Brazil and Peru, who became attractive to Japanese firms as relatively cheap labor for industries that were unattractive to the “native” Japanese (for example, manual labor in factories). Also in 1990, changes were made to the technical training program: foreign trainees were allowed to work, it was allowed to employ them to small enterprises. And in 2009, new amendments to the 1981 law made it possible to accept interns for permanent work after completing a 2-year training course. Thus, the technical internship program, which was originally intended to promote the dissemination of technical skills, has become one of the channels for the supply of labor to Japan (Vogt, 2015, p. 572).

The aggravating of the demographic situation in Japan caused further changes and the development of legislation. In 2008–2009, within the framework of special agreements on economic partnership, specialists in nursing from the countries of Southeast Asia were admitted for migration to Japan (Vogt, 2010). The reason for this was that Japan has become short of specialists in the care of elderly citizens, whose numbers are constantly growing at an accelerating rate. In 2018 was adopted a new migration facilitation program, according to which migrants will be accepted to work in 14 areas, which suffer from the shortage of workers notably (the Japanese government has indicated the labor force deficit from 2000 to 60, depending on the industry): construction, automotive, shipbuilding, agriculture, fishing, etc. It should be noted that most of the jobs offered are designated in Japan by the term “work of the three K” (*Kitsui, Kitanai, Kiken*–きつい・汚い・危険 in Japanese)—demanding, dirty, dangerous–such employment is often avoided by the local population. In addition, it is worth noting that part of the potential immigrants is to be allowed to bring their families with them, which is usually the privilege of immigrants that are considered to be high-skilled specialists.

The digitalization of the economy and the development of its innovative industries, accompanied by an increase in the need for highly qualified personnel, have become the reason for further changes in immigration legislation. In 2012, a system was introduced for immigration to Japan for highly skilled professionals (scientists, engineers, top managers, etc.), based on the points awarded to a candidate for his merits. The procedure for obtaining a permanent residency for highly skilled professionals has been simplified as well (Points-based System for Highly Skilled Foreign Professionals). This might be serving as a sign that Japan entered the competition for talented, high-quality human resources from all over the world.

Regarding the validity of identifying the main groups of migrants with relatively large shares of their total number, presented in Table 2 and Figs. 6, 7, and 8, it is important to return once again to the Japanese 1981 Immigration Control and Refugee Recognition Law with amendments up to date. This law defines 27 categories of entry visas, 18 of them allow work only in certain areas (from diplomatic to entertainment), 4 do not allow work at all (visas for students, interns (those who has 研修—*kenshuu* status), short-term visitors (for instance, tourists), and family members of persons with work permitted visa), 4 of them entitle the right to work without any restrictions (permanent resident, spouse or child of Japanese citizens, spouse or child of permanent residents, long-term resident). Based on these categories of visas, several main groups of migrants were identified with relatively large shares of the total number of registered foreigners with medium and long-term visas for 2019: permanent residents (永住者—*eijuusha*), foreigners with a residence permit, they have the right to stay in Japan indefinitely and work without restrictions, this group accounted for 27.69%; long-term residents (*x1* in calculations)—special status for foreigners who are descendants of the Japanese up to the third generation, this group occupied 6.98%; highly skilled professionals (*x2*) 11.33%; and trainees (*x3*)—13.04%; nursing and elderly care specialists—0.16%. According to H. Watanabe, some of the interns are migrants who come with a specific purpose to stay in Japan and are de facto labor force, this fact sometimes leads to a violation of migration legislation and migrants' rights (Watanabe, 2010). Trainees could stay in Japan after the end of the internship by exchanging their visa for a work visa that is subject to a guarantee of employment. As for specialists in nursing and the elderly care, they are mainly coming from the countries of Southeast Asia, now they make up an imperceptible share, similarly to recently designated relatively lower-skilled workers (Specified Skilled Worker (特定技能–*tokuteigino*) (0.0007%), however, considering the decrease in the population of Japan and aging of the population, it is reasoned to expect an increase in numbers of these groups of migrants in the future.

It is also worth mentioning that in addition to the groups indicated in Table 2, a group of special permanent residents (特別永住者—*tokubetsueijuusha*) could be also highlighted among foreigners in Japan—these are people of mainly Korean and Chinese origin who came to Japan from territories under Japan's control before the end of World War II and their descendants, 317,000 people or 11.2%. In fact, most of them were born in Japan. Another relatively large group of migrants are family members of Japanese citizens, permanent residents, and holders of visas allowing work, there are 272,000 of them or 13.21%.

All groups mentioned in the previous two paragraphs make up 83.67% of overall mid- and long-term visas holder that could be defined as immigrants.

The geographical distribution of Japan's foreign residents, who came through obtaining mid- and long-term visas, origin regions is uneven: 83.8% are citizens of Asian countries, 9.5% of South America, 2.9% of Europe (Japanese statistics also includes the countries of Central Asia in this group), 2.7% of North and Central America, 0.6% Africa, 0.5% Australia and Oceania (Fig. 10).

The most numerous foreign residents in Japan are citizens of mainland China 27.8%, followed by citizens of South Korea 16%. Other large groups include citizens of Vietnam 13.3%, Philippines 9.8%, Brazil 7.3%, Nepal 3.2%, Taiwan 2.2%, Indonesia 2.2%, US 2% and Peru 1.7%.

Among the citizens of South Korea, Taiwan, and the PRC, a significant proportion are former residents of territories once controlled by Japan, who came to Japan because of various reasons (job-seeking, etc.) before the end of World War II, and their descendants. Among the migrants from the countries of Southeast Asia, the largest group is represented by technical intern trainees. Most of the migrants from South America are people of Japanese origin, descendants of the Japanese who

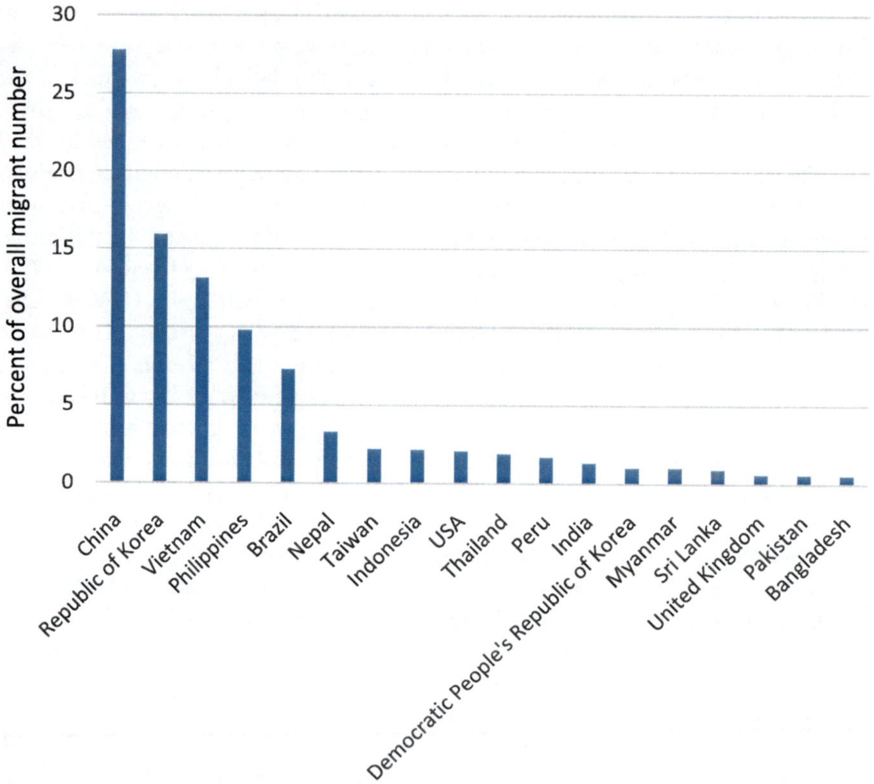

Fig. 10 Countries of origin of migrants for 2019. Compiled by the authors based on data from the Japanese Ministry of Justice (http://www.moj.go.jp/housei/toukei/toukei_ichiran_touroku.html)

migrated to these countries at the beginning of the twentieth century. It could be stated that the overwhelming majority of migrants (93.3%) are from Asia (mainly) and South America. Taking into account the fact that Central Asia is recorded in the countries of Europe, the percentage should be higher. Hence, as could be seen from Fig. 10, most numerous immigrants to Japan are citizens of the Far Eastern countries neighboring Japan and citizens of Southeast Asia.

Discussing the results of the analysis and forecast of the total number and main groups of immigrants presented in Tables 1 and 2 and Figs. 6, 7, 8, and 9, it should be emphasized that from 2009 to 2019 the number of foreigners with mid- and long-term visas increased by 31.43% (or by 676,000 people), then the number of qualified specialists increased by 111.48% (more than doubled). It could be assumed that this growth was due to the introduction of the 2012 recruitment program for highly skilled professionals and the deeper involvement of Japan in the processes of the international division of labor and economic globalization. The number of technical intern trainees has grown 5.5-fold, which illustrates Japan's increasing need for foreign workers. At the same time, the number of nursing and elderly care professionals has almost quadrupled since 2012, which also reflects Japan's growing need for foreign specialists in this area. Factors such as the emergence of new diseases (for example, the coronavirus pandemic in 2020), a further increase in the number of elderly citizens could create an ever-growing demand for foreign medical personnel in the future. The number of permanent residents has increased by 46%, which is a consequence of the fact that more and more foreigners settle in Japan for long periods, as well as because it has become easier for many high-class foreign specialists to obtain a permanent residence.

The presence of stable trends in labor migration and the high reliability of regression models allow making a forecast of both the total number and the size of the main groups of immigrants. Forecast data is obtained by substituting the forecast year (for example, 2021) into the formula of one or another trend indicated in Figs. 2, 6, 7, 8 and 9. Hence, the following forecast values could be determined (without considering the effects of the COVID-19 pandemic): 2,681,000 (total number of migrants), 390,000 (technical intern trainees), etc. From the analysis of statistical data and regression models, it could be concluded that the conditions prevailing in the demography and economy of Japan by the present time force it to be more and more open to migration. At the same time, there is a demand for both highly qualified specialists and relatively low-skilled workers.

Conclusion

Japan is still a country where the share of immigrants in the population remains insignificant, and this makes it stand out from most of the economically developed countries of the world, for example, from the countries of Western Europe or the United States. However, there are several reasons for changes in this situation. The main ones are the following: aging and shrinking of the population, globalization, increased competition for the best highly qualified specialists, the low popularity of some occupations, and a general decline in the working population. Therefore, there is no wonder why in 2009 there were 2.1 million migrants in Japan, then 10 years later their number reached 2.8 million.

This trend did not appear in recent years, however, Japan's migration policy began to gradually change back in the 80–90s of the twentieth century. At present, the increasing relevance of the above-mentioned reasons is driving Japan to more intensive expand in the reception of foreigners. The opportunity for a long-term stay in this country opens up to immigrants with different origins and backgrounds.

The largest groups of migrants are:

- highly skilled professionals (researchers, top managers, employees of transnational corporations).
- technical intern trainees who came to master the skills of blue-collar occupations, which could be applied in Japan and their home countries.
- people of Japanese origin and their descendants.

- persons who found themselves in Japan in the first half of the twentieth century, and their descendants.

By now, relatively low-skilled labor immigrants, mainly from Asian countries, prevail, and it is expected that thanks to the new migration programs, their number would grow. In addition, it is quite peculiar for the structure of migrant groups in Japan that a significant proportion of them are represented by persons who ended up in Japan before the end of World War II and their descendants. Back then, their home regions were under Japan's control. The descendants of the Japanese who migrated from Japan to South America at the beginning of the twentieth century could also be distinguished as a peculiar separate category of immigrants.

The aging population and the current demographic situation in Japan have become the reasons for the increasing demand for medical personnel and led to the emergence of a system of hiring foreign specialists in nursing and care for the elderly.

As a result of the regression analysis, it was shown that both the total number of migrants to Japan and the number of migrants in the main categories would increase with a high value of the approximation reliability (without considering the 2020 COVID-19 pandemic and consequent borders closing). According to the standardized form of the multiple regression equation, it could be stated that the category "Permanent residents" has the greatest influence on the growth of the total number of migrants. The results of the cluster analysis showed that the categories "Long-term residents", "Highly skilled professionals" and "Technical intern trainees" are the closest in terms of their dynamics. The most typical ratio of their growth in Japan over the past decade, based on the use of the simplex optimization method, was about 45%, 35%, 20% respectively.

References

Abylkalikov, S. I., & Vinnik, M. V. (2012). Economic theories of migration: Labor force and labor market. *Business Society Power, 12*, 1–19.

Brody, B. (2002). *Opening the door: immigration. ethnicity, and globalization in New York*. Routledge.

Fomin, A. A., & Kashbrasiev, R. V. (2020). Japan's involvement in international labor migration. *Rossiya v global'noi ekonomike: novye vyzovy i ugrozy, 8*, 38–48.

Immigration Control and Refugee Recognition Act. (2020). Retrieved from http://www.japaneselawtranslation.go.jp/law/detail/?id=3367&vm=&re=02&new=1

Ministry of Justice. Acceptance of New Foreign Human Resources (2020) (新たな外国人材受入れ(在留資格「特定技能」の創設等)). Retrieved from http://www.moj.go.jp/content/001291692.pdf

Ministry of Justice. Points-based System for Highly Skilled Foreign Professionals. (2020). Retrieved from http://www.moj.go.jp/ENGLISH/m_hisho06_00043.html

Ministry of Justice. (2020). 総務省 在留外国人統計. Retrieved from http://www.moj.go.jp/housei/toukei/housei05_00030.html

National Institute of Population and Social Security Research. Population Projections for Japan (2017). Retrieved from http://www.ipss.go.jp/pp-zenkoku/e/zenkoku_e2017/pp29_summary.pdf

Piggott, J., & Woodland, A. (Eds.). (2016). *Handbook of the economics of population aging*. Elsevier.

Statistics of Japan. (2020). 政府統計の総合窓口(e-Stat). Retrieved from https://www.e-stat.go.jp/stat-search/files?page=1&layout=datalist&toukei=00250012&tstat=000001018034&cycle=1&year=20190&month=12040606&tclass1=000001060399

Tikunov, V. S., Belozerov, V. S., Shchitova, N. A., Panin, A. N., & Cherkasov, A. A. (2015). Geoinformation monitoring: an instrument of spatial-temporal analysis of population migration. Herald of Moscow University. Series 5. *Geography, 2*, 33–39.

UN. (2019). *The number of international migrants reaches 272 million*. Retrieved from https://www.un.org/development/desa/en/news/population/international-migrant-stock-2019.html

UNDP. (2019). *Human Development Report 2019*. Retrieved from http://hdr.undp.org/sites/default/files/hdr2019.pdf

Vogt, G. (2010). *Care-giver migration to greying Japan. In Demographic aspects of migration* (pp. 327–348). VS Verlag für Sozialwissenschaften.

Vogt, G. (2015). Foreign workers in Japan. *The sage handbook of modern Japanese studies* (pp. 567–582).

Watanabe, H. (2010). Concerning revisions in the foreign trainee and technical intern system. *Japan Labor Review, 7*(3), 43–67.

World Bank. (2022). *GDP per capita (current US$) - Japan*. Retrieved from https://data.worldbank.org/indicator/NY.GDP.PCAP.CD?locations=JP

Assessment of Indirect Economic Benefit from the Introduction of Electricity Storage Systems

Valentina Vladimirovna Bologova, Daria Georgievna Shuvalova, Victor Aleksandrovich Menshov, Irina Gareevna Akhmetova, Oleg Pavlovich Ushchekin, and Yuliya S. Valeeva

Introduction

In recent years, the issue of energy saving has become particularly relevant, due to the change in the structure of electricity consumption, the aggravation of environmental and urban planning problems, and the desire to increase the profitability of energy enterprises. A special place in the issues of energy saving is given to the optimization of the modes of electric networks the problem of passing load peaks, complicated by the fact that electricity consumption (e/e) is a random process, which is not yet possible to accurately predict. The existing methods of accounting for the economic consequences of changes in electrical loads in modern conditions tend to become obsolete, which is associated with changes in energy markets, the complexity of the ownership structure and the variety of effects.

The problem of uneven consumption of e/e is both purely technical and economic in nature. The first is expressed in an increase in e/e losses due to uneven electricity consumption. In addition, uneven consumption of e/e complicates the operation of power plant equipment: the characteristic of the efficiency of boilers of power plants has a non-linear form, and currently there are no methods fixed in the regulatory and technical documentation (NTD) that allow to correctly consider this non-linearity (Pospelov et al., 2008). The problem of reducing the efficiency of thermal power plants, which is in the field of view of not only domestic but also foreign scientists

V. V. Bologova (✉) · D. G. Shuvalova · V. A. Menshov
National Research University "Moscow Power Engineering Institute", Russian Federation, Moscow
e-mail: BologovaVV@mpei.ru; shuvalovadg@mpei.ru; MenshovVA@mpei.ru

I. G. Akhmetova · O. P. Ushchekin · Y. S. Valeeva
Kazan State Power Engineering University, Russian Federation, Kazan

A. Rumyantseva et al. (eds.), *Challenges and Solutions in the Digital Economy and Finance*, Springer Proceedings in Business and Economics,
https://doi.org/10.1007/978-3-031-14410-3_56

(Neshumayev et al., 2018; Hübel et al., 2017), is not only local or purely temporary, but also in some cases very large-scale (Jindal & Nilakantan, 2021). At the same time, it must be remembered that the increase in fuel consumption at the power plant entails not only economic, but also environmental damage (Zeng et al., 2021). From an economic point of view, an increase in unevenness leads to an increase in capital investments in new energy facilities, an increase in the fuel component of the cost price due to an increase in specific fuel costs, an increase in operating costs at existing energy facilities. In addition, the tariffs for e/e include the costs of compensating for the proportional unevenness of the load of technical losses of e/e, which for some distribution networks can reach 70% of the total losses of e/e (Valeev et al., 2016). For manufacturers, however, the unevenness leads to a deterioration in economic indicators, which is a consequence of two factors: an increase in installed capacity (to cover peak loads) and a decrease in CIUM (because of a decrease in the coefficient of intensive use of equipment). If we present the result in figures, then for KES, a decrease in the coefficient of intensive use of equipment by 40% during a period equal to 8 h leads to an increase in the cost of electricity production by 5.6%, and for nuclear power plants this value will be even higher and will amount to 14.6% (Skret & Garievsky, 2017).

As you can see, solving the problem of passing the peaks of load schedules is very relevant both from the point of view of energy saving, and from the point of view of reliability of energy supply, and from the point of view of increasing the efficiency of energy facilities.

The paper reviewed and analyzed existing approaches to improving the efficiency of energy facilities, gave a comprehensive assessment of the result of the use of energy storage devices and carried out an average assessment of the economic result of their use. It should be noted that, considering the specifics of the economic activity of energy facilities and considering the regulation of this activity by the state, the assessment should be carried out at the level of generation, transmission, and distribution of electricity, for consumers, as well as collectively at the level of the entire system. It is important to note that the economic instruments for assessing the effect for different owners will be different.

Let's consider separately each of the above levels of the power system.

For generate:

The literature suggests various ways to cover peak loads. In magazines you can find articles devoted to the use of combined-cycle gas installations for these purposes (Aminov et al., 2013; Skret & Garievsky, 2017). Nevertheless, we must not forget, firstly, that such a method is a "palliative", not eliminating the problem itself, but only temporarily reducing its severity. For example, in the work of I. Havachi, H. Sammouda, R. The authors conducted a study of the problem of melting a phase transition material (PTM) containing paraffin in order to increase its thermal inertia (Hawachi et al., 2014). Secondly, the construction of thermal power plants based on CCGT costs hundreds of millions of rubles (Aminov et al., 2013). Thirdly, the effect of the use of CCGT is limited by the specifics of the development of power facilities in Central Russia—here CCGT is being implemented on existing reconstructed CHP plants, and the installations are forced to work according to a thermal schedule

(Skret & Garievsky, 2017). There are various technical solutions to improve the efficiency of the power plant at various stages of the technological process. Thus, the efficiency of the station can be increased by 2–3% by the introduction of pre-drying of coal (Weigl et al., 1999). For stations running on natural gas, it is possible to use solid oxide fuel cells together by compression ignition of a homogeneous mixture, which will achieve an electrical efficiency equal to 59% with a payback period of this hybrid system of only 1.5 years (Zhu et al., 2020). Cogeneration based on the Rankine cycle with an organic coolant and an absorption heat pump for coal-fired power plants can increase thermal efficiency by 9.38% (Zhang et al., 2020). Combining a photovoltaic plant with a thermal power plant is also called as a possible way to increase the efficiency of a thermal power plant (Gambini & Vellini, 2019). Along with such more classical ways of increasing the efficiency of power plants as the introduction of modernized regenerative cycles at power plants based on gas turbines (Yang, 1997), there is also a less common method of improving the efficiency of electricity production by using a variety of biomass as fuel (Cadavez & de Souza-Santos, 2021). Also, in the work of Burhan, M., Ernest, CKJ, Chun, N.K. "Electrical characteristics of concentrated photovoltaic (CPV) systems: long-term performance analysis and comparison with conventional photovoltaic systems", the transition of traditional silicon-based photovoltaic panels to concentrated photovoltaic systems (CPV) using transient solar cells and a two-axis solar tracker is considered (Burhan et al., 2016). Nevertheless, all these methods require significant reconstructions of existing power plants or even the construction of new ones.

For networks. The analysis of the graphs of active and reactive power of 32 distribution substations of 6–10 kV and 19 transformer substations of 0.38 kV located in the Moscow region showed that in 80% of cases the real tgf differs from the expected according to generally accepted data (Karchin & Myasnikova, 2017). The current degree of technical equipment of the networks does not allow to estimate the consumption of e/e in real time, completely bypassing the stage of its forecasting without the use of specialized techniques (Karchin & Myasnikova, 2017). It is becoming more and more difficult to align the load schedules of consumers using the previous methods of optimizing modes. On the other hand, it is impossible to ignore the problem of uneven loads, because incorrectly calculated loads at certain times of the year can cause malfunctions of the power system, which can lead to damage to the power system itself and to damage to consumers. An example of this is the relatively recent (August 2017) decrease in the capacity of the electric grid of the UES of the South due to a sudden and uncompensated increase in electricity consumption (Makoklyuev, 2017).

The issue of unevenness of the load schedule can be solved with the help of the introduction of industrial e/e storage devices (Bulut & Özcan, 2021). Using the example of a study of the electricity consumption of a residential building on Annikova St. in Yoshkar-Ola for 02.01.2016, it was found that stimulating the consumption of e/e at night, leading to a decrease in the unevenness of power consumption (the coefficient of unevenness of k_{ne} changed from the initial 0.44 to 0.67), entails a decrease in e/e losses by 17% (Karchin & Myasnikova, 2017), which is very significant. For the case of an even greater reduction in unevenness (with the same total daily consumption of e/e), the calculation of cost savings for the purchase of e/e was carried out, amounting to 15% (Karchin & Myasnikova, 2017). In the context of the financial benefits of installing electric power storage devices, a study by foreign scientists is noteworthy, the authors of which claim that installing a storage device with a power of 4.9 kW in one of the buildings of the Chinese Academy of Sciences with a target reduction of peak load of 16% will lead to savings in equipment modernization and maintenance costs in the amount of $108,750 (Yan et al., 2014).

The integration of storage devices into the power system has a positive effect on the operational properties of the system, allowing to achieve the following results (Gusev & Subbotin, 2019):

- maintain a stable voltage in the nodes.
- reserve power supplies.
- load generating and distribution equipment evenly.

These advantages have been repeatedly highlighted in the literature (Kononenko et al., 2014; Lushnikov, 2018) however, without any attempts to generalize estimates of their impact. It would be reasonable to describe each of the advantages quantitatively, bringing them all to a single measure – to indicators of economic efficiency. For power companies, the installation of industrial batteries reduces depreciation and operating costs for generating equipment, reduces potential compensation for the sale of substandard electrical equipment (or its under-supply in general), reduces capital investment in the construction and modernization of switchgear. Among all the costs for energy companies, the most undesirable are the costs associated with peak load issues. This can include both a decrease in the profitability of e/e production due to the operation of units in modes that are not optimal for them, and investments in the construction of new power plants specially designed for passing peaks. In foreign literature, the annual savings from the introduction of one additional kilowatt of battery power (battery) is estimated at $ 650 (Yan et al., 2014). It is obvious that for Russia such a parameter should be recalculated, especially in view of the high cost of storage devices, which does not allow to approach the issue of feasibility study without due consideration of all influencing aspects. This issue is also complicated by the fact that, as calculations and experiments show, lithium-ion batteries have a complex degradation characteristic of their parameters, depending on the number of operating cycles of perfect batteries (Atalay et al., 2020; Sarkar et al., 2019; Crawford et al., 2018; Soltani et al., 2020) the current State of charge of the battery State of charge at which the stored battery was stored for a long time electric power elements between charge-discharge cycles M. (Swierczynski et al., 2015; Kim et al., 2018), discharge depths in operating cycles (Han et al., 2014; Zhou et al., 2011). Considering all of the above, it can be concluded that the assessment of the economic efficiency of the use of energy storage due to the risks of possible unplanned early failure of batteries should be of a nature in which all aspects of the activity are accurately taken into account, as well as indicators affecting its effectiveness, for example, specific capital investments, useful life, discount rate, etc. (Fig. 1).

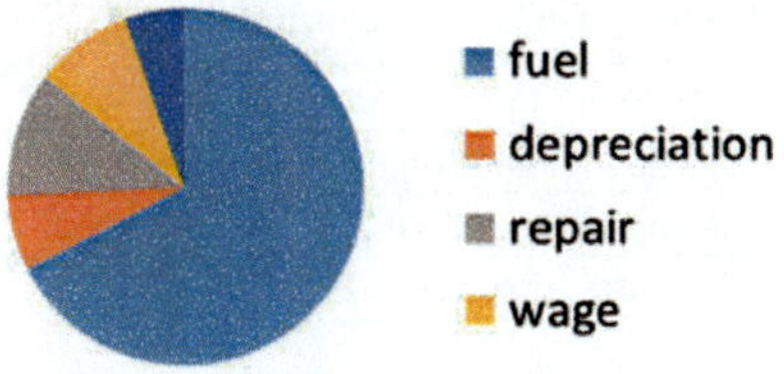

Fig. 1 Structure of the cost of electricity and heat generated by CHPPs/boilers

The savings from the introduction of the drive will be represented as the sum of three values:

1. savings on the construction of additional generating capacities (let's call it an indicator of infrastructure savings).
2. savings from an increase in the efficiency of the power plant and, consequently, a decrease in the specific fuel consumption for generation (an indicator of technological savings).
3. growth of network transmission volumes due to reduction of losses, as well as savings from the operation of peak pow-er plants (an indicator of operational savings).

The remaining part of the article contains sections, each of which is devoted to one of these values, and the conclusion.

Results and Discussions

Indicator of Infrastructure Savings

To find an indicator of infrastructure savings, it is necessary to establish the proportions of thermal and other power plants in the overall structure of capacities involved in peak coverage. For this purpose, we will use the Territorial Planning Scheme of the Russian Federation in the field of energy, from which it follows that by 2030 it is planned to build 26,220.5 MW of "maneuverable" capacity in the European part of the country: 1926.5 MW of hydroelectric power plants (7.35%) and 24,294 MW of thermal power plants (92.65%) (planning of the Russian Federation, 2016).

According to available data, in Russia, 83 GTU/CCGT with a single capacity of 30–125 MW were built in 4 combined power systems (Center, Northwest, Middle Volga, South) in 2010–2014 (Stepanova & Sushko, 2015). These 83 GTU/CCGT are divided by the authors into 3 ranges of capacities: 30–59, 60–99 and 100–125 MW, each of which in a particular unified energy system (UES) has its own average specific investments in construction. Let's find the average specific investments of GTU/CCGT of each of the capacity ranges using the formula below:

Table 1 The cost of construction of CCGT/GTU in the European part of the country

Unit power range of installations (MW)	Average unit investment, $/kW
The first range (30–59)	1438.6
Second range (60–99)	1441.0
The third range (100–125)	1206.3

$$\overline{\mathrm{K}}_{\mathrm{GTU}} = \frac{\sum_{i=1}^{m} \mathrm{K}_i^{\mathrm{GTU}} \cdot n_i^{\mathrm{GTU}\%}}{\sum_{i=1}^{m} n_i^{\mathrm{GTU}\%}} \quad (1)$$

where $\mathrm{K}_i^{\mathrm{GTU}}$—average specific investments for GTU/CCGT of a specific range of capacities in a specific ECO, $/kW/

$n_i^{\mathrm{GTU}\%}$—the specific weight of this CCGT/GTU.

The results of the calculated average unit investments can be seen in Table 1.

In total, 30 GTU/CCGT with a capacity of 30–59 MW (52.6% of the total number with a capacity of 30–125 MW), 25 with a capacity of 60–99 MW (43.9%) and 2 with a capacity of 100–125 MW (3.5%) were built in 4 OES. However, to calculate the value generalizing the average specific investments in all 4 ECO, it is more convenient to use weighting coefficients expressed not by the number of GTU/CCGT of a specific range, but by the total capacity of installations of a given range of capacities. Since the authors (Stepanova & Sushko, 2015) do not provide the aggregate capacity, we assume that each installation of its own range of capacities has a capacity equal to mid-range power. A total of 3547.5 MW of capacity were put into operation: 1335 MW of capacity of the first range units (37.6%), 1987.5 MW of the second range units (56.0%) and 225 MW of the third range units (6.4%). By this method, we obtain the weighted average unit cost of the construction of GTU/CCGT, equal to $ 1425.1/kW.

The price from rubles to dollars, according to the authors, was converted at "the average annual dollar exchange rate for the period from the beginning of 2010 to the first quarter of 2014" (Stepanova & Sushko, 2015). According to the Central Bank of Russia, the average US dollar exchange rate for the period from 01.01.2010 to 31.12.2013 was 30.6843 rubles (Website of Bank of Russia, 2022). Then, the weighted average unit cost of the construction of GTU/CCGT of the considered capacity in 2010–2013 was 43,728.2 rubles/kW. For correctness, it is necessary to consider inflation. We will rely on the price indices of the Federal State Statistics Service of the Russian Federation (Federal Service for State Statistics, 2022) in the field of producer prices for turbines. For the period 2010–2016, we will take the indices from the category "gas turbines, except turbojet and turboprop engines", for the period 2017–2019—"gas turbines, except turbojet and turboprop". The calculation of the final average value of specific investments is carried out according to the following formula:

$$\mathrm{K_{GTU}} = \overline{\mathrm{K}}_{\mathrm{GTU}} \cdot \prod_{m=1}^{n} i_{p.m} \tag{2}$$

where $i_{p.\ m}$—price index for a specific year, o.e.

The resulting value of 91,593.9 rubles/kW should be reduced to the entire range of possible capacities of GTU/CCGT. According to the authors of one article, a CCGT-based power plant with a single capacity of 450 MW in the form of a CES costs $ 1000 for each kW of power, in the form of a CHP—$ 600–1300 (Skret & Garievsky, 2017; Bondarenko et al., 2010). The article was published in 2017. The average annual exchange rate of the US dollar at that time was 58.2982 dollars per ruble (Bank of Russia, 2022). Let's take the specific cost of the powerful CCGTs we need at $ 1000/kW, which corresponds to the specific cost of the CCGT-CES or the average value of the specific cost of the CCGT-CHP. Then, in 2017 prices, one kW of a powerful CCGT cost 58298.2 rubles/kW. Recalculating this number according to Rosstat (Federal Service for State Statistics, 2012) in the prices of 2019, we get 49,982.8 rubles/kW. Unfortunately, we do not have data that allows us to establish an exact ratio between low-power and large GTU/CCGT, therefore, the final unit cost of GTU/CCGT will be obtained by us simply as an arithmetic mean, equal to 70,788.4 rubles/kW.

The final specific investments are formed not only from the cost of GTU/CCGT, but also from the specific cost of the HPP. It is determined by many factors and has a complex relationship (Avilova & Krutilova, 2017), which cannot be properly considered in the framework of the ongoing study, therefore we will be satisfied with the average unit cost of lowland hydroelectric power plants, the construction of which is reported by PJSC RusGidro (RusHydro 2022). We will take the objects from the category "new construction. Objects under construction", for which the full "costs are indicated in the prices of the corresponding years (including VAT"). As in the case of stations based on GTU/CCGT, we will calculate the unit costs of the construction of hydroelectric power plants and bring them, based on the price coefficients of "alternators (synchronous generators)", to the prices of 2019.

When calculating the average unit cost of hydroelectric power plants, small hydroelectric power plants should not be overlooked. According to the 2010 publication, 16 MPPS with a total capacity of 66.7 MW in the Rostov region cost 3.04 billion rubles (Bondarenko et al., 2010). Considering inflation for the period up to 2019, for small HPPs we have a unit price equal to 77,102.37 rubles/kW. If we calculate the average between this value and the values from Table 2, we get the average specific cost of construction generalized for HPPs in the amount of 79,290.93 rubles/kW.

Finally, using the obtained weighting coefficients for GTU/CCGT and HPP, as well as specific investments in their construction, from formula (1) we have weighted average specific investments in the construction of new peak power plants equal to 71,413.34 rubles/kW in 2019. It is more appropriate, however, to bring the amount received to annual investments. It is known that the service life of a hydrogenerator is 40 years according to GOST 5616-89. The standard in force for

Table 2 Some projects of the investment program PJSC "RusHydro" for 2014–2018 (PJSC "RusHydro", 2018)

HPP, its capacity in MW, year of completion of construction[a]	Total cost, million rubles[a]	Coef. prices	Unit cost of construction at the end of 2019[b], rub/kW
Ust-Srednekanskaya, 570. 2018	42,444.00	1.023; 1.019	77,623.15
Boguchanskaya, 2997. 2014	90,793.24	1.013; 1.014; 1,11; 1.060; 1.023; 1.019	34,763.30
Nizhne-Bureyskaya, 320. 2016	36,572.00	1.011; 1.060; 1.023; 1.019	127,674.91

[a]Data of PJSC "RusHydro"
[b]Results of our calculations

turbo generators does not stipulate their service life years according to GOST IEC 60034-3-2015, the GOST preceding it sets the minimum service life of turbo generators, also amounting to 40 years according to GOST 533-2000 (Nikolaev & Sizov, 2010). Then, the annual investment will be equal to 1785.33 rubles/(kW* year) or rounded 1800 rubles/(kW·year). This will be the indicator of infrastructure savings that we are looking for, relevant for 2019.

Indicator of Technological Savings

Operation in the variable part of the load schedule results in a decrease in the efficiency of power generation due to longer operation of generating equipment in areas of its suboptimal efficiency (Hübel et al., 2017; Skret & Garievsky, 2017; Weigl et al., 1999). Calculations show that with an eight-hour load reduction of 40%, the cost of generating electricity for CCGT increases by about 5% (Skret & Garievsky, 2017). Taking these 5% in this study as the estimated increase in the cost of e/e, we calculate the annual decrease in profitability, expressed in rubles. In 2010 The cost of electricity for a GTU-CHP located in the Middle Volga region (4 GTU with a natural gas capacity of 6.5 MW each) when operating according to an electric load schedule was 0.94 rubles/kWh (Nikolaev & Sizov, 2010). For a coal-fired thermal power plant with a capacity of 8 MW, the by-product of which is activated carbon, the cost of 1 kWh e/e in 2006–2008 prices was 0.41 rubles (Afanasyeva & Mingaleeva, 2009), or in 2010 prices—0.47 rubles. The cost of 1 kWh of electricity generated at the Novo-Irkutsk thermal power Plant (665 MW of capacity) in 2010 it was 0.56 rubles, and in the case of a hypothetical increase in the capacity of the CHP to 1425 MW due to the introduction of GTU, the cost of electricity varies in the range of 0.47–0.67 rubles/kWh (Postnikov et al., 2013). The average arithmetic cost of electricity according to these data is 0.64 rubles/kWh.

Nevertheless, it would be more correct to take the cost of electricity equal to the middle of the price range from the above data, i.e., equal to 0.68 rubles/kWh.

Inflation for the second decade of the XXI century will be considered only in the fuel component of the cost of electricity. There is an assertion that 2/3 of the cost of electricity is accounted for by the fuel consumed by the power plant (Afanasyeva & Mingaleeva, 2009). Speaking specifically about the Novo-Irkutsk CHP, fuel costs for it reach 90% of all annual station costs (Postnikov et al., 2013). Statistics on power plants of the Orenburg region for 2013 are given according to (Afanasyev & Koptsev, 2015): Sakmarskaya CHPP (460 MW)—83.5%, Orskaya CHPP-1 (245 MW)—76.1% (T Plus Group, 2022b), Kargalinskaya CHPP (320 MW)—82.3% (T Plus Group, 2022a), the geometric average for these three stations, the last two of which run on gas (Postnikov et al., 2013; Afanasyev & Koptsev, 2015)—80.6%. In general, the above-mentioned share [2/3, according to (Afanasyeva & Mingaleeva, 2009)] seems to be the most appropriate—in order to avoid overstating the cost of e/e as a result of incorrectly accounted inflation, we will focus on it, rounding it up to ten (70%). When calculating the rise in the cost of the generated electricity, using price indices for the category "natural gas (natural gas)" (Federal Service for State Statistics, 2012), we obtain for 2019 1.93 rubles/(kWh). Then a 5% increase in the cost of electricity in rubles will be equal to 0.095 rubles/(kWh or rounded 0.10 rubles/(kWh).

The indicator of infrastructure savings was expressed in terms of power (kW). The desired indicator of technological savings should also be given to it. This can be achieved if we consider the resulting increase equal to 0.10 rubles/(kWh), not as an increase in the cost of 1 kWh of electricity, but as an increase in the hourly cost of 1 kW of power. Then, as of 2019, from the introduction of one kilowatt of e/e storage, we will have savings equal to 900 rubles/ (kW year) annually.

Operational Economy Indicator

The operation of power plants erected to cover the peak load is accompanied by fuel costs, spare parts, other consumables, staff salaries, equipment maintenance, taxes, etc. For the above-considered coal-fired mini-TPP with a capacity of 8 MW, annual costs were 63.89 million rubles/year, or 7986 rubles/year for each kilowatt of capacity (Afanasyeva & Mingaleeva, 2009); for GTU-TPP with a capacity of 26 MW—169.3 million rubles/year, or 6512 rubles/year for each kilowatt of capacity (Nikolaev & Sizov, 2010). There is no doubt that the data of the end of the first decade of the XXI century should be updated due to rising prices. A valuable source in this matter is the annual reports of generating companies: PJSC Mosenergo (13 CHP, 1 low-power HPP, 1 GRES in Moscow and the Moscow region), PJSC Quadra (18 CHP, 1 GTU-CHP, 1 GRES in Smolensk, Kaluga, Tula, Ryazan, Oryol, Lipetsk Tambov, Kursk, Voronezh and Belgorod regions) and PJSC TGC-1 (12 thermal power plants and 40 HPPs in St. Petersburg, Leningrad and Murmansk regions, the Republic of Karelia). Based on the total annual costs of the above-mentioned

Table 3 Specific annual expenses large generating companies in Russia for 2019

Company	Total installed capacity of facilities, MW[a]	Specific annual expenses r[b], rub/kW
Mosenergo PJSC	12,825	13,267
PJSC "Quadra"	2939	16,864
PJSC TGC-1	6918	12,373

[a]Company data
[b]The results of our calculations

companies (Website of PJSC Mosenergo, 2019; Website of PJSC "Quadra", 2019; Website of PJSC "TGC-1", 2022), we calculate the specific annual costs, which can be seen in Table 3.

Since storage units are an alternative to peak power plants, the annual costs of which are clearly less than the costs of power plants with a high coefficient of installed capacity utilization, among the 3 values (Table 3) it would be more correct to take the lowest unit annual costs as an indicator of operational savings. Then for 2019 we have a value equal to 12,400 rubles/(kW * year).

In addition to the three components of the total indirect savings indicated in the text of this publication, it is worth paying attention to another one - reducing the cost of undersupply of electricity. The introduction of storage devices may increase the probability and (or) the duration of the e/e undersupply, the penalties for which lie in a wide range and may amount to significant amounts (Nepomnyashchy, 2010). To take this aspect into account, additional research and calculations need to be carried out, which was not the task of this work.

Conclusions

The economic consequences of technological innovations need to be assessed comprehensively and comprehensively. The effect of such innovative solutions can have systemic consequences and manifest itself among different owners both in the form of results of investment decisions and in the form of savings within the framework of operational activities. Based on the results of the research conducted in the article, the following conclusions can be drawn:

- the use of energy storage devices to cover demand at peak loads will reduce costs by an average of 12,400 rubles/(kW-year), which is the cumulative result of savings on the construction of additional generating capacities, savings from the increase in efficiency of power plants and savings from the operation of peak power plants.

- when assessing the economic effect for networks, it is necessary to consider factors acting in the opposite direction, such as changes in the depreciation component of the cost and the loss component.
- when assessing the economic effect, it is necessary to consider in each individual case an additional reduction in the cost of undersupply of electricity and the environmental component.

Based on the results of the study, it can be concluded that the economic consequences of technical and technological innovations are multifaceted, the assessment of which is an important economic task. The complexity of assessing the economic effect is due to the need to consider the existing ownership system in the energy sector with a single energy supply process.

The application of any technical solution with a complex ownership structure requires the identification of both private effects for all participants in the process and the cumulative effect for the system. Cumulative and systemic effects are important for system-forming participants in the process, such as the state. Private effects directly affect the indicators of economic efficiency of economic activity of energy companies.

Acknowledgements This research was funded by Ministry of Science and Higher Education of the Russian Federation within the framework of the state assignment № 075-03-2021-175/3 from 30.09.2021.

References

Afanasyev, V. N., & Koptsev, A. I. (2015). *Statistical study of the dynamics of the structure of costs for the production of electricity at a thermal power plant: Monograph*. Orenburg State University.

Afanasyeva, O. V., & Mingaleeva, G. R. (2009). Technical and economic indicators of a mini-thermal power plant operating on coal. *Proceedings of Akademenergo, 2*, 54–63.

Aminov, R. Z., Novikov, S. V., & Yankov, A. V. (2013). Economic evaluation of the efficiency of mono and double-block CCGT circuits when covering the daily unevenness of the electrical load schedule. News of higher educational institutions. *Problems of Energy, 7–8*, 84–92.

Atalay, S., Sheikh, M., Mariani, A., Merla, Y., Bower, E., & Widanage, W. D. (2020). Theory of battery ageing in a lithium-ion battery: Capacity fade, nonlinear ageing and lifetime prediction. *Journal of Power Sources, 478*, 229026.

Avilova, I. P., & Krutilova, M. O. (2017). Eco-oriented economic assessment of the cost of construction of infrastructure facilities on the example of hydroelectric power plants. *Bulletin of Belgorod State Technological University na VG Shukhov, 12*, 212–219.

Bank of Russia. (2022). *Dynamics of the official exchange rate of a given currency*. Retrieved from https://cbr.ru/currency_base/dynamics

Bondarenko, V. L., Gutenev, V. V., Denisov, V. V., & Kuvalkin, A. A. (2010). Small hydroelectric power plants and modernization of the drinking water supply system of the Rostov region. *Problems of Regional Ecology, 6*, 95–101.

Bulut, M., & Özcan, E. (2021). Integration of battery energy storage systems into natural gas combined cycle power plants in fuzzy environment. *Journal of Energy Storage, 36*, 102376.

Burhan, M., Chua, K. J. E., & Ng, K. C. (2016). *Electrical rating of concentrated photovoltaic (CPV) systems: Long-term performance analysis and comparison to conventional PV systems*.

Cadavez, C. C., & de Souza-Santos, M. L. (2021). Efficiency of a power generation alternative regarding the composition of feeding biomass-glycerol slurry; theoretical assessment. *Energy, 214*, 118967.

Crawford, A. J., Huang, Q., Kintner-Meyer, M. C., Zhang, J. G., Reed, D. M., Sprenkle, V. L., et al. (2018). Lifecycle comparison of selected Li-ion battery chemistries under grid and electric vehicle duty cycle combinations. *Journal of Power Sources, 380*, 185–193.

Federal Service for State Statistics. (2012). Retriveg from https://rosstat.gov.ru/

Federal Service for State Statistics. (2022). Retriveg from https://rosstat.gov.ru/

Gambini, M., & Vellini, M. (2019). Hybrid thermal power plants: Solar-electricity and fuel-electricity productions. *Energy Conversion and Management, 195*, 682–689.

Government of the Russian Federation. (2016). *The scheme of territorial planning of the Russian Federation in the field of electric power industry: Approved by the decree of the Government of the Russian Federation dated 01.08.2016 No. 1634-r "On the scheme of territorial planning of the Russian Federation in the field of energy"*. Collection of Legislation of the Russian Federation, 33, 5207.

Gusev, Y. P., & Subbotin, P. V. (2019). Development of an improved methodology for selecting parameters and locations of electric power storage systems in distribution electric networks. *Bulletin of the South Ural State University. Series: Energetika, 19*(2), 48–61.

Han, S., Han, S., & Aki, H. (2014). A practical battery wear model for electric vehicle charging applications. *Applied Energy, 113*, 1100–1108.

Hawachi, I., Sammouda, H., & Bennacer, R. (2014). Energy storage using the phase change materials: Application to the thermal insulation. *International Journal of Technology, 5*(2), 142–151.

Hübel, M., Meinke, S., Andrén, M. T., Wedding, C., Nocke, J., Gierow, C., et al. (2017). Modelling and simulation of a coal-fired power plant for start-up optimisation. *Applied Energy, 208*, 319–331.

Jindal, A., & Nilakantan, R. (2021). Falling efficiency levels of Indian coal-fired power plants: A slacks-based analysis. *Energy Economics, 93*, 105022.

Karchin, V. V., & Myasnikova, T. V. (2017). Reduction of electricity losses by reducing the unevenness of power consumption. *Bulletin of the Chuvash University, 3*, 90–98.

Kim, H. K., Kim, C. J., Kim, C. W., & Lee, K. J. (2018). Numerical analysis of accelerated degradation in large lithium-ion batteries. *Computers and Chemical Engineering, 112*, 82–91.

Kononenko, V. Y., Smolentsev, D. O., & Veshchunov, O. V. (2014). The possibilities of using network energy storage devices and their efficiency. *Proceedings of the Russian Academy of Sciences. Energy, 3*, 106–113.

Lushnikov, O. G. (2018). On the needs of the energy system in regulating capacities. In *IV International Congress REENCON-XXI of Renewable energy of the XXI century: Energy and economy. Efficiency* (pp. 126–135).

Makoklyuev, B. I. (2017). Economic aspects of the production, transmission, distribution and consumption of electricity. *Unified Grid Energy, 5*, 64–76.

Nepomnyashchy, V. A. (2010). *Economic losses from power supply disruptions to consumers*. Publishing House of MEI.

Neshumayev, D., Rummel, L., Konist, A., Ots, A., & Parve, T. (2018). Power plant fuel consumption rate during load cycling. *Applied Energy, 224*, 124–135.

Nikolaev, Y. E., & Sizov, S. V. (2010). Efficiency of operation of small thermal power plants with GTU according to thermal and electrical load schedules. *Energetika Tatarstan, 2*, 59–65.

PJSC "Quadra". (2019). *Annual report of the joint-stock company for 2019*. Retrieved from https://www.quadra.ru/aktsioneram-i-investoram/raskrytie-informatsii/godovye-otchyoty/index.php?sphrase_id=19814

PJSC "TGC-1". (2022). *Annual report of the Public Joint Stock Company "Territorial Generating Company No 1" on the results of work for 2019*. Retrieved from http://tgc1.ru/ir/reports

PJSC Mosenergo. (2019). *Presentation of financial results for IFRS for 2019 of PJSC Mosenergo*. Retrieved from https://mosenergo.gazprom.ru/investors/reports/msfo/. – Title from the screen.

PJSC RusHydro. (2018). *Investment program for 2014–2018*. Retrieved from http://www.rushydro.ru/activity/invest/investprogramm/2014;

Pospelov, A. A., Ledukhovsky, G. V., & Borisov, A. A. (2008). On taking into account the influence of the unevenness of the daily schedules of the electrical load when calculating the nominal specific fuel consumption for power units. *Bulletin of the Ivanovo State Power Engineering University, 4*, 27–29.

Postnikov, I. V., Penkovsky, A. V., Dobrovolskaya, T. V., & Yakimets, E. E. (2013). Forecasting of consumption and production of electric and thermal energy in the Irkutsk region. *Bulletin of Irkutsk State Technical University, 8*(79), 197–206.

RusHydro. (2022). Retriveg from http://www.eng.rushydro.ru/investors/presentations/IR_presentations/

Sarkar, A., Shrotriya, P., Chandra, A., & Hu, C. (2019). Chemo-economic analysis of battery aging and capacity fade in lithium-ion battery. *Journal of Energy Storage, 25*, 100911.

Skret, A. F., & Garievsky, M. V. (2017). Efficiency of the use of heating CCGTs for the regulation of variable electrical loads.

Soltani, M., Ronsmans, J., & Van Mierlo, J. (2020). Cycle life and calendar life model for lithium-ion capacitor technology in a wide temperature range. *Journal of Energy Storage, 31*, 101659.

Stepanova, E. L., & Sushko, S. N. (2015). Determination of average specific investments in the construction of combined-cycle gas plants introduced in Russia for the period 2010-2014. *Bulletin of Irkutsk State Technical University, 11*(106), 171–175.

Swierczynski, M., Stroe, D. I., Stan, A. I., Teodorescu, R., & Kær, S. K. (2015). Lifetime Estimation of the Nanophosphate $\hbox {LiFePO} _ {4}\hbox {/C} $ Battery Chemistry Used in Fully Electric Vehicles. *IEEE Transactions on Industry Applications, 51*(4), 3453–3461.

T Plus Group. (2022a). *Kargalinskaya CHP*. Retrieved from https://www.tplusgroup.ru/org/orenburg/organization/kargalinskaya-chp

T Plus Group. (2022b). *Orskaya CHPP-1*. Retrieved from https://www.tplusgroup.ru/org/orenburg/organization/orkskaya-chp1

Valeev, G. S., Dzyuba, M. A., & Valeev, R. G. (2016). Modeling of daily load schedules of 6-10 kV distribution network sections in cities and localities under conditions of limited initial information. Bulletin Of SUSU. A Series Of. *Energy, 16*(2), 23–29.

Weigl, K., Schuster, G., Stamatelopoulos, G. N., & Friedl, A. (1999). Increasing power plant efficiency by fuel drying. *Computers and Chemical Engineering, 23*, S919–S922.

Yan, X., Zhang, X., Chen, H., Xu, Y., & Tan, C. (2014). Techno-economic and social analysis of energy storage for commercial buildings. *Energy Conversion and Management, 78*, 125–136.

Yang, W. J. (1997). Reduction of specific fuel consumption in gas turbine power plants. *Energy Conversion and Management, 38*(10–13), 1219–1224.

Zeng, J., Liu, L., Liang, X., Chen, S., & Yuan, J. (2021). Evaluating fuel consumption factor for energy conservation and carbon neutral on an industrial thermal power unit. *Energy, 232*, 120887.

Zhang, H., Liu, Y., Liu, X., & Duan, C. (2020). Energy and exergy analysis of a new cogeneration system based on an organic Rankine cycle and absorption heat pump in the coal-fired power plant. *Energy Conversion and Management, 223*, 113293.

Zhou, C., Qian, K., Allan, M., & Zhou, W. (2011). Modeling of the cost of EV battery wear due to V2G application in power systems. *IEEE Transactions on Energy Conversion, 26*(4), 1041–1050.

Zhu, P., Yao, J., Qian, C., Yang, F., Porpatham, E., Zhang, Z., & Wu, Z. (2020). High-efficiency conversion of natural gas fuel to power by an integrated system of SOFC, HCCI engine, and waste heat recovery: Thermodynamic and thermo-economic analyses. *Fuel, 275*, 117883.

Estimation of Innovative Indicators as Growth Dynamics Indicator

Iana Verkhovskaia

Introduction

Global trends in socio-economic and scientific and technological development create new challenges, including in the academic environment, which on the one hand contribute to the development of new forms of activities and interactions (Hüther, 2016; Kreis, 2013), but on the other hand increase the gap and inequality, including between universities.

Modern universities, on the one hand, are participants in the process of change, and on the other—one of the sources of change (Kallio et al., 2016). Currently, many universities have implemented the principles and practices of project management, which are becoming vital for sustainable development, which also allows you to create your own university ecosystem, develop strategic partnerships and alliances and create a basis for mutual exchange of academic and commercial knowledge and experience (Lee et al., 1995; Siegel et al., 2003; Un et al., 2010), strengthening several components in the structure of an innovative society.

Thus, the development of strategies, action plans, project activities aimed at achieving strategic objectives and obtaining effective results is currently not so much the responsibility of management as a vital necessity, as it provides opportunities for the formation of various collaborations and participation in competitions and grants to attract additional resources, including government (Carayannis et al., 2000).

I. Verkhovskaia (✉)
St. Petersburg University of Management Technologies and Economics, St. Petersburg, Russian Federation

St. Petersburg State University of Industrial Technologies and Design, St. Petersburg, Russian Federation

A. Rumyantseva et al. (eds.), *Challenges and Solutions in the Digital Economy and Finance*, Springer Proceedings in Business and Economics,
https://doi.org/10.1007/978-3-031-14410-3_57

High motivational and organizational and managerial components of the university structure are among the key areas for the development and effectiveness of the university, which require high interdisciplinary competencies of the university team. Improving the technical, managerial and communication skills of the university's management and increasing practical experience in multi-functional activities makes it possible to solve emerging challenges and problems in a timely manner, including identifying root causes and eliminating risks (Dove, 1999; Lee et al., 1995). Thus, an important issue is the transformation of forms of professional interactions at the university, and consequently, the formation of new critical skills, including academic and industry skills, both among specialists and university managers (Van Looy et al., 2004).

One of the important factors in increasing the sustainability of the development of a modern university is to increase the motivation of innovative actions and interactions of specialists and organizations, exchange and accumulation of knowledge, and the use of inter-organizational ties through the development of entrepreneurial ecosystems and open innovations initiates the search for new knowledge for projects, including for public-private partnerships (Guerrero et al., 2015).

Thus, the key factor in the effectiveness of the university's development is the development of concepts of scientific and technological entrepreneurship and the formation of a system of strategic alliances, including through the development of student startups, internal entrepreneurship and open innovation.

Materials and Methods

In the context of the situation presented above, the organization of the university's innovation structure within the framework of the introduction and implementation of the ideology of project management, internal entrepreneurship and open innovation is the main goal of the diversification of management and re-engineering of organizational business processes. Thus, the purpose of the study is to analyze the innovative activity of Russian organizations and universities to assess the dynamics of development.

The theoretical and methodological basis of the research was the works of researchers in the field of sociology and sociological research, concepts of knowledge management theory and human resources management, fundamentals of management theory and organization, as well as management practice. The knowledge capital assessment of the university is based on the systematization of the types of collaborations with universities (Un et al., 2010). The data sources were the open data of Rosstat (2021) and the Global Innovation Index (WIPO, 2021). The research plan is shown in Fig. 1.

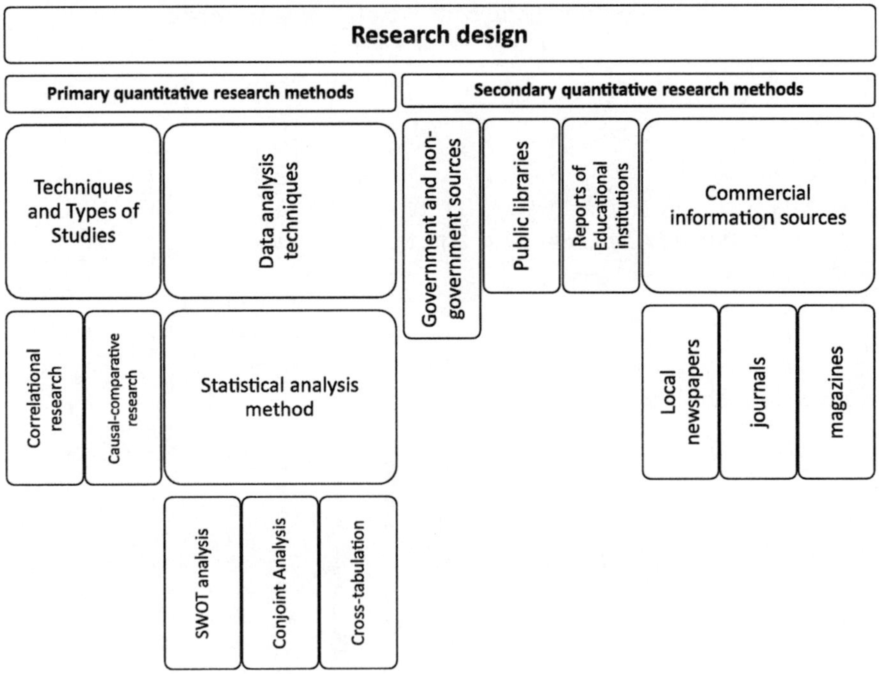

Fig. 1 Research design of estimation of innovative indicators

Results and Discussions

The analysis of Rosstat statistical data (Fig. 2) shows a general positive trend in the growth of innovative activity of Russian organizations that have carried out technological, organizational, marketing innovations since 2010—up to 39% in the Russian Federation. The most active growth of innovation activity in 2017 was observed in the construction and IT fields of activity, production and distribution of electricity, gas and water, as well as in the field of communications, the use of computer technology and the development of research and development, while indicators in the field of processing production (up to 33%) and mining fell mineral resources (up to 85%).

Since 2017, there has been a sharp increase in technological innovations in the Russian Federation, especially in the Central Federal District, Volga Federal District, North-Western Federal District, Siberian Federal District, where, after the introduction of a number of long-term programs of state support for the development of science and education, a knowledge-intensive cluster of organizations and universities has been formed R&D (Fig. 3). It is important to note that, according to Rostsat data, since 2017, the largest share of Russian organizations implementing technological innovations are in the fields of crop production, including provision of services in the field of decorative gardening and animal husbandry (growth up to

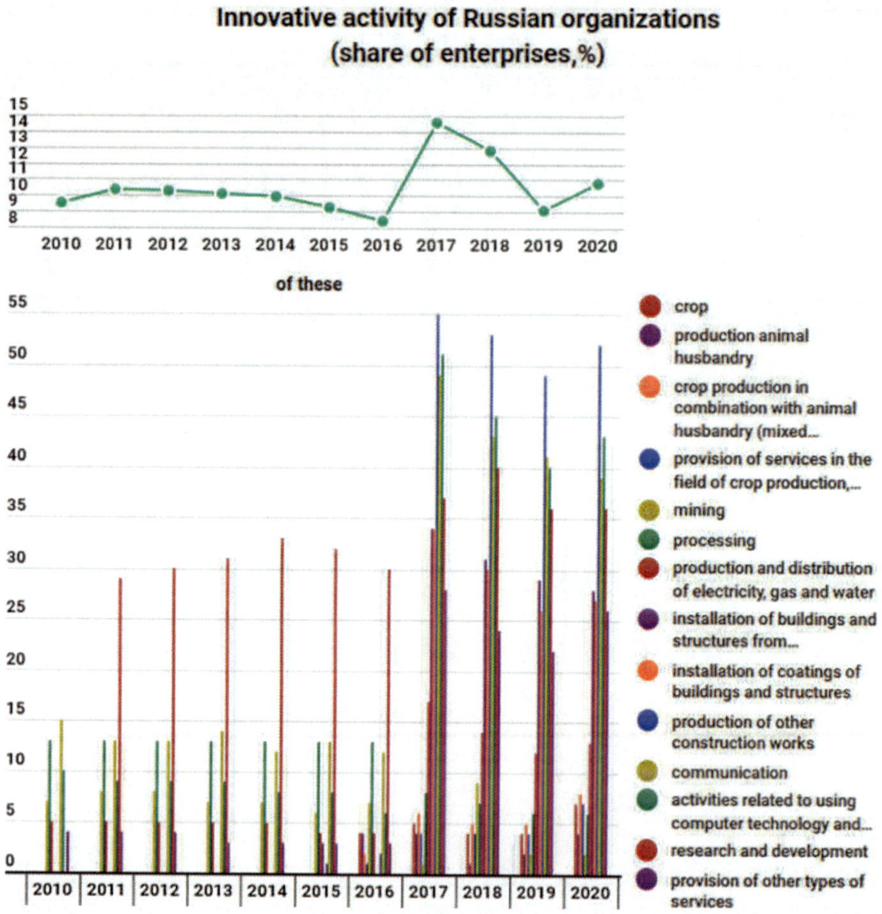

Fig. 2 Innovative activity of Russian organizations (the share of organizations in the total number of surveyed organizations) that carried out technological, organizational, marketing innovations in the reporting year by type of economic activity (compilation data by author https://infogram.com/innovative-activity-of-russian-organizations-1ho16vowl3vlx4n?live)

114%), manufacturing (growth up to 113%) and crop production (growth up to 65%) and R&D (growth up to 14%).

At the same time, according to statistics, it is important to note that the overall growth dynamics of technological innovation activity reduces the number of personnel involved in R & D (Fig. 4).

The positive dynamics of the innovative development of the Russian Federation is confirmed by the data of the Global Innovation Index (Fig. 5), where it is noted that the Russian Federation has been among the top 30 countries of the world in terms of knowledge creation since 2019 (WIPO, 2021). There are high indicators among countries with above-average incomes in the areas of human capital and research, infrastructure, business development, results in the field of knowledge and

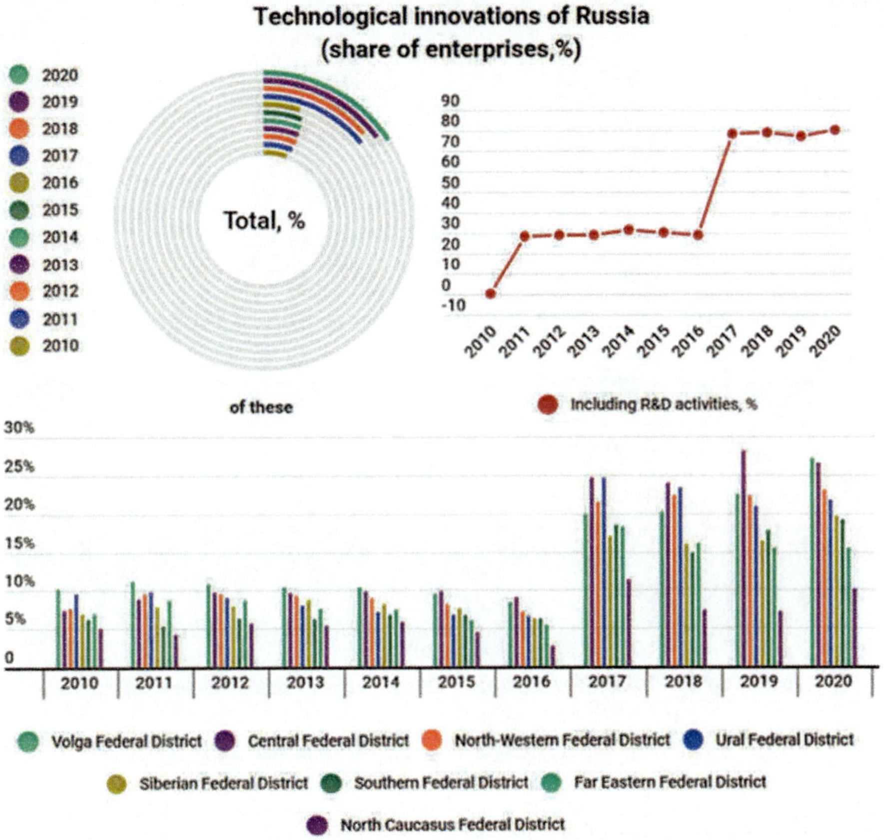

Fig. 3 The share of organizations of the Russian Federation engaged in technological innovations (compilation statistic data by author https://infogram.com/innovative-activity-of-russian-organizations-1ho16vowl3vlx4n?live)

technology, in particular, the leading positions are occupied by higher education, research and development, information and communication technologies. The Global Innovation Index is formed for 131 countries of the world by 80 indicators combined in seven areas and is calculated by two sub-indices: innovation resources (institutions, human capital and science, infrastructure, the level of market and business development) and innovation results (technology development and the knowledge economy, the results of creative activity).

According to the Global Innovation Index from 2019 to the present, the weaknesses and strengths of innovation are presented. Figures 5 and 6 shows the data of rating indicators of the strengths and weaknesses of the innovation structure of the Russian Federation. The obtained results allow us to conclude that the strengths of innovation indicators in Russia are the system of science and higher education, in which there is a high indicator of knowledge-intensive employment and results in the field of knowledge and technology (Fig. 6).

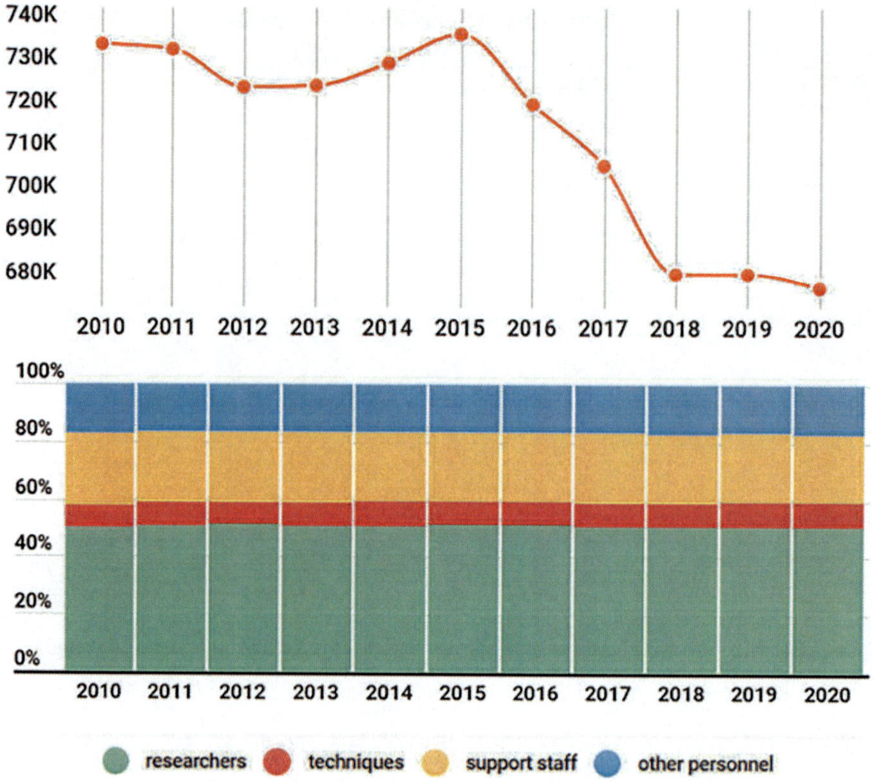

Fig. 4 Quantity and categories of staff engaged in R&D of organizations of Russian Federation (compilation data by author https://infogram.com/innovative-activity-of-russian-organizations-1ho16vowl3vlx4n?live)

Weak sides of innovative indicators in Russia are weak regulation of certification and quality, the system of additional education, the development of national feature films, etc. (Fig. 7).

Conclusion

As a result of the conducted research, the following conclusions were obtained in conclusion:

1. It is revealed that the highest innovation indicators in Russia according to GII data are associated with the system of science and higher education, knowledge-intensive employment and the results obtained in the field of knowledge and technology.

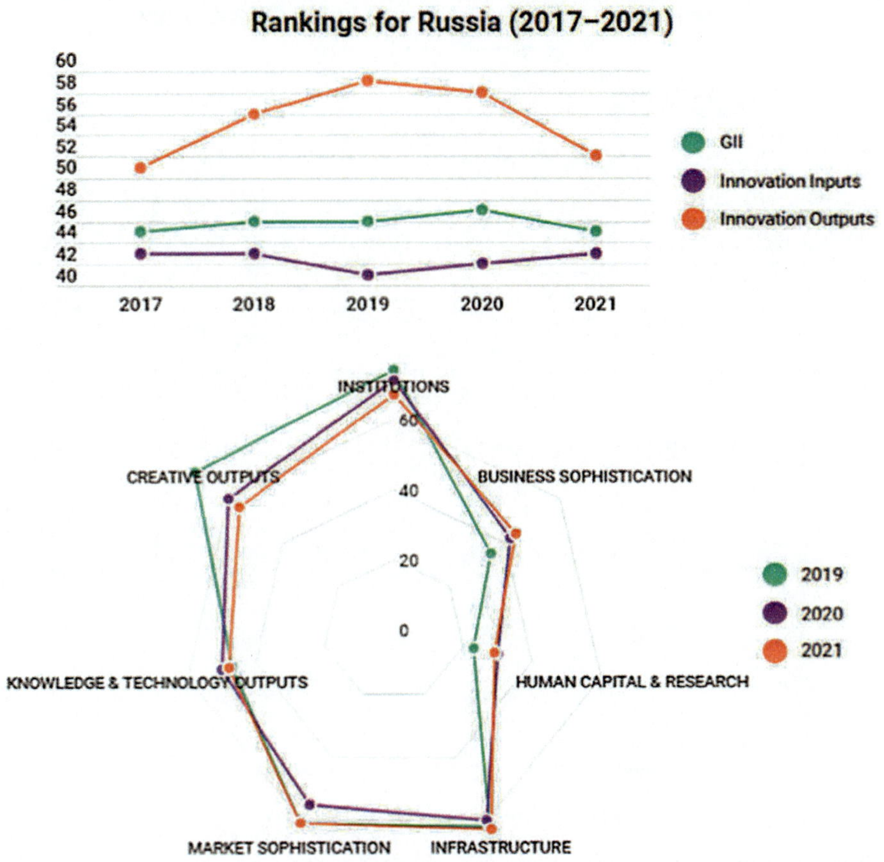

Fig. 5 Innovative activity of Russian organizations of Global Innovation Index (compilation data by author https://infogram.com/innovative-activity-of-russian-organizations-1ho16vowl3vlx4n?live)

2. It is established that there has been a reorientation of organizations towards technological innovations in the field of manufacturing, with a drop in the growth of innovation activity in this area.
3. It is established that the overall growth dynamics of technological innovation activity reduces the number of personnel involved in R&D to 4% annually.
4. Contradictions in educational preferences of managers and existing qualification deficits are identified.
5. Both "weak" and "strong" knowledge of managers of an educational organization are identified, which are not related to the profile of professional activity, the type and level of the position, or the status of the educational organization, but are related to the age, work experience of the manager and the profile of the educational organization. At the same time, in each of the 5 areas of activity, there are certain ranges of the formation of a lack of knowledge.

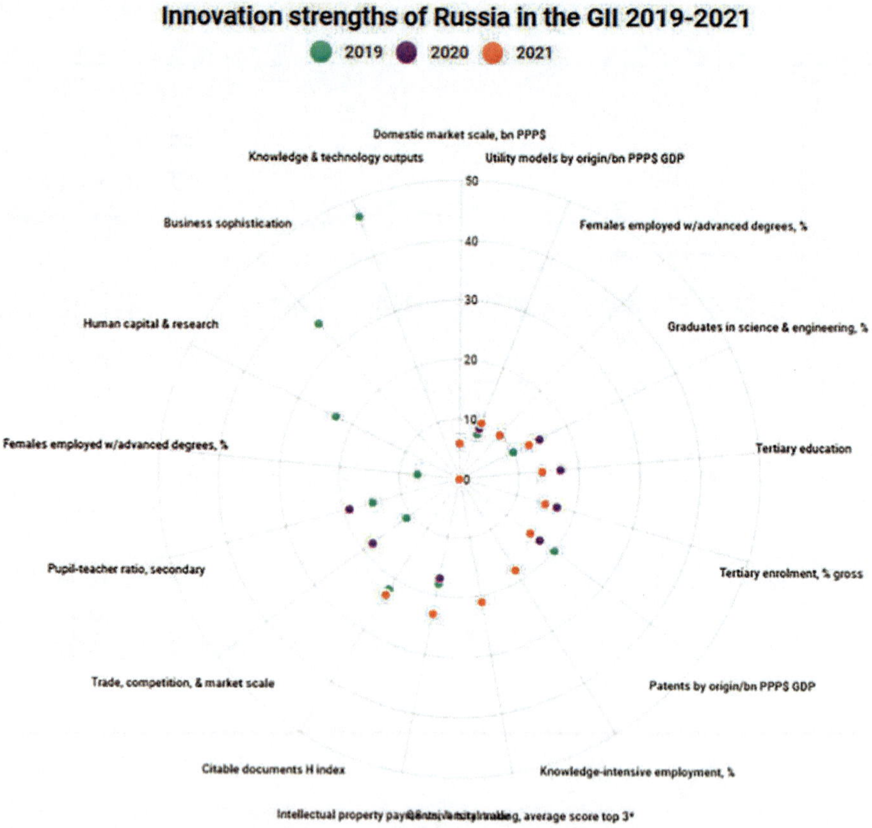

Fig. 6 Innovative strengths of Russian по результатам Global Innovation Index (compilation data by author https://infogram.com/innovative-activity-of-russian-organizations-1ho16vowl3vlx4n?live)

6. It is established that the development of professional interactions of managers on a project-functional basis can be correlated with the identified "strong" knowledge of managers.
7. It is proposed to form project teams based on the "strong" knowledge of managers and to introduce the division of projects into intra-organizational and inter-organizational.

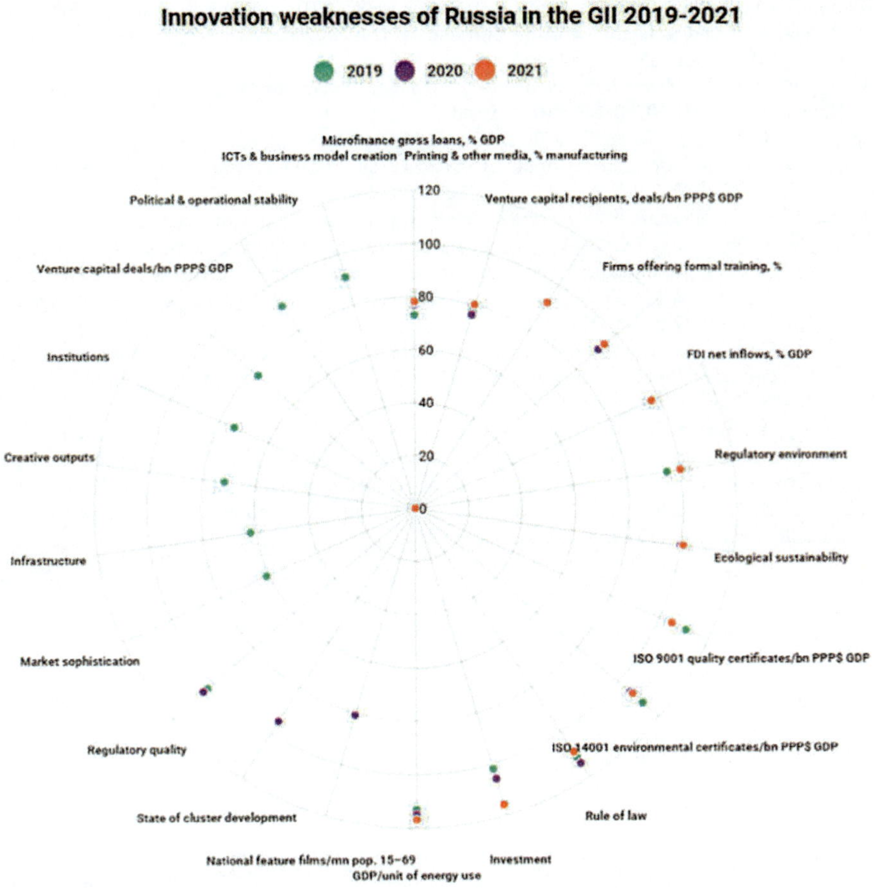

Fig. 7 Innovative weaknesses of Russian по результатам Global Innovation Index (compilation data by author https://infogram.com/innovative-activity-of-russian-organizations-1ho16vowl3vlx4n?live)

References

Carayannis, E. G., Alexander, J., & Ioannidis, A. (2000). Leveraging knowledge, learning, and innovation in forming strategic government–university–industry (GUI) R&D partnerships in the US, Germany, and France. *Technovation, 20*(9), 477–488. https://doi.org/10.1016/S0166-4972(99)00162-5

Dove, R. (1999). Knowledge management, response ability, and the agile enterprise. *Journal of Knowledge Management, 3*(1), 18–35. https://doi.org/10.1108/13673279910259367

Guerrero, M., Cunningham, J. A., & Urbano, D. (2015). Economic impact of entrepreneurial universities' activities: An exploratory study of the United Kingdom. *Research Policy, 44*(3), 748–764. https://doi.org/10.1016/j.respol.2014.10.008

Hüther, M. (2016). Digitalisation: An engine for structural change - A challenge for economic policy. B *W Policy Paper: T. 15/2016E.* titut der deutschen Wirtschaft (IW). https://www.econstor.eu/bitstream/10419/148408/1/87503487X.pdf

Kallio, K.-M., Kallio, T. J., Tienari, J., & Hyvönen, T. (2016). Ethos at stake: Performance management and academic work in universities. *Human Relations, 69*(3), 685–709. https://doi.org/10.1177/0018726715596802

Kreis, M. (2013). *Innovationsfelder der digitalen Welt: Bedürfnisse von übermorgen.* Zukunftsstudie MÜNCHNER KREIS Band.

Lee, D. M. S., Trauth, E. M., & Farwell, D. (1995). Critical skills and knowledge requirements of IS professionals: A joint academic/industry investigation. *MIS Quarterly, 19*(3), 313. https://doi.org/10.2307/249598

Rosstat. (2021). *Science, innovation and technology.* https://rosstat.gov.ru/statistics/science

Siegel, D. S., Waldman, D. A., Atwater, L. E., & Link, A. N. (2003). Commercial knowledge transfers from universities to firms: Improving the effectiveness of university–industry collaboration. *The Journal of High Technology Management Research, 14*(1), 111–133. https://doi.org/10.1016/S1047-8310(03)00007-5

Un, C. A., Cuervo-Cazurra, A., & Asakawa, K. (2010). R&D collaborations and product innovation*. *Journal of Product Innovation Management, 27*(5), 673–689. https://doi.org/10.1111/j.1540-5885.2010.00744.x

Van Looy, B., Ranga, M., Callaert, J., Debackere, K., & Zimmermann, E. (2004). Combining entrepreneurial and scientific performance in academia: Towards a compounded and reciprocal Matthew-effect? *Research Policy, 33*(3), 425–441. https://doi.org/10.1016/j.respol.2003.09.004

WIPO. (2021). *Global Innovation Index.* Russian Federation (The). https://www.globalinnovationindex.org/analysis-economy

Project Management: University Leaders' Willingness

Iana Verkhovskaia

The search for ways to develop effective, efficient, and productive management of higher education institutions in the Russian Federation is determined by the achievement of the state's priority goal in the field of education (Russian Federation Government, 2017a, b, c, 2019). Development objectives are aimed not only to meet the strategic interests of the regions (Ministry of Economic development of the Russian Federation, 2017), federal districts and the country as a whole by forming a differentiated network of educational organizations, but also to enter the international market of educational services through the development of globally competitive universities included in the first and second hundreds of world leading rankings universities (Russian Federation Government, 2017a).

The systemic changes taking place in the field of education management have recently been associated primarily with the introduction of principles and mechanisms of project management (Ministry of Education of the Russian Federation, 2017) (Russian Federation Government, 2017b) and means that the role of project activities (both "external" and "internal") increases dramatically. In fact, the development and implementation of university development projects is becoming the most important area of activity of university managers, but for this they need qualifications based on appropriate competencies. First, these are competencies expressed through labor actions of a communicative nature, the methodological basis of which is the process of knowledge exchange, their processing and appropriation at the level of the subjects of the management process. On this basis,

I. Verkhovskaia (✉)
St. Petersburg University of Management Technologies and Economics, St. Petersburg, Russian Federation

St. Petersburg State University of Industrial Technologies and Design, St. Petersburg, Russian Federation

A. Rumyantseva et al. (eds.), *Challenges and Solutions in the Digital Economy and Finance*, Springer Proceedings in Business and Economics,
https://doi.org/10.1007/978-3-031-14410-3_58

professional interactions of managers of a modern university are built (Lee et al., 1995; Kallio et al., 2016).

In addition, the management of a modern Russian university is characterized by the following contexts and challenges:

- the emergence of several strategic projects initiated by the university community and supported by the state for the development of Russian universities to strengthen their global ambitions and the formation of drivers of innovative development of regions (President of the Russian Federation, 2019a), the digital economy and the new information culture in general.
- Consolidation of universities (President of the Russian Federation (2016b), Government Decision on increasing the competitiveness of leading Russian universities among the world's leading research and educational centers, 2016a) and the emergence of a cluster of globally successful research, federal, reference universities (Project "5 in 100", Project "Reference Universities", University development projects in Concepts of "Twelve solutions for new Education" (Kuzminov & Frumin, 2018).
- "globalization" of Russian university education and the introduction of international development trends ("bolonization", "digitalization", the growing role of online education, the formation of universities as centers of innovative research and development, internationalization and growth of the "export" potential of Russian higher education, the development of "networkism" and the entry of Russian universities into international university corporations (President of the Russian Federation, On the Development of Scientific and Industrial Cooperation, 2019b) and other associations, etc.).
- Professional standardization in Russian higher education as a manifestation of global and all-Russian trends characteristic of society and the economy (President of the Russian Federation, 2019c).

All the above contributed to the development of such a direction in the university management system as strategic planning for the development of an educational organization, which is reflected in development programs and aimed at achieving strategic objectives and obtaining effective results at universities.

Currently, strategic development programs are not so much the responsibility of management as a vital necessity. Therefore, the context of the ideology of project management makes it possible to participate in various competitions and events to attract additional resources (primarily financial), which are carried out within the framework of the implementation of the State Program "Development of Education in the Russian Federation" for 2018–2025.

Naturally, in the context of the situation presented above, the organization of teamwork within the framework of the implementation of the ideology of project management is the main goal of the diversification of management and reengineering of organizational business processes at the university (Carayannis et al., 2000; Siegel et al., 2003).

Thus, the basis of the project–functional management structure is highly qualified employees capable of implementing long-term, complex projects. Thus, the high competence of specialists is the basis for effective and productive functioning of the university, which reduces the risks of erroneous management decisions, duplication of management functions and expands the possibilities of operational management of organizations (Dove, 1999; Un et al., 2010).

It is worth noting that the management faces new challenges, for example, in conducting managerial accounting of the effectiveness of professional interactions of managers to obtain additional resources for the development of the university on this basis (Van Looy et al., 2004).

It is known that information and knowledge have a cumulative character, and the infrastructure of their transmission at the levels of the due hierarchy at the university is horizontally and vertically oriented, i.e., linearly. Thus, the depth of professional experience and knowledge of each employee about the specific features of the functioning of the university have a great impact on the result, because the knowledge of top managers generalize the knowledge of managers, and those, in turn, generalize the knowledge of specialists, etc.

Thus, the research work on the identification of existing qualification deficits and the nature of their occurrence, contributing to the formation of the specifics of professional interactions of managers of a modern university in the main areas of the university's activities, is described below.

Therefore, the main research question is to determine the readiness of the manager of a modern Russian university for organizational changes.

It is worth noting that project management in the presented study will not be understood as project management in the field of education, but education management based on the results of project activities. In this case, the purpose of management activity is not to comprehensively ensure the design process, but to actualize project results by alienating project products during their dissemination.

The category of middle managers (heads and deputies of departments, departments and other divisions of the university's administrative infrastructure) was chosen by us as the subject of research for several reasons: firstly, middle managers are more accessible due to their status and position in an educational organization to participate in research procedures; secondly, it is easier for their representatives of this category to form a valid selection; thirdly, in the situation of project management, the degree of their autonomy increases in the preparation of management information of a strategic nature, since they traditionally lead the formation and real activity of university project teams implementing both "external" and "internal" strategic projects, while representatives of the top management remains, as a rule, the functions of coordination and external motivation.

Materials and Methods

The methodological basis of the research was the works of researchers in the field of sociology and sociological research, concepts of knowledge management theory and human resources management, fundamentals of management theory, organization, and practices in the field of management and mathematical modeling.

Informative Mathematical Model

If the university's management system is considered as a dynamic system with a discrete set of states arising as a result of the implementation of tasks coming from the external environment, which are represented by a sequence of events with a homogeneous flow of events. $P\{n(s, s + t) = m\}$, where $\forall s, t \geq 0, m = 0, 1, 2...$ and with their intensity $\lambda(t) = \lim_{h \to \infty} \frac{P(t,\ t+h)}{h}$, where $P(t, t + h)$ probability of occurrence of an event over a period of time $[t, t + h]$, occurring one after the other at intervals of time ($t \geq 0$) and random interval length ($h > 0$), the organization's readiness for change can be represented as a model with a special class of Markov processes describing the processes of the birth and death of the system.

This hypothesis is based on a functional-structural approach that expresses the essence of a systematic approach, namely: the university management system is viewed in the development paradigm and as a structure of objects and their connections endowed with a set of functions in relation to the structure as a whole and the relationship with the environment.

It is worth explaining that there is no one-to-one correspondence between the functions and the structure of universities, i.e., there may be several systems with the same functions, but different structures.

At the same time, the functional and structural organization adapts to the changing conditions of its existence (external environment), which causes changes in functions and, accordingly, structure.

Thus, the functional and structural system, which is the university management system, can be represented as an evolutionary type of system (Balashov E.P.) endowed with a functional load and described by a meaningful mathematical model.

Consider an ordinary stream of events, i.e. events follow one after another and do not occur together ($h \to 0$), a set of random variables $\xi_\kappa = n(t_k, t_{k+1})$, $k = 0, 1, \ldots$, m and we represent it as a function of random variables $\{\xi(t), t \geq 0\}$—a homogeneous Markov random function and with the help of a stochastic graph.

Let us present an analysis of the processes of the university management system, in the form of a system of professional interactions because of the external environment and consider the case $N < \infty$, which may be so large that the model of the system under study will be adequate.

Further, if we count $\xi(t)$ при $t \geq 0$ numerically equal to the number of members of the organization N, where N—the maximum possible number of its members, then, if the conditions of the ergodic Markov process are met with the utmost probability $\widetilde{\pi}_k = \lim_{t\to\infty} \pi_k(t), k \in E$, where E—sets of a random function $\{\xi(t), t \geq 0\}$, $E = \{x_0, x_1, \ldots\}$, $x_k = k$, $k = 0, 1, \ldots$, it follows that λ_n—the intensity of the birth of the system provided that the members in the organization n − 1, and the intensity μ_n—the intensity of death provided that the number of members, then there is a stationary distribution $\{\widetilde{\pi}_n\}_{n=0}^{N}$, and, using the homogeneous process theorem of the Markov function of the birth and death of the system, the analytical solution of the system is unique and is determined by the relations of the stationary Kolmogorov equations.

$$\begin{cases} \widetilde{\pi}_0 = \left(1 + \frac{\lambda_1}{\mu_1} + \frac{\lambda_1\lambda_2}{\mu_1\mu_2} + \ldots + \frac{\lambda_1\lambda_2\ldots\lambda_N}{\mu_1\mu_2\ldots\mu_N}\right)^{-1}, \\ \widetilde{\pi}_n = \dfrac{\lambda_n}{\mu_n}\widetilde{\pi}_{n-1}, n = 1, 2, \ldots, N. \end{cases}$$

Thus, in order to find and compare the intensities of birth and death of the system, they can be represented as $\lambda_n = n\lambda$, $\mu_n = n\lambda$, $n \leq N$ and express by $\rho = \frac{\lambda_n}{\mu_n} = \frac{\lambda}{\mu}$, by $\lambda \neq \mu$ and we get the expression $\widetilde{\pi}_0 = \left(1 + \rho + \rho^2 + \ldots + \rho^2\right)^{-1} = \frac{1-\rho}{1-\rho^{N+1}}$ and through the system of stationary Kolmogorov equations we express all probabilities through $\widetilde{\pi}_0$ and we get $\widetilde{\pi}_n = \rho\widetilde{\pi}_{n-1} = \rho\widetilde{\pi}_0, n = 1,2, \ldots ,N$..

Then on occasion $\lambda > \mu$, $\rho > 1$, that $\widetilde{\pi}_n > \widetilde{\pi}_{n-1}$. Thus, $\widetilde{\pi}_N = \max\{\widetilde{\pi}_0, \ldots, \widetilde{\pi}_N\} \to \frac{\rho-1}{\rho}$, by $N \to \infty$, a $\widetilde{\pi}_0 = \min\{\widetilde{\pi}_0, \ldots, \widetilde{\pi}_N\} \to 0$, by $N \to \infty$. Therefore, when $\lambda > \mu$ the system tends to develop and die at large N—unlikely.

Consider the case $\lambda < \mu$, то $0 < \rho < 1$ then $\widetilde{\pi}_0 > \widetilde{\pi}_1 > \ldots > \widetilde{\pi}_N$ with $\widetilde{\pi}_0 \to 1 - \rho > 0$ by $N \to \infty$, a $\widetilde{\pi}_N \to 0$ by $N \to \infty$. Then the system tends to regress and die.

And finally, if $\lambda = \mu$, то $\rho = 1$ and then $\widetilde{\pi}_0 = \widetilde{\pi}_1 = \ldots = \widetilde{\pi}_N = \frac{1}{N+1}$, that is, the distribution is uniform discrete, and the state of the system is maximally uncertain.

Let's assume that the state of the system of professional interactions of universities is represented as a system that performs the tasks of the external environment, then the following conditions for the functioning of the system can be accepted: a stream of tasks with an intensity of λ; there is no more than one person in the system at the same time N tasks to be executed; the implementation of tasks leads s independent communication channels, while the time for implementation in each case is random and distributed according to the exponential law μ; If everyone s communication channels are busy, then incoming tasks may take one of $N - s$ places in the execution waiting queue.

Thus, if we assume that the trend of development or regression of university professional interactions as a task processing system is a process $\{\xi(t), t \geq 0\}$, where $\xi(t)$—the number of tasks in the system at the moment t, then it is possible to find stationary probability distributions of its state, namely when $\xi(t) \leq s$, $s < \xi(t) \leq N$, that $n - s$.

Results and Discussions

The collection and analysis of primary statistical, regulatory, administrative, managerial data was carried out based on the study of periodicals, information and legal portals and systems publishing texts of regulatory legal acts, official information, and scientific research.

The study was conducted in three stages. The first stage consisted in a survey of respondents from 8 Federal Districts of the Russian Federation, 12 profiles of educational institutions of higher education, of which 21% are from national research, federal and reference universities and 79% from regional ones, which are represented by specialists, middle and senior managers, as well as professional teaching staff. The program requirements for achieving the representativeness of the sample were the gender, age, socio-professional composition of the subjects and their spatial localization with restrictions on professional and qualification composition.

The second stage was the processing of the received data, the analysis of the results and their interpretation. The reliability of the results of the sample study was determined with an ordinary accuracy (95% confidence probability) of generalizations, i.e., sampling error (error) of 4.32% (confidence interval of distributions at the level of 0.03–0.1) with 79% of the responses received, with a random error of sample representativeness up to 6% with a sample population of more than 5000.

The third is conducting 2 focus surveys in academic circles based on the Higher School of Management of St. Petersburg State University with the participation of expert respondents, participating respondents to interpret the results obtained and interviewing expert university leaders to obtain an expert assessment of the results obtained.

The analysis of primary questionnaire data was carried out using mathematical statistics of one-dimensional, two-dimensional descriptive analysis, the results obtained were processed using regression analysis of sociological data, as well as a quantitative content analysis of open questions.

Sample Characteristics

Representatives of national research (51%), federal (45%) and reference (4%) universities are 25% of all respondents, 84% of whom are from multidisciplinary universities of millionaire cities (57%) of the Central and Volga Federal Districts (60%), mostly women (84%). 45% of managers aged 25–39 years (43%) and aged 45–59 years (39%) with more than 8 years of experience at the university (67%)

have a PhD degree, although 49% did not consider it necessary to answer this question. 75% of respondents have a total teaching experience of over 10 years and are currently teaching (67%) under an "effective" contract (55%) combining with administrative activities (63%) as heads of various administrative levels (76%). 80% of respondents in this category took advanced training courses, 49% of whom studied according to the profile of their main activity over the past 3 years, and 43% could not remember the name of the educational program or did not consider it necessary to answer this question.

Representatives of regional universities represent 75% of all respondents, 65% of whom are from multidisciplinary and technical universities from millionaire cities (54.4%) of the Central and Northwestern Federal Districts (55%), mostly women (66%). 60% of managers aged 35–49 with more than 8 years of experience at the university (64%), 30% have a PhD degree, although 60% did not consider it necessary to answer this question. 75% of respondents have a total teaching experience of more than 10 years and currently conduct teaching activities (72%) under an "effective" contract (53%) combining with administrative activities (72%) as heads of various administrative levels (87%). 85% of respondents in this category have taken advanced training courses over the past 3 years, 38% of whom studied according to the profile of their main activity, and 51% could not remember the name of the educational program or did not consider it necessary to answer this question.

One of the tasks of analyzing the questionnaire data was to determine the request for the necessary educational programs from the point of view of university managers. The task of identifying the causes of their occurrence was also solved. Thus, 10 programs were allocated in the main areas of requirements for managers of any level: management theory, marketing, office work, work with information management systems, with information automated systems, psychology, conflictology, business and interethnic communication, social networks, and career guidance. However, the preferences received were noted only in 50% of respondents, so the focus of the search for the causes of occurrence shifted to the analysis of answers about the shortcomings of qualifications that respondents note from the management, themselves, and their subordinates. As a result, 7 directions were identified from 60% of respondents who did not coincide with the previously 10 requested programs. According to the results of the survey on the existing qualification deficits of managers, 7 programs were identified to overcome them: management theory, including strategic, knowledge of the functioning of the organization, including control, information analytics, psychology, self-management, including working time management, managerial decision-making, change management.

To identify the causes of qualification deficits, a factor analysis was carried out, according to the methodology of which 28 questions were analyzed on a semantic differential on the scale of "not changed—not much—very noticeable—it is difficult for me to answer (I do not do this)".

As a result, 5 factors with statistically reliable results were identified ($1 > p > 0.5$, $p = 0.712$) correlated with the conceptual levels of knowledge on Viig and the areas of professional activity of the manager: the first factor is associated with the strategic management of the external development of the educational organization, the second—with the strategic management of internal development, the third—with organizational processes, the fourth—with production, the fifth—with the analysis of the professional activity of the manager.

Further, the calculation of factor values for each respondent by age, by work experience, by type and level of position, by profile of professional activity, by status and profile of the university was carried out. During the calculation, both "weak" and "strong" knowledge of university managers were identified.

Using the calculation of the obtained factor values, it was determined that the qualification deficits of university managers at the time of the survey are not related to the profile of professional activity, nor the type and level of their position, nor to the status of the university, but are related to their age and work experience, as well as to the profile of the university. At the same time, in each of the identified 5 areas of professional activity, there are certain ranges of the formation of a lack of knowledge.

Qualification deficits associated with weak knowledge:

- in strategic management of external development of the organization, managers with 8–12 years of experience or from the university's legal profile.
- in strategic management of internal development of the organization—at the age of 60–64 years or with 3–8 years of work experience.
- in organizational processes—at the age of 50–54, or with 1–3 years of work experience, or from legal, socio–economic profiles of universities.
- in production processes—at the age of 40–44, 60–64 years, or with up to 1 year of work experience, or from legal, creative university profiles.
- in the analysis of professional activity—at the age of 25–29, 40–44 years, or with work experience up to 1 year, or from the socio–economic profile of universities.

Thus, the obtained factor values allow us to identify the directions of development of educational programs.

To determine the motives for the lack of knowledge, based on the obtained factor values, a cluster analysis was carried out using the cluster center analysis method. As a result, 3 groups of respondents were identified, the reasons for the formation of a lack of knowledge of which are in ensuring the basic processes of the university's activities, resource support of activities and the formation of corporate culture.

With the help of the obtained data, the calculation and analysis of the affiliation of the selected motives to each respondent was carried out.

The values calculated on the basis of data characterizing the profile of activity and the status of the university turned out to be informative, and it turned out that the motives for the occurrence of qualification deficits in managers of architectural and construction profile (100%) lie in the field of ensuring the main processes of the university's activities, but are absent in the resource support of activities (0%) and the formation of the corporate culture of the university (0%), when as in universities of culture and art (86%), national research, federal and reference universities (58%) have motives in the field of resource provision of the university's activities, but are absent in providing basic processes, and military, naval (50%), national research, federal, reference (42%)—lie in the field of forming the corporate culture of the university, but are absent in providing basic processes.

It can be stated that the obtained values allow us to formulate not only the directions of development of educational programs, but also to determine their target audience.

As noted above, systemic changes in the work of Russian universities have contributed to the formation of project management in determining and implementing the tasks of their strategic development. Therefore, in the professional activity of university managers, there is a need to combine project activities aimed at development with the fulfillment of tasks, the solution of which ensures the functioning of the organization.

Thus, professional interactions of managers are now formed based on project–functional interactions and high productivity of specialists in this field is the basis for the effective functioning of the university.

However, the effectiveness of the project depends a lot on how the project teams will be formed and productive: based on certain skills, experience, or specialization?

We believe that it is advisable to form a project team based on the previously identified "strong" knowledge of managers. Based on this, it makes sense to distribute roles within the team. At the same time, it is necessary to introduce a division into intra-organizational and inter-organizational projects of the university.

In an intra-organizational project, as well as an inter-organizational one, teams can be formed based on the directions of their activities: external and internal development, organizational and production support, as well as analytical activities, based on the calculations of "strong" relevant knowledge presented above, based on age characteristics and work experience. Project teams can also be formed based on the profile of the university.

The interaction of such project teams takes place at different stages of project preparation and execution, they can be conditionally divided into 7, which are associated with the processes of accumulation and exchange of information and knowledge within the positions and professional activities of managers in the organizational structure of the university.

For example, when implementing a university strategic development project, analysis, preparation, and implementation are the main stages of the work of project teams.

At the stage of project analysis of the external strategic development of the university, a team is formed with competencies in the field of value formulation and "strong" knowledge of the higher education system, missions, goals, capabilities of systems and schemes of their functioning, as well as methods and new approaches for making alternative decisions. Such a project team may include not only senior managers, but also highly specialized specialists from different fields.

At the next stage, the analysis of external changes is carried out and the results obtained are correlated with the strategic objectives of the development of the university, problematic issues are formed and at the next stage there is interaction with the internal development team, which implements the analysis of the external and internal development of the project.

At the stages of project implementation, it is necessary to form teams with competencies for structuring and systematizing work, working with changes.

At the next stage, there is an analysis of the project implementation, ways of further development and transfer of the acquired, accumulated knowledge to the third team for organizational support of the project, which, in turn, correlates analytical data and leads the establishment of organizational processes. The tasks of the fourth team include the organization of production processes.

As for the fifth team of analysts, they can be highly specialized representatives of different fields and form the "core" of project participants, whose purpose is to conduct an audit and examination of the project at any stage.

Conclusion

The analysis of the data of the sociological study of the management environment of higher education educational institutions found that, according to respondents, the most significant changes in the professional activities of universities of the Russian Federation occurred in internal organizational work, the number of external and internal reporting increased, the program and methodological work and educational activities changed significantly, there was an increase in the vertical ordinal communication of subordination in decision-making and decreased network interaction, while qualification requirements for specialists, researchers and managers have not changed much. This conclusion can be interpreted in the sense that the middle manager, even though objectively, as we indicated above, is included in the deep processes of essential changes, you need to react in your daily professional activity rather to external evidence of changes that do not always adequately reflect their essence.

Also, because of the conducted research, there is evidence of the active involvement of university middle managers in the above-mentioned deep changes in managerial activity. This is evidenced, first of all, by the change in the essence of their professional interactions and the increasing trend of "management", that is, the strengthening of the role of the manager in the implementation of strategies and procedures of strategic management, which are characterized, at the same time, by the restructuring of the functions of management levels, divisions and departments, the introduction of the system strategic and targeted planning based on formalized quantitative and qualitative performance criteria.

The formation of the phenomenon of "managerization" in the university management system was also determined by several indicators: statistical and functional dependence between indicators of age, seniority, and teaching activities, that is, the younger the manager, the less experience he has, the less likely it is that he conducts teaching activities.

In the process of the noted consolidation of universities within the territorial location, the management system and organizational structure tends to become more complicated. The forms and types of university management structures depend on the category, status, profile of professional activity, diversity and complexity of the tasks being solved, especially at large universities, which may include research institutes and other separate scientific units, as well as research laboratories included in the structure of departments, faculties, departments, institutes.

The development and strengthening of competition, the introduction of market incentives and mechanisms in the market of educational services of higher education leads to the formation and formation of new types of special management, which can be conditionally designated as scientific, pedagogical, administrative, and managerial management in the field of educational services.

An exhaustive list of the positions of administrative and managerial personnel and heads of educational institutions of higher education belonging to the field of higher education does not currently exist, however, there are mechanisms for regulating pedagogical, managerial, and professional activities in the field of education by introducing compliance of the positions with the qualification requirements specified in the qualification directories, and (or) professional standards.

The change in the organizational and economic foundations, the rules of functioning of the education system and the implementation of educational activities, the transition to effective project activities, the complication of the information and communication environment, contributes to the formation of quality deficits associated with the lack of the level of qualification of professional activity of managers of educational institutions of higher education.

It is worth noting that in the innovative and informational era, knowledge management has become an integral part of the strategic management of the organization. To ensure the interaction of employees and related departments as a single management system, there is a need for communication links in order to transfer information or knowledge to each other, since the basis for the formation of qualification deficits is either the insufficiency of the information received or the lack of knowledge.

Practical significance:

1. To evaluate the efficiency of the throughput capacity of the tasks, the results of which allow us to judge the current state of the system and, depending on the result, to form recommendations for eliminating the causes of resistance to external influences and the development of the system.
2. To carry out and calculate the current process of the state of university professional interactions and to perform an analysis based on the results of which it is possible to assess the current state of the system and its trend of development or regression, and, consequently, the state of the system as a process of birth or death.
3. To calculate the stationary probability distributions of the university system states and calculate the average number of task processing, the throughput capacity of the system.

Thus, the research work on identifying the readiness of university managers for organizational changes is described below.

1. To identify the ability of university managers to perceive changes,
2. Determine the types of emerging resistances.
3. Establish methods for reducing resistance.

Thus, the conducted research will make it possible to assess the personnel potential, determine the main criteria for readiness for organizational changes and ways to overcome resistance.

References

Carayannis, E. G., Alexander, J., & Ioannidis, A. (2000). Leveraging knowledge, learning, and innovation in forming strategic government–university–industry (GUI) R&D partnerships in the US, Germany, and France. *Technovation, 20*(9), 477–488. https://doi.org/10.1016/S0166-4972(99)00162-5

Dove, R. (1999). Knowledge management, response ability, and the agile enterprise. *Journal of Knowledge Management, 3*(1), 18–35. https://doi.org/10.1108/13673279910259367

Kallio, K.-M., Kallio, T. J., Tienari, J., & Hyvönen, T. (2016). Ethos at stake: Performance management and academic work in universities. *Human Relations, 69*(3), 685–709. https://doi.org/10.1177/0018726715596802

Kuzminov, Y. I., & Frumin, I. D. (2018). *Twelve solutions for new education: Report of the Center for Strategic Research and the Higher School of Economics*. National Research University "Higher School of Economics".

Lee, D. M. S., Trauth, E. M., & Farwell, D. (1995). critical skills and knowledge requirements of IS professionals: A joint academic/industry investigation. *MIS Quarterly, 19*(3), 313. https://doi.org/10.2307/249598

Ministry of economic development of the Russian Federation. (2017). *State programs*. Retrieved from http://government.ru/docs/30303/

Ministry of education of the Russian Federation. (2017). *The state program "development of education" is being transferred to project management in 2018*. Retrieved from http://government.ru/docs/30832/

President of the Russian Federation. (20 05 2016a). *The Government's decision on the issue of increasing the competitiveness of leading Russian universities among the world's leading scientific and educational centers*. Retrieved from http://government.ru/orders/selection/401/23103/

President of the Russian Federation. (2016b). *Passport of the priority project "Modern digital educational environment"*. Retrieved from http://government.ru/news/25682/

President of the Russian Federation. (2019a). *On the development of human resources in the field of science*. Retrieved from http://government.ru/news/36519/

President of the Russian Federation. (2019b). *On the development of scientific and industrial cooperation*. Retrieved from http://government.ru/news/36628/

President of the Russian Federation. (2019c). *The procedure for granting grants for state support of world-class scientific and educational centers*. Retrieved from http://government.ru/docs/36626/

Russian Federation Government. (2017a). *About the approval of the state program of the Russian Federation "development of education"*. Retrieved from http://government.ru/docs/all/115042/

Russian Federation Government. (2017b). Passport of the priority project "development of the export potential of the Russian education system". Retrieved from http://government.ru/news/28013/

Russian Federation Government. (2017c). *Rules for developing, implementing and evaluating the effectiveness of pilot state programs that are transferred to project management mechanisms.* Retrieved from http://government.ru/docs/29710/

Russian Federation Government. (2019). *Report on the implementation of the state policy in the field of education in 2018.* Retrieved from http://government.ru/news/36939/

Siegel, D. S., Waldman, D. A., Atwater, L. E., & Link, A. N. (2003). Commercial knowledge transfers from universities to firms: Improving the effectiveness of university–industry collaboration. *The Journal of High Technology Management Research, 14*(1), 111–133. https://doi.org/10.1016/S1047-8310(03)00007-5

Un, C. A., Cuervo-Cazurra, A., & Asakawa, K. (2010). R&D collaborations and product innovation. *Journal of Product Innovation Management, 27*(5), 673–689. https://doi.org/10.1111/j.1540-5885.2010.00744.x

Van Looy, B., Ranga, M., Callaert, J., Debackere, K., & Zimmermann, E. (2004). Combining entrepreneurial and scientific performance in academia: Towards a compounded and reciprocal Matthew-effect? *Research Policy, 33*(3), 425–441. https://doi.org/10.1016/j.respol.2003.09.004

Correction to: Challenges and Solutions in the Digital Economy and Finance

Anna Rumyantseva, Vladimir Plotnikov, Alexey Minin, and Hod Anyigba

Correction to:
A. Rumyantseva et al. (eds.), *Challenges and Solutions in the Digital Economy and Finance*, Springer Proceedings in Business and Economics,
https://doi.org/10.1007/978-3-031-14410-3

The author's first, middle, and last names are inadvertently mixed up in several chapters. This has now been corrected.

The updated original versions of the chapters can be found at
https://doi.org/10.1007/978-3-031-14410-3

A. Rumyantseva et al. (eds.), *Challenges and Solutions in the Digital Economy and Finance*, Springer Proceedings in Business and Economics,
https://doi.org/10.1007/978-3-031-14410-3_59